Guava

Botany, Production and Uses

Guava

Botany, Production and Uses

Edited by

Sisir Mitra

Former Dean, Faculty of Horticulture and Dean, Post-Graduate Studies
Bidhan Chandra Krishi Viswavidyalaya (State Agricultural University),
Mohanpur, West Bengal, India

CABI

CABI is a trading name of CAB International

CABI
Nosworthy Way
Wallingford
Oxfordshire OX10 8DE
UK

Tel: +44 (0)1491 832111
Fax: +44 (0)1491 833508
E-mail: info@cabi.org
Website: www.cabi.org

CABI
WeWork
One Lincoln St
24th Floor
Boston, MA 02111
USA

Tel: +1 (617)682-9015
E-mail: cabi-nao@cabi.org

A catalogue record for this book is available from the British Library, London, UK.

Library of Congress Cataloging-in-Publication Data

Names: Mitra, S. K. (Horticulturist), editor.
Title: Guava : botany, production and uses / Sisir Mitra.
Other titles: Botany, production and uses.
Description: Boston, MA, USA : CAB International, [2021] | Series: Botany, production and uses | Includes bibliographical references and index. | Summary: "Guava is considered a minor tropical fruit (together with lychee, longan, durian etc.), although it is the largest in terms of output. This is the first comprehensive book authored by an international team of experts, and covers botany, biotechnology, propagation, production, pests and diseases, postharvest, and processing"-- Provided by publisher.
Identifiers: LCCN 2020053473 (print) | LCCN 2020053474 (ebook) | ISBN 9781789247022 (hardback) | ISBN 9781789247039 (ebook) | ISBN 9781789247046 (epub)
Subjects: LCSH: Guava.
Classification: LCC SB379.G8 G83 2021 (print) | LCC SB379.G8 (ebook) | DDC 634/.421--dc23
LC record available at https://lccn.loc.gov/2020053473
LC ebook record available at https://lccn.loc.gov/2020053474

References to Internet websites (URLs) were accurate at the time of writing.

ISBN-13: 978 1 78924 702 2 (hardback)
978 1 78924 703 9 (ePDF)
978 1 78924 704 6 (ePub)

DOI: 10.1079/9781789247022.0000

Commissioning Editor: Rebecca Stubbs
Editorial Assistant: Emma McCann
Production Editor: Shankari Wilford

Typeset by SPi, Pondicherry, India
Printed and bound in the USA by Integrated Books International, Dulles, Virginia

Contents

Contributors

Siti H. Ahmad, Department of Crop Science, Faculty of Agriculture, Universiti Putra Malaysia, 43400 UPM, Serdang, Selangor, Malaysia. E-mail: sitihajarahmad@gmail.com

Martín Aluja, Red de Manejo Biorracional de Plagas y Vectores, Clúster Científico y Tecnológico BioMimic®, Instituto de Ecología A.C., Carretera Antigua a Coatepec n° 351, 91073 Xalapa, Veracruz, Mexico. E-mail: aluja@inecol.mx

Fredy H. Ballen, Tropical Research and Education Center, University of Florida Institute of Food and Agricultural Sciences (IFAS), 18905 SW 280 Street, Homestead, FL 33031, USA. E-mail: fredy.ballen@ufl.edu

Anup K. Bhattacherjee, ICAR–Central Institute for Subtropical Horticulture, Rehmankhera, Lucknow – 226101, Uttar Pradesh, India. E-mail: dadabhatu@gmail.com

Andrea Birke, Red de Manejo Biorracional de Plagas y Vectores, Clúster Científico y Tecnológico BioMimic®, Instituto de Ecología A.C., Carretera Antigua a Coatepec n° 351, 91073 Xalapa, Veracruz, Mexico. E-mail: andrea.birke@inecol.mx

Regina M.D.G. Carneiro, Embrapa Recursos Genéticos e Biotecnologia, C.P. 02372, 70849-970 Brasília, DF, Brazil. E-mail: regina.carneiro@embrapa.br

Daniel Carrillo, Tropical Research and Education Center, Department of Entomology and Nematology, University of Florida Institute of Food and Agricultural Sciences (IFAS), 18905 SW 280 Street, Homestead, FL 33031, USA. E-mail: dancar@ufl.edu

José Mauro C. Castro, Embrapa Semiárido, BR428, Km 152, 56302-970 Petrolina, PE, Brazil. E-mail: mauro.castro@embrapa.br

Po-An Chen, Department of Plant Technology Laboratories, Agricultural Technology Research Institute, 300 Taiwan. E-mail: chenpoan@mail.atri.org.tw

Rosemary J. du Preez, Agricultural Research Council, Institute of Tropical and Subtropical Crops, Mbombela, Private Bag X11208, Nelspruit 1200, South Africa. E-mail: rosedup@arc.agric.za

Edward A. Evans, Tropical Research and Education Center, University of Florida Institute of Food and Agricultural Sciences (IFAS), 18905 SW 280 Street, Homestead, FL 33031, USA. E-mail: eaevans@ufl.edu

Larissa Guillén, Red de Manejo Biorracional de Plagas y Vectores, Clúster Científico y Tecnológico BioMimic®, Instituto de Ecología A.C., Carretera Antigua a Coatepec n° 351, 91073 Xalapa, Veracruz, Mexico. E-mail: larissa.guillen@inecol.mx

Azimah Hamidon, Department of Crop Science, Faculty of Technical and Vocational, Universiti Sultan Idris, 35900 Muallim, Perak, Malaysia. E-mail: azmiah.hamidon@ftv.upsi.edy.my

Hamizah Hassan, Department of Crop Science, Faculty of Agriculture, Universiti Putra Malaysia, 43400 UPM, Serdang, Selangor, Malaysia. E-mail: misz_zah@yahoo.com

Sandeep Kumar, Division of Crop Improvement and Biotechnology, ICAR–Central Institute for Subtropical Horticulture, Rehmankhera, Lucknow – 226101, Uttar Pradesh, India. E-mail: sandeep17apr@gmail.com

Leslie R. Landrum, Natural History Collections, School of Life Sciences, Arizona State University, Tempe, AZ 85287-4108, USA. E-mail: les.landrum@asu.edu

Rodrigo Lasa, Red de Manejo Biorracional de Plagas y Vectores, Clúster Científico y Tecnológico BioMimic®, Instituto de Ecología A.C., Carretera Antigua a Coatepec n° 351, 91073 Xalapa, Veracruz, Mexico. E-mail: rodrigo.lasa@inecol.mx

Shu-Yen Lin, Department of Horticulture and Landscape Architecture, National Taiwan University, 106 Taiwan. E-mail: sylin@ntu.edu.tw

Maneesh Mishra, Division of Crop Improvement and Biotechnology, ICAR–Central Institute for Subtropical Horticulture, Rehmankhera, Lucknow – 226101, Uttar Pradesh, India. E-mail: maneeshmishra.cish@gmail.com

Ashok K. Misra, Former Project Coordinator, All Indian Co-ordinated Project on Subtropical Fruits and Head, Division of Crop Protection, Central Institute for Subtropical Horticulture, Lucknow – 226101, Uttar Pradesh, India. E-mail: misra_a_k@yahoo.co.in

Sisir Mitra, Former Dean, Faculty of Horticulture, Bidhan Chandra Krishi Viswavidyalaya, Mohanpur, West Bengal, B-12/48 Kalyani, Nadia 741235, India. E-mail: sisirm55@gmail.com

Munirah Mohamad, Department of Crop Science, Faculty of Agriculture, Universiti Putra Malaysia, 43400 UPM, Serdang, Selangor, Malaysia. E-mail: munimohd5516@gmail.com

Muthukumar, M., Division of Crop Improvement and Biotechnology, ICAR–Central Institute for Subtropical Horticulture, Rehmankhera, Lucknow – 226101, Uttar Pradesh, India. E-mail: muthukumarbt@gmail.com

Pradyot K. Pathak, KVK Murshidabad, West Bengal University of Animal and Fishery Science, Digha, Milebasa, Kalukhali, Bhagwangola-I, Murshidabad – 742135, India. E-mail: pathakpradyot@gmail.com

Pravat K. Ray, Rajendra Agricultural University, Pusa 848125, Samastipur, India. E-mail: pkraypusa@gmail.com

Marcilene F.A. Santos, Embrapa Recursos Genéticos e Biotecnologia, C.P. 02372, 70849-970 Brasília, DF, Brazil. E-mail: lenebio@gmail.com

Vinod K. Singh, ICAR–Central Institute for Subtropical Horticulture, Rehmankhera, Lucknow – 226101, Uttar Pradesh, India. E-mail: singhvk_cish@rediffmail.com

Manoj K. Soni, ICAR–Central Institute for Subtropical Horticulture, Rehmankhera, Lucknow – 226101, Uttar Pradesh, India. E-mail: manojsoni.lko@gmail.com

Margo Sulistio, Department of Horticulture and Landscape Architecture, National Taiwan University, 106 Taiwan. E-mail: d06628009@ntu.edu.tw

Nor Elliza Tajidin, Department of Crop Science, Faculty of Agriculture, University of Putra Malaysia, 43400 UPM, Serdang, Selangor, Malaysia. E-mail: elliza.tajidin@upm.edu.my

Dileep K. Tandon, ICAR–Central Institute for Subtropical Horticulture, Rehmankhera, Lucknow – 226101, Uttar Pradesh, India. E-mail: dr.dktandon@gmail.com

Chun-Ta Wu, Department of Horticulture and Landscape Architecture, National Taiwan University, 106 Taiwan. E-mail: wuct@ntu.edu.tw

Preface

The guava is native to the American tropics. The Spanish explorers took the guava to the Philippines and the Portuguese disseminated it from the Philippines to India. It spread easily and rapidly throughout the tropics because of the abundance of seeds with long viability possessed by the fruit. Although it is grown in more than 70 countries, the bulk of the production is from India, Pakistan, Brazil and Mexico. The major problems facing the guava industry are severity of wilt disease, root nematodes, high seed content of diploid commercial cultivars and postharvest management of fruit. In the last three decades, researchers have addressed many of the major production and postharvest problems.

Global sharing of information on guava was advanced by the First International Symposium on Guava held at Lucknow, India, in 2005, under the aegis of the International Society for Horticultural Science (ISHS). This has been followed by regular symposia in cities in different guava-producing countries, including Merida and Aguascalientes, Mexico (2008), Petrolina, Brazil (2012) and Cairns, Australia (2016). Proceedings of these symposia provide a rich source of scientific literature. The Working Group on Guava and other Myrtaceae Crops at its meeting in Mexico identified the need for compilation to address aspects of crop botany, orchard management, postharvest management and value addition in one volume and I have been requested by my colleagues from India, Brazil, Mexico and South Africa to lead the compilation.

This publication presents the current state of knowledge concerning the origin, history, physiology, culture and trade of guava throughout the world. The book is mainly targeted at guava researchers, teachers and academics, students, advisors and industry support personnel.

Sisir Mitra
Kalyani, India, 2020

Acknowledgements

I wish to express my appreciation and gratitude to all the authors who have given their time, without any financial reward, to contribute to this book. I am grateful to the scientists who have provided information and/or reviewed the chapters: Dr Duane P. Bartholomew, University of Hawaii at Manoa, Hawaii, USA; Dr S. Kondo, Chiba University, Japan; Dr W. Rohde, Max Planck Institute for Plant Breeding Research, Germany; Dr R.M. Khan, Central Institute for Subtropical Horticulture, Lucknow, India; Dr Cheng-Fang Heng, National Chung Hsing University, Taiwan; Dr Kamala Jayanti, Indian Institute of Horticultural Research, Bengaluru, India; Dr M.R. Dinesh, Indian Institute of Horticultural Research, Bengaluru, India; Dr P.L. Saroj, Central Institute of Arid Horticulture, India; Dr Aron Dag, The Volcani Center, Israel; Dr Ziwei Zhou, Griffith University, Australia; Dr Martha Schoeman, ARC Institute for Tropical and Subtropical Crops, South Africa; Ms Regina Cronje, ARC Institute for Tropical and Subtropical Crops, South Africa; Dr Shant Lal, G.B. Pant University of Agriculture and Technology, India; Dr Carlos A.F. Santos, Embrapa Tropical Semi-Arid, Brazil; Dr U. Lavi, The Volcani Center, Israel; Dr J. Crane, University of Florida, USA; Dr Hamide Gubbuk, Akdeniz University, Antalya, Turkey; and Dr Omayma M. Ismail, National Research Centre, Cairo, Egypt.

I also wish to thank the following for providing the photographs used in this book: Mr Erli Ropke, Frucafe, Brazil, Valfrutas, Brazil; Mr Chen Haojun, Guangxi Institute of Subtropical Crops, Guangxi, People's Republic of China; Mr Faming Zhang, Institute of Tropical and Subtropical Cash Crops, Yunnan, People's Republic of China; Mr Ahmed Salah Elsoda, Horticulture Research Institute, Agriculture Research Centre, Giza, Egypt; Salomie Willemse, ARC Institute for Tropical and Subtropical Crops, South Africa; Dr T. Matsumoto, USDA, ARS, Hilo, Hawaii, USA; Dr C.V. Pommer, Universidade Federal Rural do Semiarido, Brazil; Dr Ron Polat, The Volcani Center, Israel; Dr A. Guha Choudhury, Birsa Agricultural University, Ranchi, India; Dr G. Yang, North Carolina A & T State University, USA; Dr B.R. Jana, ICAR Research Complex for Eastern Region, Darbhanga, India; Dr Thaveesak Sangudon, Pichit Research and Development Centre, Thailand; Dr K.B.S. Gill, Punjab Agricultural University, Punjab, India; Dr K. Kishore, CHES, (ICAR-IIHR), Odisha, India; Dr A. Bhattacherjee, Central Institute for Subtropical Horticulture, Lucknow, India; Dr Pratibha, G.B. Pant University of Agriculture and Technology, India; Dr Narayan Chawda, VNR Seeds, Raipur, India; Dr Huey-Ling Lin, National Chung Hsing University, Taichung, Taiwan; Dr N.N. Truong, Southern Horticultural Research Institute, Vietnam; Dr K.K. Srivastava, Central Institute for Subtropical Horticulture, Lucknow, India; Dr Madhu Kamle, North Eastern

Regional Institute of Science and Technology, Arunachal Pradesh, India; Dr B. Ghosh, Bidhan Chandra Krishi Viswavidyalaya, Mohanpur, West Bengal, India; and Dr A. Malik, University of Agriculture, Faisalabad, Pakistan.

My special thanks to Dr P.K. Pathak, KVK, Murshidabad, West Bengal University of Animal and Fishery Science, India for his editorial assistance.

The continued support and encouragements received from various editors and assistants of CABI, particularly from Rebecca Stubbs and Emma McCann, are gratefully acknowledged.

On a personal note I thank my wife Ipsa who has unselfishly allowed me to edit this book. Without her sacrifices I could not possibly have finished.

Sisir Mitra

1 *Psidium guajava* L.: Taxonomy, Relatives and Possible Origin

Leslie R. Landrum*
Arizona State University, Tempe, Arizona, USA

1.1 Introduction

The guava (*Psidium guajava* L.) belongs to the genus *Psidium* L. of the family *Myrtaceae*. The family *Myrtaceae* has *c.*130 genera and nearly 6000 accepted species names (PoWO, 2020). Because many groups have not been studied in depth, we know that many species remain to be discovered and many names will eventually be recognized to be synonyms by future monographers. So, the numbers of species must remain approximate for now.

The family *Myrtaceae* is apparently of Gondwanan origin with centres of diversity in tropical America and Australasia and with fewer species in Africa, the Mediterranean and southern Asia (Raven and Axelrod, 1974; Thornhill *et al.*, 2015).

Several species have economic importance: *Syzygium aromaticum* (L.) Merrill & L.M. Perry (clove) and *Pimenta dioica* (L.) Merrill (allspice) are spices; *P. guajava* (guava) is the best-known tropical fruit, but other species of *Psidium* (e.g. *Psidium cattleyanum* Sabine, *Psidium friedrichsthalianum* (O. Berg) Nied.), *Acca sellowiana* (O. Berg) Burret (= *Feijoa sellowiana* (O. Berg) O. Berg; pineapple guava), *Plinia cauliflora* (Mart.) Kausel (jaboticaba), *Syzygium jambos* (L.) Alston (rose-apple) and *Syzygium malaccense* (L.) Merr. & Perry (Malay apple) are commonly cultivated; species of *Eucalyptus* L'Heritier are widely planted for timber and as ornamentals. *Myrtus communis* L., *Melaleuca* L. (including *Callistemon* R. Brown) and other genera are planted as ornamentals.

Psidium is a genus of at least 60 species and perhaps as many as 100 (McVaugh, 1968; Govaerts *et al.*, 2008), ranging from Mexico and the Caribbean to Argentina and Uruguay. The state of Bahia, Brazil is particularly rich in species of *Psidium* with 28 known so far (Landrum, 2017), about half the total for South America. A few species have been introduced as cultivated plants in the Old World and Pacific Island tropics and subtropics, and some are weedy invasives (Global Invasive Species Database, 2017). The distinguishing characters of *Psidium* are discussed in Landrum (2003) and Landrum and Sharp (1989) and are: flowers (4–)5(–6)-merous (occasionally flowers have more petals) with multiovulate locules; placenta often peltate; mature seedcoat rough or dull, covered with a pulpy layer when wet; hard portion of seedcoat (5–)8–30 cells thick at the narrowest point, with the cells thick-walled, elongate and overlapping; and a C-shaped embryo

*E-mail: les.landrum@asu.edu

 Guava: Botany, Production and Uses (ed. S. Mitra)
DOI: 10.1079/9781789247022.0001

with cotyledons much shorter than the hypocotyl. Based on small samples of two to nine species, recent molecular studies of *Myrtaceae* (Lucas *et al.*, 2007; Murillo *et al.*, 2012; Rivero *et al.*, 2012; Vasconcelos *et al.*, 2017; Flickinger *et al.*, 2020) indicate that *Psidium* may be a monophyletic group and place it in clades with such genera as *Acca* O. Berg, *Amomyrtus* (Burret) Legrand & Kausel, *Campomanesia* Ruiz & Pavon, *Legrandia* Kausel, *Mosiera* Small and *Myrrhinium* Schott and Pimenta L. These are all members of the morphologically based subtribe *Myrtinae* (i.e. those genera with embryos with relatively small cotyledons and a large hypocotyl) that appears to be a basal, paraphyletic group in the tribe.

In a new subtribal classification of the *Myrteae*, a mainly American group of berry-fruited genera, Lucas *et al.* (2019) have divided traditional paraphyletic *Myrtinae* into five subtribes that they recognize as monophyletic based mainly on molecular data. *Psidium* belongs to *Pimentinae* O. Berg in their system. Larger samples will be needed to determine which genera are the closest relatives of *Psidium* and if it is truly monophyletic.

Two genera of *Myrtaceae* that are similar to *Psidium* and sometimes confused with it are *Campomanesia* Ruiz & Pav. and *Calycolpus* O. Berg because they have generally 5-merous flowers, and seeds and embryos similar to *Psidium*. The three genera are compared in the key below. For comparison with other genera of *Myrtaceae*, see Landrum and Kawasaki (1997).

1. Ovary with (3–)6–18 locules, the locules when fertile usually 1-seeded; locular wall in fruits glandular, functioning as a false seedcoat in the fruit so that the 'seedcoat' appears to be glandular, the locules arranged in a ring in the fruit, several often without a seed inside; leaves with broadly arching lateral veins and often no clear marginal vein; bark flaky or crusty; hypocotyl swollen, much wider than the cotyledons; anthers usually with 1 gland in the connective, or none .... ***Campomanesia***.

1′. Ovary with 2–5(–6) locules, the locules when fertile with 1 to many seeds; locular wall in fruits usually not glandular, not functioning as a false seedcoat, the true seedcoat not glandular, the seeds distributed throughout the fruit, not oriented in a ring; leaves variable but often with a distinct marginal vein; bark variable, but often smooth; hypocotyl not swollen, about the same width as the cotyledons; anthers often with more than 1 gland in the connective.

2. Seedcoat dull or rough, several cells thick; cells of the hard seedcoat surface elongate, overlapping (Fig. 1.1A–E); calyx closed or nearly so in the bud, or the calyx lobes usually broader than long, more or less triangular, sometimes only evident as a sinuate margin .... ***Psidium***.

2′. Seedcoat shiny, 1 to a few cells thick; cells of hard seedcoat surface not elongate, abutting each other in a mosaic-like pattern (Fig. 1.1F and G); calyx open, the lobes often longer than broad .... ***Calycolpus***.

1.2 Geography

Psidium is naturally an American genus, although *P. guajava*, *Psidium guineense* Sw. and *P. cattleyanum* are subtropical and tropical weedy species in many other parts of the world. The greatest number of species (*c.*50) is found in South America and those of Central America and Mexico are a subset of that group. Presumably then, the Central American species are geologically recent arrivals from South America because they have not diverged from their South American relatives. The Caribbean islands are home to an unknown number of species, perhaps 20 or more, most of which are endemic to the Caribbean, and may, because of their diversity and distinctness from mainland species, represent multiple, geologically old colonizations.

The Brazilian Atlantic Coastal Forest and the adjacent cerrado and caatinga can be considered a centre of diversity for *Psidium*. We can speculate, at least, because of that diversity and endemism, that *Psidium* has a long history in the Atlantic Coastal Forest and adjacent areas. It is notable that the genus is not found in temperate south-western

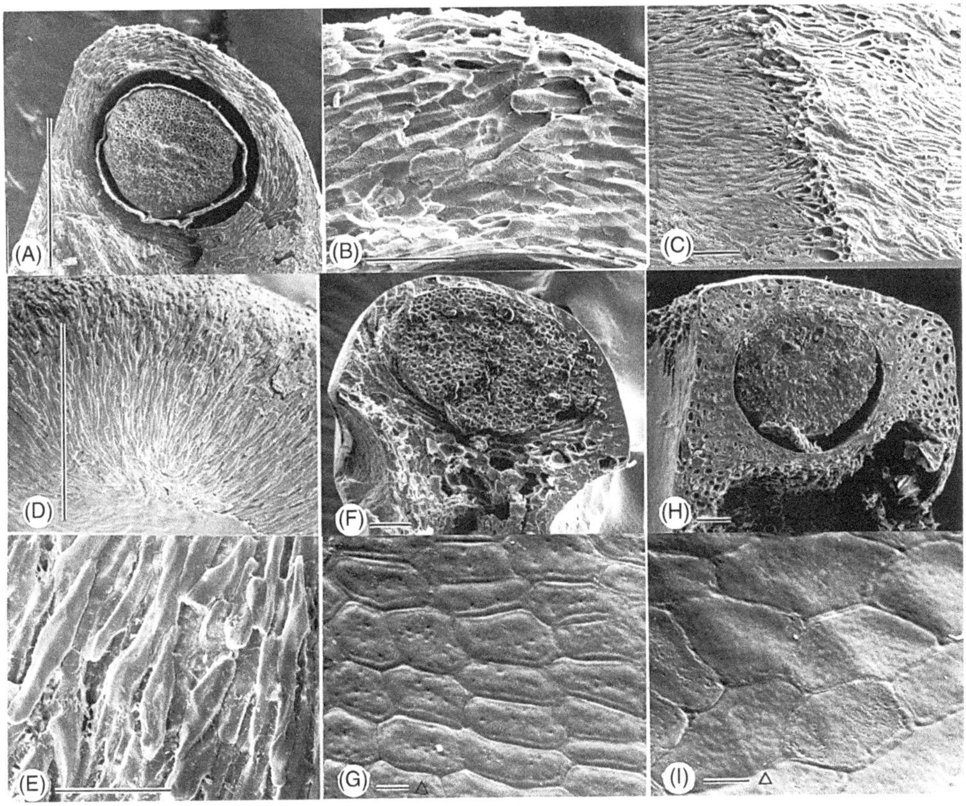

Fig. 1.1. Scanning electron micrographs of seedcoats of *Psidium*, *Calycolpus* and *Mosiera*. (A–C) *Psidium acidum*: section of seed showing cylindrical cavity and embryo (A), upper portion of seedcoat in section (B) and outer surface of seed on right and tangential section (C). (D, E) *Psidium australe*: outer seed surface from which outer pulpy covering has rotted away. (F, G) Seed of *Calycolpus moritzianus*. (H, I) *Seed of Mosiera elliptica*. Note dense overlapping, elongate cells of *Psidium* versus the mosaic pattern of non-overlapping cells in *Calycolpus* and *Mosiera*. (A–C, from *Huashikat 1311*, MO; D, E, from *Montes 851*, NY; F, G, from *Davidse and Gonzalez* 21134, MO; H, I, from *Clemente 2831*, NY.) Vertical lines = 1 mm; horizontal lines without triangle = 1/10 mm; horizontal lines with triangle = 1/100 mm. All photographs reproduced with permission from Landrum and Sharp (*Systematic Botany* 14(3), 370–376. 1989).

South America; in this respect it is similar to the large Neotropical *Myrtaceae* genera *Calyptranthes* Sw., *Eugenia* L. and *Myrcia* DC., all of which are quite diverse in the Atlantic Coastal Forest (Sobral *et al.*, 2009). *Psidium* is present but less locally diverse in the rest of Brazil, the Andean countries (excluding Chile) and the Guianas.

The Isthmus of Panama has been dated at *c.*2.8 million years old (O'Dea *et al.*, 2016). So, prior to that date, direct migration without dispersal over water barriers may have been impossible to Central America and beyond. *Psidium oligospermum* DC., at least, is clearly able to cross significant water barriers, having become established on some oceanic islands (e.g. Galapagos). Colonization of islands, especially when they are new, relatively uninhabited and with reduced biological competition, would be more likely than colonization of a continent with many species already growing there. Because of their edible fruits some species of *Psidium* in Central America may have been carried there by humans. The author does not know of any fossil evidence of *Psidium* in Central America and Mexico other than archaeological finds of

P. guajava that are about 2000 years old in the Tehuacán Valley of Mexico (Smith, 1965).

1.3 Taxonomy

The taxonomically useful morphological characters have been discussed in Landrum (2017). A few of the more important characters are summarized here.

1.3.1 Leaf venation

In *Psidium* the most common type is brochidodromous with the lateral veins (i.e. secondary veins) looping towards the apex near the margin to connect with each other to form a marginal vein that follows the margin, either as a series of arches or as a scarcely arching vein that nearly parallels the margin (e.g. *P. cattleyanum*, Fig. 1.2A). Less common is eucamptodromous venation, where the laterals diminish near the margin and no clear marginal connecting vein is evident. In some species leaves may be eucamptodromous proximally and brochidodromous distally (e.g. *P. guineense*, Fig. 1.2B) and intermediate conditions are sometimes encountered.

The tertiary veins that connect the lateral, marginal and midveins may have a dendritic pattern (Fig. 1.2A) or a ladder-like pattern (the latter found in eucamptodromous leaves only; Fig. 1.2B). The dendritic pattern may seem to be without clear direction or may seem to arise from the marginal vein.

Species of the *P. guajava* complex have, to varying degrees, the eucamptodromous venation with ladder-like tertiary veins. This pattern is well developed in *P. guajava*, *Psidium rutidocarpum*, and usually in *P. guineense*. In *Psidium guyanense*, *Psidium nutans* and *Psidium rostratum*, this pattern may be less pronounced or lacking.

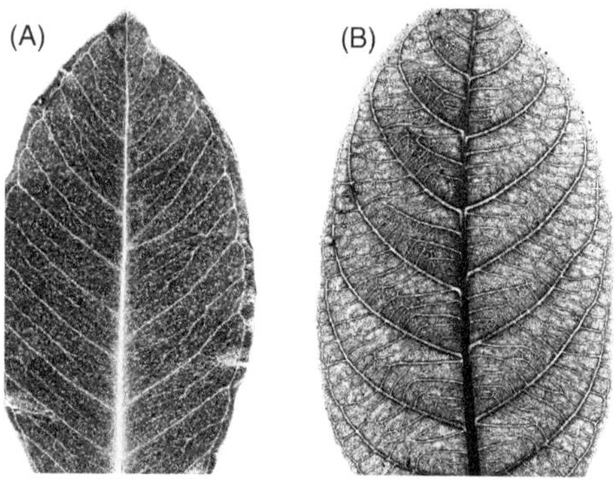

Fig. 1.2. Scanned leaves of (A) *Psidium cattleyanum* and (B) *Psidium guineense*. *P. cattleyanum* has brochidodromous venation with a marginal vein that connects the lateral (secondary) veins and parallels the margin; tertiary veins between the laterals are dendritic. *P. guineense* illustrates eucamptodromous venation without a clear marginal vein for most of the leaf; in this specimen the ladder-like connecting tertiary veins link between the laterals; this is the typical venation of the *P. guajava* complex. Photograph by L.R. Landrum.

1.3.2 Twig shape

Young twigs vary from terete or compressed, to quadrangular, or 4-winged. *P. guajava* and other species of the complex often have quadrangular or 4-winged twigs.

1.3.3 Flower size

A good measure of flower size in *Psidium* that is a convenient way to compare species is the length of the style. In *Psidium* the style varies from 3 to 23 mm long. In the *P. guajava* complex lengths are usually between 10 and 15 mm. Other species complexes (*Psidium acidum* complex, *Psidium grandifolium* complex and *Psidium acutangulum* complex) have similarly large flowers but are distinct because of other characteristics. Other measures of flower size are stamen number, locule number or ovule/locule number, but these are less easily seen.

1.3.4 Calyx

The calyx structure is especially important in *Psidium*. For convenience, calyx morphology may be divided into two types: (i) bowl-like, with the globe of the corolla clearly visible in the closed flower bud (Fig. 1.3C); and (ii) closed, hiding the globe of the closed corolla completely or enclosing it except for a terminal pore (Fig. 1.3A and G). The amount of closure of the calyx is variable between and sometimes within species.

In *P. guajava* and its relatives the calyx is closed in the bud or has a terminal pore with no clear lobes before the flower bud opens.

1.3.5 Ovary and ovules

Psidium ovaries are 2–5(–6)-locular. In *Psidium* the number of ovules per locule

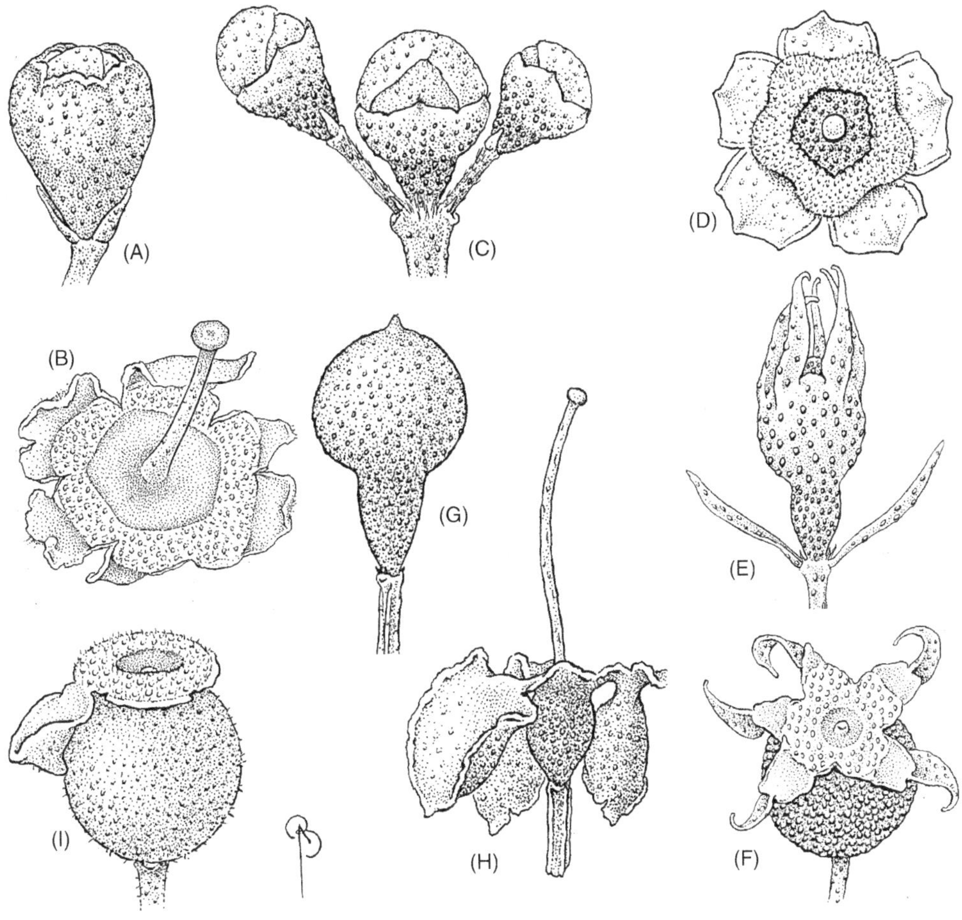

Fig. 1.3. *Psidium* flowers before and after anthesis. *Psidium cattleyanum*: (A) closed bud before anthesis with apical pore; (B) apical view of flower after anthesis showing tears in staminal ring. *Psidium occidentale*: (C) 3-flowered dichasium with closed buds; (D) view from above after anthesis, tears forming between lobes but not penetrating the staminal ring. *Psidium appendiculatum*: (E) a nearly closed calyx before anthesis with a flange-like apical appendage on each lobe; (F) after anthesis, tears forming between lobes penetrating the staminal ring. *Psidium acidum*: (G) bud with completely closed calyx; (H) calyx tears irregularly at anthesis, the staminal ring, not visible in this drawing, is not penetrated by tears at anthesis. *Psidium brownianum*: (I) side view of immature fruit showing persistent calyptra and no tears in staminal ring. (A, from *Rossato et al. 4855*, ASU; B, from *Folli 4925*, ASU; C, D, from the isotype *Rubio and Quelal 659*, ASU; E, from *Proença et al. 1445*, ASU; F, from *Filgeiras and Lopes 2406*, ASU; G, H, from *Perea et al. 2098*, ASU; I, from *Stannard et al. H515615*, ASU.) All illustrations by Bobbi Angell.

varies from as few as 3 to over 250. Numbers below 10 and more than 100 are relatively rare. The placenta often protrudes as two lamellae that form a peltate structure (Figs 1.7G and 1.13E below). The number of rows of ovules on the edge of a lamella varies from 1 to about 4. In the *P. guajava* complex, the locules are 3–5(–6), the placenta is often peltate and the ovules number 50 to about 200, except in *P. rutidocarpum* with 25–40 ovules per locule.

1.3.6 Fruits and seeds

The seeds of *Psidium* are unique among the *Myrteae* because of their dense cell structure (Landrum and Sharp, 1989; Fig. 1.1A–E). The cells of the seedcoat are elongate, with little or no lumen, are closely packed together and in a few to several layers (Fig. 1B). The seed surface is not a smooth, shiny mosaic of non-overlapping cells as in most other genera of subtribe *Myrtinae* with hard seeds (e.g. *Calycolpus*, Fig. 1.1F and G; *Mosiera*, Fig. 1.1H and I), but rather a rough or dull surface when dry and a pulpy layer when wet. The very hard, dense seedcoat is hard to break and this characteristic may be related to fruit predators. The germinating embryo emerges via a pore in the hard seedcoat covered by a plug-like operculum (Rotman, 1976; Fig. 1.4A). The operculum is found in several other genera with hard seedcoats of the subtribe *Myrtinae* in the broad traditional sense.

In *Psidium* the number of seeds in a fruit varies from 1 to 300, but the ranges for a particular species are much smaller. The size of seeds varies from *c.*2.5 to 12 mm long. Seed morphology is often important. Seeds may be approximately reniform with uniformly rounded surfaces (e.g. *Psidium*

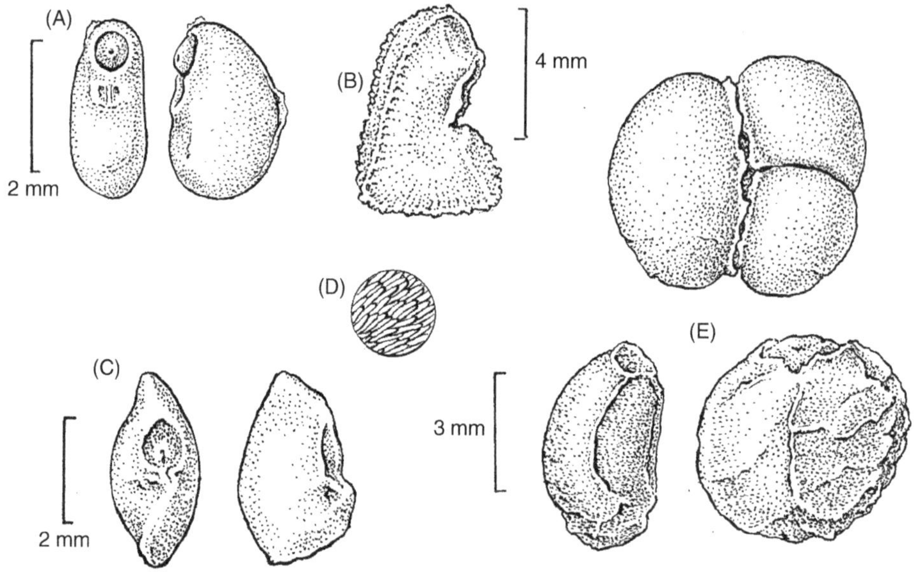

Fig. 1.4. Seeds of *Psidium*. (A) Two views of seed of *Psidium firmum* with rounded edges and generally smooth surface. (B) Angular seed of *Psidium riparium* with rough surface. (C) Two views of seed of *Psidium myrsinites* with compressed sides. (D) Close-up of cell structure of a general *Psidium* seed. (E) Top: three seeds of a fruit with rounded smooth surface towards fruit wall and flat rough surface where two seed abut. Bottom: two views of a seed showing internal rough surface. (A, from *Irwin et al. 9189*, MICH; B, from *da Silva et al. 4200*, ASU; C, from *Azevedo et al. 1098*, ASU; E, from *Pereira s.n.*, ASU.) All illustrations by Bobbi Angell.

firmum, Fig. 1.4A). In this case the author believes the seeds mature in the fruit without abutting the fruit wall or other seeds. When seeds develop tightly packed within the fruit, they may have nearly flat sides where they abut other seeds and rounded sides that are adjacent to the fruit wall (e.g. *Psidium myrtoides*, Fig. 1.4E) or are isolated in the pulp of the fruit. In one mainly Amazon group (*P. acutangulum* complex) the seeds are tightly packed in the fruit and are often angular, C- or L-shaped, or irregularly shaped, with few rounded surfaces (e.g. *Psidium riparium*, Fig. 1.4B).

In *P. guajava* and its relatives, the seeds are relatively small (*c.*2–5 mm long) and numerous (often over 100). *P. rostratum* is an exception with seeds 8–11 mm long and few (4–12). Seeds of the complex are rounded (never angular) but sometimes have flattened sides.

1.4 Chromosome Number and Genome Size

Chromosome number and genome size are not subjects the author has worked on himself, but they have been researched by others and should be valuable in better understanding the evolution of *Psidium*. The common chromosome number in *Myrtaceae* is $2n = 22$, but polypoidy is found in *Eugenia* and *Psidium* (Rye, 1979). Chromosome numbers for *Psidium* have been reported by Atchison (1947), Costa and Forni-Martins (2006, 2007), Costa *et al.* (2008), Chakraborti *et al.* (2010), Machado (2016) and Marques *et al.* (2016). In *P. cattleyanum* polyploidy seems especially common with levels up to $2n = 12x = 132$ (Machado, 2016). Genome size can be a measure of ploidy level in *Psidium* and may prove to be a valuable tool in assessing ploidy level in many species of *Psidium* (Costa *et al.*, 2008; Marques *et al.*, 2016). Based on the author's own studies, hybridization appears to be frequent in *Psidium*; hybridization coupled with polyploidy may explain some of the confusing variation in *Psidium* in such groups as the *P. grandifolium* complex (Landrum, 2005) and in *P. guineense*.

1.5 Phytochemistry and Medicinal Uses

P. guajava is known around the world for its medicinal properties and has been frequently studied for its chemical components and their effects. Pérez Gutiérrez *et al.* (2008) offer an excellent summary of these subjects complete with an illustrated appendix of the known chemical components of *P. guajava*. The lesser-known relatives of *P. guajava* have not been studied so frequently but a few recent contributions are cited here: *P. guineense* (Fernandes *et al.*, 2012); *P. acutangulum* (Wen *et al.*, 2011; Houël *et al.*, 2015); *P. friedrichsthalianum* (Flores *et al.*, 2013); and *P. cattleyanum* (Medina *et al.*, 2011). Further studies of the medicinal potential of other *Psidium* species should prove rewarding. Essential oils are the most frequently studied compounds in *Psidium*. Commonly several essential oils are found in a single individual, but a few will be much more abundant than the others. Among the more common dominant essential oils in *Psidium* are α-pinene, α-selinene, γ-selinene, 1,8-cineole, β-pinene, β-caryophyllene, β-bisabolene and *p*-cymene (Tucker *et al.*, 1995; Silva *et al.*, 2003). There seems to be considerable variation within species as to which oils dominate and whether or not essential oils will be taxonomically important is still unclear.

Flavonoid chemistry has proved helpful in distinguishing between *P. guajava*, on the one hand (myricetin absent), and *P. guineense* (and its suspected hybrid with *P. guajava*), on the other (myricetin present) (Landrum *et al.*, 1995). Flavonoid chemistry may prove useful in other studies of hybridization.

1.6 What are the Closest Relatives of *Psidium guajava*?

Knowing which of the 60 or more species of *Psidium* are the closest relatives of *P. guajava* would be useful in deducing where *P. guajava* originated and what species might provide useful genetic materials for the improvement of the species. As a working hypothesis, the present author proposes that six species belong to the *P. guajava* complex: *P. guajava, P. guineense, P. guyanense* Pers., *P. nutans* O. Berg, *P. rostratum* McVaugh and *Psidium rutidocarpum* Ruiz & Pav. This analysis will be based mainly on morphology, but it is worth noting that in a molecular phylogenetic study of *Mosiera*, including 12 species of *Psidium*, Salywon (2003) found that the closest relative of *P. guajava* to be *P. guineense*. He found *P. grandifolium* DC. to be sister to the *P. guajava–P. guineense* clade. Other species of what the present author considers to be members of the *P. guajava* complex were not included. Salywon (2003) used internal transcribed spacer (ITS) sequences only but his study has been the one to include the most species of *Psidium* so far.

1.7 Method for Development Key to Species Complexes

Taxonomic studies often proceed slowly and that has been the case with the study of *Psidium*. For each species, a description has been prepared in a standardized format that includes vegetative and reproductive features and examination of multiple specimens, usually 20 to 100 or more. In a few cases species are known from one or a few collections only. Unfortunately, these descriptions were sometimes written years apart from each other. After descriptions for the great majority of the mainland species (but not species restricted to Caribbean islands) had been completed, they were checked for consistency. Concurrently a database of characters was constructed using the descriptions. This allowed for the efficient comparison of all species or subgroups of species. Usually potential groups of species were perceived without using the character database, but with the database it was possible to find characters that supported these taxonomic concepts.

It is worth mentioning that the important values when comparing species are the ranges of characters rather than averages. This key is not meant to be useful in the identification of species. It does not use some of the most easily observable characters (e.g. indumentum density, degree of calyx closure) because these vary across more than one complex. For identification it will be better to use regional keys.

1.8 Key to the Species Complexes of *Psidium*

1. Flowers large: style usually 10–15 mm long (shorter in *P. grandifolium* complex); stamens 200–800; ovary locules usually 3–5; ovules per locule usually 20–200; calyx closed in most species (open in some species of *P. grandifolium* and *P. acutangulum* complexes); young twigs frequently 4-winged to quadrangular.

2. Leaves brochidodromous, with a clear marginal vein from the base to the apex.

3. Seeds rounded or with some flat sides; southern Mexico to Peru including upper Amazon basin (some species cultivated); locules usually 3–5; anthers often with a few to several co-equal glands; twigs always 4-winged; peduncle sometimes 4-winged; calyx always closed .... ***P. acidum* complex**: *P. acidum, P. friedrichsthalianum, Psidium guayaquilense* and *Psidium montanum*.

3′. Seed angular; mainly Amazon basin (one species in Paraná river basin); locules usually 2–4; anthers with a terminal gland and often with smaller gland below; twigs 4-angled or terete; peduncle not winged; calyx open in some species .... ***P. acutangulum* complex**: *P. acutangulum, Psidium densicomum, Psidium kennedyanum, Psidium maribense, P. riparium* and *Psidium striatulum*.

2′. Leaves eucamptodromous (entirely without a clear marginal vein), or eucamptodromous proximally and brochidodromous distally (with a clear marginal vein for part of the leaf).

4. Placenta barely protruding into locule; locule walls sometimes incomplete; calyx open or with a terminal pore in the bud (rarely closed); anthers up to about 1 mm long; seeds mostly less than 85, 3–5(–6) mm long; tertiary veins connecting lateral veins in a dendritic pattern; shrubs and subshrubs of open areas (campo, cerrado, savannahs) .... ***P. grandifolium* complex:** *Psidium australe, P. grandifolium, Psidium missionum, Psidium ratterianum* and *Psidium suffruticosum.*

4′. Placenta protruding into locules as a peltate structure; locule walls complete; calyx closed or with a terminal pore in the bud; anthers often over 1 mm long; seeds sometimes few but often over 100, 2.5–5(–11) mm long; tertiary veins commonly connecting lateral veins in a ladder-like pattern, less often in a dendritic pattern; shrubs and trees of forested areas, sometimes in riparian or disturbed habitats .... ***P. guajava* complex**: *P. guajava, P. guineense, P. guyanense, P. nutans, P. rostratum* and *P. rutidocarpum.*

1′. Flowers small: style usually 3–8 mm long; stamens usually less than 200; ovary locules usually 2–3; ovules per locule usually 3–50; calyx closed or open and bowl-like; young twigs terete to compressed (not known to be 4-winged or quadrangular).

5. Flowers cauliflorous; eastern Bahia and Espirito Santo, Brazil .... ***Psidium cauliflorum* complex**: *P. cauliflorum* and *Psidium grazielae.*

5′. Flowers not cauliflorous; widespread.

6. Calyx with apical appendages (not always in *P. oligospermum*), appearing closed or nearly so in bud, tearing into nearly regular lobes at anthesis (or usually irregularly in *P. oligospermum*) .... ***P. oligospermum* complex**: *Psidium appendiculatum, Psidium glaziovianum, P. oligospermum* and *Psidium schenckianum.*

6′. Calyx without apical appendages, open or closed in bud.

7. Shrubs and subshrubs of open areas (campo, cerrado, savannahs); calyx open .... ***Psidium salutare* complex**: *Psidium laruotteanum* and *P. salutare.*

7′. Shrubs or trees of forests and open habitats; calyx open or closed.

8. Andean species with small flowers (styles 3–6 mm long); dichasial inflorescences common (these sometimes aggregate into panicles); calyx open .... ***Psidium pedicellatum* complex**: *Psidium fulvum, P. pedicellatum* and *Psidium occidentale.*

8′. Species of eastern South America (mainly Brazil) with small or large flowers; dichasial inflorescence occasional; calyx open or closed. Miscellaneous species without clear affinities.

1.9 Key to Species of *Psidium guajava* Complex

1. Lateral veins usually 12–20 pairs; tertiary veins clearly ladder-like; leaves frequently more than 2.6 times as long as wide.

2. Leaves usually 3–4 times as long as wide, narrowly lanceolate-elliptic, tapering from mid-leaf or below, with an acuminate apex; inner surface of calyx densely covered with reddish brown or whitish hairs; immature fruit with a few longitudinal ridges; ovules per locule up to about 40; endemic to eastern Peru; not cultivated .... ***P. rutidocarpum.***

2′. Leaves usually less than 3 times as long as wide, mostly elliptic-oblong, not tapering from below mid-leaf, usually with an acute to obtuse apex; inner surface of calyx glabrous to pubescent, the hairs whitish; immature fruit smooth; ovules per locule usually more than 90; widespread in subtropical and tropical regions; frequently cultivated .... ***P. guajava.***

1′. Lateral veins usually 4–10 pairs; tertiary veins ladder-like or dendritic; leaves usually less than 2.6 times as long as wide.

3. Seeds 4–12, 8–11 mm long; closed calyx often with a rostrate apex; ovules per locule up to *c.*26; stamens 500 or more; anthers 0.6–1 mm long, with 0 or 1 gland; endemic to north-western

Peru and western Ecuador .... *P. rostratum*.

3′. Seeds 27–300, 3–5 mm long; closed calyx without a rostrate apex; ovules per locule 50 or more; stamens up to *c*.400; anthers 1–3 mm long, usually with a few to several glands; widespread species.

4. Young growth hirtellous, the hairs mainly less than 0.1 mm long; closed bud often with an apical pore clearly exposing a portion of the corolla; leaves elliptic, the apex acute to acuminate; eastern Amazon basin .... *P. guyanense*.

4′. Young growth glabrous to pubescent, the hairs mainly over 0.5 mm long; closed bud normally without an open apical pore exposing the corolla; leaves elliptic-oblong, elliptic or obovate, the apex obtuse, rounded or acute; widespread.

5. Leaves, twigs and flowers usually abundantly pubescent; tertiary veins usually predominantly ladder-like; calyx closed completely or nearly closed and with 5 minute lobes at the apex; disturbed habitats or occasionally cultivated ...... *P. guineense*.

5′. Leaves, twigs and flowers glabrous or very sparsely pubescent; tertiary veins often predominantly reticulate, but ladder-like veins common; calyx nearly closed and with 5 minute lobes at the apex; habitats frequently wet .... *P. nutans*.

1.10 *Psidium guajava* L., Sp. Pl. 470. 1753. TYPE: 'Habitat in India', Cultivated Plant from Hortus Cliffortianus (LECTOTYPE: BM-628598)

George Clifford was governor of the Dutch East India Company and hired Carl Linnaeus to describe plants growing in his garden in Bennebroek, Netherlands. Clifford and Linnaeus probably thought that *P. guajava* was native to the East Indies, rather than America.

P. guajava has been described as a 'new' species several times. Some of the more commonly found synonyms are listed below. A more complete list with citations can be found in Landrum (2017).

- *Psidium cujavus* L.
- *Psidium pomiferum* L.
- *Psidium pyriferum* L.
- *Psidium cujavillus* Burm.
- *Psidium angustifolium* Lamarck
- *Psidium sapidissimum* L.
- *Psidium pumilum* Vahl
- *Psidium aromaticum* Blanco
- *Psidium fragrans* Macfad.

SHRUB or TREE up to *c*.12 m high, subglabrous to densely appressed pubescent on young growth and lower leaf surfaces, the trunk smooth, light brown to light grey-green, with large flaky scales; *hairs* whitish, yellowish or silvery, up to *c*.0.7 mm long, erect or appressed; *young twigs* quadrangular, slightly to strongly winged, often sulcate (at least when dry), densely to moderately appressed pubescent, the older twigs at first scaly with longitudinal striations or fibres, eventually smooth with irregular scales falling as patches. LEAVES elliptic, oblong, elliptic-oblanceolate, elliptic-obovate or lanceolate, 4.5–14 cm long, 2.4–7.5 cm wide, 1.6–4 times as long as wide, densely to sparsely appressed pubescent below, subglabrous except for puberulent midvein above; *apex* acute, acuminate, to rounded; *base* rounded to slightly cordate; *petiole* 2–5 mm long, 1–2 mm thick, channelled, densely pubescent to subglabrous; *venation* brochidodromous distally to eucamptodromous proximally, the midvein impressed above, prominent below, the lateral veins 9–22 prominent pairs, ascending at angle of *c*.45°, nearly straight, curving towards apex near the margin and connecting with the next lateral, the marginal vein not clearly present or arching between the laterals, the tertiary veins connecting the laterals in a ladder-like to reticulate pattern; *blades* coriaceous to submembranous, drying yellow-green, grey-green, to dark reddish brown. FLOWER BUDS subfusiform to pyriform, 9–14 mm long, sometimes strongly constricted near the midpoint, the hypanthium narrowly campanulate, barrel-shaped or fusiform, 4–6 mm long, the distal portion of bud more or less ovoid, sometimes strongly so with a conical apex, 4.5–9.5 mm long; *indumentum pattern of buds* with peduncles, hypanthium and bracteoles sparsely to moderately appressed pubescent, the calyx without glabrous to sparsely pubescent (usually less densely covered than that hypanthium), the calyx within glabrous or densely pubescent, the petals, disk and style glabrous; *peduncles* 1–3-flowered, 1–3.5 cm long, 1–1.5 mm thick, terete; *bracteoles* linear to narrowly triangular, 2–5 mm long. CALYX closed, tearing irregularly as the bud opens, persisting or falling in *c.*3 parts; *petals* obovate to elliptic, 13–22 mm long; *disk* 4–6 mm across; *stamens* 280–720, 7–15 mm long; *anthers* 0.7–1 mm long, with 1–7(–10) glands; *style* 10–15 mm long; *ovary* 3-6-locular; *ovules* 90–180 per locule, multiseriate. FRUIT globose to pyriform, 2–6(–8) cm long, green to yellow without, with pink, yellow or white flesh, aromatic; *seeds* numerous, subreniform, 3–4 mm long, more or less smooth, the seedcoat *c.*0.25 mm thick. $2n$ = 22, 44 (Figs 1.5 and 1.6A).

PHENOLOGY. Flowering mainly in spring months; fruiting throughout year but mainly in summer months.

HABITAT AND DISTRIBUTION. Disturbed areas such as roadsides, pastures and frequently cultivated, from near sea level to 1000 m. Widely distributed as a cultivated and

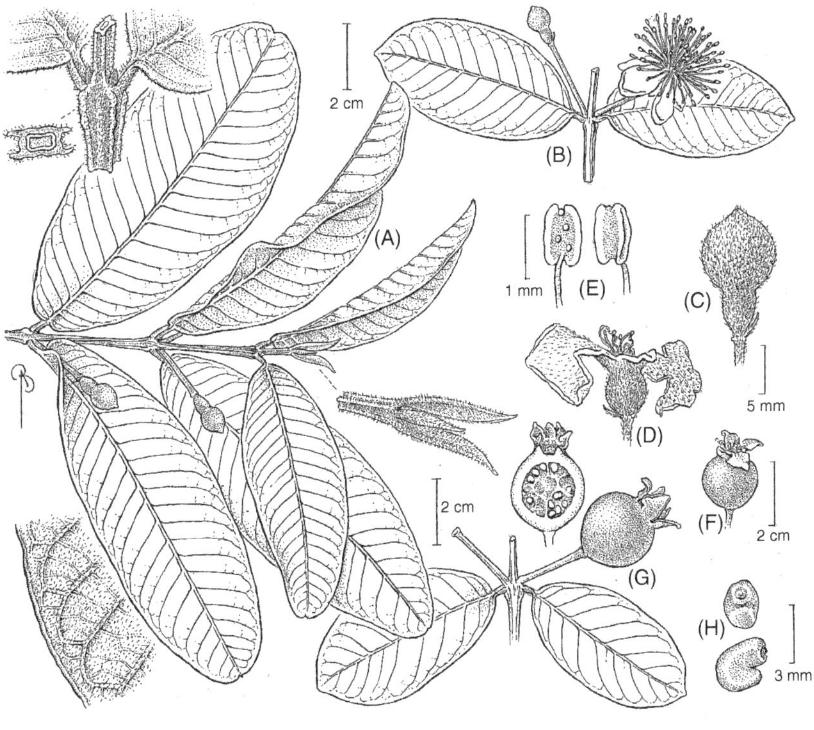

Fig. 1.5. *Psidium guajava*. (A) Branch with flower buds, including close-ups of node showing wings on twigs (upper left) and growing tip with two decussate pairs of immature leaves (right), and ladder-like tertiary veins (lower left). (B) Node with open flower and closed bud. (C) Closed bud with one persistent bracteole. (D) Flower after anthesis with irregularly torn calyx. (E) Two views of anther with multiple glands. (F) Fruit. (G) Node with fruit attached and longitudinal section of fruit showing seeds. (H) Two views of a seed. (A, from fresh material from Tempe, Arizona, USA, unknown origin; B, E, from *Sanders 8615*, ASU0004830; C, D, F, from *Landrum 6301*, ASU0004836; G, H, from *Landrum 6343*, ASU0004869.) All illustrations by Bobbi Angell.

Fig. 1.6. (A) *Psidium guajava* and (B) *Psidium rutidocarpum*. (A) Cultivated plant from Tempe, Arizona, USA (photograph by L.R. Landrum). (B) Wild-growing plant from Estación Biológica Huampal, Parque Nacional Yanachaga-Chemillen, Oxapama, Peru (photograph by Rocio Rojas).

escaped-weedy species in tropical and subtropical regions around the world.

COMMON NAMES. Goiaba (Portuguese); guayaba (Spanish); guava (English); gobaya (French Guiana); bayabas (Philippines).

DISTINGUISHING FEATURES. Calyx closed in flower bud or open only as a terminal pore, tearing irregularly as the bud opens, usually in 2 or 3 parts; lateral veins usually more than 10 pairs; hairs on lower leaf surface appressed, whitish or silvery.

P. guajava is frequently confused with similar *P. guineense*; they have been hypothesized to hybridize (Landrum *et al.*, 1995). They are contrasted in the key below.

1. Lateral veins usually 9–22 pairs; young twigs quadrangular, more or less winged; indumentum of lower leaf surface appressed, whitish, yellowish or silvery; calyx usually tearing into 2 or 3 parts; anthers 0.7–1 mm long, usually with less than 10 glands .... ***P. guajava***.

1′. Lateral veins 5–10 pairs; young twigs more or less terete or compressed (some vigorous shoots sometimes 4-winged); indumentum of lower leaf surface more or less erect, reddish brown, or less often appressed, whitish or greyish; calyx usually tearing into 5 parts; anthers 1–3 mm long, often with more than 10 glands .... ***P. guineense***.

The origin of cultivated *P. guajava* is unknown, but various interesting clues exist. The original habitat may have been riparian areas with periodic drought because roadsides with occasional abundant water, and disturbed areas, such as pastures, are where the species thrives presently.

The earliest archaeological remains known to the author that are thought to be of *P. guajava* come from two South American sites. The oldest is from Teotônio, Rondônia, Brazil (5000–9000 cal. BP), a locality especially good for fishing along the Madeira River (a tributary of the Amazon River) with evidence of human habitation as early as 9000 years ago. Evidence of other edible plants includes remains of squash (*Cucurbita* L. sp.), beans (*Phaseolus vulgaris* L.), manioc (*Manihot esculenta* Cranz) and pequiá (*Caryocar* L. sp.) (Watling *et al.*, 2018). The second site is Caral, Supe River valley, Peru along a river valley of arid coastal Peru and may have been cultivated there as early as 4000 years ago along with plants such as squash, beans, camote (*Ipomoea batatas* (L.) Lam.) and cotton (*Gossypium barbadense* L.), but not maize (*Zea mays* L.) (Shady Solis *et al.*, 2001). These two South American sites are quite different in climate and separated by the Andes mountain range, but geographically separated by only 1500 km. So, for the present, this part of South America seems like a likely area of origin for cultivated *P. guajava*.

P. rutidocarpum, an endemic species of eastern central Peru, appears to be a close relative of *P. guajava*. It lives between these two archaeological sites, which lends support to the hypothesis that this region gave rise to cultivated *P. guajava*.

In Central America and Mexico, the earliest archaeological find of *P. guajava* is about 2000 years old in the Tehuacán Valley of Mexico (Smith, 1965). The earliest records of peanut (*Arachis hypogaea* L., another

South American cultivated plant) in Mexico are also from the Tehuacán Valley and of the same approximate age (Smith, 1965).

By the time of European contact, *P. guajava* was widely cultivated in the Caribbean region and various cultivars had been selected according to Fernández de Oviedo y Valdez (1851, vol. 1, p. 304) who wrote his account in the early 1500s.

It is interesting that 'goiaba' is the common name frequently used for this species in Brazil, a variant of 'guayaba' reported by Fernández de Oviedo y Valdez (1851) and the name frequently used in Spanish-speaking countries. Other species of *Psidium* in Brazil are usually called 'araçá' (Legrand and Klein, 1977), a name from Guarani language. So, it is possible that *P. guajava* is a relatively recent arrival in much of Brazil.

In conclusion, *P. guajava* likely originated in South America because the large majority of *Psidium* species live there. Based on archaeological evidence and the related species *P. rutidocarpum*, Peru and the western Brazilian Amazon are hypothesized as being its area of domestication.

1.11 Gallery of *Psidium* Species

Several species of *Psidium* have been mentioned in this chapter as close relatives of *P. guajava* or as cultivated species of other complexes. Some are frequently cultivated and may be of interest to readers of this book. Drawings and/or photographic images are provided for the following species: *P. guajava* (Figs 1.5 and 1.6A), *P. rutidocarpum* (Fig. 1.6B), *P. guineense* (Figs 1.7 and 1.8) and *P. rostratum* (Fig. 1.9), all of the *P. guajava* complex. Cultivated species of other complexes are: *P. acidum* (Figs 1.10 and 1.11), *P. friedrichsthalianum* (Fig. 1.12) and *P. cattleyanum* (Figs 1.13 and 1.14).

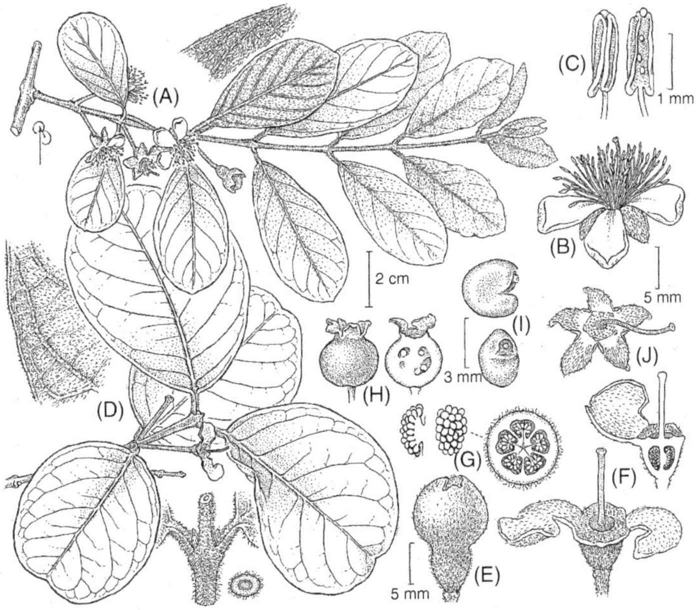

Fig. 1.7. *Psidium guineense.* (A) Branch with flowers and flower bud; detail of lower leaf surface (form with appressed hairs on lower leaf surface). (B) Open flower. (C) Anthers with glands. (D) Branch with old flower; detail of lower leaf surface and ladder-like tertiary veins (form with erect spreading hairs). (E) Closed bud just beginning to open. (F) Flowers after anthesis with irregularly opening calyx. (G) Cross section of ovary showing 5 locules; detail of placentation and ovules. (H) Fruit, whole and sectioned. (I) Two views of seed. (J) Flower after anthesis showing calyx tearing in 5 nearly equal lobes. (A, B, from *Landrum 8804*, ASU0008042; C, from *Landrum 5676*, ASU0004988; D–I, from fresh material grown from seeds from Chiapas, Mexico; J, from *Nee 39697*, ASU0007532.) All illustrations by Bobbi Angell.

Fig. 1.8. *Psidium guineense*. Plant cultivated at Tempe, Arizona, USA; grown from seeds collected in Chiapas, Mexico. Bars = 3 cm. Photographs by L.R. Landrum.

Fig. 1.9. *Psidium rostratum*. (A) Flower buds. (B) Twig with fruit and insert showing seeds. (C) Trees. (D) Bark of trunk. All photos of plants growing wild. (A, *Cornejo 8777*, Cerro Seco Biological Reserve, Manabi, Ecuador; B–D, *Cornejo 8829*, Bosque Protector Cerro Blanco, Guayas, Ecuador.) All photographs by X. Cornejo.

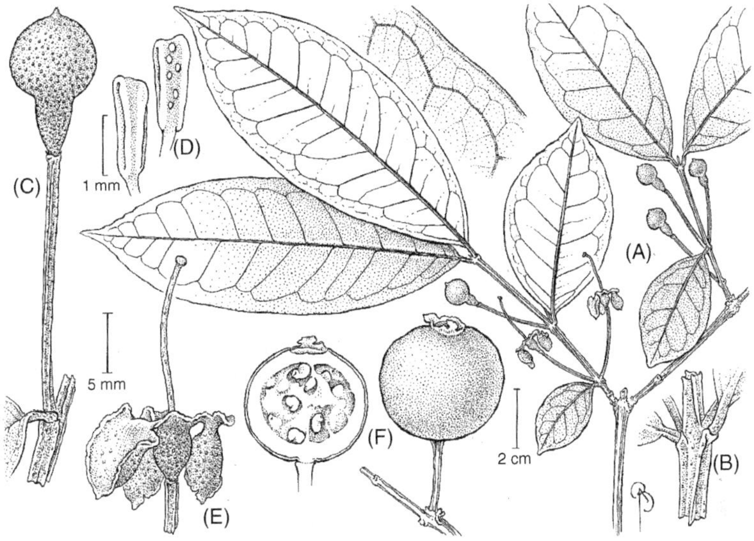

Fig. 1.10. *Psidium acidum*. (A) Flowering branch. (B) Detail of winged twig. (C) Closed flower bud. (D) Anthers with multiple glands. (E) Flower after anthesis. (F) Fruit, whole and sectioned. (A–E, from *Perea et al. 2098*, ASU0005139; F, from *Ceron 3634*, ASU0005129.) All illustrations by Bobbi Angell.

Fig. 1.11. *Psidium acidum*. (A) Habit. (B) Bark. (C) Fruits, with insert of fruit showing persistent calyx. (D) Leaves and winged twig. (All from *Orejuela & E. Trujillo 3004*, Centro Demostrativo Agroforestal Guacayaco, Piamonte, Cauca, Colombia.) All photographs by Edwin Trujillo.

Fig. 1.12. *Psidium friedrichsthalianum*. (A) Flower branches with fruits. (B) Open flower. (C) Fruits. (A, C, cultivated plant (*L.R. Landrum 6555*) from Heredia, Santo Domingo, Costa Rica; B, cultivated plant from Tempe, Arizona, USA.) All photographs by L.R. Landrum.

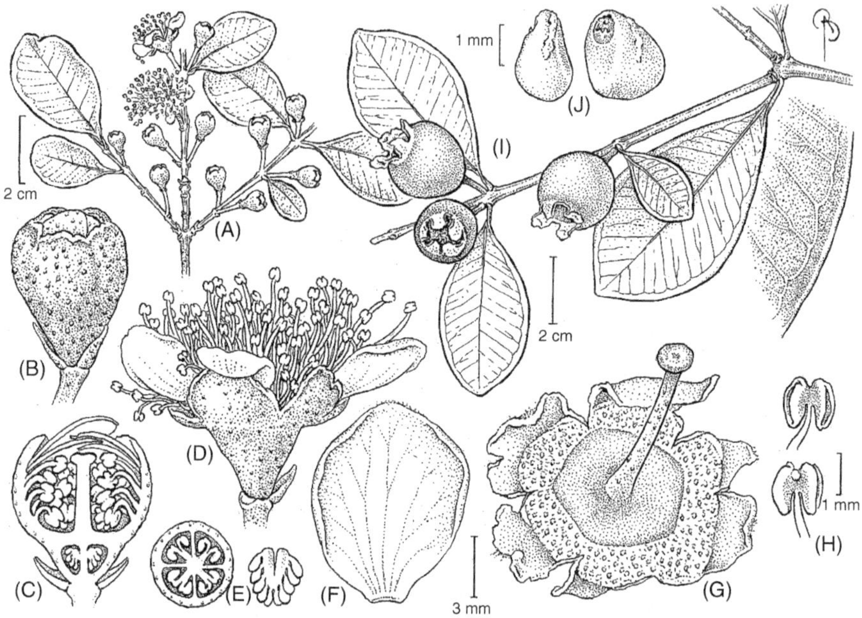

Fig. 1.13. *Psidium cattleyanum.* (A) Twig at beginning of anthesis. (B) Flower bud. (C) Longitudinal section of flower bud. (D) Opening flower. (E) Cross-section of ovary and extracted placenta with ovules. (F) Petal. (G) Apical view of flower after anthesis showing tears in calyx cutting into staminal ring. (H) Two views of stamen and anther with single terminal gland. (I) Fruiting twig. (J) Seeds. (A–C, from *Rossato et al. 4855*, ASU0006118; D, from photograph of live specimen; E–H, from *Folli 4925*, ASU0006103; I, from *Baitello 414*, ASU0006091; J, from *Carvalho et al. 6859*, ASU0006121.) All illustrations by Bobbi Angell.

Fig. 1.14. *Psidium cattleyanum*. (A) Branches with young fruits, from a cultivated plant at Universidade Federal de Santa Catarina, Florianópolis, Brazil (photograph by Marla Ibrahim). (B–E) Flower buds, open flower and young fruit at different stages of development, from a cultivated plant at Tempe, Arizona, USA (photographs by L.R. Landrum). (F) Plant in fruit growing wild at Kōke'e State Park, Kauai, Hawaii, USA (photograph by D. Wolkis).

References

Atchison, E. (1947) Chromosome numbers in the Myrtaceae. *American Journal of Botany* 34, 159–164.

Chakraborti S., Sinha, S. and Sinha, R. (2010) Chromosome number and karyotype analysis of wild guava *Psidium guineense* Sw. – a new report from Tripura, India. *Indian Journal of Science and Technology* 3, 925–927.

Costa, I.R. and Forni-Martins, E.R. (2006) Chromosome studies in Brazilian species of *Campomanesia* Ruiz and Pávon and *Psidium* L. (Myrtaceae Juss.). *Caryologia* 1, 7–13.

Costa, I.R. and Forni-Martins, E.R. (2007) Karyotype analysis in South American species of Myrtaceae. *Botanical Journal of the Linnean Society* 155, 571–580.

Costa, I.R., Dornelas, M.C. and Forni-Martins, E.R. (2008) Nuclear genome size variation in fleshy-fruited Neotropical Myrtaceae. *Plant Systematics and Evolution* 276, 209–217.

Fernandes, T.G., de Mesquita, A.R.C., Randau, K.P., Franchitti, A.A. and Ximenes, E.A. (2012) *In vitro* synergistic effect of *Psidium guineense* (Swartz) in combination with antimicrobial agents against methicillin-resistant *Staphylococcus aureus* strains. *Scientific World Journal* 2012, 138237.

Fernández de Oviedo y Valdez, G. (1851) *Historia General y Natural de las Indias, Islas y Tierrafirme del Mar Océano*, Vol. 1, ed. D.J. Amador de Los Rios. Real Academia de La Historia, Madrid.

Flickinger, J.A., Jestrow, B., Oviedo Prieto, R., Santiago-Valentín, E., Sustache-Sustache, J. *et al.* (2020) A phylogenetic survey of Myrtaceae in the Greater Antilles with nomenclatural changes for some endemic species. *Taxon* 69(3), 448–480.

Flores, G., Dastmalchi, K., Wu, S.-B., Whalen, K., Dabo, A.J. *et al.* (2013) Phenolic-rich extract from the Costa Rican guava (*Psidium friedrichsthalianum*) pulp with antioxidant and anti-inflammatory activity. Potential for COPD therapy. *Food Chemistry* 141, 889–895.

Global Invasive Species Database (2017) Available at: http://www.iucngisd.org/gisd/ (accessed May 2017).

Govaerts, R., Sobral, M., Ashton, P., Barrie, F., Holst, B.K. *et al.* (2008) *World Checklist of Myrtaceae*. Kew Publishing, Royal Botanic Gardens, Kew, UK.

Houël, E., Fleury, M., Odonne, G., Nardella, F., Bourdy, F. *et al.* (2015) Antiplasmodial and anti-inflammatory effects of an antimalarial remedy from the Wayana Amerindians, French Guiana: Takamalaimë (*Psidium acutangulum* Mart. ex DC., Myrtaceae). *Journal of Ethnopharmacology* 166, 279–285.

Landrum, L.R. (2003) A revision of the *Psidium salutare* complex (Myrtaceae). *Sida* 20(4), 1449–1469.

Landrum, L.R. (2005) A revision of the *Psidium grandifolium* complex (Myrtaceae). *Sida* 21(3), 1335–1354.

Landrum, L.R. (2017) The genus *Psidium* (Myrtaceae) in the State of Bahia, Brazil. *Canotia* 13, 1–101.

Landrum, L.R. and Kawasaki, M.L. (1997) The genera of Myrtaceae in Brazil: an illustrated synoptic treatment and keys. *Brittonia* 49, 508–536.

Landrum, L.R. and Sharp, W.P. (1989) Seed coat characters of some American Myrtinae (Myrtaceae): *Psidium* and related genera. *Systematic Botany* 14, 370–376.

Landrum, L.R., Clark, W.D., Sharp, W.P. and Brendecke, J. (1995) Hybridization between *Psidium guajava* and *P. guineense* (Myrtaceae). *Economic Botany* 49(2), 153–161.

Legrand, C.D. and Klein, R.M. (1977) *Psidium. Flora Ilustrada Catarinense* [MIRT.], 684–724.

Lucas, E., Harris, S., Mazine, F., Belsham, S.R., Nic Lughadha, E.M. *et al.* (2007) Suprageneric phylogenetics of Myrteae, the generically richest tribe in Myrtaceae (Myrtales). *Taxon* 56, 1105–1128.

Lucas, E.J., Holst, B., Sobral, M., Mazine, F.F., Nic Lughadha, E.M. *et al.* (2019) A new subtribal classification of tribe Myrteae (Myrtaceae). *Systematic Botany* 44(3), 560–569.

Machado, R.M. (2016) Geographic distribution and karyotype analysis in cytotypes of *Psidium cattleyanum* Sabine (Myrtaceae). M.Sc. dissertation, Universidade Estadual de Campinas, São Paulo, Brazil.

McVaugh, R. (1968) The genera of American Myrtaceae – an interim report. *Taxon* 17, 354–532.

Marques, A.M., Tuler, A.C., Carvalho, C.R., Carrijo, T.T., Ferreira, M.R.S. and Clarindo, W.R. (2016) Refinement of the karyological aspects of *Psidium guineense* (Swartz, 1788): a comparison with *Psidium guajava* (Linnaeus, 1753). *Comparative Cytogenetics* 10(1), 117–128.

Medina, A.L., Haas, L.I.R., Chaves, F.C., Salvador, M., Zambiazi, R.C. *et al.* (2011) Araçá (*Psidium cattleyanum* Sabine) fruit extracts with antioxidant and antimicrobial activities and antiproliferative effect on human cancer cells. *Food Chemistry* 128, 916–922.

Murillo, A.J., Ruiz, P.E., Landrum, L.R., Stuessy, T.F. and Barfuss, M.H.J. (2012) Phylogenetic relationships in *Myrceugenia* (Myrtaceae) based on plastid and nuclear DNA sequences. *Molecular Phylogenetics and Evolution* 62, 764–776.

O'Dea, A., Lessios, H.A., Coates, A.G., Eytan, R.I., Restrepo-Moreno, S.A. *et al.* (2016) Formation of the Isthmus of Panama. *Science Advances* 2, e1600883.

Pérez Gutiérrez, R.M., Mitchell, S. and Solis, R.V. (2008) *Psidium guajava*: a review of its traditional uses, phytochemistry, and pharmacology. *Journal of Ethnopharmacology* 117(1), 1–27.

PoWO (2020) Plants of the World Online. Facilitated by the Royal Botanic Gardens, Kew. Available at: http://www.plantsoftheworldonline.org/ (accessed 24 June 2020).

Raven, P.H. and Axelrod, D.I. (1974) Angiosperm biogeography and past continental movements. *Annals of the Missouri Botanical Garden* 61, 539–673.

Rivero, G., Salazar, G., Pacheco, D., Sánchez, A., Quirós, M. and Sthormes, G. (2012) Relaciones filogenéticas entre especies de *Psidium* (Myrtaceae) presentes en el occidente de Venezuela a partir de secuencias de ADN nuclear (ITS) y plastidial (trnH-psbA). *Interciencia* 37(11), 838–844.

Rotman, A. (1976) Revisión del género *Psidium* en la Argentina. *Darwiniana* 20, 418–444.

Rye, B.L. (1979) Chromosome number variation in variation in the Myrtaceae and its taxonomic implications. *Australian Journal of Botany* 27, 547–573.

Salywon, A.M. (2003) A monograph of *Mosiera* (Myrtaceae). PhD dissertation, Arizona State University, Tempe, Arizona.

Shady Solis, R., Haas, J. and Creamer, W. (2001) Dating Caral, a preceramic site in the Supe Valley on the central coast of Peru. *Science* 292, 723–726.

Silva, J.D.d., Luz, A.I.R., Silva, M.H.L.d., Andrade, E.H.A., Zoghbi, M.B. and Maia, J.G.S. (2003) Essential oils of the leaves and stems of four *Psidium* spp. *Flavour and Fragrance Journal* 18, 240–243.

Smith, C.E. (1965) The archeological record of cultivated crops of New World origins. *Economic Botany* 19(4), 322–334.

Sobral, M., Lucas, E., Landrum, L. and Soares-Silva, L. (2009) Myrtaceae. In: Stehmann, J.R., Campostrini Forzza, R., Salino, A., Sobral, M., da Costa, D.P. and Yoshino Kamino, L.H. (eds) *Plantas da Floresta Atlântica*. Jardim Botânico do Rio de Janeiro, Rio de Janeiro, Brazil, pp. 352–366.

Thornhill, A.H., Ho., S.Y.W., Külheim, C. and Crisp, M.D. (2015) Interpreting the modern distribution of Myrtaceae using a dated molecular phylogeny. *Molecular Phylogenetics and Evolution* 93, 29–43.

Tucker, A.O., Maciarello, M.J. and Landrum, L.R. (1995) Volatile leaf oils of American Myrtaceae. III. *Psidium cattleianum* Sabine, *P. friedrichsthalianum* (Berg) Niedenzu, *P. guajava* L., *P. guineense* Sw., and *P. sartorianum* (Berg) Niedenzu. *Journal of Essential Oil Research* 7, 187–190.

Vasconcelos, T.N.C., Proença, C.E.B., Ahmad, B., Aguilar, D.S., Aguilar, R. *et al.* (2017) Myrteae phylogeny, calibration, biogeography and diversification patterns: increased understanding in the most species rich tribe of Myrtaceae. *Molecular Phylogenetics and Evolution* 109, 113–137.

Watling, J., Shock, M.P., Mongeló, G.Z., Almeida, F.O., Kater, T. *et al.* (2018) Direct archaeological evidence for Southwestern Amazonia as an early plant domestication and food production centre. *PLoS ONE* 13(7), e0199868. https://doi.org/10.1371/journal.pone.0199868

Wen, L., Haddad, M., Fernández, I., Espinoza, G., Ruiz, C. *et al.* (2011) Actividad antinfúngica de cuatro plantas usadas en la medicina tradicional peruana. Aislamiento de 3′-formil-2′,4′,6′-trihidroxidihidrochalcona, principio activo de *Psidium acutangulum*. *Revista de la Sociedad Quimica del Peru* 77, 199–204.

2 Production and Trade

Fredy H. Ballen* **and Edward A. Evans**
University of Florida IFAS, Homestead, Florida, USA

2.1 Introduction

Mostly unknown outside their production areas, minor tropical fruits are considered novelty products sold in international and ethnic markets, or by premium retailers. Because of their unique nutritional attributes and flavour profiles, minor tropical fruits are increasingly receiving more attention from consumers worldwide.

In terms of volume, guava is the most important minor tropical fruit crop. For example, during the period 2015–2017, average production of guava was 6.75×10^6 t (Mt), accounting for 24.48% of the total volume of minor tropical fruits, followed by lychee (14.15%) and longan (14.02%).

Guava is indigenous to the American tropics; it has become naturalized in tropical and subtropical regions throughout the world. Early Spanish and Portuguese colonizers were the first to transport the fruit to other parts of the world. Asia and warmer parts of Africa quickly adopted guava as a commercial crop. Guava gained widespread acceptance in Asia, which has become the major production region, accounting for over 80% of the global guava production during the period 2015–2017.

The bulk of the production remains in the guava-producing regions due to strong domestic demand; only a small percentage of the total production reaches international markets. Demand for guava is increasing in both emerging and developed economies as familiarity with the fruit expands worldwide. Given the intense competition among major tropical fruits in international markets, guava represents an interesting option to increase farm income, export revenue and diversify fruit exports.

Regular supply of the fruit to import markets is not an issue because the fruit is available year-round from tropical and subtropical production areas. The main issues are the lack of promotional activities to expand the export market, improvements in preharvest and postharvest practices and wider distribution to improve pricing.

This chapter provides an overview of guava fruit world production, exports, imports and consumption trends, with the focus on US and European markets.

*E-mail: fredy.ballen@ufl.edu

DOI: 10.1079/9781789247022.0002

2.2 Guava Production, Trade and Consumption

2.2.1 Area harvested and world production

There are four different pulp colours of guava: white, red, pink and yellow. Red- and white-pulped are the main cultivars grown worldwide for the fresh and processed markets. Guava fruit is rich in dietary fibre; vitamins A and C; folic acid; and the dietary minerals, potassium, copper and manganese. Guavas are considered excellent sources of antioxidant phytochemicals, including ascorbic acid, carotenoids, antioxidant dietary fibre and polyphenolics, which act as chemoprotective agents against degenerative diseases and have antimutagenic effects and antiviral effects, among others (Rani and Vijayanchali, 2017).

The main constraint in guava production is insect pest management; there have been reports of approximately 80 insect pests, but only a few of them, mainly as the fruit fly, cause significant economic loss. For example, India, the major global guava producer, has experienced crop losses in the range of 16 to 40% due to fruit flies (Gundappa *et al.*, 2018).

Because of the informality of production, several guava-producing countries do not collect production data[1] on a consistent basis. In an effort to quantify production, average global guava production[2] for the period 2015–2017 was estimated at 6.75 Mt; Asia is the main producing region, accounting for 81.74% of the global production, followed by South America (8.50%), Africa (6.45%) and Central America and the Caribbean (3.31%) for the period 2015–2017 (Altendorf, 2018).

Production information for some of the major guava producers with official data for the period 2009–2018 is presented in Table 2.1. While not comprehensive due to data unavailability, the information presented provides important insights in identifying key producers and illustrating production trends. It represents detailed information for about 84% of the aggregated global production reported by Altendorf (2018) for the period 2015–2017.

Global guava production rates have varied widely. For India, by far the largest guava producer, production has grown by 79%, from 2.27 Mt in 2009 to 4.05 Mt in 2018 (NHB, 2020). Guava production in Pakistan has fluctuated slightly, ranging from 0.51 Mt in 2010 to 0.59 Mt in 2018 (Ministry of National Food Security & Research, Government of Pakistan, 2018). Brazil has significantly increased its guava production by 95%, from 0.30 Mt in 2009 to 0.58 Mt in 2018 (SIDRA, 2020). Guava production in Mexico has fluctuated from 0.29 Mt in 2009 to 0.31 Mt in 2018 (SIAP, 2020). After a downward trend from 2009 to 2013, guava production in Indonesia has increased, from 0.18 Mt in 2014 to 0.23 Mt in 2018 (Statistics Indonesia, 2019). Guava production in Thailand has doubled, from 0.11 Mt in 2009 to 0.23 Mt in 2018 (OAE, 2012; DOAE, 2013, 2018).

Increases in global guava production are due to expanded harvested areas and/or higher productivity. Gains in guava production in India are due to a combination of increases in both harvested area and productivity. India's guava harvested area has increased by 30%, from 204,000 ha in 2009 to 265,000 ha in 2018 (Table 2.2), and average yield has

Table 2.1. Production data for selected fresh guava producers, 2009–2018 ($\times 10^3$ t). From official sources.

	Year									
Country	2009	2010	2011	2012	2013	2014	2015	2016	2017	2018
India	2,270	2,572	2,462	2,510	3,198	3,668	3,994	4,048	3,826	4,054
Pakistan	512	509	524	557	525	527	537	523	548	586
Brazil	297	324	343	345	350	359	424	421	458	579
Mexico	289	305	291	295	298	303	294	309	325	312
Indonesia	220	205	212	208	182	182	196	207	200	231
Thailand	114	100	95	100	256	197	199	193	250	225
Total	3,702	4,015	3,927	4,015	4,809	5,236	5,644	5,701	5,607	5,987

Table 2.2. Harvested area for selected guava producers, 2009–2018 ($\times 10^3$ ha). From official sources.

	Year									
Country	2009	2010	2011	2012	2013	2014	2015	2016	2017	2018
India	204	220	205	220	236	268	246	255	260	265
Pakistan	62	62	62	67	66	66	65	70	70	65
Brazil	15	16	16	15	15	16	18	17	19	22
Mexico	22	22	21	21	21	20	21	22	22	22
Thailand	6	6	6	6	9	6	6	6	8	7
Total	310	326	310	329	346	377	356	370	380	381

increased by 35.4%, from 11.13 t ha^{-1} in 2009 to 15.30 t ha^{-1} in 2018. Brazil and Mexico have managed to increase their productivity without a significant expansion in area; average yield has increased by 35.35 and 5.86%, respectively. Thailand is a special case, where harvested area has increased slightly, while fruit yield has doubled as result of productivity gains, from 17.67 t ha^{-1} in 2009 to 31.29 t ha^{-1} in 2018.

2.2.2 Global trade – exports and imports

While only a very small percentage of the global production lands on international markets, the overall trend in exports is encouraging. Export data are not comprehensive because official information for some countries is unavailable. The information presented herein covers exports for some of the major producers based on available official data (Table 2.3). Mexico is a key guava supplier in international markets. Its share of export volume has increased by 152%, from 4306 t in 2009 to 10,850 t in 2018. Mexico's share of exported domestic production has increased from 1.5% in 2009 to 3.48% in 2018 (SIAVI, 2020).

Next in importance in global trade is Thailand, whose export volume has grown at an annual rate of 34%, from 2001 t in 2009 to 8117 t in 2018. Thailand's share of exported domestic production has increased from 1.75% in 2008 to 3.6% in 2018 (Thailand Trading Report, 2020).

Despite being the dominant global guava producer, fresh guava exports from India are minimal, where exports have fluctuated from a high of 1691 t in 2009 to a low of 301 t in 2011. On average, about 0.04% of India's domestic guava production is destined for international markets (APEDA, 2020).

Brazil's guava exports have increased by 9.15%, from 153 t in 2009 to 167 t in 2018. On average, Brazil's share of exported domestic production is about 0.04% (AGROSTAT, 2020).

Along with the increase in export volume, export value has also increased (Table 2.4). The export value of Mexican guavas has fluctuated from a low of US$8.34 million in 2009 to a high of US$24.66 million in 2017. Unit prices for Mexican guava exports have fluctuated from a low of US$1678 t^{-1} in 2014 to a high of US$2034 t^{-1} in 2016 (SIAVI, 2020). For Thailand, guava export value has trended upwards, from a low of US$0.93 million in 2009 to a high of US$6.45 million in 2018. Unit prices for Thai guava exports have fluctuated from a low of US$465 t^{-1} in 2009 to a high of US$795 t^{-1} in 2018 (Thailand Trading Report, 2020).

The export value of Brazilian guavas has fluctuated from a low of US$0.28 million in 2012 to a high of US$0.50 million in 2015. Unit prices for Brazilian guava exports have fluctuated from a low of US$1961 t^{-1} in 2009 to a high of US$2708 t^{-1} in 2013 (AGROSTAT, 2020). India's fresh guava export value has fluctuated from a low of US$0.14 million in 2011 to a high of US$1.09 million in 2016. On a per unit basis, India's fresh guava export value has fluctuated from a low of US$390 t^{-1} in 2009 to a high of US$701 t^{-1} in 2014 (APEDA, 2020).

Import information is not comprehensive as official data from most of the countries are unavailable.[3] Overall, there is an upward trend in fresh guava imports

Table 2.3. Fresh guava export volumes from major producers with official data, 2009–2018 (t). From official sources.

	Year									
Country	2009	2010	2011	2012	2013	2014	2015	2016	2017	2018
Mexico	4,306	5,531	5,203	6,752	7,604	8,051	9,297	11,959	12,340	10,850
Thailand	2,001	2,426	3,043	2,896	4,069	4,985	5,901	6,360	6,625	8,117
India	1,691	516	301	1,382	1,180	970	908	1,914	1,408	1,230
Brazil	153	147	137	120	144	171	204	172	143	167
Total	8,151	8,621	8,684	11,150	12,996	14,177	16,310	20,405	20,516	20,363

Table 2.4. Fresh guava export value from major producers based on official data, 2009–2018 (million US$). From official sources.

	Year									
Country	2009	2010	2011	2012	2013	2014	2015	2016	2017	2018
Mexico	8.34	9.86	10.08	11.56	14.05	13.51	15.99	24.33	24.66	21.90
Thailand	0.93	1.16	1.58	1.61	3.08	3.70	4.01	4.19	4.42	6.45
India	0.66	0.24	0.14	0.66	0.65	0.68	0.60	1.09	0.91	0.86
Brazil	0.30	0.33	0.30	0.28	0.39	0.44	0.50	0.40	0.34	0.40
Total	10.22	11.59	12.10	14.10	18.17	18.33	21.11	30.00	30.34	29.60

Table 2.5. Fresh guava import volume for selected destinations, 2009–2018 (t). From official sources.

	Year									
Country/ Region	2009	2010	2011	2012	2013	2014	2015	2016	2017	2018
USA	2,728	5,361	4,768	5,907	4,694	6,078	6,343	8,111	9,582	9,022
Singapore	1,909	2,033	2,696	2,697	2,456	2,467	2,871	3,075	3,234	4,110
Myanmar	2	11	3	19	1,296	2,183	2,690	3,062	3,124	2,258
Malaysia	52	214	264	54	43	56	101	64	89	1,570
Europe	207	196	159	247	223	199	251	202	253	288
Total	4,898	7,816	7,890	8,923	8,713	10,982	12,256	14,514	16,282	17,248

(Table 2.5). The USA is the main importer of fresh guavas, with fruit imports fluctuating from a low of 2728 t in 2009 to a high of 9582 t in 2017, and then dropping to 9022 t in 2018 (USDA-FAS, 2020). Singapore is also an important destination for fresh guava; its share of imports has grown from 1909 t in 2009 to 4110 t in 2018. Guava imports to Myanmar have grown significantly from just 2 t in 2009 to 2258 t in 2018. Guava imports to Malaysia have been below the level of 270 t for most of the period 2009–2018; then, in 2018, fruit imports reached 1570 t. Guava imports to Europe have fluctuated from a low of 159 t in 2011 to a high of 288 t in 2018.

2.2.3 US production

In the USA, guava was first introduced to Florida from Cuba in 1847 (Popenoe, 1920). US guava production takes place in the states of Florida, California, Hawaii and Texas. Florida is the main guava producer accounting for 65.19% of the total area in 2017, followed by California (23.46%), Hawaii (11.25%) and Texas (0.10%), respectively. Total area for guava production in the USA is minor (Table 2.6). Guava planted area reached a high of 702 ha in 2012 before dipping to 421 ha in 2017. Total non-bearing area, representative of new plantings, also decreased from 282 ha to just 41 ha that same year.

Demand for major tropical fruits like mango, papaya and pineapple in the US market seem to have reached a plateau; this may signal opportunities for minor tropical fruits such as guava. Figure 2.1 illustrates the US per capita consumption of papaya and guava for the period 2010–2018. While per capita consumption of the major tropical fruits began to wane in 2017, per capita consumption for guava grew slightly from about 0.02 kg in 2009 to 0.03 kg in 2018.

2.2.4 US imports

The US fresh guava import market opened in 2008, when commercial imports of Mexican guava were granted market access conditioned on the fruit being irradiated with a minimum absorbed dose of 400 Gy (American Shipper, 2008). Irradiation uses gamma rays to kill harmful bacteria and mould, along with live pests (e.g. weevils and fruit flies). In some cases, importers may be compensated for irradiation costs due to the lack of competitive substitutes (USDA-ERS, 2011).

Table 2.6. US bearing and total guava area, 2002–2017 (ha). From USDA-NASS (2020).

	2002	2007	2012	2017
Bearing area	388	257	420	380
Non-bearing area	103	100	282	41
Total area	491	357	702	421

Fresh guava imports in the USA have increased considerably since 2010; with an average annual growth rate of 9.32%, imports rose from 5361 t in 2010 to 9861 t in 2019. The value of such imports grew from US$11.2 million to US$22.1 million during this period, denoting an annual growth rate of 10.81%. As seen in Fig. 2.2, Mexico is the largest supplier of fresh guava to the US market, accounting for roughly 94% of total imports volume in the period 2017–2019.

Although US market demand for guava continues to rise, domestic guava supply is not expected to increase significantly in the near future. Therefore, fruit imports from other countries are expected to expand beyond Mexico. In October 2019, the USDA (US Department of Agriculture) authorized the importation of fresh guava from Taiwan to the USA (USDA-APHIS, 2020). Fresh guava imports from Taiwan are granted market access subject to bagging of fruit intended for export and phytosanitary treatments (cold treatment or irradiation) to meet US standards (US Federal Register, 2019).

In addition to the fresh fruit market, guavas also enter the USA as processed products in the form of paste/purée, jams, dried fruit, and other prepared or preserved types (Fig. 2.3). The volume of these imports increased from 11,771 t in 2010 to 16,091 t in 2019, representing an overall growth of 37%. Meanwhile their value increased by 31%, from US$12.6 million to

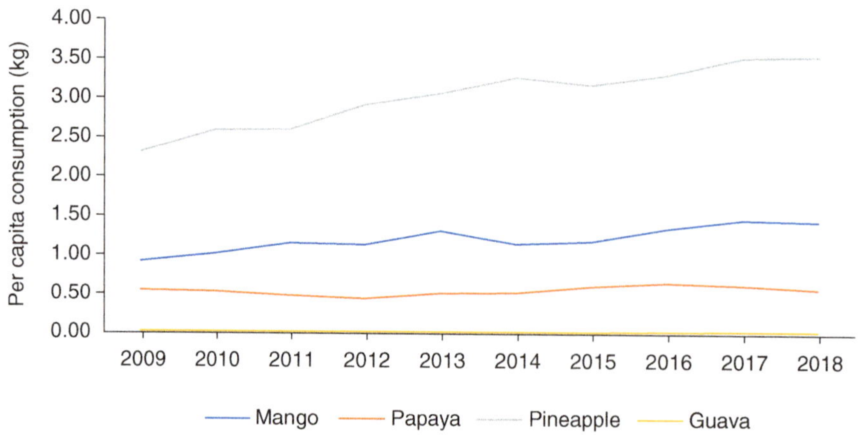

Fig. 2.1. US per capita consumption of selected tropical fruits, 2010–2018. From USDA-ERS (2020).

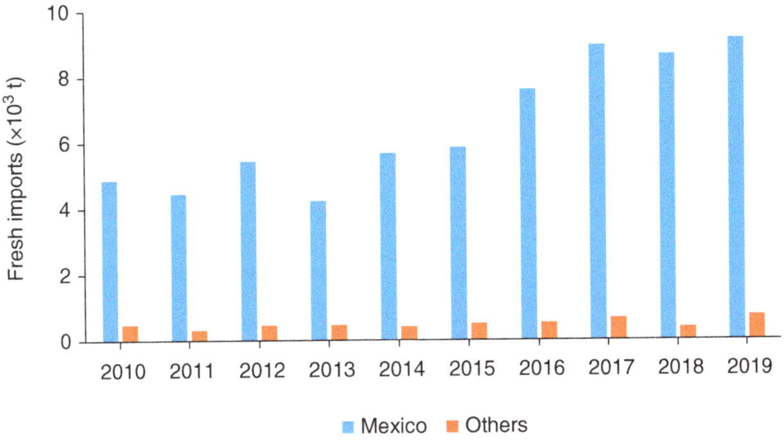

Fig. 2.2. US fresh guava imports, 2010–2019. From USDA-FAS (2020).

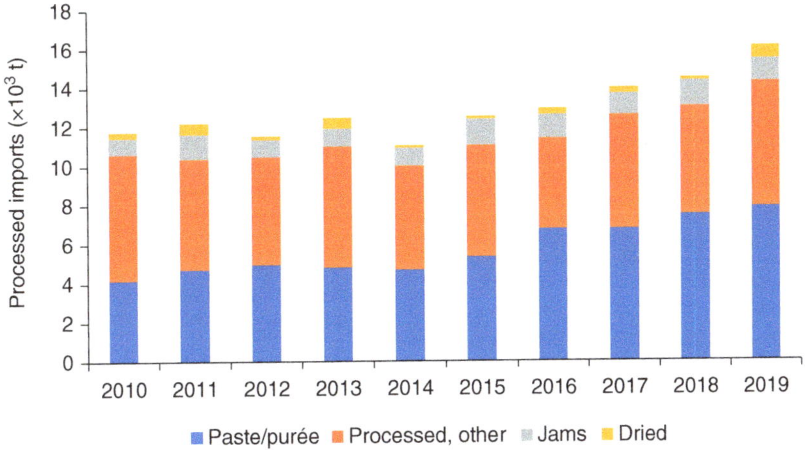

Fig. 2.3. US processed guava imports, 2010–2019. From USDA-FAS (2020).

US$16.5 million. Guava paste comprises the bulk of the processed imports, accounting for a roughly 50% share of total guava imports in the period 2017–2019, followed by prepared or preserved guavas (39.8%), jams (8.0%) and dried guavas (2.5%). While Mexico provides the vast majority of fresh guava imports, Brazil is the largest supplier of processed guavas, delivering 54% of total processed imports between 2017 and 2019.

Average import prices during the period 2015–2019 for the top three US fresh guava suppliers are presented in Table 2.7. Mexico has been the lowest-cost supplier for most of the period 2015–2019, with the exception of 2019 when guavas from India suddenly dipped to US$1.06 kg^{-1}.

Guavas from Thailand are by far the most expensive, with prices remaining above US$5 kg^{-1} for the period 2015–2019. The guava variety contributes largely to its price, as Thai guavas are deemed more desirable due to their larger size and crunchy texture similar to an apple (Crane *et al.*, 2014).

2.2.5 US market wholesale prices

San Francisco and New York City were selected as representative guava markets of the US east and west coasts. The majority of domestic guava supply comes from the US states of California and Florida, and the bulk of imported guavas comes from Mexico. While retail prices are not available for either city, terminal market prices for each market are shown in Figs 2.4 and 2.5, respectively.

Wholesale average prices for the period 2017–2019 for domestic and imported guavas in the San Francisco market are shown in Fig. 2.4. Guavas from California have exhibited higher prices than those imported from Mexico. Prices for California guavas fluctuate depending on the season, while prices for Mexican imported guavas remain more constant due their year-round availability. California guava prices in the period 2017–2019 averaged a low of US$2.69 kg^{-1} in the month of October. Mexican guava prices experienced their low in the same month, at US$2.62 kg^{-1}. The highest average price for California guavas was US$6.61 kg^{-1} in the month of April while for Mexican guavas it was US$4.03 kg^{-1} in January.

Wholesale average prices for the period 2017–2019 for domestic and imported guavas in the New York City market are shown in Fig. 2.5. As is the case with the San Francisco market, guavas imported from Mexico sell for lower prices than those acquired domestically from Florida. During the winter season especially, Florida guavas exhibit higher prices while in the summer and autumn seasons prices notably drop but never quite contend with Mexican imports. The average price for Florida guavas in the period 2017–2019 reached a high of US$7.45 kg^{-1} in February and a low of US$4.06 kg^{-1} in August. Mexican imports averaged a high price of US$3.81 kg^{-1} in May and a low of US$3.49 kg^{-1} in March.

Table 2.7. Average import prices for the top three US fresh guava suppliers, 2015–2019 (US$ kg^{-1}). From USDA-FAS (2020).

	Year				
Country	2015	2016	2017	2018	2019
Mexico	1.52	1.62	1.94	2.19	2.14
Thailand	6.05	6.14	5.98	6.75	5.14
India	4.57	3.97	2.82	3.95	1.06

2.2.6 European market

European agricultural trade import statistics show the aggregate quantities and values for mangos, guavas and mangosteens. While the information[4] shown in Table 2.8 is not comprehensive due to data limitations, it provides some insights in terms of

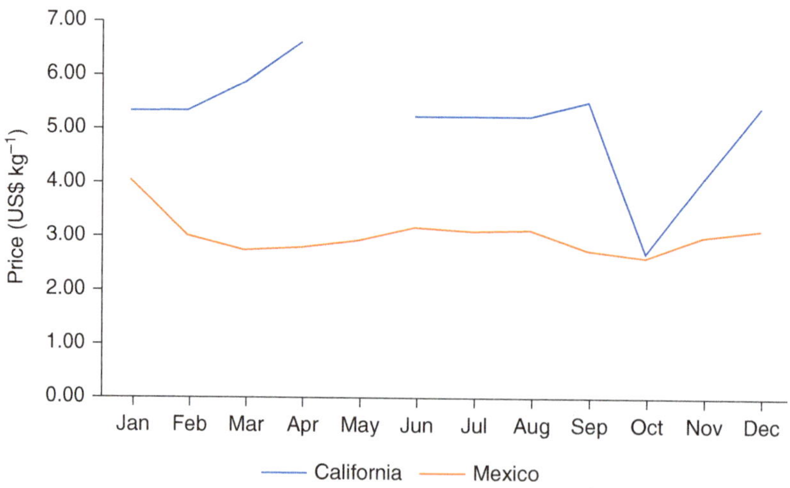

Fig. 2.4. Average terminal market prices for fresh guavas in San Francisco market, 2017–2019. From USDA-AMS (2020).

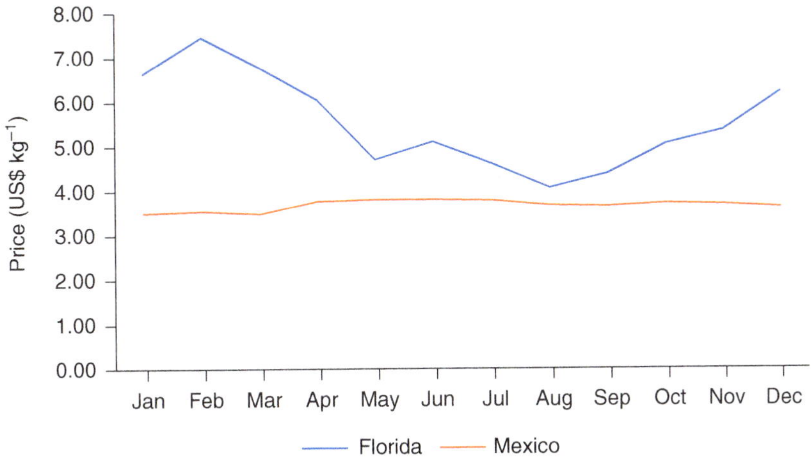

Fig. 2.5. Average terminal market prices for guavas in New York City market, 2017–2019. From USDA-AMS (2020).

Table 2.8. European guava-importing countries, 2009–2018 (t). From APEDA (2020), SIAVI (2020), AGROSTAT (2020) and Thailand Trading Report (2020).

	Year									
Country	2009	2010	2011	2012	2013	2014	2015	2016	2017	2018
UK	44	61	39	26	50	65	98	104	97	107
Netherlands	37	15	17	131	94	41	68	28	92	86
France	60	65	66	50	51	62	56	44	37	38
Portugal	24	16	16	10	9	16	17	17	14	19
Spain	20	17	10	9	5	7	8	7	5	7
Others	21	22	11	21	15	8	4	3	8	31
Total	207	196	159	247	223	199	251	202	253	289

identifying the major European guava importers and visualizing fruit trade trends. In Europe, the UK is the main guava importer, ethnic origin seems to be an important demand driver; for instance Indian- and Pakistani-born citizens account for 1.2% (694,000) and 0.9% (482,000) of the UK population, respectively (GOV.UK, 2018). The Netherlands is also an important guava destination market; however, the Netherlands is the European port of entry and the fruit is re-exported to other places in Europe. In contrast to the UK, fresh guava imports in France, Portugal and Spain have trended downwards.

Consistent with the increase in export volume to Europe, export value has also increased, albeit at a slower rate (Table 2.9). Fresh guava exports value to the UK reached US$139,000 in 2018. UK unit export value has trended downwards, from US$2308 t^{-1} in 2014 to US$1299 t^{-1} in 2018. Fruit export value to the Netherlands has remained unchanged at US$1000 t^{-1} during the period 2009–2018. Export value for fresh guavas to the French market has trended downwards, and unit export value has decreased gradually from US$3118 t^{-1} in 2013 to US$2289 t^{-1} in 2018.

2.2.7 Market outlook

The market outlook for fresh guavas in export markets seems promising as demand

Table 2.9. European fresh guava import values, 2009–2018 (thousand US$). From APEDA (2020), SIAVI (2020), AGROSTAT (2020) and Thailand Trading Report (2020).

Country	Year									
	2009	2010	2011	2012	2013	2014	2015	2016	2017	2018
UK	68	95	54	36	109	150	191	170	140	139
Netherlands	59	37	46	133	118	84	113	69	151	163
France	125	147	156	133	159	160	129	104	87	87
Portugal	35	29	24	14	12	29	43	44	37	52
Spain	27	28	27	22	14	22	20	15	12	20
Others	29	49	34	52	25	7	8	7	9	33
Total	342	386	341	390	437	453	505	409	436	494

for major tropical fruits in developed markets like the USA is starting to wane, although some challenges persist. From the consumer's side, there are encouraging signs; income growth, increasing health awareness and evolving dietary preferences will help to strengthen the demand for the fruit in both the domestic and international markets.

There is an expanding market for the fruit from emerging economies in Asia, particularly Singapore and Malaysia, where imports have grown significantly during the last 10 years. China is also another important market to consider in Asia; however, there are no available data to visualize market prospects.

Demand prospects from the developed economies, particularly the USA and Europe, remain strong; demographic factors, such as ethnicity, will continue to drive the demand for the fruit in the near future. Because of increasing health awareness and the desire to improve dietary habits/patterns, promotional activities to create market awareness about the nutritional and health benefits of the fruit will go a long way to expand the market and attract new consumers.

On the supply side, there are also both opportunities and challenges. Guava might be a good alternative for smallholder farms to increase income compared with staple crops. Major producers of the fruit are positioned to benefit the most from an increase in demand in the international markets, as they do not need to make substantial investments in new plantings. While area increases are feasible, it would be far easier to increase productivity at the farm level; as an example, Thailand has doubled its average yield without substantial increases in harvesting area.

Additional work is needed to improve quality standards, which are very strict in international markets. At this time, organic guava production is still in the early stages; demand for organic fruit may benefit from an increase in demand for conventionally grown guava. Fruit fly damage is the major production-limiting factor; nevertheless, this problem is easily manageable through fruit bagging, which is an environmentally friendly and cost-effective practice. Fruit bagging and phytosanitary treatments (irradiation or cold treatment) are required to have market access in international markets.

Due to the seasonality in guava production and given that fruit production occurs in tropical and subtropical areas, the opportunity to coordinate a regular flow of the fruit to international markets exists. An increase in the frequency of adverse weather effects in some of the producing areas has the potential to cause significant disruption in the production and supply, resulting in price volatility.

Given the highly perishable nature of the fruit, improvements in transportation and logistics will go a long way to improve fruit quality and gain consumer acceptance. Economies of scale from production to retail will be key to improve pricing and to increase the demand for guavas.

Notes

[1] FAO (Food and Agriculture Organization of the United Nations) reports aggregated production and trade data for mango, guava and mangosteen.
[2] Based on official information, industry sources and secondary sources.
[3] Imports for Singapore, Myanmar, Malaysia and Europe are compiled from the listed exporters' destinations.
[4] European guava import data are compiled from the export destinations for Brazil, India, Mexico and Thailand.

References

AGROSTAT (2020) Estatisticas de Comércio Exterior do Agronegócio Brasileiro. Ministério da Agricultura, Pecuária e Abastecimento, Brasil. Available at: http://indicadores.agricultura.gov.br/agrostat/index.htm (accessed 1 June 2020).

Altendorf, S. (2018) Minor tropical fruits: mainstreaming a niche market. *Food Outlook*, July 2018. Food and Agriculture Organization of the United Nations, Rome. Available at: http://www.fao.org/fileadmin/templates/est/COMM_MARKETS_MONITORING/Tropical_Fruits/Documents/Minor_Tropical_Fruits_FoodOutlook_1_2018.pdf (accessed 24 May 2020).

American Shipper (2008) USDA to allow guava fruit imports from Mexico. Available at: https://www.freightwaves.com/news/usda-to-allow-guava-fruit-imports-from-mexico (accessed 5 June 2020).

APEDA (2020) India Export Statistics. Agricultural & Processed Food Products Export Development Authority, New Delhi. Available at: https://agriexchange.apeda.gov.in/indexp/genReport_combined.aspx#content (accessed 5 April 2021).

Crane, J., Evans, E.A. and Garcia, S. (2017) Cost estimates of establishing and producing Thai guavas in Florida, 2014. Document No. FE998. University of Florida IFAS Extension, Gainesville, Florida. Available at: https://edis.ifas.ufl.edu/fe998 (accessed 15 May 2020).

DOAE (2013) Agricultural Production. Department of Agricultural Extension, Ministry of Agriculture and Cooperatives, Bangkok.

DOAE (2018) Agricultural Production. Department of Agricultural Extension, Ministry of Agriculture and Cooperatives, Bangkok.

GOV.UK (2018) UK population by ethnicity. People born outside the UK. Available at: https://www.ethnicity-facts-figures.service.gov.uk/uk-population-by-ethnicity/demographics/people-born-outside-the-uk/latest#place-of-birth-uk-or-non-uk-by-ethnicity (accessed 29 June 2020).

Gundappa, B., Balaji Rajkumar, M., Singh, S. and Rajan, S. (2018) Pests of guava. In: Omkar (ed.) *Pests and Their Management*. Springer, Singapore, pp. 491–516. Available at: https://link.springer.com/chapter/10.1007/978-981-10-8687-8_15 (accessed 25 June 2020).

Ministry of National Food Security & Research, Government of Pakistan (2018) Agricultural Statistics of Pakistan 2017–18. Available at: http://www.mnfsr.gov.pk/frmDetails.aspx# (accessed 25 June 2020).

NHB (2020) Horticultural Statistics at a Glance. National Horticulture Board, Ministry of Agriculture and Farmers Welfare, Government of India. Available at: http://nhb.gov.in/Statistics.aspx?enc=i3aXhtkJwc/n3rCHOr1FVp4BttTNWlLSQ8DhVptPrAbUppswYCodsFDUK1EY4Ru6yxB1yyjqgJ6NwxLqpANwXQ== (accessed 10 May 2020).

OAE (2012) Agricultural Statistics of Thailand, 2012. Office of Agricultural Economics, Ministry for Agricultural and Cooperatives, Bangkok. Available at: http://www.oae.go.th/assets/portals/1/files/ebook/yearbook55.pdf (accessed 10 May 2020).

Popenoe, W. (1920) *Manual of Tropical and Subtropical Fruits*. Macmillan, New York.

Rani, D.J. and Vijayanchali, S.S. (2017) Phytochemical, antioxidant activity and lycopene analysis of red guava fruits. *Journal of Research, Extension and Development* (6), 25–30. https://doi.org/10.2139/ssrn.3345542

SIAP (2020) Estadística de Producción Agrícola. Servicio de Información Agroalimentaria y Pesquera, Gobierno de México. Available at: http://infosiap.siap.gob.mx/gobmx/datosAbiertos.php (accessed 20 May 2020).

SIAVI (2020) Sistema de Información Arancelaria via Internet. Secretaría de Economía, Gobierno de México. Available at: http://www.economia-snci.gob.mx (accessed 15 June 2020).

SIDRA (2020) Produção Agrícola Municipal. Sistema IBGE de Recuperação Automática, Brasil. Available at: https://sidra.ibge.gov.br/tabela/5457 (accessed 15 May 2020).

Statistics Indonesia (2019) Agriculture and Mining: Production of Horticulture (Dynamic). BPS – Statistics Indonesia, Jakarta. Available at: https://www.bps.go.id/site/pilihdata (accessed 16 June 2020).

Thailand Trading Report (2020) Exports of Thailand Classified by Commodity. Available at: http://www.ops3.moc.go.th/hs/export_commodity/ (accessed 20 June 2020).

USDA-AMS (2020) Terminal Market Averages. US Department of Agriculture, Agricultural Marketing Service, Washington, DC. Available at: https://www.ams.usda.gov/market-news/fruits-vegetables (accessed 3 June 2020).

USDA-APHIS (2020) USDA Issues Notice of Decision to Authorize the Importation of Fresh Guava Fruit from Taiwan into the Continental United States. US Department of Agriculture, Animal and Plant Health Inspection Marketing Service, Washington, DC. Available at: https://www.aphis.usda.gov/aphis/newsroom/stakeholder-info/sa_by_date/2019/sa-10/taiwan-guava-fruit (accessed 11 June 2020).

USDA-ERS (2011) Irradiation of produce imports: small inroads, big obstacles. *Amber Waves*, 16 June. US Department of Agriculture, Economic Research Service, Washington, DC. Available at: https://www.ers.usda.gov/amber-waves/2011/june/irradiation-of-produce-imports/ (accessed 1 June 2020).

USDA-ERS (2020) Fruit and Tree Nuts Yearbook Tables. Fruits: Supply and Utilization. US Department of Agriculture, Economic Research Service, Washington, DC. Available at: https://www.ers.usda.gov/data-products/fruit-and-tree-nuts-data/fruit-and-tree-nuts-yearbook-tables/ (accessed 4 June 2020).

USDA-FAS (2020) Global Agricultural Trade System. US Department of Agriculture, Foreign Agricultural Service, Washington, DC. Available at: https://apps.fas.usda.gov/gats/default.aspx (accessed 2 June 2020).

USDA-NASS (2020) Statistics by State. US Department of Agriculture, National Agricultural Statistics Service, Washington, DC. Available at: https://www.nass.usda.gov/Statistics_by_State/ (accessed 10 June 2020).

US Federal Register (2019) Decision to Authorize the Importation of Fresh Guava from Taiwan into the Continental United States. Available at: https://www.federalregister.gov/documents/2019/10/17/2019-22648/decision-to-authorize-the-importation-of-fresh-guava-from-taiwan-into-the-continental-united-states (accessed 15 June 2020).

3 Composition and Processing

Anup K. Bhattacherjee* and Dileep K. Tandon
ICAR–Central Institute for Subtropical Horticulture, Lucknow, Uttar Pradesh, India

3.1 Introduction

Guava fruit is an excellent source of ascorbic acid or vitamin C (about 4–5 times higher than orange), niacin or vitamin B_3, carotenoids or provitamin A (pink- or red-pulped), pantothenic acid or vitamin B_5, as well as potassium, phosphorus, magnesium and calcium. Besides vitamins and minerals, the fruits also contain dietary fibre, antioxidant phenolic compounds, sugars and organic acids (McCook-Russell *et al.*, 2012). Guava leaf is rich in essential oils, carotenoids and some polyphenols, while the seeds possess fatty acids and minor amounts of ascorbic acid. The plant parts of guava (leaf, bark, fruit, etc.) are commonly used as folk medicine in numerous regions of the world (North America, Central America, Asia and Africa) to cure human ailments like diarrhoea, fever, caries, dysentery, diabetes, gastroenteritis and hypertension, and for pain relief and wounds (Anand *et al.*, 2016; Naseer *et al.*, 2018). The anticancer and antioxidant activities of guava fruits are believed to be due to the presence of phenolic compounds and ascorbic acid.

3.2 Composition

The composition of guava fruit varies according to its development stage, pulp colour and variety, and with environmental conditions. There is a lot of variation observed in fruit weight. The average fruit weight varies from 150 to 300 g for Indian cultivars. The fruits grown in Thailand and Taiwan are usually much bigger in size (400–550 g). The fruit contains about 65–75% pulp, 25–35% seed core and a negligible amount of peel, and it is eaten with peel whether raw or ripe. The seed/pulp ratio varies from 1:20 to 1:40 depending on cultivar. The moisture content in fruit generally ranges between 75 and 85%. The pink/red-pulped cultivars usually have bold seeds. Due to its delicious taste and exotic flavour, guava is mostly consumed fresh in many countries.

3.2.1 Minerals

Minerals are of great importance in our daily diet, although they comprise only 4–6% of human body weight. Some minerals are

*E-mail: dadabhatu@gmail.com

 Guava: Botany, Production and Uses (ed. S. Mitra)
DOI: 10.1079/9781789247022.0003

macroelements that are required in amounts greater than 100 mg day^{-1}, representing 1% or less of body weight. These essential macroelements include calcium (Ca), magnesium (Mg), phosphorus (P), potassium (K), sodium (Na), sulfur (S) and chlorine (Cl). Zinc (Zn), iron (Fe), copper (Cu), manganese (Mn), selenium (Se), iodine (I) and molybdenum (Mb) are essential micro (trace) elements and are required in amounts less than 100 mg day^{-1}, making up less than 0.01% of body weight (Imelouane *et al.*, 2011). Guava fruits are particularly rich in potassium, phosphorus, magnesium and calcium besides containing iron, manganese, sodium and zinc (Table 3.1). The copper and zinc contents were reported to be higher in winter-season fruits, while iron and manganese contents were higher in rainy-season fruits (Chauhan *et al.*, 1986). 'Allahabad Safeda' guava contained good amounts of copper (147 µg 100 g^{-1}), iron (1502 µg 100 g^{-1}) and zinc (247 µg 100 g^{-1}), while 'Banarsi Surkha' was reported to be rich in manganese (383 µg 100 g^{-1}). The concentrations of minerals vary significantly with tree age and canopy position of fruit where it is grown. Fruits from a 15-year-old guava (cultivar 'Allahabad Safeda') tree collected from middle canopy possessed maximum copper (1.43 mg kg^{-1}), manganese (0.55 mg kg^{-1}) and iron (3.10 mg kg^{-1}), while fruits from a 20-year-old tree had maximum magnesium (17.37 mg kg^{-1}, upper canopy) and zinc (0.57 mg kg^{-1}, middle canopy) (Asrey *et al.*, 2007). The presence of 18 mg calcium, 256 mg potassium, 9 mg magnesium and 23 mg phosphorus was observed in a single guava fruit without seed (90 g) by Murdock (2002). Vora *et al.* (2018) have observed that red guava pulp contained much higher amounts of sodium, calcium and phosphorus compared with white guava pulp, although white guava pulp contained a significantly higher amount of potassium than red guava pulp. Guava seeds also possess minerals like calcium, potassium, magnesium, sodium, phosphorus, sulfur, iron, manganese and zinc. Uchôa *et al.* (2008) reported magnesium content of 0.19 mg 100 g^{-1} in fresh guava seeds. Guava-seed powder, obtained after pulp processing, contained iron as the major mineral (13.8 mg 100 g^{-1}) followed by zinc (3.31 mg 100 g^{-1}) (Uchôa-Thomaz *et al.*, 2014).

3.2.2 Sugars

The fruits are usually rich in carbohydrates. Carbohydrates are sugars that exist in fruits generally in two main forms: (i) simple; and (ii) complex. The difference between a simple and a complex carbohydrate lies in the number of sugar molecules they possess and how quickly they are broken down and metabolized. Usually simple sugars are monosaccharides (glucose, fructose, galactose, etc.) and disaccharides (sucrose, maltose, lactose, etc.). Any complex carbohydrate that is made up of more than two simple sugars is referred to as a polysaccharide, such as starch, pectin, cellulose, etc. Sugars are one of the major components of guava fruit and responsible for the fruit's sweetness. The sugars found in fruit are mainly reducing (glucose and fructose) and non-reducing (sucrose) sugars. Reducing sugars comprise more than 80% of the total sugars content in fruit. Fructose is the major sugar identified in guava, followed by glucose and sucrose (Mowlah and Itoo, 1982; Wilson *et al.*, 1982). Fructose level was highest among the three sugars (fructose, glucose and sucrose) monitored during maturity, while glucose content increased between 10 and 12 weeks of fruit development (Yusof and Mohamed, 1987). It was observed by El Bulk *et al.* (1997) that the contents of total sugars, fructose, glucose and sucrose were in the range of 13.7–30.6, 5.6–7.7, 1.9–8.0 and 6.2–7.8 mg 100 ml^{-1} juice, respectively, during development and ripening of guava fruits. Total soluble solids (TSS) and total sugars increased in peel and pulp of both white- and pink-pulped guavas with decrease in pulp firmness during ripening, because of enzymatic breakdown of starch into sugars, and peel contained more total sugars than pulp (Bashir and Abu-Goukh, 2003).

3.2.3 Dietary fibre and pectin

Dietary fibre is increasingly considered an essential aspect of good nutrition. Intake of dietary fibre alters the water content, viscosity and microbial mass of the intestinal contents, resulting in changes in the rate

Table 3.1. Proximate composition and nutritional value of guava fruits (per 100 g serving).

Proximate composition/ Nutritional value	Indian guava (Singh *et al.*, 2003/Kamath *et al.*, 2008)	Bangladeshi guava (Uzzaman *et al.*, 2018)	US guava (USDA, 2019/ Murdock, 2002[a])
Energy	66 mg/77–86 g	77–86 g	285 kJ (68 kcal)/46 kcal
Moisture	~80%	~80%	78%/80.8 g
Carbohydrates	9.1–17 mg	9.1–17 mg	14.32 g/11 g
Total sugars	6–9%	–	8.92 g
Dietary fibre	0.9–1.0 g	0.9–1.0 g	5.4 g/5 g
Protein	0.3–1.5%/0.1–0.5 g	0.1–0.5 g	2.55 g/1 g
Fat	0.03–0.1%/0.43–0.7 g	0.43–0.70 mg	0.95 g/1 g
Ash	9.5–10 g	9.5–10 mg	1.39 g
Vitamins			
Vitamin A equivalents	200 IU	–	31 µg/624 IU/71 RE[b]
β-Carotene	0.046 mg	0.046 mg	374 µg
Lycopene	–	–	5204 µg
Vitamin B_1	0.03–0.04 mg	0.03–0.04 mg	0.067 mg
Vitamin B_2	0.6–1.068 mg	0.6–1.068 mg	0.04 mg
Vitamin B_3	35–40 IU	35–40 IU	1.084 mg/1 mg
Vitamin B_5	–	–	0.451 mg
Vitamin B_6	–	–	0.11 mg
Vitamin B_9	–	–	49 µg
Vitamin C	150–350 mg	–	228.3 mg/165 mg
Vitamin K	–	–	2.2 µg/2.6 µg
Vitamin E	–	–	0.73 mg/1 mg
Vitamin G_4	36–50 mg	36–50 mg	–
Minerals			
Calcium	0.01–0.02%/17.8–30 mg	17.8–30 mg	18 mg/18 mg
Iron	1.0–1.8%/200–400 IU	200–400 IU	0.26 mg
Magnesium	–	–	22 mg/9 mg
Potassium	–	–	417 mg/256 mg
Phosphorus	0.01–0.04%/0.3–0.7 mg	0.3–0.7 mg	40 mg/23 mg
Manganese	–	–	0.15 mg
Sodium	–	–	2 mg
Zinc	–	–	0.23 mg
Copper	–	–	0.23 mg
Selenium	–	–	0.6 µg
Fatty acids			
Total saturated	–	–	0.272 g
Total monounsaturated	–	–	0.087 g
Total polyunsaturated	–	–	0.401 g
Cholesterol	–	–	0 mg
Amino acids			
Tryptophan	–	–	0.022 g
Threonine	–	–	0.096 g
Isoleucine	–	–	0.093 g
Leucine	–	–	0.171 g
Lysine	–	–	0.072 g
Methionine	–	–	0.016 g
Phenylalanine	–	–	0.006 g
Tyrosine	–	–	0.031 g
Valine	–	–	0.087 g
Arginine	–	–	0.065 g
Histidine	–	–	0.022 g
Alanine	–	–	0.128 g

Continued

Table 3.1. Continued.

Proximate composition/ Nutritional value	Indian guava (Singh *et al.*, 2003/Kamath *et al.*, 2008)	Bangladeshi guava (Uzzaman *et al.*, 2018)	US guava (USDA, 2019/ Murdock, 2002[a])
Aspartic acid	–	–	0.162 g
Glutamic acid	–	–	0.033 g
Glycine	–	–	0.128 g
Proline	–	–	0.078 g
Serine	–	–	0.075 g

[a]Per 90 g serving of fruit (without seed).
[b]RE, retinol equivalents.

and ease of passage through the intestine (Elleuch *et al.*, 2011). Dietary fibre improves glucose tolerance by delaying the transport of carbohydrates into the small intestine, thereby reducing the risk of heart disease and easing constipation (Anderson *et al.*, 1994). There are two types of dietary fibre available in food materials: (i) soluble; and (ii) insoluble. Fruits contain mostly soluble dietary fibre. Pectin is a nutritionally important but complex polysaccharide in plant cell walls with applications in food, pharmaceuticals and a number of other industries. Pectin also serves as a gelling and stabilizing polymer in diverse food and speciality products, has positive effects on human health and multiple biomedical uses (Mohnen, 2008). It can be used to reduce blood cholesterol levels and gastrointestinal disorders. In the food industry, pectin is used to prepare jams, jellies and frozen foods as well as in low-calorie foods as a fat/sugar substitute (Thakur *et al.*, 1997). Guava fruit is rich in pectin too.

Both pulp and peel fractions of guava contain high amounts of dietary fibre (48.55–49.42%, dry matter basis) and could be used to obtain antioxidant dietary fibre, an item with the properties of dietary fibre and antioxidant compounds combined in a single natural product (Jiménez-Escrig *et al.*, 2001). Total dietary fibre and pectin contents in red-pulped guava were reported to be higher (7.2 and 1.04 g 100 g^{-1}, respectively) than in white-pulped guava (4.0 and 0.77 g 100 g^{-1}, respectively) and were listed highest among nine tropical fruits (green and ripe) grown in Florida, USA (Mahattanatawee *et al.*, 2006). Guava fruits grown in Queensland, Australia contained 18% of daily fibre intake per 100 g of fruit (Fanning *et al.*, 2008). Dietary fibre content of 'Sardar', 'Allahabad Safeda' and 'Apple Guava' was recorded as 4.90, 4.22 and 4.84%, respectively (Shukla and Shukla, 2017). It was also noticed that the pectin content in these cultivars ranged between 1.08% ('Apple Guava') and 1.23% ('Sardar').

Guava pectic substances, fractionated into water-, oxalate- and alkali-soluble fractions in two canned guava cultivars and in calcium chloride ($CaCl_2$)-treated fresh fruits, indicated the gradual conversion of protopectin into soluble pectin which diffused into the syrup (El Tinay *et al.*, 1979). Similarly, the conversion of protopectin into soluble pectin during maturity and the subsequent increase in soluble pectin during fruit development were reported in Vietnamese guava (Yusof and Mohamed, 1987). The crude pectin content in Indian guava fruits was found to be significantly higher in winter-season crops compared with rainy-season crops; however, the pectin quality in terms of methoxyl content, anhydrogalacturonic acid content, degree of esterification and jelly grade was better in rainy-season fruits (Dhingra *et al.*, 1983). Total pectin content of guava fruit grown was found to be higher at mature stage (3.40%) than in ripe stage (0.67%) (Chyau *et al.*, 1992).

3.2.4 Protein and fat

The fruits are generally poor in protein and fat. Winter-season guava contained higher amounts of protein and fat than rainy-season guava (Chauhan *et al.*, 1986). 'Allahabad Safeda' guava had protein content of 1.24% while 'Sardar' guava had 1.21% during the winter season, while during the rainy season 'Sardar' contained a protein content of 0.91%. The fat content did not vary much among the cultivars during both seasons and 'Sardar' contained 0.16 and 0.18% fat in rainy and winter season, respectively. Total protein in pulp and peel of white- and pink-pulped guava fruits increased gradually up to the full-ripe stage and declined thereafter (Bashir and Abu-Goukh, 2003). The crude protein content in Indian guava fruits (white- and pink-pulped) varied from 0.69 to 0.94% (Joshi, 2016). The amount of protein varied from 1.38 to 1.68 mg 100 g^{-1} among the 128 guava accessions collected from four different regions of Kenya (Chiveu *et al.*, 2019). The US National Nutrient Database (USDA, 2019) showed that USA-grown common guava fruits contained 2.55 g of protein and 0.95 g of fat per 100 g. Among the lipids, total saturated, total monounsaturated and total polyunsaturated fatty acid contents were 0.272, 0.087 and 0.401 g 100 g^{-1}, respectively. Similarly seventeen amino acids, which are the building blocks of protein, were detected in guava fruit, among them glutamic acid content was the maximum (0.333 g 100 g^{-1}) and phenylalanine content was minimum (0.006 g 100 g^{-1}). Six fatty acids (lauric, oleic, myristic, palmitic, stearic and linoleic acids) were detected in guava-seed oil of which linoleic acid (65–75%) content was reported as maximum (Kobori and Jorge, 2005). Unsaturated fatty acids constituted the major portion (86%) compared with saturated fatty acids (14%). A protein content of 9.2 g 100 g^{-1} was obtained during the isolation of protein from guava seed (Fontanari *et al.*, 2008). Guava-seed protein has functional properties similar to those of other seeds that have been used as food ingredients and may be an alternative source of protein for future use in processed foods. Guava-seed powder (cultivar 'Paluma'), obtained from pulp processing in Brazil, contains high amounts of protein (11.19 g 100 g^{-1}) and total lipids (13.93 g 100 g^{-1}) (Uchôa-Thomaz *et al.*, 2014). The lipid profile of seed powder showed a predominance of unsaturated fatty acids (87.06%) especially linoleic acid and oleic acid. Eleven different kinds of fatty acids, namely lauric acid (0.07%), myristic acid (0.10%), palmitic acid (8.00%), heptadecanoic acid (0.07%), stearic acid (4.48%), oleic acid (9.42%), linoleic acid (77.35%), arachidic acid (0.12%), gondoic acid (0.14%), linolenic acid (0.15%) and behenic acid (0.10%), were recorded in guava-seed powder. Palmitic and stearic acids were the major saturated fatty acids detected.

3.2.5 Vitamins

Vitamins are organic molecules, but essential nutrients required by the human body in small amounts to function properly and stay healthy. The vitamins are either water- or fat-soluble. Vitamins A, D, E and K are fat-soluble while C and B complex are water-soluble. Fruits are excellent sources of vitamins. Guava fruit is one of the richest sources of vitamin C (ascorbic acid). It contains generally three times the amount of recommended daily intake of vitamin C (which is 80 mg day^{-1}). However, the vitamin C content depends on the fruit ripeness and cultivar as well as location. Plenty of reports are available on profiling of ascorbic acid in fresh guava as well as its processed products. Ascorbic acid is a well-known antioxidant, having the ability to quench reactive oxygen species, keep membrane-bound α-tocopherol (vitamin E) in a reduced state, act as a cofactor for some enzymes by keeping metal ions in a reduced state, can play a role in stress management and appears to be a substrate for oxalate and tartarate biosynthesis (Davey *et al.*, 2000; Arrigoni and de Tullio, 2002). It also helps in the formation of connective tissues like teeth, blood vessels and bones, prevents many diseases such as scurvy, anaemia, common cold and haemorrhagic disorders, and aids wound healing.

The average ascorbic acid content (mg 100 g^{-1} pulp) of several Indian cultivars has been estimated by many researchers and the ascorbic acid content of some cultivars was reported as follows: 'Allahabd Safeda' (329), 'Behat Coconut (98), 'Sardar' (250), 'Kerala'

(268), 'Chittidar' (234), 'Red Fleshed' (226), and 'Apple Colour' (220) (Tandon *et al.*, 1983b; Chauhan *et al.*, 1986; Pandey and Singh, 1998; Reddy *et al.*, 1999; Patel *et al.*, 2013).

White-pulped guavas contain higher ascorbic acid (84.5–250.8 mg 100 g^{-1}) than pink-pulped guavas (70.0–190.7 mg 100 g^{-1}) grown in Sudan. Ascorbic acid contents were estimated as 130 and 112 mg 100 g^{-1} in white-pulped (cultivar 'Pansithong') and red-pulped guava (cultivar 'Samsi'), respectively, in Thailand (Thuaytong and Anprung, 2011). Among the two cultivars of pink-pulped guava widely produced in Malaysia, 'Semenyih' contained a much higher concentration of ascorbic acid (202 mg 100 g^{-1}) compared with 'Sungkai' (135 mg 100 g^{-1}) and peel contained higher ascorbic acid than fruit (Musa *et al.*, 2015). In Pakistan, the highest concentration of ascorbic acid (234.6 mg 100 g^{-1}) was recorded in 'Hong Kong' cultivar, while the lowest (114.6 mg 100 g^{-1}) was in 'Ruby × Supreme' (Ghani *et al.*, 2016). Guava accessions collected from the eastern region of Kenya showed highest ascorbic acid content (128.1 mg 100 g^{-1}), compared with 43.7 mg 100 g^{-1} in fruits sampled from the Rift Valley region (Chiveu *et al.*, 2019). Guava-seed powder also possessed a considerable amount of vitamin C (87.44 mg ascorbic acid per 100 g) as mentioned by Uchôa-Thomaz *et al.* (2014).

Pink/red-pulped guava contains a considerable amount of lycopene, a major compound from the carotenoids group (vitamin A). Other vitamins present in minute amounts are niacin (vitamin B_3), folate (vitamin B_9), pantothenic acid (vitamin B_5), thiamine (vitamin B_1), riboflavin (vitamin B_2) and vitamin K (trace amount) (Table 3.1). Carotenoids like β-carotene, lycopene, lutein and zeaxanthin can act as the most efficient singlet oxygen quencher in biological systems without the production of any oxidizing products (Das *et al.*, 2012). Lycopene is the most powerful antioxidant from the carotenoids group, helping in decreasing lipid peroxidation and hydroxyl radical formation by quenching singlet oxygen (Dimascio *et al.*, 1989). Various functions and actions have been attributed to carotenoids, making determination of their concentrations in food items highly desirable. Pink-pulped guava cultivar 'IAC-4', sampled from São Paulo state, Brazil, contained 53.4 ± 6.3 μg of lycopene and 3.7 ± 0.7 μg of β-carotene per gram, respectively, where other pigments from the carotenoids group, namely ζ-carotene, γ-carotene, 5,6,5′,6′-diepoxy-β-carotene, zeinoxanthin and a trihydroxy-5,8-epoxy-β-carotene, were also detected at trace levels (Padula and Rodriguez-Amaya, 1986). Guava fruits of unknown varieties, collected from the states of Pernambuco and Ceará in Brazil, had lycopene and β-carotene concentrations of 53.4 ± 14.1 and 47.0 ± 15.7 μg g^{-1} and 11.9 ± 5.2 and 5.5 ± 2.3 μg g^{-1}, respectively. In the state of Rio de Janeiro, Brazil, pink/red-pulped ripe guavas have a considerable amount of lycopene (44.8–60.6 μg g^{-1}), a principal pigment of the carotenoids group, much higher than papaya (17.7–28.6 μg g^{-1}) (Wilberg and Rodriguez-Amaya, 1995). However, they have less β-carotene (3.02–5.84 μg g^{-1}), major provitamin A, than mango (8.2–28.7 μg g^{-1}) but higher than papaya (0.80–1.76 μg g^{-1}). Sixteen carotenoids from the pulp of Brazilian, red-pulped guavas were isolated using ultraviolet–visible (UV-VIS) spectroscopy, liquid chromatography (LC)–mass spectrometry (MS) and nuclear magnetic resonance spectroscopy (NMR) by Mercadante *et al.* (1999). The identified carotenoids were phytofluene; (all-*E*)-, (9*Z*)-, (13*Z*)- and (15*Z*)-β-carotene; (all-*E*)-γ-carotene; (all-*E*)-, (9*Z*)-, (13*Z*)- and (15*Z*)-β-lycopene; (all-*E*,3*R*)-β-cryptoxanthin; (all-*E*,3*R*)-rubixanthin; (all-*E*,3*S*,5*R*,8*S*)-cryptoflavin; (all-*E*, 3*R*,3′*R*,6′*R*)-lutein; and (all-*E*,3*S*,5*R*,6*R*,3′*S*,5′*R*, 8′*R*)- and (all-*E*,3*S*,5*R*,6*R*,3′*S*,5′*R*,8′*S*)-neochrome. In Brazil, some other researchers mentioned lycopene content in pink guava varieties as 35, 644.9 or 6999.3 μg g^{-1} depending on cultivar and fruit maturity (Oliveira *et al.*, 2011; Sousa *et al.*, 2011; Silva *et al.*, 2014). Guava fruits grown in Thailand possessed 714.5 μg of ascorbic acid, 0.15 μg of lutein, 0.02 μg of β-cryptoxanthin, 0.12 μg of β-carotene, 0.34 μg of total carotenoids, 2.40 μg of α-tocopherol and 1.12 mg of gallic acid equivalents per gram of total phenols, but no lycopene was detected (Isabelle *et al.*, 2010). Lycopene content in two pink-pulped guavas ('Sungkai' and 'Semenyih') grown in Malaysia was reported to be 40.5 and 52.8 mg kg^{-1}, respectively (Musa *et al.*, 2015). Among the ten seedling progenies of 'Apple Colour' guava, lycopene content varied from 2.66 (AC Selection 2/11) to 3.61 mg 100 g^{-1} (AC Selection 6/10) (Marak and Mukunda, 2007).

Although no investigation has been done for the estimation of other vitamins in guava, especially from the B-complex group, a few reports are available regarding vitamin contents in Indian, Bangladeshi and American guavas (average value) as presented in Table 3.1.

3.2.6 Phenolic compounds

Phenolic compounds (polyphenols) are secondary metabolites of plants widely distributed in various plant parts and plant-based food products. According to the nature of their carbon skeleton, phenolic compounds can be classified into four broad groups: (i) phenolic acids like cinnamic acid, ellagic acid, cafffeic acid, chlorogenic acid, *p*-coumaric acid, ferulic acid, etc.; (ii) flavonoids which include flavans such as gallic acid, catechin, epicatechin, epigallocatechin, etc., flavonols such as kaempferol, quercetin, etc. and anthocyanins such as cyanidin, delphinidin, malvidin, peonidin, etc.; (iii) stilbenes such as resveratrol, piceatannol, etc.; and (iv) lignans, namely sesamin, matairesinol, pinoresinol, lariciresinol, secoisolariciresinol, etc. (Lee *et al.*, 2003). These chemicals have the capacity to quench lipid peroxidation, prevent DNA oxidative damage, scavenge free radicals and prevent inhibition of cell communication (Cao and Cao, 1999). The antioxidant activity of phenolic compounds is due to the reactivity of the hydroxyl group in the phenol moiety of the aromatic ring, which has the ability to scavenge free radicals via hydrogen or electron donation (Shahidi *et al.*, 1992). Several *in vitro* studies have suggested that they have the ability to alter numerous cellular processes like gene expression, apoptosis, platelet aggregation and intercellular signalling, with anticarcinogenic and anti-atherogenic implications (Duthie *et al.*, 2003). Therefore, phenolic compounds can be considered a positive quality of fruits, although they are not considered vital nutrients for humans (Rickman *et al.*, 2007). Since there are hundreds of phenolic compounds identified, many researchers report composite total phenolics or total polyphenols (TP) values. The amounts of bioactive phenolic compounds in fruits depend on numerous factors such as cultivar, degree of maturity, agroclimatic conditions, soil composition, geographic location, storage conditions, etc., which explains the discrepancy between data generated from different researchers for the same fruit.

Guava fruit has high concentrations of extractable polyphenols (2.62–7.79%) containing protocatechuic acid, quercetin, ferulic acid, gallic acid and caffeic acid in pulp and peel fractions (Jiménez-Escrig *et al.*, 2001). Free ellagic acid and glycosides of apigenin and myricetin were detected in guava (Koo and Mohamed, 2001). In fruits, the polyphenols are mostly flavonoids and present mainly as glycoside and ester forms (Fleuriet and Macheix, 2003). Red-pulped guava fruit have ellagic acid conjugates, flavone glycosides and gallic acid conjugates as phenolic compounds (Mahattanatawee *et al.*, 2006). The contents of ellagic acid and quercetin were determined as 10.50 and 9.20 mg g^{-1} dry extract, respectively, by high-performance liquid chromatography (HPLC) in Costa Rican guava fruit (Flores *et al.*, 2013). Chen *et al.* (2015) identified gallic acid, galangin, kaempferol, homogentisic acid and cyanidin-3-glucoside in the peel, seeds and pulp of guava and that the amount of phenolics was high in seeds and peel compared with pulp. Myricetin, quercetin, luteolin, kaempferol and isorhamnetin were characterized as flavonoids in two pink-pulped cultivars ('Semenyih' and 'Sungkai') in Malaysia, where kaempferol was the dominant flavonoid in both cultivars (Musa *et al.*, 2015). Total flavonoids content was reported to be significantly higher in skin and pulp of 'Semenyih' than in 'Sungkai' fruit. Ellagic acid, quercetin, myricetin, isorhamnetin, quercitrin and 1-*O*-*trans*-cinnamoyl-β-D-glucopyranose were detected as polyphenols in guava (Flores *et al.*, 2015). Similarly, quercetin, gallic acid and tannin were identified as phenolic compounds from residual pulp of pink guava. Total polyphenols content of processing residues of pink-pulped guava ranged between 0.03 and 0.12 mg g^{-1} (Sukeksi and Sarah,

2016). Six phenolic compounds, namely gallic acid, (+)-catechin, (–)-epicatechin, quercetin, luteolin and kaempferol, were detected in guava fruit, among which (+)-catechin (391.93 mg kg^{-1}) was the most abundant (Fu *et al.*, 2016). Simultaneous estimation of 13 bioactive phenolic compounds (gallic acid, caffeic acid, chlorogenic acid, *p*-coumaric acid, syringic acid, ellagic acid, vanillic acid, ferulic acid, *trans*-cinnamic acid, rutin, catechin, quercetin and kaempferol) in fruit by HPLC–pulsed amperometric detection (PAD) has been conducted by dos Santos *et al.* (2017). They recorded high concentrations (dry weight basis) for ellagic acid (5.72–30.60 mg 100 g^{-1}), gallic acid (1.27–5.62 mg 100 g^{-1}), rutin (5.02–45.02 mg 100 g^{-1}) and catechin (1.89–13.07 mg 100 g^{-1}) as phenolic compounds. Similarly, 60 phenolic compounds were characterized by HPLC–MS in pink-pulped guava peel and pulp and classified as ellagitannins, flavones, flavonols, flavanols, proanthocyanidins, anthocyanidins, dihydrochalcones, phenolic acid derivatives, stilbenes, acetophenones and benzophenones (Rojas-Garbanzo *et al.*, 2017). Using HPLC coupled with tandem MS, Chiari-Andréo *et al.* (2017) have identified ten phenolic compounds (most of them in ester form) in Brazilian guava. HPLC–MS has recently been employed to detect 69 phenolic compounds in peel, pulp and seed of guava by Liu *et al.* (2018), where flavonoids, hydrolysable tannins, phenolic acid derivatives and benzophenones were identified as the predominant phenolic compounds. They reported that peel of white-pulped guava was a superior source of antioxidant compounds and could be exploited for food composition. The chapter authors have identified and quantified seven phenolic compounds, namely gallic acid, chlorogenic acid, ellagic acid, caffeic acid, catechin, epicatechin and *p*-coumaric acid, in six guava cultivars at edible ripe stage using HPLC–PAD. 'Shweta', 'Lalit' and 'Lalima' guavas were found superior in polyphenols content (Bhattacherjee and Dikshit, 2019).

Guava leaves also contain considerable amounts of phenolic compounds (Metwally *et al.*, 2010). Five flavonoid compounds, namely quercetin, quercetin-3-*O*-α-L-arabinofuranoside, quercetin-3-*O*-β-D-arabinopyranoside, quercetin-3-*O*-β-D-glucoside and quercetin-3-*O*-β-D-galactoside, were isolated from Egyptian guava leaves. Gallocatechin, guaijaverin, quercetin and leucocyanidin were isolated as polyphenols from guava leaves long back (Seshadri and Vasishta, 1965). Using gas chromatography (GC)–MS and HPLC characterization, Afzal *et al.* (2019) have identified quercetin, vanillic acid, syringic acid, *m*-coumaric acid and cinnamic acid as phenolic compounds in leaves, among which *m*-coumaric acid was the predominant compound.

3.2.7 Organic acids

Organic acids are found in almost all fruits and also are available in the human body. Organic acids not only strongly influence the organoleptic properties of fruits and their processed products, particularly flavour, aroma and colour, but also are responsible for fruit sourness or acidity (Kader, 2008). Organic acids also indirectly affect the phenolic metabolism by altering pH and act as precursors of phenolics and flavour components (Galdon *et al.*, 2008). Organic acids represent intermediates of major carbon metabolism in plant cells and are involved in various biochemical pathways such as glycolysis, the tricarboxylic acid (TCA) cycle, photorespiration, the glyoxylate cycle and the photosynthetic C_4 cycle, among others. They also display unexpected roles in controlling whole plant cell physiology by regulating a broad range of basic cellular processes like the modification of cellular pH or the redox state and help in maintaining acid–alkali balance in the human body (Drincovich *et al.*, 2016). Estimation of organic acids in fruit is important for characterization of different genotypes or when investigating the influence of maturity or agronomic factors on fruit quality (Flores *et al.*, 2012). Organic acids can be used as a helpful index of authenticity in fruit products because they exert an important influence on the sensory attributes (Kelebek, 2010).

The literature on the estimation of organic acids in guava is very scanty. Ascorbic acid can also be considered an organic acid because of its mild acidic nature. Six non-volatile organic acids in guava fruits (lactic, malic, citric, ascorbic, galacturonic and succinic acids) were isolated and identified using thin-layer chromatography (TLC) and gas–liquid chromatography (GLC) (Chan *et al.*, 1971). In cultivated guavas citric and malic acids are present in almost equal amounts and lactic acid in a much lesser amount; however, in wild guava, citric acid is the predominant acid with lesser amounts of malic and lactic acids (Chan *et al.*, 1971). HPLC analysis revealed the presence of glycolic, malic, ascorbic and citric acids in five Florida guava cultivars, where citric acid was the major acid in all five cultivars, and traces of fumaric acid were also detected in guava fruits (Wilson *et al.*, 1982). Five organic acids, namely citric, ascorbic, tartaric, oxalic and malic acids, were identified and quantified by HPLC in six guava cultivars ('Allahabad Safeda', 'Sardar', 'Lalit', 'Shweta', 'Dhawal' and 'Lalima') grown in Lucknow, India, where citric acid was the major acid and malic acid the minor one (Bhattacherjee and Dikshit, 2019).

3.2.8 Volatile components

Volatile components are actually some secondary metabolites like essential oils from terpenes, alkaloids, saponins, etc. and some antioxidant phytochemicals like flavonoids, tannins and glycosides. Several studies have been published on estimation of volatile components of guava fruit. More than 500 volatile compounds have already been isolated from guava fruits whose concentrations vary at different stages of maturity and ripening (Chen *et al.*, 2007). Forty volatile compounds were identified in Venezuelan guava including 2-methylpropyl acetate, myrcene, hexyl acetate, benzaldehyde, ethyl decanoate, β-caryophyllene, α-humulene and α-selinene, which has a guava-like aroma (MacLeod and Troconis, 1982). One hundred and fifty-four volatile constituents were identified from guava fruit using GC, GC–MS and GC–Fourier transform infra-red (FTIR), with lipid peroxidation products such as C_6 aldehydes and alcohols predominating (Idstein and Schreier, 1985). Hashinaga *et al.* (1987) have estimated 85 volatile components in guava fruit and leaves during maturation. The major components in immature fruit were ethyl acetate, isobutyl alcohol, β-caryophyllene and α-humulene, whereas in ripe fruit they were ethyl acetate, ethyl butyrate and ethyl caproate. Nishimura *et al.* (1989) analysed the volatile constituents of guava fruits and canned purée; a total of 122 volatile compounds were identified, with C_6 compounds being the major ones in fresh fruits. Chyau and Wu (1989) determined volatile compounds in inner and outer pulp-peel of guava fruit in Taiwan and observed that inner pulp-peel was especially rich in ethyl acetate, other ethyl esters and C_6 aldehydes, while (*Z*)-ocimene, β- and γ-caryophyllene existed in larger amounts in the outer pulp-peel. Ethyl esters may play a significant role in the characteristic sweet and very pleasant flavour of guava. Vernin *et al.* (1991) have investigated the composition of volatile compounds in guava fruits from Egypt, where GC–MS analysis confirmed the existence of 2-hex-3-enyl acetate (11.0%) and its corresponding alcohol (7.5%), pentan-2-one (9.15%), cinnamyl alcohol (10.2%), 3-phenylpropyl acetate (5.0%) and its corresponding alcohol (3.5%) among 132 compounds. Ekundayo *et al.* (1991) identified 25 volatile compounds in Nigerian guava where β-caryophyllene and oxygen-containing sesquiterpenes were the typical ones. Chyau *et al.* (1992) have identified 34 volatile components in mature and ripe guava fruits in Taiwan using GC, GC–MS and GC–FTIR. The major constituents in mature fruit were 1,8-cineole, (*E*)-2-hexenal and (*E*)-3-hexenal, whereas in ripe fruits, ethyl hexanoate and (*Z*)-3-hexenyl acetate were the major volatile components.

One hundred and seventy-three volatile compounds were sensorially characterized by sniffing–GC in Costa Rican guava, with β-caryophyllene, α-terpineol, α-pinene, α-selinene, β-selinene, δ-cadinene, 4,11-selinadiene and α-copaene being the major constituents

(Pino *et al.*, 2002). The presence of many aliphatic esters and terpenic compounds may contribute to the unique flavour of this fruit. GC–MS estimation of the aroma profile in commercial guava essence and fresh guava purée yielded a total of 51 principal components (Jordán *et al.*, 2003). The major difference between the aromas of the commercial guava essence and fresh fruit purée could be due to the presence of acetic acid, 3-hydroxy-2-butanone, 3-methyl-1-butanol, 2,3-butanediol, 3-methylbutanoic acid, (*Z*)-3-hexen-1-ol, 6-methyl-5-hepten-2-one, limonene, octanol, ethyl octanoate, 3-phenylpropanol, cinnamyl alcohol, α-copaene and an unknown component. (*E*)-2-Hexenal seemed to be more significant to the aroma of the commercial guava essence than to the aroma of the fresh fruit purée. Chen *et al.* (2006) have characterized 64 volatile components in Taiwanese guava fruit using GC–MS and identified α-pinene, 1,8-cineole, β-caryophyllene, nerolidol, globule, C_6 aldehydes, alcohols and esters as the principal components. They have opined that the presence of C_6 aldehydes and esters, terpenes and 1,8-cineole may contribute to the unique flavour of guava fruit. Soares *et al.* (2007) have determined volatile chemical compounds in white-pulped guava at three stages of maturity using headspace GC–MS. The aldehydes like (*E*)-2-hexenal and (*Z*)-3-hexenal were the predominant compounds in immature fruits and those in their intermediate stage of maturation, whereas in mature fruits, esters like (*Z*)-3-hexenyl acetate and (*E*)-3-hexenyl acetate and sesquiterpenes like caryophyllene, α-humulene and β-bisabollene were the principal components. Volatile compounds in guava were analysed by the solid-phase microextraction (SPME)/GC–MS method. The major constituents identified in white- and red-pulped guavas were cinnamyl alcohol, ethyl benzoate, β-caryophyllene, (*E*)-3-hexenyl acetate and α-bisabollene (Thuaytong and Anprung, 2011).

Two triterpenoids, 20β-acetoxy-2α,3β-dihydroxyurs-12-en-28-oic acid (guavanoic acid) and 2α,3β-dihydroxy-24-*p*-*Z*-coumaroyloxyurs-12-en-28-oic acid (guavacoumaric acid), along with six known compounds, namely 2α-hydroxyursolic acid, jacoumaric acid, isoneriucoumaric acid, asiatic acid, ilelatifol D and β-sitosterol-3-*O*-β-D-glucopyranoside, have been isolated from the leaves of *Psidium guajava* and their structures were determined through spectroscopic methods (Begum *et al.*, 2002a). Two other triterpenoids, guajavolide (2α,3β,6β,23-tetrahydroxy-urs-12-en-28,20β-olide) and guavenoic acid (2α,3β,6β,23-tetrahydroxy-urs-12,20(30)-dien-28-oic acid) – along with one known triterpene, oleanolic acid – were isolated from fresh guava leaves in Pakistan (Begum *et al.*, 2002b). Five constituents including one new pentacyclic triterpenoid (guajanoic acid) and four known compounds (β-sitosterol, uvaol, oleanolic acid and ursolic acid) have been isolated from the leaves of guava. The new constituent has been characterized as 3β-*p*-*E*-coumaroyloxy-2α-methoxyurs-12-en-28-oic acid through two-dimensional (2D) NMR techniques and chemical transformations (Begum *et al.*, 2004). Some investigations showed that the major constituents of the essential oil are α-pinene, limonene, menthol, terpene lactate, isopropyl alcohol, longicyclene, β-caryophyllene, (*E*)-nerolidol, β-bisabolene, caryophyllene oxide, β-copanene, farnesene, humulene, selinene, cardinene, curcumene and 1,8-cineole (Chen and Yen, 2007; Satyal *et al.*, 2015; Soliman *et al.*, 2016). As a result of GC–MS analysis, four compounds were identified from leaf oil of *P. guajava*, with cryptonine (56%), prenol (26%) and dehydro benzophenanthridine (20%) being named as the major constituents of the essential oil. Among the four compounds, flavanone-2,2′-ene (4.2%) has been identified as the minor component present in the essential oil of guava leaf (Joseph *et al.*, 2010). Four new triterpenoids, psiguanins A–D (**1**–**4**), along with 13 known ones, were isolated from the leaves of *P. guajava*. The structures of the new compounds were determined as 2α,3β-dihydroxy-taraxer-20-en-28-oic acid (**1**), 2α,3β,12α,13β-tetrahydroxy-urs-28-oic acid (**2**), 2α,3β,12β,13β-tetrahydroxy-urs-28-oic acid (**3**) and 2α,3β,12β,13α-tetrahydroxy-urs-28-oic acid (**4**), respectively, on the basis of comprehensive spectroscopic methods and molecular modelling calculations (Shao *et al.*, 2012). Among them, compound **4** was characterized as an unusual ursane-type triterpenoid with *cis*-fused C/D ring system. Seventeen components were identified and quantified in the essential oil from guava leaves in Brazil using the GC method and mostly comprised hydrocarbon monoterpenes, oxygenated monoterpenes, hydrocarbon sesquiterpenes and oxygenated sesquiterpenes (Silva *et al.*, 2016). Sesquiterpene compounds represented the highest concentration (62.0%) and monoterpenoids and monoterpenes the lowest (1.1 and 2.2%, respectively). Among the compounds identified, *trans*-caryophyllene (17.1%), α-humulene (16.2%), aromadendrene (12.2%), α-selinene (11.3%) and selin-11-en-4α-ol (9.92%) were the predominant components. It was observed that collection time of guava leaves within a day did not affect the essential oil composition qualitatively, but it did so quantitatively for some major essential oil components whose concentrations varied significantly with collection time. The authors have concluded that essential oil composition is directly influenced by the oscillation of temperature and humidity throughout the day. GC–MS analysis of guava-leaf extract in Pakistan showed the presence of 33 chemical components among which caryophyllene oxide (22.70%), caryophyllene (15.65%), α-cubebene (8.02%), *cis*-calamenene (6.21%) and α-humulene (5.90%) were the major constituents (Afzal *et al.*, 2019). Recently a total of 54 chemical compounds, most of them terpenoids and their derivatives, were detected using GC–MS which is representing 98.17% of the total oil. The dominating ingredients in the leaf oil were iso-caryophyllene (33.53%), veridiflorene (13.00%), farnesene (11.65%), D,L-limonene (9.84%), δ-cadinene (1.75%), α-copaene (2.80%), α-humulene (3.74%), aromadendene (1.70%) and τ-cadinol (0.08%) (Weli *et al.*, 2019).

3.2.9 Pharmacological value

Guava, commonly known as 'poor man's apple of the tropics', has a long history of traditional pharmacological uses for a wide range of diseases. The fruit and its juice are freely consumed for their great taste and many health benefits. Much of the traditional uses have also been validated by scientific investigations. Both leaf and fruit are safe for human consumption without any side-effects as proven by toxicity studies in mice and other animal models as well as some controlled human studies. A number of chemicals isolated from plants like quercetin, guaijaverin, flavonoids and galactose-specific lecithins have shown promising activity in many human trials (Kamath *et al.*, 2008). Extensive studies on various plant parts of guava have shown potent therapeutic activities of its major components indicating antidiarrhoeal, antihypertensive, hepatoprotective, antioxidant, antimicrobial, hypoglycaemic and antimutagenic activities (Table 3.2). Phenolic compounds in guava help in curing cancerous cells and prevent premature skin ageing. The terpenes (caryophyllene oxide and *p*-selinene) present in fruit and leaf produce relaxation effects. Guava leaves contain many compounds which can act as fungistatic and bacteriostatic agents. Quercetin is considered the most active antioxidant in guava leaves responsible for spasmolytic activity. The ethyl acetate extract of leaves can stop the germ infection and thymus production. The ethanolic extract of guava can increase the quality and quantity of sperm and can be utilized in treating infertile males. Guava possesses antiviral, anti-inflammatory, antiplaque and radio-protective activities (Naseer *et al.*, 2018). Quite a significant amount of work has been done on the pharmacological and biological activities and possible application of chemical compounds isolated from various parts of the guava plant.

3.3 Processing and Value Addition

Guava fruit is mostly consumed fresh because of its excellent taste, exotic flavour and lower prices. However, the fruit is highly perishable and its shelf-life is very poor (2–3 days) under ambient conditions. Due to inappropriate handling, transportation and storage, approximately 25% of fruits are spoiled

Table 3.2. Therapeutic uses of guava plant in various parts of the world. From Kamath *et al.* (2008).

Region/Country	Usage
Amazonia	For diarrhoea, dysentery, menstrual disorders, stomach ache, vertigo
Brazil	For anorexia, cholera, diarrhoea, digestive problems, dysentery, gastric insufficiency, inflamed mucous membranes, laryngitis, swelling of mouth, skin problems, sore throat, ulcers, vaginal discharge
Cuba	For colds, dysentery, dyspepsia
Ghana	For coughs, diarrhoea, dysentery, toothache
Haiti	For dysentery, diarrhoea, epilepsy, itch, piles, scabies, skin sores, sore throat, stomach ache, wounds, as an antiseptic and astringent
India	For anorexia, cerebral ailments, childbirth, cholera, convulsions, epilepsy, nephritis, jaundice
Malaya	For dermatitis, diarrhoea, epilepsy, hysteria, menstrual disorders
Mexico	For deafness, diarrhoea, itch, scabies, stomach ache, swelling, ulcer, worms, wounds
Peru	For conjunctivitis, cough, diarrhoea, digestive problems, dysentery, oedema, gout, haemorrhages, gastroenteritis, gastritis, lung problems, premenstrual syndrome, shock, vaginal discharge, vertigo, vomiting, worms
Philippines	For sores, wounds, as an astringent
Trinidad	For bacterial infections, blood cleansing, diarrhoea, dysentery

before they reach the consumer. The fruits could better be utilized by employing adequate processing technologies and preserving the fruit into different value-added products like pulp, juice, jelly, jam, nectar, blended ready-to-serve (RTS) beverages, toffee, leather, cheese, canned slices, dehydrated products, etc. This would not only provide higher prices to growers and more employment, but also make the nutritious fruit products available throughout the year. The selection of a guava variety for processing depends on characters like pulp percentage and its colour, along with contents of sugars, acid, pectin and ascorbic acid. The demand for coloured-pulped (red or pink) fruits and their processed products is higher than for white-pulped guava fruit.

3.3.1 Pulp/purée

Guava fruits can be processed and preserved in the form of pulp or purée. The pulp content not only varies from cultivar to cultivar but also depends on geographical location. Indian cultivars with high pulp content include ‘Hisar Safeda’ (97.39%), ‘Allahabad Safeda’ (94.10%) and ‘Sardar’ (92.86%) (Pandey and Singh, 1999; Pandey *et al.*, 2016; Tiwari *et al.*, 2016). Fully mature firm ripe guava fruits were found to be best for pulp preparation.

The pulp is extracted from fruits by blending the cut pieces of fruits with water (up to 20%) and filtering out the seeds. The pulp is heated to 75–80°C and preserved with 1000 ppm sulfur dioxide (SO_2) as potassium metabisulphite (KMS) in airtight containers for up to 12 months under ambient conditions. For preservation of pulp from pink-pulped cultivars, a combination of KMS and sodium benzoate (500 ppm) can be used (Singh *et al.*, 2003). Generally two types of pulp extraction method are followed: (i) cold extraction methods; and (ii) hot extraction methods. In the cold extraction process, the fruit pieces are fed into a motor-driven machine which grates and centrifuges the juice. The juice and pomace are mixed together and passed through a 60-mesh stainless-steel sieve in order to get pulp. In the hot extraction method, small fruit pieces are heated to 95 ± 2°C with occasional stirring for about 20 min. The whole mass is cooled and passed through a 60-mesh stainless-steel sieve to obtain pulp (Murari and Verma, 1989). Although pulp recovery is 5–8% higher in the hot method of pulp extraction, it results in development of pink/brown pigment which further increases upon processing to nectar and during storage. However, the cold pulping method retains the pulp's original colour irrespective of cultivar. The merits of cold extraction were found to be superior quality compared with hot extraction and helping to increase amounts of reducing and non-reducing sugars in stored guava pulp (Harnanan *et al.*, 1980). The cold extraction method was suggested by many researchers to extract pulp from ripe fruits of cultivars 'Sardar', 'Allahabad Safeda', 'Lalit', 'Shweta' and 'Gorakh Bilas Pasand' (Bons and Dhawan, 2006, 2013; Kadam *et al.*, 2012; Tiwari *et al.*, 2016), while the hot extraction method (90°C) was also used for extracting pulp from fully mature and ripe fruits from cultivars 'Lalit' and 'Sardar' (Yadav *et al.*, 2017). Different types of chemical preservatives like KMS, potassium sorbate, sodium benzoate, etc. at various concentrations and in combination with heating or without heating can be used for the preservation of guava pulp. In most of the studies, KMS was adjudged the best preservative for this purpose in different concentrations (500 to 2000 ppm). The pulp then can be stored in various packaging materials, namely ordinary and food-grade polyethylene pouches/bags, glass jars/bottles, polyvinyl chloride (PVC) containers, food-grade plastic jars, etc., and at different temperatures like room (15–30°C), low (2–6°C) and freezing (–18 to –20°C) (Kalra and Revathi, 1981; Tandon *et al.*, 1983a; Tandon and Kalra, 1984; Sandhu *et al.*, 2001; Jain and Asati, 2004; Bons and Dhawan, 2006, 2009, 2013).

A storage study of guava pulp (cultivars 'Allahabad Safeda' and 'Sardar') in three types of container at room (20–25°C) and cold temperatures (4–6°C) for up to 45 days revealed that porcelain containers were inferior to glass and PVC containers for storage of guava pulp. Although ascorbic acid content was higher in 'Sardar' compared with 'Allahabad Safeda', the former was found to be slightly less suitable for bulk storage (Kalra and Revathi, 1981). During storage, protein content, acidity and TSS showed an upward trend and SO_2 level declined and under refrigeration pulp could be stored much longer (6 months) than at room temperature. It was indicated that 500 ppm SO_2 was sufficient to check the deterioration of guava pulp, prepared from fruits of 'Allahabad Safeda' and 'Sardar', during storage at room temperature (26–30°C) for up to 60 days, while 1000 ppm SO_2 was required for longer duration of storage in PVC containers (Tandon *et al.*, 1983a). Guava pulp prepared from 'Allahabad Safeda' could be stored at low temperature (6 ± 1°C) for up to 60 days with good overall acceptability (Jain *et al.*, 2011). The best organoleptic score was obtained for stored guava pulp (90 days) preserved at low temperature (–20°C), followed by addition of KMS (0.1%), sodium benzoate and KMS (0.05% each) and potassium sorbate (0.1%) (Yadav *et al.*, 2017).

Preserved guava pulp could be utilized for the preparation of various products like juice, squash, RTS beverages, nectar, leather, etc. The juice obtained from 15% stored pulp, 12°Brix TSS and 0.3% acidity was quite acceptable organoleptically and could be stored well in bottles for up to 1 year at ambient temperature (Tandon *et al.*, 1983a). RTS drinks and leather could be prepared from 6-months stored pulp of 'Banarsi Surkha' and 'Allahabad Safeda' with good acceptability (Sandhu *et al.*, 2001). In Brazil, two types of guava paste were prepared from red guava pulp (cultivar 'Paluma') – conventional sweet guava paste and light sweet guava paste. Conventional sweet guava paste was prepared by mixing 50% guava pulp with 50% sugar, 1% pectin, 0.5% citric acid, 0.02% sodium benzoate and 0.02% potassium sorbate. Light sweet guava paste comprised 65% red guava pulp, 35% sugar, 2% pectin, 0.5% citric acid, 0.02% sodium benzoate, 0.02% potassium sorbate and cyclamate/saccharine (1:1). Quality of light sweet guava paste was found to be superior to that of conventional sweet guava paste as it contained higher amounts of bioactive compounds like phenolics, lycopene and ascorbic acid (Freda *et al.*, 2018).

Guava fruits are usually macerated into purée first and then further processed to different products. For the preparation of guava purée, fully mature firm ripe guava fruits are macerated and finished by a rotor crusher, a paddle pulper and a paddle finisher, all lined up in sequence. The pulper removes seeds and fibrous material and forces the remainder of the product through a perforated stainless-steel screen. The residual stone cells may be ground by passing the finished pulp through a mill as it improves the mouth feel (Jagtiani *et al.*, 1988). Guava purée is preserved by: (i) freezing to −29°C and storing at −18°C (frozen guava purée); (ii) canning (canned guava purée); (iii) aseptic packaging; or (iv) dehydration. Changes in sensory quality in aseptically packaged guava purée in pre-sterilized, low-oxygen-permeability, laminated bags were noticed during storage. Aseptic processing of guava purée had no effect on its ascorbic acid content and flavour but did cause a significant loss of colour. After 6 months of ambient storage, ascorbic acid loss was small (30%) and slight colour and flavour changes occurred (Chan and Cavaletto, 1982). The deaeration of aseptically packaged guava purée helped in retention of ascorbic acid during storage for up to 6 months (Chan and Cavaletto, 1986). Guava purée was also prepared using a flash vacuum-expansion process and compared with traditionally prepared reference products. The vacuum-expanded purée had alcohol-insoluble residue (AIR) levels higher than the reference products. The vacuum-expanded purée exhibited consistencies and apparent viscosities higher than those of the reference products which were related to their AIR contents (Brat *et al.*, 2002).

3.3.2 Juice

Tropical fruit juices have become important in recent years due to the overall increase in 'natural fruit' juice consumption as an alternative to the traditional caffeine-containing beverages like coffee, tea or carbonated soft drinks (Jagtiani *et al.*, 1988). Guava juice may be obtained from either fresh fruits or stored pulp. From fresh fruit, juice is extracted by squeezing guava pieces through a hydraulic filter press. It can be made from pulp by diluting it with water and filtering. The recovery of juice may be increased by treating the pulp with pectic enzymes (Singh *et al.*, 2003). Cultivars suitable for juice preparation must have soft flesh and good colour and flavour. 'Banarasi Surkha' and 'Allahabad Safeda' both have soft pulp, give better recovery of juice and can be considered as better cultivars for the juice industry (Singh and Dhawan, 1983). The highest loss of ascorbic acid during the guava juice-making process occurred due to peeling (6%) followed by exhausting (4.5%), and processing led to an overall decrease in ascorbic acid of 20.4% for juice (Jawaheer *et al.*, 2003). Storage of juice at low temperature (4°C) did not significantly reduce the ascorbic acid content over time. The consumption of either one fresh guava daily (~250 g) or one glass of juice daily (200 ml), even after storage, still satisfies the recommended dietary allowance for vitamin C (which is 90 mg day^{-1} for men and 75 mg day^{-1} for women) for healthy non-smoking adults. Ascorbic acid degradation in guava juice was substantially reduced by refrigerated storage temperature and time, and the lowest rate constant (k = 3.0–5.8 × 10^{-2} day^{-1}) was obtained in juice samples stored at 5°C which were also found to be healthier after 7 days of storage (Sinchaipanit *et al.*, 2015). Shrivastava *et al.* (2017) have investigated the effect of various concentrations of ginger extract and sodium benzoate as preservative on guava juice (cultivar 'Apple Colour') during storage at room temperature and observed that, quality-wise, the juice treated with 800 ppm sodium benzoate and 1% ginger extract was the best during 90 days of storage.

Fruit juices are usually cloudy with colloidal suspensions. The colloidal particles which cause turbidity in juice carry flavour substances and bioactive antioxidant molecules. Pink-pulped guava fruits also contain large amounts of lycopene which remains in the structural tissue during pressing. The use of pectic enzymes along with fining agents in fruit processing is essential to get better juice yield, improve filtration rate and produce clear juice of good quality for the concentration process

(Brasil *et al.*, 1995). Good yield of guava juice (84.70%) could be obtained by clarifying the juice with 600 ppm of pectic enzyme along with silica sol and gelatine as fining agents at 45°C for 120 min. The pressed juice was cloudy and pink in colour but after addition of fining agents and filtration a clear juice was visible with a light yellow colour. Time and temperature used for enzyme treatments significantly affect the yield and quality of cloudy juice. Increasing exposure time enhances the yield as well as causes a reduction in ascorbic acid content of the juice due to oxidation (Imungi *et al.*, 1980). However, an increase in ascorbic acid content (10.6%) in guava juice was obtained by treating guava pulp with 400 ppm of pectic enzyme at 45°C for 90 min. Enzymes are usually not used in guava juice production for better clarification with good yield. Commercial preparations of enzymes like pectinase, arabinase and cellulase may benefit production of guava juice. Pectinase assists in pectin hydrolysis which causes a reduction in pulp viscosity as well as a significant increase in juice yield. Arabinase and cellulase convert arabinan and cellulose to soluble sugars and increase the soluble solids content. Arabinase also helps in eliminating the turbidity of juice caused by arabinan, which is visible only after 3–4 weeks of storage (Askar *et al.*, 1992).

Guava juice can further be processed and utilized in the form of concentrate, beverages, drinks and powder. A good-quality beverage should have 25% juice, 0.5% acidity and 20% TSS. Guava juice was extracted after treating with 0.5% enzyme at room temperature for 16 h using a hydraulic press and concentrated under vacuum at 50–55°C in a glass evaporator (Sandhu and Bhatia, 1985). Clear guava juice could be concentrated successfully up to 45–48°Brix, although enzymatic treatment of pulp resulted in a considerable reduction of pectin content. RTS drink prepared by reconstituting the concentrate (40–45°Brix) was comparable in colour, flavour and cloudiness to that prepared from freshly extracted juice. Partially clarified guava juice concentrate was prepared from aseptically prepared single-strength guava purée (5.5°Brix) by treatment with pectinase enzyme (2 h at 50°C), extraction of juice with a rack-and-cloth press and vacuum concentration to 23°Brix; the ascorbic acid content of the concentrate was 867 mg 100 ml^{-1}, a significant increase from 209 mg 100 g^{-1} in the purée (Hodgson *et al.*, 1990). For better clarity, guava juice prepared using ultrafiltration was clearer, with 89.6% transmission compared with 82.8% for plate-and-frame filtered juice. However, plate-and-frame filtered juice retained more soluble solids, contained 5.8% more ascorbic acid than the ultrafiltered juice and had higher flux rates at all times (Chopda and Barrett, 2001).

3.3.3 Squash

Squash is concentrated syrup containing fruit pulp, sugar and acidulant used in beverage making. Usually the pulp content in squash varies from 25 to 40%. A squash recipe containing 25% pulp and 45% TSS with 1.0% acidity was found to be most ideal (Pandey and Singh, 1998). The squash prepared from 'Sardar' guava was found to be better than those from 'Allahabad Safeda', 'Apple Colour' and 'Sangam'. Increase in non-enzymatic browning (NEB) and decrease in ascorbic acid content were observed during storage of the squash. The squash quality was found organoleptically acceptable for up to 6 months (Pandey, 2004). Five red-pulped guava cultivars ('Arka Kiran', 'Arka Rashmi', 'Lalit', 'Hybrid H-17-16' and 'Punjab Pink') were evaluated for preparation and storage of squash (Ravi *et al.*, 2018). The squash was prepared using 25% pulp, 40–50% TSS and 0.3% acidity. The squash prepared from 'Punjab Pink' retained maximum ascorbic acid after 90 days of storage, while lycopene retention was maximum in squash prepared from 'Arka Kiran'. 'Hybrid H-17-16' scored maximum in overall acceptability and 'Arka Kiran' and 'Arka Rashmi' had minimum microbial load during storage. The contents of ascorbic acid, anthocyanins and total phenols decreased during storage. Mango–guava blended squashes at various proportions have been prepared and their storage study indicated that the squash obtained from mango–guava blend 70:30 and preserved with 0.1% KMS and 0.05% carboxymethyl cellulose was found best after 90 days of storage at room temperature on the basis of physicochemical analysis and sensory evaluation (Nooruddin *et al.*, 2019).

3.3.4 Nectar

Guava nectar is a juice or non-carbonated non-alcoholic beverage made from guava pulp, purée, clarified juice or juice concentrate by mixing with water, sugar and citric acid. Other flavour additives may be added to it and cloudy or clear nectar may be obtained. The former type of nectar is more popular than the latter in most countries (Singh, D. *et al.*, 2007). There are a number of uses for guava nectar ranging from mixed drinks to straight consumption and it is especially popular in tropical regions (Usha Kumari *et al.*, 2017). A number of workers have prepared guava nectar from different cultivars using numerous combinations of fruit pulp, TSS and acidity. The best and ideal nectar could be prepared using the combination of 15 to 20% pulp, 12 to 20% TSS and 0.2 to 0.35% acidity (Kerure and Kjedkar, 1982; Singh and Dhawan, 1983; Kalra and Tandon, 1984; Khurdiya and Sagar, 1991; Choudhary and Dikshit, 2006; Choudhary *et al.*, 2008; Dubey *et al.*, 2011; Bal *et al.*, 2014). Among these, the combination of 20% pulp, 17% TSS and 0.3% acidity is generally used for the preparation of nectar.

Pink-pulped guavas are found particularly more suitable for the preparation of nectar. The inner pulp is sieved and blended with sugar syrup of 15°Brix and 0.25% acidity to obtain a guava nectar with delicious taste and aroma (Sharma *et al.*, 1999). The production of nectar from fresh pink guava fruits reduced titratable acidity, lycopene and ascorbic acid contents in nectar, while pH and TSS increased significantly. Guava nectar from pink-pulped cultivar 'Lalit' was prepared with different ratios of pulp and TSS and it was noticed that the treatments having 20% pulp and 15°Brix TSS and 16% pulp and 17°Brix TSS were better for nectar preparation considering changes in chemical constituents as well as sensory attributes after 8 months of storage at room temperature (Bal *et al.*, 2014). In Taiwan, cloudy guava nectar having 25% juice, 11°Brix of sugar and 0.2% acidity at pH 3.8 is available (Chen *et al.*, 1994). During the ambient storage of processed, white-pulped, cloudy guava nectar, NEB played an important role in the deterioration in quality while ascorbic acid and tannins were involved in discoloration. Chen and Wu (1993) have concluded that the reduction in pH by the addition of citric acid in the formulation of nectar might reduce the browning rate effectively.

Blending of guava nectar with other nutritious fruit juices/nectars is also possible, although Kalra and Tandon (1984) have noticed that blends of guava and mango nectar were inferior in quality to guava or mango nectar alone. Value-added nutraceutical beverages were prepared from guava blended with aloe vera and roselle at 70:25:5 juice ratio following standard recipes (Kumar *et al.*, 2012).

3.3.5 Ready-to-serve beverages

RTS beverages are one of the popular drinks made from guava pulp or juice. The demand for fruit beverages is mainly based on their nutritive value, flavour, aroma and colour. The beverages are a good source of vitamins, minerals, carbohydrates, amino acids, flavonoids and phenolic compounds, and many more (Chandel *et al.*, 2018). Numerous workers have prepared RTS beverages from several guava cultivars using different combinations of fruit pulp, TSS and acidity or by blending it with other fruit juices. RTS beverages can be prepared using the recipe of 10 to 15% pulp/juice (10–20% for blended RTS), 11 to 15% TSS and 0.2 to 0.4% acidity (Pandey and Singh, 1999; Tiwari, 2000; Singh, P. *et al.*, 2007; Bons and Dhawan, 2009; Jakhar *et al.*, 2013; Bhat and Singh, 2014; Abhangrao *et al.*, 2017). Among these, the recipe of 10% pulp, 11% TSS and 0.3%

acidity is abundantly used for beverage preparation. Pink-pulped guavas are better suited for RTS beverage preparation due to their attractive colour (Joshi *et al.*, 2017) and pink-pulped 'Lalit' was found most suitable with high acceptability. From 100 kg of red-pulped guava, 247 litres of RTS beverage could be prepared (Bhuvaneswari and Tiwari, 2007). After evaluating the varietal suitability ('Sardar', 'Allahabad Safeda', 'Apple Colour' and 'Sangam') and storage stability of recipes for commercial preparation of guava RTS beverages, the recipe containing 10% pulp, 11% TSS and 0.25% acidity was found ideal (Pandey and Singh, 1999). Fresh guava beverage obtained with 10% guava pulp, 11% TSS and 0.3% acidity scored maximum organoleptically, followed by beverage made from 10% pulp, 11% TSS and 0.4% acidity (Abhangrao *et al.*, 2017). Varietal evaluation for preparation and storage of RTS beverage revealed that 'Sardar' guava showed maximum overall acceptability after 60 days of storage followed by 'Apple Guava' and 'Red Fleshed'. RTS drink with 10.11°Brix TSS and 0.3% acidity had many health benefits and could be stored for 6 months with good overall acceptability (Rashid *et al.*, 2018).

Guava juice has a strong flavour and taste with good nutritional quality, but many times an unattractive colour. For diversified products guava pulp and juice are used to prepare RTS beverages by blending with other fruit juices like pear, apple, mango, papaya, aonla, pineapple, jamun, bael, etc. The RTS beverage prepared from a blend of 90% pineapple juice and 10% guava juice was rated best with 88% overall acceptability rating followed by the blend of 80% pineapple juice and 20% guava juice (rating 85%) after 3 months of storage. RTS beverage obtained from 70% guava and 30% pineapple juice and having combination of 10% pulp, 11% TSS and 0.2% acidity had highest acceptability after 105 days of storage at ambient temperature, although it could be stored for up to 120 days with acceptable score (Singh, P. *et al.*, 2007).

Guava–aonla (80:20) blended beverage having 12°Brix TSS was found to have the best organoleptic score after 90 days of storage. It was noticed that quality parameters such as TSS, titratable acidity, ascorbic acid, total sugars, reducing sugars and tannins contents did not change much in the RTS beverage but NEB increased with prolongation of the storage period (Mall and Tandon, 2007).

3.3.6 Jam, jelly and preserve

Jelly is one of the popular products prepared from slightly underripe fresh guava fruits with plenty of pulp and less seeds. The fruits are cut into small pieces or slices and boiled with an equal amount of water for 30–35 min using a low flame. The material is filtered through a strainer/muslin cloth and clear juice is obtained which is used for jelly making. The residual pulp is used for the preparation of guava cheese. After testing the juice for pectin content by either drop test or chemically, it is mixed with sugar (0.75 kg kg^{-1} for pectin-rich juice or 0.50 kg kg^{-1} for low-pectin juice) and boiled in a shallow vessel. Approximately 6–7 g of citric acid solution in water is added per kilogram of juice 15 min after boiling. Boiling is continued until the temperature reaches 105.5°C or until it gives a sheeting test. The hot jelly is poured into sterilized glass bottles/jars and stored in a cool dry place. The jelly should have an attractive purplish red colour, a pleasant aroma and good taste (Singh *et al.*, 2003). Guava jam consists of 45% guava purée or pulp, 55% sugar, 0.5% pectin and 0.4% citric acid. All the ingredients are mixed and cooked until the TSS content of the product reaches above 65%. The jam is bottled while hot in airtight containers and it may be kept for more than a year without any preservative. Guava preserve is the product of appropriate processing of the fruit edible parts with added sugars, water, pectin (0.5–1.5%), pH adjusted between 3.0 and 3.4, permitted colour and food additives until adequate consistency is reached with assured product stability (Kanwal *et al.*, 2016).

Guava varieties for making jelly should be rich in pectin and acid with thick pulp. 'Sardar' guava has been found rich in pectin, acid and ascorbic acid with strong flavour and higher productivity. Jelly made from this cultivar was found to be the best in flavour, colour and taste (Singh and Dhawan, 1983). The pink-pulped 'Beaumont' gave jelly of better quality than white-pulped or other red-pulped cultivars (Ramanjaneya, 1983). Effects of addition of potassium sorbate and different types of packaging (polypropylene, metallic and cellophane film) on the quality of guava preserve during storage at 19.6°C temperature with 76.2% humidity were evaluated by Menezes *et al.* (2011). They concluded that the addition of potassium sorbate was not viable to store guava preserve for 5 months since it caused an increase in TSS and a decrease in water activity, and the economically most viable packaging should be used, as the packaging factor did not influence the stability of the product during storage.

3.3.7 Toffee/candy

Fruit toffees naturally are very nutritious as they possess most of the nutraceuticals of the fruit from which they are prepared. Guava toffee has a good taste and may be compared with chocolate. For the preparation of guava toffee, 1.5 kg of sugar and 125 g of butter are added to 1.0 kg of pulp and the mixture is heated to obtain a thick mass. One teaspoonful salt (5 g), citric acid (2 g) and edible colour are added to it before the toffee mixture is spread in flat, greased trays. After cooling, the desired size of cubes is cut and wrapped in butter paper.

For the preparation of guava candy, the same method as for the preparation of guava toffee can be utilized with the addition of required quantities of sugar, butter, milk powder and citric acid, but no salt is needed here. The final strength of the sugar syrup may be raised to 68–70°Brix and the candies are dried in an oven or dehydrator at 55–60°C for 5–6 h beyond sticky condition. A cooking temperature range of 85–95°C and cooking period of 120 min were found to be optimal for candy preparation from guava cultivars 'Allahabad Safeda' and 'Sardar' (Kishore *et al.*, 2016). Candy was reported to be more nutritious than other processed products, having 0.28–0.42% protein, 3.87–5.21% fibre, 85.09–87.23% carbohydrates, 7.72–9.39 mg of ascorbic acid per 100 g and 3.20–4.90 mg of total carotenoids per 100 g (in pink-pulped cultivars) (Joshi *et al.*, 2019).

3.3.8 Dehydrated products

There are two types of dehydrated products of guava available on the market: (i) osmotically dehydrated slices; and (ii) powder. Guava slices or chunks can be dehydrated by air drying, osmotic dehydration and/or osmovac dehydration. In the case of air drying, guava slices are blanched in boiling water for 4 min followed by sulfuring for 20 min and air drying at 71°C until final moisture content reaches 6–7%, which usually takes about 15 h (Campbell and Campbell, 1983). To prevent browning during the drying process, some treatments like chemical blanching with 0.1% KMS + 2% $CaCl_2$ at 100°C for 3 min, sulfiting with 1% KMS for 5 min, or sulfuring with 2 g of sulfur per kilogram of fruit slices for 4 h, can be applied. For osmotically dehydrated product, guava slices are heated in an equal weight of 70°Brix sugar syrup containing 0.1% KMS at 90°C for 3 min. The slices are left in the same syrup for 24 h, drained, spread in a net tray, and dried at 80°C for 1 h and then at 60–65°C for 7–8 h. Slices are packed in airtight containers. In osmovac dehydration, the guava slices are submerged in 70°Brix syrup for 5–6 h. They are then dried under vacuum until a final moisture content of 2% is attained (Adsule and Kadam, 1995). The influence of sugars (sucrose, glucose, fructose and their mixture) along with chemical preservatives (KMS and potassium sorbate) and antioxidants (citric acid and ascorbic acid) on the nutritional quality of osmotically dehydrated, intermediate-moisture, sweetened guava slices (final moisture content 25%) has been studied during storage for 90 days (Ayub *et al.*, 2005). Ascorbic acid content reduced from 293.9 to 44.5 mg 100 g^{-1} during the dehydration process. Treatments like sucrose + glucose (7:3), ascorbic acid and KMS, and sucrose + glucose (7:3), citric acid and KMS, were found to be the most effective sweetener combination in maintaining the nutrient stability of intermediate-moisture guava slices. Osmo-dried guava slices were prepared by dipping the slices in different concentrations of sugar syrup containing 0.05% KMS and 0.1% citric acid for varying time periods and temperatures (Sagar and Kumar, 2007). During the osmosis process, water loss and solid gain increased with the increase in sugar concentration and temperature of the solution. The optimum solid gain (13.1%), water loss (34.2%) and mass reduction (21.1%) in slices were observed in 60°Brix sugar syrup at 60°C, which were found to be optimum parameters to produce better osmo-dehydrated guava slices. The use of lower temperatures and shorter treatment times can diminish the loss of volatiles with respect to the fresh samples. The lowest total volatiles loss occurred at 30 and 40°C for up to 3 h under pulsed vacuum or atmospheric pressure (Pino *et al.*, 2008). The slight water activity reduction promoted by the process may provide stable products with good nutritional and sensorial quality and with characteristics similar to those of the fresh products. Four different drying conditions, namely cabinet drier, vacuum oven drier, low-temperature drier and solar drier, were evaluated to prepare osmo-dried guava slices (Kumar and Sagar, 2014). Vacuum oven drying was found superior as it holds maximum nutrients like ascorbic acid, acidity, sugar and water removal and moisture ratio along with the best sensory score.

Powder production through various drying processes has enormous benefits in terms of volume reduction, decrease in packaging and transportation costs, nutrient retention and extension of shelf-life. Guava powder can be prepared either by grinding the dehydrated slices or by using different drying techniques like sun drying, hot air drying, cabinet drying, foam mat drying, double drum drying, freeze drying, spray drying, convection oven drying, microwave oven drying, etc. Powder prepared by freeze or spray drying is far superior in nutritional quality and sensory attributes than that prepared by other drying procedures. Powder can be used in the preparation of guava juice, RTS beverages, milkshakes or *shrikhand*. Guava powder can become an efficient alternative for storage of the fruit because the reduction in water activity is directly related to the decline in chemical and enzymatic reactions responsible for the deterioration of foods. A milkshake, prepared by mixing 1.5 g of guava powder and 16 g of sugar with 100 ml of milk, had good colour, aroma and taste (Ahire, 1989). Powder was made from clarified guava juice using freeze drying, spray drying and tunnel drying methods. The freeze-dried product had superior quality, but the spray-dried powder was stable and more economical (Chopda and Barrett, 2001). Instant guava drink powder was prepared by dehydrating the concentrated guava juice (10.5°Brix) using freeze, spray and tunnel drying techniques (Mahendran, 2010). Guava powder could not be obtained satisfactorily by spray drying without additives (maltodextrin) because of the hygroscopic and thermoplastic nature of the product. Although the freeze-dried product had superior nutritional and sensory qualities, spray-dried powder was stable and more economical to obtain free-flowing guava powder with good stability. Response surface methodology (RSM) was used to optimize the spray drying process for the preparation of guava powder. The inlet air temperature exerted maximum influence on moisture and ascorbic acid content, while the maltodextrin concentration showed similar influence on solubility and dispersibility (Mahendran, 2010). The recommended optimum spray drying conditions for making guava powder are 185°C inlet air temperature and 7% maltodextrin concentration (Patil *et al.*, 2014). The spray-dried guava powder contained a higher amount of ascorbic acid compared with commercial fruit juice powder and was also found to be free-flowing without any physical alterations, viz. caking, stickiness, collapse and crystallization. Spray drying of aqueous guava extract (cultivar 'Allahabad Safeda') with 7% maltodextrin increased the proportion of ascorbic acid (118.6 mg 100 g^{-1}) in guava powder and also showed effective antimicrobial activity against *Shigella* sp. (minimum inhibitory concentration (MIC) = 11 mg ml^{-1}), *Escherichia coli* (MIC = 8 mg ml^{-1}) and *Candida* spp. (MIC < 1 mg ml^{-1}). The proximate and physiochemical properties showed good quality and enhanced solubility of guava powder compared with plain guava extract. The spray-dried guava powder may be a good source of natural antioxidants and profoundly increase the use of guava in value addition and dietary intake (Chauhan *et al.*, 2014). Drying temperature and drying time were the key parameters directly affecting the ascorbic acid concentration and colour of guava powder. Freeze drying was the best technique to preserve maximum ascorbic acid content and natural colour of guava, but it was quite expensive. The chemical composition of oven- and freeze-dried guava powders for future use as antioxidant-rich flavour enhancers was evaluated. Oven drying was found a viable option for the production of a functional ingredient that would improve the phenolic content of cereal foods while adding desirable guava flavour (Nunes *et al.*, 2016). Spray-dried, pink-pulped guava powder produced with 20% maltodextrin was found to have better quality than freeze-dried powder in terms of lowest moisture content (2.17%), lowest water activity (0.33), highest glass transition temperature (215°C), less electricity and time consumption, and moderate retention of lycopene and ascorbic acid (Shishir *et al.*, 2018).

3.3.9 Leather/bar

Fruit leather is an intermediate-moisture food also called 'fruit roll', 'fruit bar' or 'fruit sheet' commercially and developed by dehydration of fruit paste into a leathery sheet. The destruction of the original fruit structure by puréeing and restructuring it into a dehydrated sugar–acid–pectin gel called 'fruit leather' provides an attractive coloured product. Fruit leather is a dried sheet of fruit pulp having a soft, rubbery texture and sweet taste. Guava emits a pleasant sweet aroma, is refreshing and acidic in flavour, besides being a rich source of pectin; its pulp shows compatibility and suitability for blending and making mixed fruit products like jam, jelly, candy, leather, etc. (Jain *et al.*, 2011). Leather can be consumed as a confectionery or cooked to give a sauce. The method for the preparation of guava leather was standardized from preserved pulp (cultivar 'Banarsi Surkha') with 0.4% acidity and 20% TSS followed by drying in a cabinet drier at 50 ± 5°C for 4 h to a moisture content of about 29–30% (Sandhu *et al.*, 2001). The product wrapped in butter paper and packed in polyethylene bags was found to be acceptable for up to 3 months under ambient conditions. Microbiologically all packed leather samples were found to be safe for consumption after storage. Organoleptically, aluminium foil-packed leather is most suitable for storage followed by vegetable parchment paper. Guava leather is high in protein, fat, crude fibre and ash content (Ashaye *et al.*, 2005).

Guava fruit bar was prepared by mixing guava juice with maltodextrin, sucrose, soluble starch, wheat flour, pectin and anti-browning agent and drying at 50°C in a cross flow, hot air drier to final moisture content of 14–15% (Vijayanand *et al.*, 2000). Guar gum was used to improve the thickness of the product. On the basis of physicochemical analysis and sensory evaluation, the treatments of guava pulp with sucrose/glucose (7:3) and guar gum (0.25%), and of guava pulp with sucrose/glucose (0:10) and guar gum (0.25%), were found adequate.

3.3.10 Cheese

Guava cheese is a semi-solid concentrated fruit slice that is widely consumed as a snack in many countries, produced by dehydration of fruit pulp. Better-quality cheese is prepared from fresh fruits rather than the leftover pulp from jelly preparation. After the extraction of pulp from firm ripe guava fruits, the required quantities of sugar, citric acid and butter are added to 1 kg of pulp and the mixture is cooked until it becomes very thick. Small quantities of common salt and permitted colour may be added towards the end. The whole product is allowed to set and then cut into pieces of attractive shape, wrapped in cellophane or butter paper, and stored in a dry clean place (Sharma *et al.*, 1999). Cheese can also be prepared from the leftover guava pulp obtained during jelly preparation. Guava cheese was prepared from rainy- and winter-season fruits of three cultivars ('Allahabad Safeda', 'Banarsi Surkha' and 'Sardar') by adding 1.5 kg of sugar, 125 g of butter, 2 g of citric acid and 1 tablespoon of salt per kilogram of pulp extracted (Singh *et al.*, 1983).

3.3.11 Minimally processed/fresh-cut fruit

Minimal processing of fruits is the process that makes them ready to eat without losing their freshness, with good quality and degree of sanitization after eliminating non-edible parts like rinds, stems and seeds, followed by cutting, washing, classification, sanitization, centrifugation, packaging and storage, possibly including low levels of irradiation and whitening (Ahvenainem, 1996). Minimal processing makes the fruits more susceptible to spoilage compared with the whole fruit, causing an increase in respiration rate, generating more chemical reactions leading to faster degeneration and to increased water evaporation with much quicker wilting or withering. The combination of storage temperature (5°C), modified-atmosphere packaging (MAP) and the osmotic dehydration process helps to maintain the quality of minimally processed guavas during storage. MAP in polyethylene terephthalate (PET) containers has a strong influence on colour preservation and weight loss of the guavas (Pereira *et al.*, 2004). 'Paluma' guava fruits were peeled or not, cut in halves with seeds removed and packaged in polystyrene (PS) trays covered with PVC film or in a PET container with a lid. Packed minimally processed fruits were stored at low (5 and 10°C) and ambient (22.6°C) temperatures for 12 days. Weight loss was affected by storage temperature but not by peeling, with fruits packaged in PS trays having a greater weight loss. At low temperatures, product could be stored for 8 days with low microbial growth and no coliforms, but products stored at ambient temperature showed short shelf-life (3–4 days) with rapid spoilage (Durigan *et al.*, 2005). Chitosan coating can help to maintain the quality of fresh-cut guava fruit (cultivar 'Klom-Sali') for up to 7 days when packed in foam trays, enclosed with PVC film and preserved at 4°C (Thommohaway *et al.*, 2007). Minimally processed guava (cultivar 'Paluma') packed in PS–PVC can be preserved for 6 days when stored at 3°C after sanitizing the fruits with dehydrated sodium dichloroisocyanurate compound and cutting them into halves (Lima *et al.*, 2010). This treatment was also found more efficient to control autochthonous aerobic microbia (aerobic mesophilic microorganisms). Nasution *et al.* (2015) have mentioned that with suitable additives, aloe vera gel has potential as an edible coating for fresh-cut guava because of its ability to prolong the shelf-life and maintain the characteristics of the fruit for a longer period at 5°C. Fresh-cut guava coated with aloe vera, ascorbic acid and potassium sorbate was the most acceptable organoleptically with maximum ascorbic acid content (190.0 ± 14.1 mg 100 g^{-1}). Minimally processed guava (cultivar 'Lalit') fruit can retain quality attributes and bioactive phytochemicals when packed in airtight low-density polyethylene pouches, compared with whole guava fruits, during storage at 10°C for 15 days, thereby offering convenience and more choice to consumers (Chethan Kumar *et al.*, 2017). For fresh-cut guava fruits (cultivars 'Kimju' and 'Pan Srithong'), hot water treatment at 50°C for 10 or 30 min could be used to maintain quality and minimized microbial load when packed in a foam tray, wrapped with PVC film and stored for 6 days at 10°C (Poubol *et al.*, 2018). Development of browning was observed on the surface of fruits of both cultivars during storage which increased with the increase in temperature and time.

3.3.12 Canned slices

Canned slices are another important but less popular product of guava. The fruit is preserved by canning as halves or quarters, with or without the seed core. Fully ripe but firm fruits are lye peeled, peeled with a knife, or unpeeled, and cut into desired pieces. The cut pieces are dipped in 1–2% brine solution for 5 min to prevent browning and then canned in hot sugar syrup of 40°Brix containing 0.25% citric acid. The high sugar concentration of the syrup reduces the water activity and enhances the product's shelf-life. Canned slices often have a taste and aroma better than those of the fresh fruit. Loss of ascorbic acid during canning amounts to 19.4% (Sharma *et al.*, 1999). In Sudan, canning of guava in sugar syrup was done by the conventional method with double seaming and heated by steam. Due to canning, ascorbic acid content declined from 568.8 to 131.3 mg 100 g^{-1}. Acidity and reducing sugars also showed a decreasing trend, while total sugars, pH and TSS increased (Mudawi *et al.*, 2016). Organoleptically, colour and flavour of canned guava were found to be perfect and texture also remained good.

3.3.13 Fermented beverages

Two types of fermented beverages can be prepared from guava fruit: (i) wine (guava-juice wine and guava-pulp wine); and (ii) cider (with comparatively low alcohol percentage). Mature and ripe guavas with their high composition of fermentable reducing sugars such as glucose and fructose could serve as substrates for fruit wine production using wine yeast (*Saccharomyces cerevisiae*). High wastage of this fruit especially at the peak of its production season necessitates the need for alternative preservation and postharvest technologies towards its value addition that can reduce the level of postharvest losses besides increasing diversity of wine (Minh *et al.*, 2019). Guava wine may prove to be a quality wine with alcohol (stimulant) and high concentrations of ascorbic acid and phenolic compounds (antioxidants) besides being economically viable to farmers especially during market glut. The treatment of guava pulp with pectinase enzyme increased the final yield of wine. Guava-pulp wine is prepared in the same way as guava-juice wine. When the TSS content reaches 10°Brix, the pomace is removed and 10% more sugar is added to the fermenting materials and the mixture is allowed to ferment further (Bardiya *et al.*, 1974). Guava fruit could be converted into wine of acceptable quality with or without the addition of grape-grown yeast as reported by Shankar *et al.* (2006). Upon fermentation, TSS, reducing sugars and total sugars decreased as the sugars were converted into alcohol and carbon dioxide; whereas alcohol, total acidity, fixed acidity and volatile acidity increased, and pH decreased, as various acids were formed as by-products during fermentation. Also the increase in aroma and flavour of guava wine, produced from guava pulp at 1:4 dilution with water, and ethanol production with supplementation of nitrogen and phosphorus in the 'must' were highlighted. Yu and Zhang (2008) have optimized initial sugar concentration as 20% for guava wine production and stated that filtration and pasteurization of young guava wine could help to obtain a clear wine. A slight increase in acidity and alcohol and decrease in total phenol content were noticed during refrigerated storage of wine for 60 days (Kumar, 2009).

Guava juice required 'chaptalization' to adjust its Brix and make a perfect wine out of it (Kocher and Pooja, 2011). The chaptalized juice (must) was treated with pectinase or a combination of enzymes and fermented with traditional yeasts at a temperature range of 22 to 30°C and inoculums size of 6 to 11%. The addition of nitrogen and phosphorus through diammonium hydrogen orthophosphate (DAHP) improved ethanol production and quality parameters of the wine. Ageing and racking of guava wine also improved its sensory characteristics. Two different strains of *S. cerevisiae*, NCIM

3095 and NCIM 3287, were evaluated for the production of guava fruit wine and it was found that both *S. cerevisiae* strains were capable for guava wine production, but strain NCIM 3287 gave better results (Sevda and Rodrigues, 2011). Optimized fermentation parameters for *S. cerevisiae* NCIM 3287 were reported to be 25°C fermentation temperature, 4.0 pH, 0.06% diammonium phosphate, 100 ppm SO_2 and 6% inoculums level for better-quality guava wine. Wine was prepared from overripe guava fruit (cultivar 'Safeda') juices by using yeast *S. cerevisiae* var. HAU 1 after optimization of the fermentation process by adjusting TSS and pH. It was observed that wine prepared from juices having 15°Brix TSS and 4.0 pH yielded higher alcohol content (10.65%) (Younis *et al.*, 2014). Guava wine from 'Punjab Pink' cultivar prepared using the optimal fermentation attributes 25°Brix TSS, 25°C, 9% inoculums size and DAHP supplementation of 300 mg 100 ml^{-1} had an ethanol production of 13.8% (v/v) in 6 days (Nikhanj and Kocher, 2015). Sensory analysis of guava wine, obtained through ethanolic fermentation by *S. cerevisiae* MTCC 11815, revealed that wine prepared from untreated (without pectinase treatment) guava must was of standard quality as compared with superior quality of wine prepared from pectinase-treated must at 30 days of storage. However, both wines were superior as reflected by sensory scores after 90 days of storage (Nikhanj *et al.*, 2017).

Cider is legally defined to be a beverage made 'wholly or partly from the fermented juice of apples'. However, other fruits can also be utilized for cider preparation and such ciders are named by adding the name of the fruit before the word 'cider'. Guava cider from cultivar 'Safeda' was prepared and stored up to 12 months at ambient temperature (Pandey, 2004). TSS and acidity of the cider decreased and alcohol content increased continuously up to 5 months of storage, thereafter it remained constant. Regular decrease in ascorbic acid content and increase in NEB during storage also occurred. Improvement in organoleptic quality made the cider acceptable for up to 12 months. Six acidic accessions of guava (Acc. Nos 95, 26, 40, 33, 61 and 84), having low table value due to high acidity and low pectin content, were screened for cider preparation (Garg *et al.*, 2007). Highest ascorbic acid content (11.5 mg 100 ml^{-1}) was recorded in Acc. No. 33 followed by Acc. No. 40 (10.0 mg 100 ml^{-1}). Organoleptic evaluation during 45 days of storage revealed that cider prepared from Acc. No. 40 was best as it had better sugar–acid blend with good flavour retention, followed by Acc. No. 33.

3.4 Conclusion

Guava fruit is not only a rich source of vitamins and antioxidants but also a good source of minerals. Guava is useful in food and many other commercial and industrial applications. Because of its flavour and nutritional quality, guava has the potential to become a commercially important tropical fruit crop not only for fresh consumption but also for processing into value-added products. Despite this, the utilization of guava as a commercially important tropical fruit has not been exploited to its full potential. Recently, several advanced technologies have been developed for value addition and there is immense scope for diversified value-added products of guava including blending with other fruits. The products like guava powder, wine/cider, toffee/candy, juice and jelly have immense potential to attract consumers and expand the marketability of the fruit in future. Development of low-cost processing technologies, value addition through extension of shelf-life and processing of marketable surplus into value-added products, and utilization of food industries' waste/by-products are the key areas upon which to focus in order to increase the share of value-added products in Indian exports. Therefore, it is evident that there is a lot of scope for setting up guava processing industries in the country.

References

Abhangrao, A.K., Naidu, A.K. and Yadlod, S.S. (2017) Effect of recipes and cultivars on long storage of guava RTS. *Journal of Pharmacognosy and Phytochemistry* 6(5), 993–998.

Adsule, R.N. and Kadam, S.S. (1995) Guava. In: Salunkhe, D.K. and Kadam, S.S. (eds) *Handbook of Fruit Science and Technology, Production, Composition, Storage and Processing*. Marcel Dekker Inc., New York, pp. 419–433.

Afzal, M., Iqbal, R., Mahmood, Z., Zeshan, B. and Wattoo, J.I. (2019) Study of GC–MS and HPLC characterized metabolic compounds in guava (*Psidium guajava* L.) leaves. *Pakistan Journal of Agricultural Sciences* 56(3), 709–713.

Ahire, A.A. (1989) Studies on the development of preserved products from guava (*Psidium guajava* L.). MSc thesis, Mahatma Phule Agricultural University, Rahuri, India.

Ahvenainem, R. (1996) New approaches in improving the shelf life of minimally processed fruit and vegetables. *Trends in Food Science & Technology* 7(6), 179–197.

Anand, V., Manikandan, K.V., Kumar, S. and Pushpa, H.A. (2016) Phytopharmacological overview of *Psidium guajava* Linn. *Pharmacognosy Journal* 8, 314–320.

Anderson, J.M., Smith, B.M. and Gustafson, N.J. (1994) Health benefits and practical aspects of high-fiber diets. *American Journal of Clinical Nutrition* 59(5 Suppl.), 1242S–1247S.

Arrigoni, O. and de Tullio, M.C. (2002) Ascorbic acid: much more than just an antioxidant. *Biochimica et Biophysica Acta* 1569, 1–9.

Ashaye, O.A., Babalola, S.O., Babalola, A.O., Aina, J.O. and Fasoyiro, S.B. (2005) Chemical and organoleptic characterization of pawpaw and guava leathers. *World Journal of Agricultural Sciences* 1(1), 50–51.

Askar, A., El-Samahy, S.K. and El-Salem, N.A.A. (1992) Production of instant guava drink powder. *Confructa-Studien* 36(5–6), 154–161.

Asrey, R., Pal, R.K., Sagar, V.R. and Patel, V.B. (2007) Impact of tree age and canopy position on fruit quality of guava. *Acta Horticulturae* 735, 259–262.

Ayub, M., Zeb, A., Ullah, J. and Khattak, M.M.A.K. (2005) Effect of various sweeteners on chemical composition of guava slices. *Sarhad Journal of Agriculture* 21(1), 131–134.

Bal, L.M., Ahmad, T., Senapati, A.K. and Pandit, P.S. (2014) Evaluation of quality attributes during storage of guava nectar cv. Lalit from different pulp and TSS ratio. *Journal of Food Processing & Technology* 5(5), 329.

Bardiya, M.C., Kundu, B.S. and Tauro, P. (1974) Studies on fruit wines, 1. Guava wines. *Haryana Journal of Horticultural Sciences* 3(4), 140–146.

Bashir, H.A. and Abu-Goukh, A.A. (2003) Compositional changes during guava fruit ripening. *Food Chemistry* 80(4), 557–563.

Begum, S., Hassan, S.I., Siddiqui, B.S., Shaheen, F., Ghayur, M.N. and Gilani, A.H. (2002a) Triterpenoids from the leaves of *Psidium guajava*. *Phytochemistry* 61(4), 399–403.

Begum, S., Hassan S.I. and Siddiqui, B.S. (2002b) Two new triterpenoids from the fresh leaves of *Psidium guajava* (L.). *Planta Medica* 68, 1149–1152.

Begum, S., Hassan, S.I., Ali, S.N. and Siddiqui, B.S. (2004) Chemical constituents from the leaves of *Psidium guajava*. *Natural Product Research* 18(2), 135–140.

Bhat, F.M. and Singh, R. (2014) Preparation, quality evaluation and shelf life studies of whey–guava beverage. *World Journal of Agricultural Sciences* 10(3), 141–145.

Bhattacherjee, A.K. and Dikshit, A. (2019) Profiling of ascorbic acid, phenolic compounds and organic acids in guava varieties by HPLC–PDA. In: *Progressive Horticulture Conclave (PHC) – 2019 Futuristic Technologies in Horticulture, ICAR-IISR, Lucknow, India, 8–10 December 2019*, p. 110 (abstract).

Bhuvaneswari, S. and Tiwari, R.B. (2007) Pilot scale processing of red flesh guava RTS beverage. *Journal of Horticultural Sciences* 2(1), 50–52.

Bons, H.K. and Dhawan, S.S. (2006) Effect of heating/freezing with added chemical preservation on pulp preservation of guava (*Psidium guajava* L.). *Haryana Journal of Horticultural Sciences* 35(1), 22–25.

Bons, H.K. and Dhawan, S.S. (2009) Studies on the preparation of ready-to-serve beverage from stored guava pulp. *Beverage & Food World* 36(4), 41–42.

Bons, H.K. and Dhawan, S.S. (2013) Studies on preservation of guava pulp. *Indian Journal of Horticulture* 70(3), 452–454.

Brasil, I.M., Maia, G.A. and Figueiredo, R.W. (1995) Physical-chemical changes during extraction and clarification of guava juice. *Food Chemistry* 54(4), 383–386.

Brat, P., Olle, D., Reynes, M., Alter, P., Brillouet, J.M. *et al.* (2002) Preparation of tropical fruit purees by flash vacuum-expansion. *Acta Horticulturae* 575, 535–541.

Campbell, B.A. and Campbell, C.W. (1983) Preservation of tropical fruits by drying. *Proceedings of the Florida State Horticulture Society* 96, 229–231.

Cao, Y.H. and Cao, R.H. (1999) Angiogenesis inhibited by drinking tea. *Nature* 398(6726), 381.

Chan, H.T. Jr and Cavaletto, C.G. (1982) Aseptically packaged papaya and guava puree: changes in chemical and sensory quality during processing and storage. *Journal of Food Science* 47(4), 1164–1169.

Chan, H.T. Jr and Cavaletto, C.G. (1986) Effects of deaeration and storage temperature on quality of aseptically packaged guava puree. *Journal of Food Science* 51(1), 165–168.

Chan, H.T. Jr, Brekke, J.E. and Chang, T.U. (1971) Nonvolatile organic acids in guava. *Journal of Food Science* 36(2), 237–239.

Chandel, N., Kurrey, V.K., Minz, R.R. and Thakur, O. (2018) Guava nectar as a refreshing beverage: an overall review. *Plant Archives* 18(1), 1163–1169.

Chauhan, A.K., Singh, S., Singh, R.P. and Kumar, A. (2014) Determination of antioxidant capacity, total phenolics and antimicrobial properties of spray-dried guava extract for value-added processing. *Journal of Food Processing & Technology* 5(9), 368.

Chauhan, R., Kapoor, A.C. and Gupta, O.P. (1986) Note on the effect of cultivar and season on the chemical composition of guava fruits. *Haryana Journal of Horticultural Sciences* 15(3–4), 228–230.

Chen, C.C. and Wu, J.S.B. (1993) Effect of heating time, storage temperature, heavy metal ions presence, pH reduction, and cysteine addition on non-enzymatic browning in guava nectar. *Food Science (Shih Pm Kb Hsueh) (Taiwan)* 20, 43.

Chen, H.C., Sheu, M.J., Lin, L.Y. and Wu, C.M. (2007) Nutritional composition and volatile compounds in guava. *Fresh Produce* 1(2), 132–139.

Chen, H.S., Sheu, M.J., Lin, L.Y. and Wu, C.M. (2006) Characterization of volatiles in guava (*Psidium guajava* L. cv. Chung-Shan-Yueh-Pa) fruits from Taiwan. *Journal of Food and Drug Analysis* 14(4), 398–402.

Chen, H.Y. and Yen, G.C. (2007) Antioxidant activity and free radical-scavenging capacity of extracts from guajava leaves. *Food Chemistry* 43, 686–694.

Chen, M.L., Lee, C.Y. and Wu, J.S.B. (1994) An evaluation of possible mechanisms for nonenzymatic browning in guava nectar during storage. *Food Science (Shih Pm Kb Hsueh) (Taiwan)* 21, 293–303.

Chen, Y., Zhou, T., Zhang, Y., Zou, Z., Wang, F. and Xu, D. (2015) Evaluation of antioxidant and anticancer activities of guava. *International Journal of Food and Nutritional Safety* 6(1), 1–9.

Chethan Kumar, P., Garg, N., Shukla, D.K., Oberoi, H.S. and Yadav, K. (2017) Comparative evaluation of quality attributes and shelf life of minimally processed guava vis-à-vis whole guava (*Psidium guajava* L.) fruits during storage. *Indian Journal of Agricultural Sciences* 87(9), 1246–1251.

Chiari-Andréo, B.G., Trovatti, E., Marto, J., Almeida-Cincotto, M.G.J., Melero, A. *et al.* (2017) Guava: phytochemical composition of a potential source of antioxidants for cosmetic and/or dermatological applications. *Brazilian Journal of Pharmaceutical Sciences* 53(2), e16141.

Chiveu, J., Naumann, M., Kehlenbeck, K. and Pawelzik, E. (2019) Variation in fruit chemical and mineral composition of Kenyan guava (*Psidium guajava* L.): inferences from climatic conditions, and fruit morphological traits. *Journal of Applied Botany and Food Quality* 92, 151–159.

Chopda, C.A. and Barrett, D.M. (2001) Optimization of guava juice and powder production. *Journal of Food Processing and Preservation* 25(6), 411–417.

Choudhary, M.L. and Dikshit, S.N. (2006) Studies on shelf life of guava beverages. *Current Agriculture* 30(1/2), 75–78.

Choudhary, M.L., Dikshit, S., Shukla, N.N. and Saxena, R.R. (2008) Evaluation of guava varieties and standardization for nectar preparation. *Journal of Horticultural Sciences* 3, 161–163.

Chyau, C.C. and Wu, C.M. (1989) Differences in volatile constituents between inner and outer flesh-peel of guava (*Psidium guajava* L.) fruit. *Lebensmittel-Wissenschaft und Technologie* 22, 104–106.

Chyau, C.C., Chen, S.Y. and Wu, C.M. (1992) Differences of volatile and non-volatile constituents between mature and ripe guava (*Psidium guajava* Linn) fruits. *Journal of Agricultural and Food Chemistry* 40, 846–849.

Das, L., Bhaumik, E., Raychaudhuri, U. and Chakraborty, R. (2012) Role of nutraceuticals in human health. *Journal of Food Science and Technology* 49(2), 173–183.

Davey, M.W., van Montagu, M., Inze, D., Sanmartin, M., Kanellis, A. *et al.* (2000) Plant L-ascorbic acid: chemistry, function, metabolism, bioavailability and effects of processing. *Journal of the Science of Food and Agriculture* 80, 825–860.

Dhingra, M.K., Gupta, O.P. and Chundawat, B.S. (1983) Studies on pectin yield and quality of some guava cultivars in relation to cropping season and fruit maturity. *Journal of Food Science and Technology* 20(1), 10–13.

Dimascio, P., Kaiser, S. and Sies, H. (1989) Lycopene as the most effective biological carotenoid singlet oxygen quencher. *Archives of Biochemistry and Biophysics* 274, 532–538.

Dina, O.M.A., Ahmed, A.R. and Babikir, E.B. (2014) Physicochemical and nutritional value of red and white guava cultivars grown in Sudan. *Journal of Agri-Food and Applied Sciences* 2(2), 27–30.

dos Santos, W.N.L., Sauthier, M.C.S., dos Santos, A.M.P., Santana, D.A., Azevedo, R.S.A. and Caldas, J.C. (2017) Simultaneous determination of 13 phenolic bioactive compounds in guava (*Psidium guajava* L.) by HPLC-PAD with evaluation using PCA and neural network analysis (NNA). *Microchemical Journal* 133, 583–592.

Drincovich, M.F., Voll, L.M. and Maurino, V.G. (2016) Editorial: on the diversity of roles of organic acids. *Frontiers in Plant Science* 7, 1–2.

Dubey, S., Banafar, R.N.S. and Sahu, G.D. (2011) Effect of storage period on biochemical composition of guava nectar. *Environment and Ecology* 29(4), 1911–1917.

Durigan, J.F., Mattiuz, B.H., Lima, M.A., Epiphanio, R.D.V. and Biscegli, C.I. (2005) Minimally processed guava fruits *(Psidium guajava* L.). *Acta Horticulturae* 682, 1953–1960.

Duthie, G.G., Gardner, P.T. and Kyle, J.A.M. (2003) Plant polyphenols: are they the new magic bullet? *Proceedings of the Nutrition Society* 62, 599–603.

Ekundayo, O., Ajani, F., Seppaenen-Laakso, T. and Laakso, I. (1991) Volatile constituents of *Psidium guajava* L. (guava) fruit. *Flavour and Fragrance Journal* 6, 233–236.

El Bulk, R.E., Babiker, E.E. and El Tinay, A.H. (1997) Changes in chemical composition of guava fruits during development and ripening. *Food Chemistry* 59(3), 395–399.

Elleuch, M., Bedigian, D., Roiseux, O., Besbes, S., Blecker, C. and Attia, H. (2011) Dietary fibre and fibre-rich by-products of food processing: characterisation, technological functionality and commercial applications: a review. *Food Chemistry* 124(2), 411–421.

El Tinay, A.H., Saeed, A.R. and Bedri, M.F. (1979) Fractionation and characterization of guava pectic substances. *International Journal of Food Science & Technology* 14(4), 343–349.

Fanning, K., Murray, D., Stanley, R. and Netzel, M. (2008) The health benefits of tropical fruit grown in Queensland, Australia. In: *Proceedings of the Tropical Fruits in Human Nutrition and Health Conference, Gold Coast, Australia, 8-11 November 2008*. State of Queensland, Department of Employment, Economic Development and Innovation, Brisbane, Australia, pp. 202–214.

Fleuriet, A. and Macheix, J.J. (2003) Phenolic acids in fruits and vegetables. In: Rice-Evans, C.A. and Packer, L. (eds) *Flavonoids in Health and Disease*. Marcel Dekker Inc., New York, pp. 1–41.

Flores, G., Dastmalchi, K., Wu, S.B., Whalen, K., Dabo, A.J. *et al.* (2013) Phenolic-rich extract from the Costa Rican guava (*Psidium friedrichsthalianum*) pulp with antioxidant and anti-inflammatory activity. Potential for COPD therapy. *Food Chemistry* 141, 889–895.

Flores, G., Wu, S.B., Adam, N. and Kennelly, E.J. (2015) Chemical composition and antioxidant activity of seven cultivars of guava (*Psidium guajava*) fruits. *Food Chemistry* 170, 327–335.

Flores, P., Hellin, P. and Fenoll, J. (2012) Determination of organic acids in fruits and vegetables by liquid chromatography with tandem-mass spectrometry. *Food Chemistry* 132, 1049–1054.

Fontanari, G.G., Jaco, M.C., Souza, G.R., Batistuti, J.P., Neves, V.A. *et al.* (2008) DSC studies on protein isolate of guava seeds *Psidium guajava*. *Journal of Thermal Analysis and Calorimetry* 93(2), 397–402.

Freda, S.A., Krumreich, F.D., Rutz, J.K., Hartwig, N. and Zambiazi, R.C. (2018) Bioactive compounds during processing and storage of sweet guava (conventional and light). *International Food Research Journal* 25(3), 1181–1188.

Fu, L., Lu, W.Q. and Zhou, X.M. (2016) Phenolic compounds and *in vitro* antibacterial and antioxidant activities of three tropic fruits: persimmon, guava, and sweetsop. *BioMed Research International* 2016, 4287461.

Galdon, B.R., Rodriguez, C.T., Rodriguez, E.R. and Romero, C.D. (2008) Organic acid contents in onion cultivars (*Allium cepa* L.). *Journal of Agricultural and Food Chemistry* 56(15), 6512–6519.

Garg, N., Yadav, P., Goel, N., Bhattacharjee, A.K., Rajan, S. and Kumar, R. (2007) Screening of acidic guava accessions for cider preparation. *Acta Horticulturae* 735, 647–650.

Ghani, A., Hameed, T., Hussain, M., Ikram, M., Imran, M. *et al.* (2016) Comparative analysis of ascorbic acid concentration in guava varieties collected from four different tehsils of district Bhakkar. *International Journal of Basic and Applied Chemical Sciences* 6(1), 38–40.

Harnanan, S.W., Bains, G.S. and Singh, K.K. (1980) Studies on processing of pink and white fleshed guava varieties for pulp. *Punjab Horticulture Journal* 20(3/4), 179–189.

Hashinaga, F., Shima, Y. and Itoo, S. (1987) Production of volatile components of guava during maturation. *Bulletin of the Faculty of Agriculture, Kogoshima University* 37, 59–64.

Hodgson, A.S., Chan, H.T. Jr, Cavaletto, C.G. and Perera, C.O. (1990) Physical-chemical characteristics of partially clarified guava juice and concentrate. *Journal of Food Science* 55(6), 1757–1761.

Idstein, H. and Schreier, P. (1985) Volatile constituents from guava (*Psidium guajava* L.) fruit. *Journal of Agricultural and Food Chemistry* 33(1), 138–143.

Imelouane, B., Tahri, M., Elbastrioui, M., Aouinti, F. and Elbachiri, A. (2011) Mineral contents of some medicinal and aromatic plants growing in eastern Morocco. *Journal of Materials and Environmental Sciences* 2(2), 104–111.

Imungi, J.K., Scheffeldt, P. and Saint-Hislare, P. (1980) Physical-chemical changes during extraction and contraction of clear guava juice. *Lebensmittel-Wissenschaft und Technologie* 13, 248–251.

Isabelle, M., Lee, B.L., Lim, M.T., Koh, W.P., Huang, D. and Ong, C.N. (2010) Antioxidant activity and profiles of common fruits in Singapore. *Food Chemistry* 123, 77–84.

Jagtiani, J., Chan, H.T. and Sakai, W.S. (eds) (1988) Guava. In: *Tropical Fruit Processing*. Academic Press, New York, pp. 9–44.

Jain, P.K. and Asati, V.K. (2004) Evaluation of guava cultivars for pulp preparation. *Journal of Food Science and Technology* 41(6), 684–686.

Jain, P.K., Jain, P. and Nema, P.K. (2011) Quality of guava and papaya fruit pulp as influenced by blending ratio and storage period. *American Journal of Food Technology* 6(6), 507–512.

Jakhar, M.S., Vaish, P.K. and Pathak, S. (2013) Studies on the standardization and preservation of guava (*Psidium guajava* L.) and Barbados cherry (*Malpighia glabra* L.) blended ready-to-serve beverage. *Progressive Horticulture* 45(1), 95–99.

Jawaheer, B., Goburdhun, D. and Ruggoo, A. (2003) Effect of processing and storage of guava into jam and juice on the ascorbic acid content. *Plant Foods for Human Nutrition* 58, 1–12.

Jiménez-Escrig, A., Rincón, M., Pulido, R. and Saura-Calixto, F. (2001) Guava fruit (*Psidium guajava* L.) as a new source of antioxidant dietary fiber. *Journal of Agricultural and Food Chemistry* 49(11), 5489–5493.

Jordán, M.J., Margaría, C.A., Shaw, P.E. and Goodner, K.L. (2003) Volatile components and aroma active compounds in aqueous essence and fresh pink guava fruit puree (*Psidium guajava* L.) by GC–MS and multidimensional GC/GC–O. *Journal of Agricultural and Food Chemistry* 51(5), 1421–1426.

Joseph, B., Mini Priya, R., Helen, P.A.M. and Sujatha, S. (2010) Bio-active compounds in essential oil and its effects of antimicrobial, cytotoxic activity from the *Psidium guajava* (L.) leaf. *Journal of Advanced Biotechnology* 9(10), 10–14.

Joshi, H. (2016) Nutritional evaluation of different varieties of guava (*Psidium guajava* L.) and their preserved products. M.Sc. thesis, Punjab Agricultural University, Ludhiana, India.

Joshi, H., Kochhar, A. and Boora, R.S. (2017) Organoleptic and nutritional evaluation of value added products developed from new varieties of white and pink-fleshed guavas. *Chemical Science Review and Letters* 6(24), 2108–2113.

Joshi, H., Kochhar, A. and Boora, R.S. (2019) Organoleptic characteristics and nutritive value of candy developed from new varieties of guava. *International Journal of Chemical Studies* 7(2), 2124–2127.

Kadam, D.M., Kaushik, P. and Kumar, R. (2012) Evaluation of guava products quality. *International Journal of Food Science and Nutrition Engineering* 2(1), 7–11.

Kader, A.A. (2008) Flavor quality of fruits and vegetables. *Journal of the Science of Food and Agriculture* 88(11), 1863–1868.

Kalra, S.K. and Revathi, G. (1981) Storage studies on guava (*Psidium guajava* L.) pulp. *Indian Food Packer* 35(6), 29–33.

Kalra, S.K. and Tandon, D.K. (1984) Guava nectars from sulphited pulp and their blends with mango nectar. *Indian Food Packer* 38, 74–77.

Kamath, J.V., Rahul, N., Ashok Kumar, C.K. and Lakshmi, S.M. (2008) *Psidium guajava* L.: a review. *International Journal of Green Pharmacy* 2(1), 9–12.

Kanwal, N., Randhawa, M.A. and Iqbal, Z. (2016) A review of production, losses and processing technologies of guava. *Asian Journal of Agriculture and Food Sciences* 4(2), 96–101.

Kelebek, H. (2010) Sugars, organic acids, phenolic compositions and antioxidant activity of grapefruit (*Citrus paradisi*) cultivars grown in Turkey. *Industrial Crops and Products* 32, 269–274.

Kerure, Y.E. and Kjedkar, D.M. (1982) Studies on ready-to-serve beverages. *International Food Abstract of Technical Papers* 7, 1.

Khurdiya, D.S. and Sagar, V.R. (1991) Note on processing and storage of guava nectar. *Indian Journal of Horticulture* 48(1), 19–21.

Kishore, N., Madhavi, T., Edukondalu, L. and Gayathri, R.L. (2016) Development and chemical evaluation of guava candies. *International Journal of Agriculture Sciences* 8(58), 3272–3277.

Kobori, C.N. and Jorge, N. (2005) Characterization of some seed oils from fruits for utilization of industrial residues. *Ciência e Agrotecnologia* 29(5), 108–114.

Kocher, G.S. and Pooja (2011) Status of wine production from guava (*Psidium guajava* L.): a traditional fruit of India. *African Journal of Food Science* 5(16), 851–860.

Koo, M.H. and Mohamed, S. (2001) Flavonoid (myricetin, quercetin, kaempferol, luteolin and apigenin) content of edible tropical plants. *Journal of Agricultural and Food Chemistry* 49, 3106–3112.

Kumar, P.S. and Sagar, V.R. (2014) Drying kinetics and physico-chemical characteristics of osmo-dehydrated mango, guava and aonla under different drying conditions. *Journal of Food Science and Technology* 51, 1540–1546.

Kumar, R. (2009) Utilization of guava (*Psidium guajava* L.) fruit for wine making. M.Sc. thesis, Chaudhary Charan Singh Haryana Agriculture University, Hisar, India.

Kumar, S.N.S., Sreenivas, K.N., Shankarappa, T.H. and Ravindra, V. (2012) Standardization of recipe for value added nutraceutical beverages of guava blended with aloe vera and roselle. *Environment and Ecology* 30(3B), 995–1001.

Lee, K.W., Kim, Y.J., Kim, D.O., Lee, H.J. and Lee, C.Y. (2003) Major phenolics in apple and their contribution to the total antioxidant capacity. *Journal of Agricultural and Food Chemistry* 51, 6516–6520.

Lima, M.S., Pires, E.M.F., Maciel, M.I.S. and Oliveira, V.A. (2010) Quality of minimally processed guava with different types of cut, sanification and packing. *Ciência e Tecnologia de Alimentos, Campinas* 30(1), 79–87.

Liu, X., Yan, X., Bi, J., Liu, J., Zhou, M. *et al.* (2018) Determination of phenolic compounds and antioxidant activities from peel, flesh, seed of guava (*Psidium guajava* L.). *Electrophoresis* 39(13), 1654–1662.

McCook-Russell, K.P., Nair, M.G., Facey, P.C. and Bowen-Forbes, C.S. (2012) Nutritional and nutraceutical comparison of Jamaican *Psidium cattleianum* (strawberry guava) and *Psidium guajava* (common guava) fruits. *Food Chemistry* 134(2), 1069–1073.

MacLeod, A.J. and Troconis, N.G. (1982) Volatile flavor components of guava. *Phytochemistry* 21, 1339–1342.

Mahattanatawee, K., Manthey, J.A., Luzio, G., Talcott, S.T., Goodner, K. and Baldwin, E.A. (2006) Total antioxidant activity and fibre content of select Florida-grown tropical fruits. *Journal of Agricultural and Food Chemistry* 54, 7355–7363.

Mahendran, T. (2010) Physico-chemical properties and sensory characteristics of dehydrated guava concentrate: effect of drying method and maltodextrin concentration. *Tropical Agricultural Research & Extension* 13(2), 48–54.

Mall, P. and Tandon, D.K. (2007) Development of guava–aonla blended beverage. *Acta Horticulturae* 735, 555–560.

Marak, J.K. and Mukunda, G.K. (2007) Studies on the performance of open pollinated seedling progenies of guava cv. 'Apple Colour'. *Acta Horticulturae* 735, 79–84.

Menezes, C.C., Borges, S.V., Ferrua, F.Q., Vilela, C.P. and Carneiro, J.D.S. (2011) Influence of packaging and potassium sorbate on the physical, physicochemical and microbiological alterations of guava preserves. *Ciência e Tecnologia de Alimentos, Campinas* 31(3), 674–680.

Mercadante, A.Z., Steck, A. and Pfander, H. (1999) Carotenoids from guava (*Psidium guajava* L.): isolation and structure elucidation. *Journal of Agricultural and Food Chemistry* 47(1), 145–151.

Metwally, A.M., Omar, A.A., Harraz, F.M. and Sohafy, S.M.E. (2010) Phytochemical investigation and antimicrobial activity of *Psidium guajava* L. leaves. *Pharmacognosy Magazine* 6(23), 212–218.

Minh, N.P., Pham, V.T., Tre, T.T., Kieu, T.T., Nhu, N.T.H. and Van, T.T.C. (2019) Different factors affecting Guava (*Psidium guajava*) wine fermentation. *Journal of Pharmaceutical Sciences & Research* 11(4), 1458–1463.

Mohnen, D. (2008) Pectin structure and biosynthesis. *Current Opinion in Plant Biology* 11, 266–277.

Mowlah, G. and Itoo, S. (1982) Guava (*Psidium guajava* L.). Sugar component and relation to enzymes at stages of fruit development and ripening. *Journal of Japanese Society for Food Science and Technology* 29(8), 472–476.

Mudawi, H.A., Abdalatif, F.Y. and Saifedin, M.K. (2016) Evaluation of canned papaya (*Carica papaya*) and guava (*Psidium guajava* L.) fruits in Sudan. *International Journal of Food Science and Nutrition* 1(5), 28–31.

Murari, K. and Verma, R.A. (1989) Studies on the effect of varieties and pulp extraction methods on the quality of guava nectar. *Indian Food Packer* 43, 11–15.

Murdock, H.D. (2002) *Encyclopedia of Foods, A Guide to Healthy Nutrition*. Academic Press, San Diego, California.

Musa, K.H., Abdullah, A. and Subramaniam, V. (2015) Flavonoid profile and antioxidant activity of pink guava. *Science Asia* 41, 149–154.

Naseer, S., Hussain, S., Naeem, N., Pervaiz, M. and Rahman, M. (2018) The phytochemistry and medicinal value of *Psidium guajava* (guava). *Clinical Phytoscience* 4, 32.

Nasution, Z., Ye, J.N.W. and Hamzah, Y. (2015) Characteristics of fresh-cut guava coated with aloe vera gel as affected by different additives. *Kasetsart Journal (Natural Science)* 49, 111–121.

Nikhanj, P. and Kocher, G.S. (2015) Fermentative production of guava-wine (*Psidium guajava* L.) using *S. cerevisiae* MTCC 11815. *Current Nutrition & Food Science* 11(1), 21–30.

Nikhanj, P., Kocher, G.S. and Boora, R.S. (2017) Fermentative production of guava wine from pectinase treated and untreated juice of 'Punjab Pink' cultivar of *Psidium guajava* L. *Agricultural Research Journal* 54(2), 244–247.

Nishimura, O., Yamaguchi, K., Mihara, S. and Shibamoto, T. (1989) Volatile constituents of guava fruit (*Psidium guajava* L.) and canned puree. *Journal of Agricultural and Food Chemistry*, 37, 139–142.

Nooruddin, M., Wahab, S., Muhammad, A., Bilal, H., Din, M.U. *et al.* (2019) Development and quality evaluation of mango and guava blended squash during storage. *Pure and Applied Biology* 8(2), 1182–1190.

Nunes, J.C., Lago, M.G., Castelo-Branco, V.N., Oliveira, F.R., Torres, A.G. *et al.* (2016) Effect of drying method on volatile compounds, phenolic profile and antioxidant capacity of guava powders. *Food Chemistry* 197, 881–890.

Oliveira, D.S., Aquino, P.P., Ribeiro, S.M.R., Proença, R.P.C. and Pinheiro-Sant'ana, H.M. (2011) Vitamin C, carotenoids, phenolic compounds and antioxidant activity of guava, mango and papaya ceasa coming from the state of Minas Gerais. *Acta Scientiarum Health Sciences* 33, 89–98.

Padula, M. and Rodriguez-Amaya, D.B. (1986) Characterisation of the carotenoids and assessment of the vitamin A value of Brasilian guavas (*Psidium guajava* L.). *Food Chemistry* 20, 11–19.

Pandey, A.K. (2004) Study about the storage stability of guava beverages. *Progressive Horticulture* 36(1), 142–145.

Pandey, A.K. and Singh, I.S. (1998) Physico-chemical studies on utilization of guava cultivars. *Progressive Horticulture* 30(1&2), 73–75.

Pandey, A.K. and Singh, I.S. (1999) Studies on preparation and preservation of guava ready-to-serve beverage. *Indian Journal of Horticulture* 56(2), 130–132.

Pandey, D., Pandey, A.K. and Yadav, S.K. (2016) Evaluation of newly developed guava cultivars & selections under Lucknow conditions. *Indian Journal of Horticulture* 73(3), 334–338.

Patel, R.K., Maiti, C.S., Deka, B.C., Deshmukh, N.A. and Nath, A. (2013) Changes in sugars, pectin and antioxidants of guava (*Psidium guajava*) fruits during fruit growth and maturity. *Indian Journal of Agricultural Sciences* 83(10), 1017–1021.

Patil, V., Chauhan, A.K. and Singh, R.P. (2014) Optimization of the spray-drying process for developing guava powder using response surface methodology. *Powder Technology* 253, 230–236.

Pereira, L.M., Rodrigues, A.C.C., Sarantópoulos, C.I.G.L., Junqueira, V.C.A., Cunha, R.L. and Hubinger, M.D. (2004) Influence of modified atmosphere packaging and osmotic dehydration on the quality maintenance of minimally processed guavas. *Journal of Food Science* 69(4), FEP172–FEP177.

Pino, J.A., Marbot, R. and Vázquez, C. (2002) Characterization of volatiles in Costa Rican guava [*Psidium friedrichsthalianum* (Berg) Niedenzu] fruit. *Journal of Agricultural and Food Chemistry* 50, 6023–6026.

Pino, J.A., Panadés, G., Fito, P., Chiralt, A. and Ortega, A. (2008) Influence of osmotic dehydration on the volatile profile of guava fruits. *Journal of Food Quality* 31, 281–294.

Poubol, J., Techavuthiporn, C. and Kanlayanarat, S. (2018) Guava fruit treated with hot water on microbiological quality of fresh-cut 'Kimju' and 'Pan Srithong' guava. *International Food Research Journal* 25(3), 903–907.

Ramanjaneya, K.H. (1983) Studies on some factors influencing quality of guava jelly. *Mysore Journal of Agricultural Sciences* 17(4), 408.

Rashid, R., Bhat, A., Dayal, A., Sood, M. and Sharma, S. (2018) Studies on storage stability of guava RTS. *The Pharma Innovation Journal* 7(5), 230–233.

Ravi, G.K., Jholgiker, P., Thippanna, K.S., Kumar, B.N.P. and Pattepur, S. (2018) Evaluation of red fleshed guava (*Psidium guajava* L.) varieties for their processing potential. *International Journal of Current Microbiology and Applied Sciences* 7(11), 3475–3483.

Reddy, N.N., Gangopadhyay, K.K., Rai, M. and Kumar, R. (1999) Evaluation of guava cultivars under rainfed sub-humid region of Chhotanagpur plateau. *Indian Journal of Horticulture* 56(2), 135–140.

Rickman, J.C., Barrett, D.M. and Bruhn, C.M. (2007) Nutritional comparison of fresh, frozen and canned fruits and vegetables. Part 1. Vitamins C and B and phenolic compounds. *Journal of the Science of Food and Agriculture* 87, 930–944.

Rojas-Garbanzo, C., Zimmermann, B.F., Schulze-Kaysers, N. and Schieber, A. (2017) Characterization of phenolic and other polar compounds in peel and flesh of pink guava (*Psidium guajava* L. cv. 'Criolla') by ultra-high performance liquid chromatography with diode array and mass spectrometric detection. *Food Research International* 100(3), 445–453.

Sagar, V.R. and Kumar, P.S. (2007) Processing of guava in the form of dehydrated slices and leather. *Acta Horticulturae* 735, 579–585.

Sandhu, K.S. and Bhatia, B.S. (1985) Physico-chemical changes during preparation of fruit juice concentrate. *Journal of Food Science and Technology* 22(3), 202–206.

Sandhu, K.S., Singh, M. and Ahluwalia, P. (2001) Studies on processing of guava into pulp and guava leather. *Journal of Food Science and Technology* 38, 622–624.

Satyal, P., Paudel, P., Lamichhane, B. and Setzer, W.N. (2015) Leaf essential oil composition and bioactivity of *Psidium guajava* from Kathmandu, Nepal. *American Journal of Essential Oils and Natural Products* 3(2), 11–14.

Seshadri, T.R. and Vasishta, K. (1965) Polyphenols of the leaves of *Psidium guajava* – quercetin, guaijaverin, leucocyanidin and amritoside. *Phytochemistry* 4(6), 989–992.

Sevda, S.B. and Rodrigues, L. (2011) Fermentative behavior of *Saccharomyces* strains during guava (*Psidium guajava* L.) must fermentation and optimization of guava wine production. *Journal of Food Processing and Technology* 2, 118–127.

Shahidi, F., Janitha, P.K. and Wanasundara, P.D. (1992) Phenolic antioxidants. *Critical Reviews in Food Science and Nutrition* 32, 67–103.

Shankar, S., Dilip Babu, J. and Reddy, Y.N. (2006) Fermentation of guava pulp with grape-grown yeast (*Saccharomyces cerevisiae* var. *ellipsoideus*) for wine production. *Indian Journal of Horticulture* 63(2), 171–173.

Shao, M., Wang, Y., Huang, X.J., Fan, C.L., Zhang, Q.W. *et al.* (2012) Four new triterpenoids from the leaves of *Psidium guajava*. *Journal of Asian Natural Products Research* 14(4), 348–354.

Sharma, S., Rajat, K., Prasad, R. and Vasudevan, P. (1999) Biology and potential of *Psidium guajava*. *Journal of Scientific and Industrial Research* 58, 414–421.

Shishir, M.R.I., Taip, F.S., Saifullah, M., Yong, S.Y., Aziz, N.A. and Talib, R.A. (2018) Changes in quality attributes of pink guava (*Psidium guajava*) powder with respect to different drying techniques and maltodextrin concentrations. *International Food Research Journal* 25(4), 1625–1632.

Shrivastava, P., Prashad, V.M., Kumar, J., Mishra, P.K. and Singh, R.K. (2017) Preservation of guava (*Psidium guajava* L.) juice with sodium benzoate and ginger extract. *Progressive Horticulture* 49(1), 53–58.

Shukla, R. and Shukla, Y.K. (2017) Studies of different guava cultivars (*Psidium guajava* L.) for nutritional and livelihood security suited to degraded soils. *The Asian Journal of Horticulture* 12(1), 91–95.

Silva, E.A.J., da Silva, V.P., Alves, C.C.F., Alves, J.M., Souchie, E.L. and Barbosa, L.C.A. (2016) Harvest time on the content and chemical composition of essential oil from leaves of guava. *Ciência Rural, Santa Maria* 46(10), 1771–1776.

Silva, L.M.R., Figueiredo, E.A.T., Ricardo, N.M.P.S., Vieira, I.G.P., Figueiredo, R.W. *et al.* (2014) Quantification of bioactive compounds in pulps and by-products of tropical fruits from Brazil. *Food Chemistry* 143, 398–404.

Sinchaipanit, P., Ahmad, M. and Twichatwitayakul, R. (2015) Kinetics of ascorbic acid degradation and quality changes in guava juice during refrigerated storage. *Journal of Food and Nutrition Research* 3(8), 550–557.

Singh, G., Misra, A.K., Haseeb, M., Tandon, D.K. and Pathak, R.K. (2003) *The Guava*. Extension Bulletin No. 17. ICAR-CISH, Lucknow, India.

Singh, I.S. and Dhawan, S.S. (1983) Potentiality of various fruits for processing industry. *Indian Food Packer* 37, 47–53.

Singh, D., Wangchu, L., Bhatnagar, P. and Moond, S.K. (2007) Emerging vistas in post harvest paradigm of guava. *Indian Journal of Arid Horticulture* 2(2), 45–54.

Singh, P., Shukla, A., Singh, R. and Singh, A.K. (2007) Utilization of guava juice by value addition through blended beverages. *Acta Horticulturae* 735, 639–645.

Singh, R., Kapoor, A.C. and Gupta, O.P. (1983) Effect of cultivars, seasons and storage on the nutritive value and keeping quality of guava cheese. *Indian Food Packer* 37, 71–75.

Soares, F.D., Pereira, T., Marques, M.O.M. and Monteiro, A.R. (2007) Volatile and non-volatile chemical composition of the white guava fruit (*Psidium guajava* L.) at different stages of maturity. *Food Chemistry* 100, 15–21.

Soliman, F.M., Fathy, M.M., Salama, M.M. and Saber, F.R. (2016) Comparative study of the volatile oil content and antimicrobial activity of *Psidium guajava* L. and *Psidium cattleianum* Sabine leaves. *Bulletin of Faculty of Pharmacy, Cairo University* 54, 219–225.

Sousa, M.S.B., Vieira, L.M., Silva, M.J.M. and Lima, A. (2011) Nutritional characterization and antioxidant compounds in pulps residues of tropical fruits. *Ciência e Agrotecnologia* 35, 554–559.

Sukeksi, L. and Sarah, M. (2016) Characterizations and extraction of polyphenols from residual pulp of pink guava as source of antioxidants. *ARPN Journal of Engineering and Appied Sciences* 11(8), 5209–5216.

Tandon, D.K. and Kalra, S.K. (1984) Chemical evaluation of stored guava pulp in polyethylene pouches. *Indian Food Packer* 38(5), 57–59.

Tandon, D.K., Kalra, S.K., Kulkarni, J.K. and Chadha, K.L. (1983a) Chemical and microbiological evaluation of stored guava pulp in PVC containers. *Journal of Food Science and Technology* 20(3), 118–120.

Tandon, D.K., Kalra, S.K., Singh, H. and Chadha, K.L. (1983b) Physico-chemical characteristics of some guava varieties. *Progressive Horticulture* 15(1–2), 42–44.

Thakur, B.R., Singh, R.K. and Handa, A.K. (1997) Chemistry and uses of pectin – a review. *Critical Reviews in Food Science and Nutrition* 37(1), 47–73.

Thommohaway, C., Kanlayanarat, S., Uthairatanakij, A. and Jitareerat, P. (2007) Quality of fresh-cut guava (*Psidium guajava* L.) as affected by chitosan treatment. *Acta Horticulturae* 746, 449–454.

Thuaytong, W. and Anprung, P. (2011) Bioactive compounds and prebiotic activity in Thailand-grown red and white guava fruit (*Psidium guajava* L.). *Food Science and Technology International* 17(3), 205–212.

Tiwari, A., Pal, A.K., Singh, S.P., Singh, S., Singh, B.K. and Singh, P. (2016) Evaluation of guava cultivars for quality pulp production. *Research in Environment and Life Sciences* 9(11), 1406–1408.

Tiwari, R.B. (2000) Studies on blending of guava and papaya pulp for RTS beverage. *Indian Food Packer* 54(2), 68–72.

Uchôa, A.M.A., Costa, J.M.C., Maia, G.A., Silva, E.M.C., Carvalho, A.F.F.U. and Meira, T.R. (2008) Parametros fisico-quimicos: teor de fibra bruta e alimentar de pos alimenticios obtidos de residues de frutas tropicais. *Segurança Alimentar e Nutricional* 15(2), 58–65.

Uchôa-Thomaz, A.M.A., Sousa, E.C., Carioca, J.O.B., Morais, S.M., Lima, A. *et al.* (2014) Chemical composition, fatty acid profile and bioactive compounds of guava seeds (*Psidium guajava* L.). *Food Science and Technology, Campinas* 34(3), 485–492.

USDA (2019) Guavas, common, raw. FoodData Central Search Results. USDA National Nutrient Database for Standard Reference Legacy Release, April 2018. US Department of Agriculture, Agricultural Research Service, Washington, DC. Available at: https://fdc.nal.usda.gov/fdc-app.html#/food-details/173044/nutrients (accessed 25 May 2020).

Usha Kumari, K., Rishitha, G., Prasad, K.R. and Kumar, P.S. (2017) Value added products of guava. *Agriculture Update* 12, 2171–2177.

Uzzaman, S., Akanda, K.M., Mehjabin, S. and Parvez, G.M.M. (2018) A short review on a nutritional fruit: guava. *Toxicology Research* 1(1), 1–8.

Vernin, G., Verin, E., Vernin, C. and Metzger, J. (1991) Extraction and GC–MS–SPECMA data bank analysis of the aroma of *Psidium guajava* L. fruit from Egypt. *Flavour and Fragrance Journal* 6, 143–148.

Vijayanand, P., Yadav, A.R., Balasubramanyam, N. and Narasimham, P. (2000) Storage stability of guava fruit bar prepared using a new process. *Lebensmittel-Wissenschaft und Technologie* 33, 132–137.

Vora, J.D., Mankame, G. and Madav, P. (2018) Biochemical and nutritional assessment of guava (*Psidium guajava*). *IOSR Journal of Biotechnology and Biochemistry* 4(5), 1–7.

Weli, A., Al-Kaabi, A., Al-Sabahi, J., Said, S., Hossain, M.A. and Al-Riyami, S. (2019) Chemical composition and biological activities of the essential oils of *Psidium guajava* leaf. *Journal of King Saud University – Science* 31(4), 993–998.

Wilberg, V.C. and Rodriguez-Amaya, D.B. (1995) HPLC quantitation of major carotenoids of fresh and processed guava, mango and papaya. *Lebensmittel-Wissenschaft und Technologie* 28, 474–480.

Wilson, C.W., Shaw, P.E. and Campbell, C.W. (1982) Determination of organic acids and sugars in guava (*Psidium guajava* L.) cultivars by high-performance liquid chromatography. *Journal of the Science of Food and Agriculture* 33(8), 777–780.

Yadav, S.K., Sarolia, D.K., Pilania, S., Gora, J.S. and Singh, D.K. (2017) Organoleptic evaluation of preserved guava pulp during storage. *International Journal of Current Microbiology and Applied Sciences* 6(6), 950–955.

Younis, K., Siddiqui, S., Jahan, K. and Dar, M.S. (2014) Production of wine from over ripe guava (*Psidium guajava* L. cv. Safada) and ber (*Ziziphus mauritiana* L. cv. Umran) fruits using *Saccharomyces cereviciae* var. HAU 1. *IOSR Journal of Environmental Science, Toxicology and Food Technology* 8(1), 93–96.

Yu, H. and Zhang, X. (2008) Development of guava fruit wine. *China Brew* 13, 36.

Yusof, S. and Mohamed, S. (1987) Physico-chemical changes in guava (*Psidium guajava* L.) during development and maturation. *Journal of the Science of Food and Agriculture* 38(1), 31–39.

4 Propagation

Sisir Mitra[1]* and Pravat K. Ray[2]
[1]*Former Dean, Bidhan Chandra Krishi Viswavidyalaya, Mohanpur, West Bengal, India;*
[2]*Rajendra Agricultural University, Samastipur, Bihar, India*

4.1 Introduction

Guava (*Psidium guajava* L.) is a fast-growing tropical and subtropical species adapted to a wide range of environmental conditions. It is a precocious and prolific reproducer with seed dispersal aided by avian and mammalian vectors (CABI, 2020). Guava trees have solitary hermaphrodite flowers, and their morphology is such that they favour self-pollination. However, the extent of cross-pollination in guava carried out mainly by bees and other insects has been estimated to range from 25.7 to 41.3% (Alves and Freitas, 2007; BPI, 2011; Pereira *et al.*, 2016). The proportion of cross-pollination may vary among genotypes with the environment of the production area and the availability and efficiency of the pollinating agents. Therefore, propagation by seed features individuals that are genetically heterogeneous, which can be observed between orchards and plants of the same orchards established by planting seedlings. Hence, seed propagation is not recommended in commercial orchards intended for high quality and productivity. Various asexual propagation techniques are available, but the levels of adoption vary greatly in different guava-producing areas of the world. Considering the high level of orchard management practices currently being used at different places, plant propagation plays a key role in providing healthy, good-quality planting materials free from diseases for establishing orchards with inherent capability to produce high-quality fruits and better yields. This chapter deals with various propagation methods used in the multiplication of guava and progress made so far in this field.

4.2 Seed

Guava can be propagated easily from seed. However, seed propagation should be avoided as owing to the heterozygous nature of the tree and cross-pollination, the seedling-raised plants are never true to type. Apart from a longer juvenile phase the seedlings differ widely in yield and physico-chemical characteristics of the fruits (Mitra and Sanyal, 2004). In many countries where this crop was grown earlier using seedlings, several commercial cultivars were obtained by selecting promising seedlings from open pollination. Also, on the other hand, the continuous use of seeds virtually eliminated some cultivars in Brazil,

*E-mail: sisirm55@gmail.com

DOI: 10.1079/9781789247022.0004

such as 'Guanabara', whose genetic characteristics are no longer the same as the originally selected material (Pereira *et al.*, 2016). Nevertheless, seedlings are required to be grown essentially in the case of raising hybrid seedlings for breeding new cultivars and producing rootstocks for grafting.

Although guava seeds retain their viability for a considerable period, it is better to sow seeds immediately after extraction from fruits. The viability of seeds can be extended if they are washed, dried and stored in an airtight container in a dry cool place. At 8°C and low humidity guava seeds retain their viability for approximately 1 year (CABI, 2020). The seeds can retain nearly 50% viability for up to 18 months when they are stored with or without charcoal in a sealed tin container or a glass or polyethylene jar (Chacko and Singh, 1968). Seeds could be stored for 104 days when they were pre-soaked for 24 h in ferulic acid at 10^{-3} M concentration (Mitra and Bose, 1990). Large fruit of 'Sardar' guava had the highest seed weight (2.093 g per fruit), maximum number of seeds (141.4 seeds per fruit) and showed the highest germination (70%) compared with medium and small fruits (Vijayakumar *et al.*, 1991).

The seedcoat of guava is hard and causes delay in germination. Acid scarification and boiling of seeds in hot water for 5 min shortened the time required for germination without any adverse effect on germination (Hayes, 1974). A combination of scarification in sand for 15 min and germination in sand gave the best germination (98.15%) and seeds germinated in 8 to 11 days (Tavares *et al.*, 1995). Soaking of seeds in water for 12 h or in hydrochloric acid for 3 min (Singh and Soni, 1974) or in water for 36 h (Pandey and Singh, 2000) resulted in 90% germination of seeds of 'Allahabad Safeda' guava. Scarification of seeds with sulfuric acid greatly increased germination, recording 98% germination (Essien, 2004). Soaking of seeds in 500 ppm 2-chloroethylphosphonic acid (ethephon) followed by 300 ppm naphthalene acetic acid (NAA) (Sinha *et al.*, 1973) or in gibberellic acid (GA_3) at 1000 ppm (Kalyani *et al.*, 2014) or 3000 ppm solution (Chandra and Govind, 1990) or 48 h stratification + 300 ppm GA_3 (Narayan, 2017) increased germination and seedling growth.

The germination percentage of seeds decreased if the medium had high salinity. The salinity tolerance limit (50% reduction in germination) was reported as 6.45 dS m^{-1} (Hooda and Yamdagni, 1991). Increase in salinity depressed the radical length, hypocotyl length and number of secondary roots (Kaul *et al.*, 1988).

4.2.1 Raising seedlings

The seedlings are raised in the nursery and then transferred into polyethylene bags for use as rootstock. The soil of the seedbed should be sterilized to eliminate any infection with nematodes or other pathogens. Seeds should be sown 1–2 cm deep and 8–10 cm apart in a well-drained sandy loam or similar soil. There must be intimate contact between the seed and soil for water absorption by the embryo and the endosperm. More than 90% of fresh seeds germinate in 15–20 days. Immediately following germination, partial shade is desirable, especially during hot weather. Watering should be carried out using a sprinkling can or with a fine spray head on a hose at low pressure. When the seedlings are 10–15 cm tall, they are normally transferred to polyethylene bags of 0.1 mm gauge and 25 cm × 30 cm size, filled with soil and compost (1:1 v/v). After transplanting, they should be grown in partial shade and watered daily for 3 months. They are then gradually hardened under full sun before being grafted. Seedlings may be budded or grafted when stem diameters are 15–20 mm, with greater diameter being especially suitable for budding (Nakasone and Paull, 1998).

4.3 Stem Cuttings

Although cuttings are inexpensive and simple to prepare, they can be unreliable. Success depends on good temperature and humidity control for rooting, the type of wood, the stage of growth and the time of year. In general, guava hardwood cuttings were observed to be hard to root (Dhua *et al.*, 1982; Mitra and Bose, 1996). Semi-hardwood cuttings rooted better than hardwood cuttings and cuttings with two nodes and four leaves showed a better response to rooting. Rooting of greenwood cuttings, with two to four leaves retained and treated with root-inducing compounds and rooted in intermittent mist, has made it possible to produce large quantities of plants in a relatively short time for commercial orchard development (Paxton *et al.*, 1980). Apical-shoot cuttings at juvenile stage of 12 cm in length, with two to four nodes carrying two to four leaves and treated with auxin, showed good rooting under a low-plastic tunnel (Akram *et al.*, 2017).

Maintaining high relative humidity (RH) was considered very important for high success in rooting. Mist and fog systems are mostly used for propagating guava by cuttings. Treatment with auxins increases the proportion of cuttings that root, hastens root initiation and increases the number and quality of roots (Dhua *et al.*, 1982; Mitra and Bose, 1996; Akram *et al.*, 2017).

Indole butyric acid (IBA) is more effective than NAA and indole acetic acid (IAA) in inducing rooting of guava cuttings. IBA at 200–3000 mg l^{-1} is used for treatment of semi-hardwood and softwood cuttings. Debnath and Maiti (1990) reported 83.3, 73.3 and 73.3% rooting of softwood cuttings for cultivars 'Baruipur', 'Sardar' and 'Harijha', respectively, by using IBA at 2500 mg l^{-1}. The same concentration of IBA (2500 mg l^{-1}) caused 98% rooting of hardwood cuttings with bottom heat (30 ± 2°C) in the propagation bench (Prasad *et al.*, 1988). Softwood cuttings of guava cultivar 'Paluma' showed best rooting by prolonged soaking (14 h) of cuttings in IBA solution at 200 mg l^{-1} (Pereira *et al.*, 1991).

Preconditioning treatment of mother plants such as etiolation, localized blanching, girdling or wounding before taking cuttings improved rooting (Mitra and Pathak, 2018). Spraying the stock plants twice with ethrel ((2-chloroethyl)phosphonic acid) at 100 ppm before 30 days and 20 days of taking cuttings and treatment with IBA at 3000 ppm and *p*-hydroxybenzoic acid at 200 ppm at planting time caused 92.8% rooting with an average of 15.1 roots per cutting compared with 53.3% rooting and 6.6 roots per cutting in the control (Dhua *et al.*, 1982). Mitra (1996) reported maximum rooting of leafy semi-hardwood cuttings when mother plants were etiolated for 10 days + 10 days of normal exposure after removing the etiolation cover and treated with IBA at 3000 mg l^{-1} followed by girdling (20 days before taking the cutting) and treated again with IBA at 3000 mg l^{-1} at planting.

Experimental evidence indicates that there exists a synergistic relationship between phenolic substances and auxin in root regeneration from cuttings. Reddy and Majumder (1978) reported that treatment with rutin, *o*-coumaric acid, quercetin, umbelliferone and syringic acid at 2000 ppm, each in combination with IBA at 5000 mg l^{-1}, gave 87% rooting. Dhua *et al.* (1982) obtained 93.3% rooting in semi-hardwood cuttings under mist with *p*-hydroxybenzoic acid (200 ppm) and IBA at 5000 mg l^{-1}. Addition of phenolic substances improved root characteristics (number of roots per cutting and length of roots).

Cuttings are usually 15–20 cm long and 0.8–1.5 cm wide. Thicker cuttings root less readily and have fragile roots. Green woody cuttings rooted better than dark grey ones. The cuttings should have at least three nodes, with one node in the rooting medium. The basal cut should be at an angle (slant), making it somewhat easier to cut, more convenient to insert into the rooting medium, less likely to damage the xylem vessels while doing insertion and exposing more cut surface area for water absorption to prevent desiccation. Furthermore, making the cut just below a node provides a greater concentration of root sowing to higher concentration of phytohormones particularly root-promoting auxins. Cuttings are buried 3–4 cm into the medium to reduce drying out and to give support. Several reports are available regarding use of rooting media, which include sand, silt and topsoil (Akram *et al.*, 2017), sand (Dhua *et al.*, 1982; Mitra, 1996; Kareem *et al.*, 2013), vermiculite (Pereira *et al.*, 1991; Gautam *et al.*, 2010), coco peat (Rani *et al.*, 2018), coco peat–perlite (1:1 v/v) (Sardoei, 2014) and silt (Qadri *et al.*, 2018). The rooted cuttings showed nutrient deficiency symptoms after 45 to 50 days when vermiculite or sand was used as rooting medium, because these media had no nutrients except the ones transferred through irrigation water. This indicated the need for exogenous application of nutrients when sphagnum moss–grit (1:1) and sand are used as rooting medium (Prasad *et al.*, 1988) and for development of a stronger root system (Gautam *et al.*, 2010).

There seems to be agreement on the time of year at which the cuttings should be taken. In India, cuttings are best taken in April following the spring flush, or in August–September when it is warm and humid (Mitra and Bose, 1996). Rahman *et al.* (1991) reported best success for planting in mid-August and there was no rooting response between the end of December and mid-March. Gautam *et al.* (2010) stated that the rooting potential of cuttings (82.0 to 90%) was marginally affected by seasonal variation under a polyhouse with misting facility and maintaining the environmental conditions (humidity 75 ± 10%, temperature 35 ± 5°C). However, during extremely hot (June) and cold (January) months, inhibition in root formation (10 to 20%) was recorded due to non-availability of actively growing tips of terminal shoot cuttings.

Rooted cuttings (15–20 cm long) should be transplanted into perforated polybags filled with a mixture of sand, soil and leaf mould (1:1:1) and should be kept in partial shade under a net house with 50 to 75% shading intensity (depending on weather conditions) for a month. Sprinklers should be used to irrigate young plants under net house conditions. These plants should be gradually exposed to less humidity under an open nursery before transplanting them to the field (Gautam *et al.*, 2010).

4.4 Root Cuttings

Guava can form dense thickets which displace native vegetation and is reported as an invasive weed in many countries. The balance between its valuable fruit production and its invasive potential requires careful monitoring (CABI, 2020). Vegetative reproduction from root suckers is common and impedes mechanical methods of control (Cronk and Fuller, 1995). Use of root suckers/cuttings is the oldest method of asexually propagating guava (BPI, 2011). About 12–20 cm long roots are cut and induced to sprout by placing them flat on the bed and covering them with about 5 cm of fine soil, which must be kept moist to promote sprouting (BPI, 2011). According to Pereira *et al.* (2016), this method was used successfully in the past. The roots are cut at approximately 0.5 to 1 m from the trunk of a mature tree. Shoots that develop from the chopped roots are removed with their roots and planted in 5-litre plastic bags. However, with this method there is the possibility of guava wilt disease (GWD) entrance through wounds in the roots and thus it is not recommended.

4.5 Air Layering

Air layering or marcotting has a long history in many guava-growing countries of the world and is the most widely used method for propagating guava. It involves the production of a plant *in situ* from aerial branches. Shoots selected for air layering should be 1.0 cm in diameter and preferably from the previous year's growth. A ring of bark of about 3.0 cm long is removed, the exposed tissue is enclosed in a ball of moist sphagnum moss or similar, which is wrapped in polythene film to reduce water loss (Fig. 4.1). When roots appear on the stem, the air layer is cut from the mother plant and planted on its own roots under shade.

Several factors influence success and establishment in the field, including maturity and thickness of the stem, time of layering, position of the shoot in the crown, wrapping materials, rooting media, application of growth regulators, care in the nursery and pruning of the shoot after separation from the mother plant.

The rainy season was found more favourable than spring and winter for air layering. In India, the best time for air layering is

Fig. 4.1. Air layering on tree. Photograph courtesy of Dr B. Ghosh.

June–July when the atmospheric humidity is very high (Sharma *et al.*, 1991; Mitra and Bose, 2001). Application of auxins (IBA and NAA) has been found to increase the percentage of success in air layers. Auxins are used in the form of lanolin paste or powder and are applied evenly above the upper portion of the cut surface. Treatment with IBA between 3000 and 10,000 mg l^{-1} is used to increase rooting success and improve root quality (Bhujbal, 1972; Patel *et al.*, 1989; Sharma *et al.*, 1991; Singh *et al.*, 1995; Singh and Jain, 1996).

Some reports suggested that a mixture of IBA and NAA was more effective than either of the two auxins used alone (Bhandari and Kologi, 1960; Sharma *et al.*, 1974; Patel *et al.*, 1996). Anandhanambi *et al.* (2016) reported 91.68% rooting of air layers by using *Azospirillum* in combination with IBA and NAA each at 3000 mg l^{-1}. The treatment also caused higher number of roots and longest roots. Etiolation of shoots for 15–60 days prior to preparing the layers and use of auxins were found to improve rooting success (Bhatt and Chundawat, 1982; Kumari *et al.*, 2017).

The preferred medium for air layers is moist sphagnum moss. For best results, the moss should be soaked in water for 24 h prior to its use. A quantity of about two handfuls of the moss, with the excess water squeezed out, is placed around the stem to enclose the cut surface. However, Singh and Jain (1996) suggested use of soil and moss (1:1) when etiolated shoots were used. Rymbai *et al.* (2012) suggested use of coco peat and sphagnum moss (1:1) as medium instead of sphagnum moss alone. The medium is then wrapped with a piece of polythene film to cover the moss completely. The two ends should be twisted to make sure that no water can seep inside and that the medium does not dry out, then be tied off with a piece of string. The use of tightly secured polythene eliminates the need for frequent hand watering. Black polythene wrappers have been found to be better than white wrappers in respect of rooting and subsequent growth of layers (Singh *et al.*, 1995; Patel *et al.*, 1996).

The air layer is removed when a good root system has developed and at least six to eight roots have turned from white to creamy brown. Roots normally appear after 3–4 weeks in the rainy season, with layers ready for removal after 7–8 weeks. Once sufficient roots have formed, the branch is cut below the root ball. Extreme care should be taken during this stage since the roots are weak and fragile. The plastic wrapping should be removed carefully. The new plants are generally planted in 15 cm × 15 cm × 18 cm polythene bags. The potting mix should be pushed tightly against the roots, covering them to at least 7–10 cm, and watered. Care should be taken not to allow the bags to dry out. The air layers should be kept in shade, watered regularly and protected from strong winds and desiccating atmosphere.

4.6 Stooling

This technique has immense potential in rapid multiplication of guava plants. Air-layered plants are grown for 3–5 years at close spacing (0.75 cm × 0.75 cm) and headed back at 15 cm above the ground level in spring to induce new growth. A ring of bark of 1.5–2 cm is removed from towards the base of the newly emerged shoots in July and IBA at 5000 mg l^{-1} in lanolin is applied. The basal part of the shoots is then covered up with soil for roots to form. The rooted shoots are separated in September and planted in the nursery (Sinha *et al.*, 1993). Lal *et al.* (2007) obtained maximum rooting (96.67%) of stools, average number of roots per shoot (46.93), average root length (8.45 cm) and survival (75%) after transplant in the field by treatment with IBA at 7500 mg l^{-1} in guava cultivar 'Sardar'. Using NAA in combination with IBA reduced rooting compared with IBA alone (Bhagat *et al.*, 1998; Lal *et al.*, 2007). Treating the ringed portion of shoots with paclobutrazol at 2500 ppm also showed higher rooting and root quality of 'Seedless', 'Sardar' and 'Allahabad Safeda' guava (Singh, 1998). Dutta and Mitra (1991) obtained 100% rooting of shoots of guava cultivar 'Harijha' in etiolated shoots treated with IBA (5000 mg l^{-1}) and *p*-hydroxybenzoic acid (200 mg l^{-1}). High rooting success was observed when performed between 1 July and 16 July (83.78–90.73%) which decreased to 63.79–68.04% in August and 52.00–57.90% in September (Patil *et al.*, 2016). A 4.5 m × 4.5 m nursery bed can produce at least 300 rooted shoots every year with 100% survival. The shoots can be reused for multiplying plants every year (Majumder and Mukherjee, 1968).

Among the different methods of rootstock multiplication, stooling has immense potential for clonal propagation of rootstock. Saroj and Pathak (1994) reported that treatment with IBA + NAA (each at 7500 mg l^{-1}) induced the best rooting in *Psidium chinensis*, *Psidium cujavillus*, *Psidium molle* and *Psidium cattleyanum* shoots with a high degree of establishment (82–99%) of rooted shoots under field conditions. However, the same treatment produced only a few roots in *Psidium friedrichsthalianum* and this shy rooting was attributed to its complete failure in establishment in the field. Mishra *et al.* (2007) evaluated rooting performance of seven *Psidium* species, namely *P. chinensis*, *Psidium guineese*, *P. cujavillus*, *P. molle*, *P. cattleyanum*, *Psidium araca* and *P. friedrichsthalianum*. Heading back of plants in different guava species was done to retain 15 cm height from the ground level. Urea and single superphosphate at 150 g and 300 g, respectively, were applied in each bed to energize the headed-back plants. Sprouting was initiated during March in all headed-back plants. Shoots at 5.0 cm above from their origin were girdled, removing surrounding bark to a length of 2.5 cm, and IBA at 5000 mg l^{-1} was applied in lanolin paste. Afterwards, these treated shoots were properly mounded with compost-enriched soil. Stool beds were irrigated at an interval of 10 days. After 2 months, shoots were separated. The treatment caused 100% rooting of *P. chinensis*, followed by *P. cujavillus* (84.44%). Moderate rooting was observed in *P. guineese* (52.28%) and *P. cattleyanum* (33.33%). Rooting was low in *P. friedrichsthalianum* (8.88%), while there was no rooting in *P. molle* and *P. araca*.

4.7 Grafting

Grafting has long been practised in guava. However, the success rate is variable. Approach grafting is reported to be in practice in India since long back (Cheema and Deshmukh, 1927;

Mitra and Bose, 1990). This technique gives a considerably high percentage of success (about 95%) but it is more cumbersome and labour-intensive and requires a great deal of skill and experience (Mitra and Bose, 1990; Sinha *et al.*, 1993). The grafts are likely to perform better in the field under biotic or abiotic stresses as the process of joining the scion with the rootstock gives the resulting plant a certain characteristic of the rootstock – for example, hardiness, drought tolerance, dwarfness or disease resistance. For these reasons, the grafted plants are often preferred by the growers and thus nurserymen see a marked advantage in grafting over more convenient propagation techniques to justify the time and cost involved in grafting.

The side wedge method of grafting is a more popular technique (Nakasone and Paull, 1998) in guava propagation than approach grafting. Rapid and successful propagation through wedge grafting is possible throughout the year even in extreme climatic conditions such as severe cold (Singh *et al.*, 2005). Singh *et al.* (2007) standardized the wedge grafting technique at Central Institute for Subtropical Horticulture, Lucknow, India. Raising rootstock in a polythene bag is recommended, as it gives better establishment of plants in the field on account of an undisturbed taproot system (Singh *et al.*, 2005). Seedlings 6–8 months old having a stem diameter of 0.5 to 1.0 cm are suitable for wedge grafting. Singh *et al.* (2007) used scion shoots 15–18 cm long of 0.5–1.0 cm thickness with three or four healthy buds for grafting. Selected scion shoots were defoliated on the mother plant, about 5 to 7 days prior to detachment. At the same time, the apical growing portions of selected shoots were also beheaded, which helps in forcing the dormant buds to swell.

After selection of the material, the rootstock (seedling) was headed back, leaving a 15–18 cm long stem above the polyethylene bag. The beheaded rootstock was split open about 4.0 to 4.5 cm deep through the centre cut end of the rootstock with a grafting knife. A wedge-shaped cut, slanting from both sides (4.0–4.5 cm long) was made on the lower portion of the scion shoot. The scion stick was inserted into the split of the stock and pressed properly so that the cambium tissues of rootstock and scion should come in contact with each other. The combination of stock and scion was then tied with the help of a 150-gauge, 2 cm wide and 25–35 cm long polyethylene strip. Using this wedge grafting method, Singh *et al.* (2007) evaluated the success in guava cultivars 'Allahabad Safeda' and 'Sardar' under a greenhouse as well as in the open field (Table 4.1). Significantly higher success of grafts (64.56–94.33%) was recorded under greenhouse conditions compared with field conditions (51.30–70.63%) in both cultivars. However, the highest success rate was obtained in greenhouse (88.63–94.33%) as well as in open field conditions (66.6–78.63%) when grafting was carried out during November to February. The temperature range of 20 to 26°C and RH of 70 to 80% were found to be most conducive for better (>70%) success. However, Mukherjee and Singh (1965) observed higher success in veneer grafting in March, May or June, while Bhandary and Mukherjee (1970) obtained higher success (85%) in veneer grafting in July than in March, April and June. Gotur *et al.* (2017) reported that the wedge grafting in August gave better results in the polyhouse (69.88%) and open field (67.12%) conditions. The most successful grafts in August could be due to the optimal temperature and high humidity that prevail during this period, which resulted in the successful bonding of the layers of cambium of stocks and scion, the first formation of calluses and the beginning of the subsequent growth. A high success rate in softwood grafting and graft survival was observed in 35% shaded houses (68.80 and 87.19%, respectively), followed by 50% shade (58.00 and 79.13%, respectively) (Manga and Jholgiker, 2017).

The timing for grafting depends on the species/genotypes and the technique used. Several studies reveal that, for the best results, grafting needs to be performed only at certain specific times when both weather conditions and the physiological stage of the scion are optimum (Singh *et al.*, 2010; Syamal *et al.* 2012; Rani *et al.* 2015; Nanditha *et al.* 2017; Vanaja *et al.*, 2017; Singh

Table 4.1. Seasonal effect on success of wedge grafting in different months under greenhouse as well as open field conditions for guava cultivars 'Sardar' and 'Allahabad Safeda' (%). Adapted from Singh *et al.* (2007).

	Greenhouse		Open condition		Mean	
Month	'Allahabad Safeda'	'Sardar'	'Allahabad Safeda'	'Sardar'	'Allahabad Safeda'	'Sardar'
January	89.10	89.86	63.60	69.70	76.30	71.70
February	92.00	94.33	74.76	70.20	83.38	82.20
March	77.83	93.90	58.50	56.76	68.16	75.33
April	75.56	70.90	54.86	54.83	65.21	62.86
May	70.56	80.73	53.86	51.30	62.21	66.01
June	69.50	64.56	54.73	50.46	62.11	57.51
July	70.70	69.90	58.66	60.13	64.68	65.01
August	76.00	78.56	55.40	65.00	65.70	71.78
September	71.23	77.30	56.73	68.40	63.98	72.85
October	75.60	75.90	65.90	65.20	70.75	70.55
November	89.00	88.63	69.60	74.46	79.30	81.54
December	90.20	90.00	77.00	78.63	83.60	84.31
Mean	78.10	77.53	59.85	59.59	68.97	67.46
Critical difference ($P = 0.05$)	2.17	6.70	6.28	8.51	7.84	1.77

et al., 2018). Conditions for doing grafting in India usually become congenial with the onset of the monsoon in June and can be extended up to August. It has also been suggested that the scion wood be harvested freshly, preferably as close to the time of grafting as possible or at least on the same day of grafting. Use of scions stored in a refrigerator at low temperature produces poor results. Furthermore, only good-quality, healthy scions and rootstocks completely free from insects, diseases or any physical deformity are selected for grafting.

Several grafting techniques, such as cleft (Fig. 4.2), saddle and tongue (El-Taweel *et al.*, 2015) and veneer (Mukherjee and Singh, 1965; Bhandary and Mukherjee, 1970), have been tried with varying degrees of success in guava. Wedge grafting with polycap showed the highest sprouting (96.08%) under polyhouse conditions in November, compared with February (93.95%) and July (91.13%) (Joshi *et al.*, 2014). Casagrande *et al.* (2020) explored the feasibility of omega bench grafting in two guava cultivars using rootstocks from root-knot nematode-tolerant strain 'Tailandesa' raised by stem cuttings. They observed that the grafting was not successful as the tissues presented fast oxidation, dehydration and graft death. However, immersion of the grafts into a solution containing 2.0% citric acid and 1.0% ascorbic acid controlled the oxidation process and resulted in 28.6 and 37.5% success, respectively, in guava cultivars 'Paluma' and 'Pedro Sato'. The best-quality scion wood usually comes from shoots grown the previous season (Pereira *et al.*, 2016). A modified veneer grafting method (Fig. 4.3) has been developed and commercialized by VNR nursery, Raipur, Chhattisgarh, India which showed more than 95% success (N. Chawda, Chhattisgarh, 2020, personal communication). In this method, scions are prepared with two cuts, one long and one small on each side of the scion. If the rootstock diameter is different from that of scion, the scion is inserted towards the side of the rootstock. When both stock and scion diameters are similar, the scion is inserted in the centre of the rootstock.

According to Singh and Singh (2018), wedge grafting in the month of August gave better results in polyhouse (69.88%) as well as open field conditions (67.12%). Manga *et al.* (2017) suggested covering the grafted portion with a polytube cap for 1 month or until the graft sprouted. High humidity environments (mist house) showed higher survivability of successful grafts. This not only prevents desiccation of the tissues at the scion and stock interface but

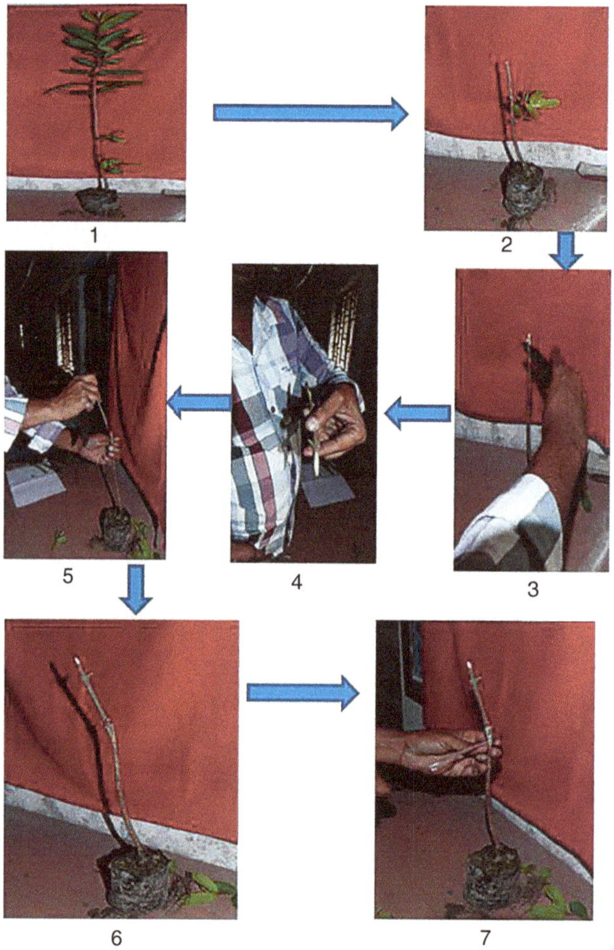

Fig. 4.2. Cleft grafting. 1, Rootstock; 2, preparation of rootstock for grafting; 3, insertion in rootstock; 4, preparation of scion; 5, inserting scion in rootstock; 6, stock and scion after operation; 7, wrapping with tape. Photograph courtesy of Dr B. Ghosh.

also favours rapid callus tissue development leading to formation of a better graft union.

4.8 Budding

Budding is a form of grafting in which the scion contains a single bud and a small section of bark with or without wood. It is preferred to scion-shoot grafting as the practice of budding is more convenient and faster. Since each bud on the scion shoot is a potential plant, several plants can be raised from a single scion-shoot. Several techniques of budding such as forkert, shield, patch, chip and 'T' are in use in guava. Selection of a proper rootstock is the prerequisite of achieving better success in budding (Mitra and Sanyal, 2004).

Forkert budding in February–March using buds from the previous season's growth on 1-year-old stock gave 92% success, while it was 32% in shield budding (Srivastava, 1962a) One hundred per cent bud-take was obtained by forkert and patch budding in the rainy season (Srivastava, 1962b). Above 80%

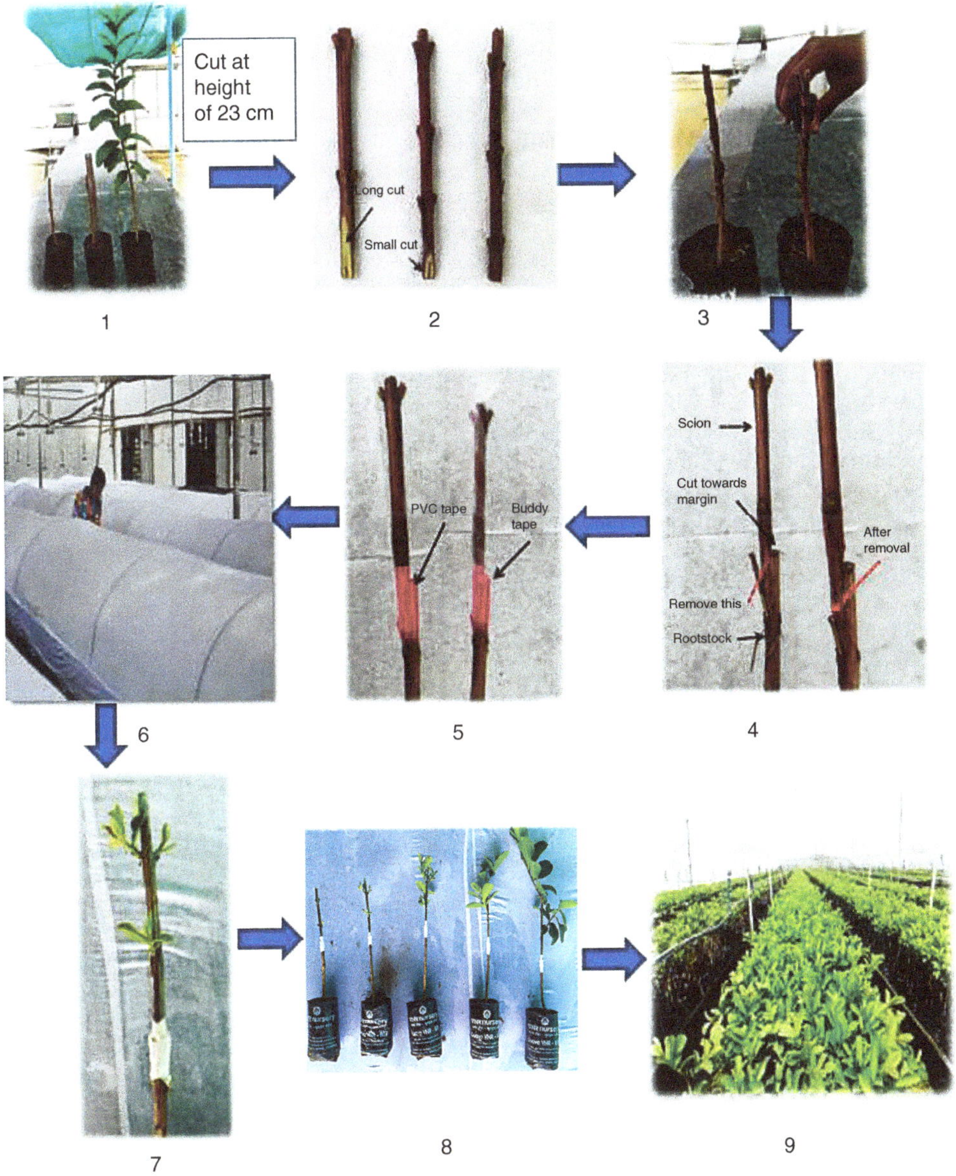

Fig. 4.3. Modified veneer grafting. 1, Rootstock; 2, preparation of scion; 3, preparation of rootstock for grafting and making insertion on rootstock; 4, inserting scion in rootstock; 5, stock and scion wrapping with tape after operation; 6, heeling of grafted plants; 7, initiation of new buds (15–20 days after grafting); 8, different stages after grafting (days after grafting, left to right: 15–20; 20–25; 30–35; 40–45; 55–60); 9, growth of plants at nursery. Photograph courtesy of Dr Narayan Chawda, VNR Seeds, India.

success could be obtained in patch budding when performed in May (Kaundal *et al.*, 1987) which might be due to higher summer temperature that favoured callus formation.

For rootstock, seedlings of about 1 year of age, uniform and active in growth are selected. Patch budding is most satisfactory when a vigorously growing plant with

1.25–2.5 cm stem diameter is used as stock. The trees from which buds are taken should be highly vegetative with lush succulent growth to permit easy separation of dormant buds from leaf axils of mature twigs of the scion variety. A patch of approximately 1.0 to 1.5 cm with a bud should be taken for better success. A similar, 1.0–1.5 cm long patch is removed from the rootstock and the bud is fitted into the remaining portion of the stock seedling. The bud should be fitted at a height of nearly 15 cm above ground level. A polythene strip is used for keeping the bud close to the stock. When the bark adheres tightly to the wood, budding is usually successful. After about 2–3 weeks of budding the polythene strip can be opened to examine the success. In successful cases, about one-third of the shoot of the rootstock can be removed for forcing the growth of buds. The remaining two-thirds can be removed after 3 weeks of the first cutting, leaving about 2–3 cm above the bud (Rajan, 2013).

Babu *et al.* (2009) evaluated patch and shield budding using white-fleshed ('Hybrid-1', 'Selection 11', 'Allahabad Safeda') and red-fleshed ('Selection 1' and 'Hybrid-4') genotypes. Patch budding was found superior over shield budding with regard to bud-take and survival percentage. Budding on the third week of February showed highest success for 'Hybrid-1' (30.0%) and on the fourth week of February for 'Allahabad Safeda' (85%). Patch budding performed during 15–21 August showed highest per cent success (92.07%) recorded after 90 days of propagation (Singh and Singh, 2018).

In chip budding, a downward 45° angle is made below the bud on a scion. A second cut is made about 12–15 mm above the bud, inward and downward in a transverse manner, meeting the first cut. The bud chip is removed and placed near the base of the stock between two nodes where an identical cut has been made. It is essential that the cambium of the stock and scion match properly. The graft is usually wrapped in waxed cloth or budding tape. Chip budding, however, was reported to be less successful than patch budding (Bhatt *et al.*, 2013).

'T' budding is done when both the rootstock and the scion are actively growing and when the buds can be easily removed. Plump axillary or side buds are cut from the middle of firm, young shoots. A T-shaped cut is made in the bark of the stock, the bud inserted, and rubber bands, soft string or tape used to hold the stock together. The top of the rootstock above the graft is removed when the bud grows, and the plants are transplanted after completion of at least one flush. While evaluating the different methods of budding (patch, chip and 'T') in different months (June, July and August), Bhatt *et al.* (2013) observed very low success rate in 'T' budding (13.33 to 23.33%) compared with chip (38.33 to 49.99%) and patch (53.99 to 73.33%). Good success with patch budding was attributed to a greater area between the matrix of the stock and scion compared with 'T' and chip budding.

4.9 Rootstock

Seedlings of 1–2 years old are generally used as rootstocks. These may be the same or a different cultivar. However, seedling rootstocks can lead to considerable variability in tree growth and performance. The effects of rootstock cultivar on fruit yield and quality have been documented (Gill and Chahil, 2013; Gill *et al.*, 2014).

The cultivar 'Sardar' produced mean highest yield (average of rainy and winter season) on 'Portugal' rootstock (76.27 kg per tree) followed by on 'Gutaniwala' (62.18 kg per tree), compared with 40.86 kg per tree on 'Pear Shaped'. The average fruit weight was 143.50 g on 'Sindhajli' compared with 114.20 g on 'Red Fleshed'. 'Sindhajli' rootstock caused maximum total soluble solids (TSS) content (9.80°Brix) of fruit of 'Sardar' guava, while the rootstock 'Chittidar' caused the maximum vitamin C content of fruit (152.71 mg 100 g^{-1}) and it was least (88.85 mg 100 g^{-1}) when 'Behat Coconut' was used as a rootstock (Gill and Chahil, 2013). 'Portugal' was reported as the most invigorating rootstock which produced smooth bud union, increased vegetative growth and gave significantly higher mean fruit yield in comparison to other rootstocks (Gill *et al.*, 2014). Rootstock 'Riverside Vermelha' was considered a good potential rootstock in Brazil, exhibiting earliness, vigour and growth rate (Vasconcelos and Cardoso, 1997).

GWD caused by *Nalanthamala psidii* is a serious disease occurring in guava-producing countries all over the world. The Chinese guava (*P. friedrichsthalianum*) is reported to be resistant to GWD and a compatible rootstock for guava with dwarfing effect (Edward and Shanker, 1964; Pereira *et al.*, 2016). Rootstocks raised from seeds of scion cultivars 'Red Fleshed', 'Behat Coconut' and 'Banarasi Surkha' were found highly susceptible to wilt disease with scion cultivars 'Sardar' and 'Allahabad Safeda', while in combination with 'Portugal' rootstock, the plants were totally free from wilt symptoms (Gill and Chahil, 2009).

The genus *Psidium* is composed of approximately 150 species of evergreen trees and shrubs (Paull and Duarte, 2012) but only a few have been tested so far for possible use as rootstocks. Mitra and Bose (1990) stated that different *Psidium* species, namely *P. cujavillus*, *P. molle*, *P. cattleyanum* and *P. guineense*, can also be used as rootstock for *P. guajava*. Trees on *Psidium pumilum* had a dwarfing effect and *P. cujavillus* produced bigger but non-uniform and rough-skinned fruits (Teaotia and Phogat, 1971). Based on anatomical studies, Saroj *et al.* (1977) categorized *P. chinensis* as dwarf, *P. molle* as semi-dwarf and *P. cujavillus* and *P. cattleyanum* as vigorous rootstock. The fruit yield of cultivar 'Allahabad Safeda' on *P. cattleyanum* was reported to be higher but TSS and reducing sugar contents of fruit were higher on *P. cujavillus* (Singh *et al.*, 1976). Cattley (*P. cattleyanum*) guava is resistant to nematode *Meloidogyne enterolobii* (Robaina *et al.*, 2015).

In general, guava is diploid ($2n = 2x = 22$), but triploid, tetraploid or hexaploid ($2n = 3x/4x/6x = 33/44/66$) species/cultivars also exist in nature. In Costa Rica, a widely grown strain of *P. friedrichsthalianum* is hexaploid ($2n = 66$) (Hirano and Nakasone, 1969). Although taxonomic proximity is a general prerequisite for successful graft-take and long-term survival of the grafted (composite) plant, seedling rootstocks of some more *Psidium* species need to be tried as rootstock to impart biotic or abiotic hardiness. In some cases, graft incompatibility between different species has been resolved using a suitable interstock (Pereira *et al.*, 2016).

A potentially dwarf rootstock 'Aneuploid No. 82' has been identified for guava at the Indian Agricultural Research Institute, New Delhi, through a selection made out of 48 different aneuploid seedling rootstocks that could induce substantial dwarfing of guava cultivar 'Allahabad Safeda'. The overall yield of plant was highest in 'Aneuploid No. 82' compared with 'Allahabad Safeda' grafted on 'Allahabad Safeda', which indicated a strong potential for its being used as a dwarfing rootstock on a commercial scale in increasing the production and profitability of guava orchards (Sharma *et al.*, 1992). 'Aneuploid No. 82' was found the best in imparting dwarfness in commercial cultivar 'Allahabad Safeda' and was released as 'Pusa Srijan' in 2004.

4.9.1 Rootstock–scion interaction

Grafting has been widely practised for a long time to multiply plants and preserve inherent qualities of the scion used, but rootstocks exert notable influence on the scion and modify its characteristics. In general, phenotypic characteristics of the grafted plants like fruit quality, resistance to pests and pathogens, tolerance to adversity and stress, and other physiological disorders are influenced and as such the technique has been widely used to improve resistance to pests and diseases in many fruit crops. In recent years, increasing effort has been made to reveal the mechanisms that control graft-induced changes in the scion characteristics (Li *et al.*, 2013; Paultre *et al.*, 2016) and to determine how the union takes place and what macromolecules are transferred between scions and rootstocks in the grafted plants (Wang *et al.*, 2017).

When the cambium of the scion joins fully with that of the rootstock, intact cells divide and proliferate into calli, which eventually differentiate into vasculature and plasmodesmata forms (Melnyk and Meyerowitz, 2015). Although the detailed molecular mechanisms underlying this process require further research, some studies have found that hormones such as auxin, cytokinin and gibberellic acid play a pivotal role in regulating stock–scion interactions (Aloni *et al.*, 2010). After cell walls fuse in the graft union, plasmodesmata stretch in small groups over the spaces of the inner cell wall, interconnecting the protoplasts of contiguous cells (Kollmann and Glockmann, 1985). Heterogeneous cells then interdigitate through the plasmodesmata (Melnyk and Meyerowitz, 2015). The plasmodesmata provide tunnels for small molecules and even selectively permit the movement of macromolecules, such as proteins and nucleic acids. Additionally, vascular reconstruction at the graft union enables macromolecules to be transported (Harada, 2010).

Earlier, botanists believed that plant hormones are responsible for these interactions due to their roles in regulating plant vegetative growth and reproduction. Since many plant hormones are highly mobile, they can be translocated easily in the graft chimeras. According to existing evidence, hormonal signalling is involved in root–shoot interactions, including graft-union formation, scion–rootstock communication, and plant growth and development (Aloni *et al.*, 2010).

According to Taller *et al.* (1998) and Tsaballa *et al.* (2013), heritability of graft-induced phenotypic changes suggests that regulatory processes underlying the scion–rootstock communication also involve a genetic component. Taller *et al.* (1998) detected several random amplification of polymorphic DNA (RAPD) markers in the graft-induced variants and found the same bands in the rootstock cultivar but not in the scion. They suggested that the genetic changes caused by grafting were attributable to direct DNA uptake through the vascular bundles. Stegemann and Bock (2009) observed that plasmodesmata formation and re-establishment of vascular bundles provide transport channels for horizontal gene transfer (HGT) during formation of the graft union.

Liu (2006) suggested that, in grafting, mRNA molecules derived from the stock cells could possibly be reverse transcribed into complementary DNA (cDNA) capable of being integrated into the genome of the scion cells, resulting in heritable changes in the scion. Fuentes *et al.* (2014) found that nuclear genome transfer between scion and stock has occurred, producing new fertile and stable allopolyploid species at a considerable rate. Thus, grafting could lead to a direct transfer of the entire nuclear and plastid genomes across the graft junction. Zhou and Liu (2015) are of the view that entire nuclear genomes can be transferred between plant cells and across the graft junction, resulting in the formation of a new allopolyploid species.

Wang *et al.* (2017) reported that changes in small RNA (sRNA) abundance after grafting were important factors for initiating graft-induced changes in gene expression. They tried to examine the phenomena and molecular mechanisms underlying graft-induced phenotypic variation in anatomy, morphology and production and suggested a model by which macromolecules, including RNA, protein and even DNA, are transported between scions and rootstocks via vascular tissues. Harada (2010) showed that vascular reconstruction at the graft union enables macromolecules to be transported.

A successful graft is based on vascular reconnection. The possible molecular mechanisms that lead to vascular reconnection have been discussed in detail in recent reports (Goldschmidt, 2014; Melnyk *et al.*, 2015; Zhou and Liu, 2015; Wang *et al.*, 2017; Sharma and Zheng, 2019). Most of the evidence demonstrates that molecular signalling plays an important role in the process of graft formation (Miyashima *et al.*, 2011; Melnyk *et al.*, 2015). It is possible that graft compatibility is associated with the molecular signalling and sharing of genes and/or mobile factors to regulate vascular reconnection. Melnyk *et al.* (2015) demonstrated complex communication involved in graft-junction reconnection. They revealed potential signals involving auxin-related genes that regulate graft formation and vascular reconnection.

According to Fuentes *et al.* (2014) and Melnyk and Meyerowitz (2015), entire nuclear genomes are transferred between plant cells of scion and stock, leading to the formation of new alloploid-derived different species. Heterografting experiments also provided evidence for long-distance transcript or RNA–protein complex trafficking, which plays significant regulatory roles in the response to developmental processes. Phytohormones have also been found to be crucial for plant growth. With the exception of straightforward translocation of hormones during scion–stock interactions, the regulation of hormone levels, through the trafficking events, is a complex mechanism. Hormonal signals, auxin in particular, are believed to play an important role in the wound healing and vascular regeneration within the graft-union zone (Lin *et al.*, 2007). Incomplete and convoluted vascular connections impede the vital upward and downward whole-plant transfer routes. Long-distance protein, mRNA and sRNA graft-transmissible signals currently emerge as novel mechanisms which regulate nutritional and developmental root/top relationships and may play a pivotal role in influencing scion characteristics (Goldschmidt, 2014).

The majority of current research has been dedicated to using rootstocks to influence shoot phenotypes, but the root changes induced by scions have been seldom discussed, probably due to the important role that scions play in horticultural practices and also the relative difficulty in observing root phenotype changes given that they are below ground. This aspect needs to be investigated as it may throw additional light on the stock–scion relationship.

4.10 *In vitro* Propagation

Fast plant regeneration is the major outcome of micropropagation or plant tissue culture where somatic embryogenesis (SE) and organogenesis are frequently experimented for the regeneration of genetically identical plants. The process by which somatic cells develop into plants is called somatic or asexual embryogenesis while the process of organ formation under the influence of several hormones *in vitro* is termed organogenesis.

4.10.1 Organogenesis

Successful plantlet formation from somatic tissue of a mature guava plant was first reported by Amin (1987) and Jaiswal and Amin (1987). The best response of shoot multiplication rate was reported on Murashige and Skoog (1962) (MS) medium supplemented with 4.5 μM benzyl adenine (BA) alone. However, the survival and response of shoot-tip explants were not satisfactory. Therefore, they have looked for suitable alternative explant sources. Rapid clonal propagation of guava through *in vitro* shoot proliferation from nodal explants of mature trees was reported (Amin and Jaiswal, 1987; Jaiswal and Amin, 1987). Most of the researchers used actively growing shoot tips or nodal segments as explants (Prakash and Tiwari, 1996; Singh *et al.*, 2001; Mishra *et al.*, 2007a; Xiaomei and Guochen, 2011).

Nodal explants should be collected during early spring as they show better culture establishment, profuse sprouting and less contamination (Prakash and Tiwari, 1996; Singh *et al.*, 2001; Bisen, 2004). Explants of 1.0–3.0 cm size and for surface sterilization, hydrogen peroxide (H_2O_2), silver nitrate ($AgNO_3$), mercuric chloride ($HgCl_2$), sodium hypochlorite (NaOCl) are used (Fitchet-Purnell, 1990; Pirela-Fuenmayor and Magollon-Montero, 1997; Bajpai *et al.*, 2007; Mishra *et al.*, 2007b; Kadam *et al.*, 2017).

During *in vitro* propagation, the problem of browning or blackening of explants in culture medium due to leaching of phenolics, microbial contamination and tissue recalcitrance was reported (Ahmad *et al.*, 2016). The problem of media browning could be reduced by pre-soaking the explants for 5 h in a solution of ascorbic acid (250 mg l^{-1}), citric acid (300 mg l^{-1}), polyvinylpyrrolidone

(500 mg l^{-1}) or charcoal (1000 mg l^{-1}). Charcoal was found more effective than other treatments (Zamir *et al.*, 2007; Mangal *et al.*, 2008; Ahmad *et al.*, 2016). Treatment with NaOCl at 5–10% followed by $HgCl_2$ reduced microbial contamination (Ali *et al.*, 2007; Usman *et al.*, 2012). Coating the explants at their cut ends with a solution of commercial silicon in diethyl ether completely inhibited phenol-based browning (Youssef *et al.*, 2010).

Different basal media, namely MS, Woody Plant Medium (WPM) (Lloyd and McCown, 1981) and BS (Gamborg *et al.*, 1968), have been tried for culture initiation and MS, WPM, Modified Blueberry Medium (MBM) (Zimmerman and Broome, 1980) and Olive Medium (OM) (Rugini, 1984) for shoot proliferation (Meghwal *et al.*, 2010).

The different media, however, showed no significant variations in bud sprouting. Maximum bud sprouting was noted in ½ MS medium (66.66%). Shoot proliferation was best on WPM (Meghwal *et al.*, 2010). Amin and Jaiswal (1987) observed better shoot multiplication in MBM than MS medium and Papadatou *et al.* (1990) reported maximum proliferation of shoot tips in OM.

Proliferation and shoot multiplications from explants on different media were induced with various growth regulators. BA was mostly used at 1.0 to 3.0 mg l^{-1} (Fitch-et-Purnell, 1990; Papadatou *et al.*, 1990; Pirela-Fuenmayor and Mogollon-Montero, 1997; Mishra *et al.*, 2007b; Rai *et al.*, 2009). Mangal *et al.* (2008) reported maximum shoot proliferation by using BAP 2.0 mg l^{-1} in combination with GA_3 0.1 mg l^{-1} and phyloroguсinol 100 mg l^{-1} and Kadam *et al.* (2017) with BA 1.0 mg l^{-1} and GA_3 0.25 mg l^{-1} on MS medium. However, Yang and Lu (2007) obtained maximum callus production by using 2,4-dichlorophenoxyacetic acid (2,4-D) at 1.6 mg l^{-1} in both MS medium and WPM.

Auxins like IAA and IBA are used for root induction and considered an important factor for adventitious root formation from *in vitro*-raised shoots. IBA at 0.2 to 10.0 mg l^{-1} has been used for most induction by different researchers (Papadatou *et al.*, 1990; Mishra *et al.*, 2007b; Rai *et al.*, 2009; Kadam *et al.*, 2017). A combination of IBA and NAA each at 0.2 mg l^{-1} has been suggested by Pirela-Fuenmayor and Mogollon-Montero (1997) and NAA (0.2 mg l^{-1}), IBA (0.2 mg l^{-1}) and activated charcoal (1.5 mg l^{-1}) by Zamir *et al.* (2007), Mangal *et al.* (2008) and Butt *et al.* (2013).

The survival rate of plantlets is significantly influenced by the potting mixture. Better establishment was reported in pots containing non-sterile garden soil and compost (Amin and Jairwal, 1987), activated peat-based compost (Papadatou *et al.*, 1990), soil, sand and farmyard manure (1:1:1 v/v) (Prakash and Tiwari, 1996), autoclaved coconut husk (Mishra *et al.*, 2007b), sand and garden soil (3:1 v/v) (Rai *et al.*, 2009) and soil and vermicompost (1:1 v/v) (Kadam *et al.*, 2017).

Liu and Yang (2011) developed a protocol for micropropagation from mature guava plants (Fig. 4.4). Apical-shoot explants about 5–7 cm in length were collected from 10-year-old, greenhouse-growing, elite mature 'Beaumont' guava. Various disinfection methods and plant growth regulators were tested *in vitro*. The most effective method involved treating explants in a 15% bleach solution for 20 min followed by culturing in MS medium with polyvinylpyrrolidone at 250 mg l^{-1}. This method maximized the percentage of bud breakage (53.3%), while producing the minimum browning rate (18.3%) for the explants. The best observed proliferation rate (71.2%) occurred on MS medium supplemented with 4.4 μM BA, 4.65 μM kinetin and 0.54 μM NAA. It produced the highest mean number of shoots (2.2). Shoots were then rooted (65%) when dipped in 4.9 μM IBA solution for 1 min and rooted plantlets survived (100%) after acclimatization in a greenhouse.

4.10.2 Embryogenesis

Several attempts have been made in the past to multiply plants of commercial guava varieties successfully through embryogenesis (Gaffoor and Alderson, 1994; Vilchez *et al.*, 2002; Akhtar, 2010; Meghwal *et al.* 2010; Xiaomei and Guochen, 2011; Rai *et al.*, 2012; Kamle and Baek, 2017). Nowadays SE

Fig. 4.4. *In vitro* proliferation of guava. (A) Stock guava plants in the greenhouse. (B) Nodal section browning after sterilization. (C) New shoots break out from healthy nodal sections. (D) Shoots proliferated. (E) Elongated shoots. (F) Rooted shoots by the medium method (medium with IBA). (G) Rooted shoots by the dipping method. (H) Guava plantlets acclimatized into the soil for 2 weeks. (I) Guava plantlets acclimatized into the soil for 10 weeks. IBA, indole butyric acid. From Liu and Yang (2011), with permission.

has become one of the most desired pathways in the regeneration of plants via tissue culture because it bypasses the necessity of time-consuming and costly manipulation of individual explants, which is a problem with organogenesis. Two patterns of SE are recognized: (i) direct embryogenesis, where the embryo develops directly on the explant; and (ii) indirect embryogenesis, in which the embryo arises from a callus (Bhatia and Bera, 2015). SE allows the utilization of somatic embryos as synthetic seeds.

There is a considerable demand for the establishment of successful and efficient regeneration protocols via SE in guava. Plants regenerated through SE could be more useful than plants obtained through organogenesis because, in most cases, somatic embryos are of single-cell origin and have a low frequency of chimeras and a high number of regenerations. Kamle and Baek (2017) described a schematic representation of the SE process for guava (Fig. 4.5). Akhtar (2010) established an embryogenic protocol for plant regeneration of guava using 10-week post-anthesis, zygotic embryo explants. Somatic embryogenesis was induced on MS medium containing 3% (w/v) sucrose, 0.8% (w/v) agar and various concentrations of 2,4-D by continuous treatment of the zygotic embryo explants. Somatic embryos appeared as globular structures at the end of the third week from culture initiation, and heart-shaped, cotyledonary-stage and torpedo-stage embryos appeared within the next few weeks.

Rai *et al.* (2012) and Kamle *et al.* (2013) found that SE helps to study plant differentiation, totipotent cell expression level and has also been extensively used for the modification at genetic level for woody plants.

Fig. 4.5. Schematic representation of the somatic embryogenesis process for guava. (A) Guava fruit. (B) Zygotic embryo. (C) Embryo induction. (D) Embryo proliferation. (E) Embryo maturation and conversion. (F) Plantlet development. (G) Plant maturation. (H) *In vitro* acclimatization in coco peat. (I) Polyhouse hardening. (J) Field establishment. From Kamle and Baek (2017) with permission.

Explants from other sources, such as leaf disks, internodes, mesocarp and petals, were also explored for embryogenesis; however, only embryos from induced mesocarp were developed (Chandra and Mishra, 2007; Chandra *et al.*, 2007; Butt *et al.*, 2013). Embryos from immature zygotes were attained after 8–10 weeks of embryogenesis in cultivar 'Banarasi' (Rai *et al.*, 2010). Different genotypes of guava show variation in frequency, intensity and efficiency of somatic embryo maturation and germination. Somatic embryos greater than 8 weeks old were not converted well, while plantlets were normal (Kamle *et al.*, 2013). In general, embryogenesis is a sequential series of dynamic processes that include cell division and growth, and the elaboration of differentiation programmes leading to cell fate specification. Maximum conversion of plantlets is achieved 8 weeks after somatic embryo induction.

In most species studied in which growth regulators are needed for induction of SE, auxins and cytokinins are key factors determining the embryogenic response. This is probably due to their participation in the regulation of cell cycling, division and differentiation (Fehér *et al.*, 2003). The auxin 2,4-D, in isolation or combined with other growth regulators, mainly cytokinins, has been used for the induction of SE via tissue culture of seeds and zygotic embryos for several species (Fehér *et al.*, 2003). Rai *et al.* (2012) and Kamle *et al.* (2013) succeeded in inducing embryogenesis in MS medium with a mixture of 2,4-D and some other auxins or cytokinins. Ascorbic acid and L-glutamine, which are sources of nitrogen, actually induced SE, while polyethylene glycol and L-proline accelerated the maturity of somatic embryos (Rai *et al.*, 2012). Among the different sources of carbon, 5–6% sucrose was better for inducing and maturing of somatic embryos, while the addition of glucose, fructose, maltose, sorbitol and mannitol in the medium had reducing effects (Rai *et al.*, 2008). The agar (0.7–0.8% w/v) mixed solidified medium showed a better initiation of embryos compared with the liquid medium. The embryo germination can be enhanced by sucrose (3%) and reducing salts (half) in the medium (Rai *et al.*, 2007). A rapid multiplication rate, successful acclimatization of plantlets in soil, uniform genetic fidelity of progeny and cost-effectiveness are the key considerations for developing a successful SE protocol (Kamle *et al.*, 2017). Today, the overall scenario is that the development in the clonal propagation of guava is being promoted strongly by the usage of organogenesis and SE in elite cultivars. The commercial exploitation of the technologies is still in its initial stages.

The decisions to adopt a specific propagation method, or combination of methods, for production of improved planting materials will depend on a number of factors including the planting materials, resources and time frame available; as well as the farmer's need and ability to invest in getting standard high-quality planting materials. Since these factors will vary tremendously both within and between regions, it is not possible to provide a simple recommendation here on the most appropriate technique. In many South American countries, rooting herbaceous stem cuttings under intermittent mist is the common method of propagation (Pereira *et al.*, 2016). In the Indian subcontinent the crop is propagated mainly by air layering, but grafting is also practised where skilled manpower is available. In African countries both cutting and air layering are employed according to resources and facilities available.

4.11 Conclusions

Guava can be propagated by both sexual and asexual methods. Since plants raised through seed do not come true to type, sexual propagation is restricted mainly to use in breeding programmes and production of rootstocks. Among different methods of vegetative propagation, shoot cuttings, air layering, budding and grafting are used with varying degrees of success in different parts of the world where guava is grown commercially. Shoot cuttings of the commercial cultivars are hard to root. They do not root satisfactorily unless treated with auxins like IBA or IAA and placed under mist. Air layering with moist sphagnum moss as wrapping gives fairly good success. Treating the layers with auxin (IBA) at the time of wrapping improves rooting and enhances their success percentage appreciably. This is the most common method of propagation in many countries. Among various methods of grafting, wedge and modified veneer grafting have produced the better results. Likewise, forkert budding in February–March is reported to produce higher (>90%) success. Factors like varieties, time of grafting, method, growing conditions, maturity of the scion, age of rootstock, etc. influence the success and survivability of the grafts. Propagation in the rainy season starting from the latter part of June to August has shown the best growth and survival of grafts. Only seedling rootstocks have been tried so far and information on performance of standard scions on the clonal rootstocks is lacking.

Rootstocks have the potential to alter tree size, canopy growth, productivity, fruit quality, and resistance to biotic and abiotic stresses. Detailed genetic and physiological studies on the rootstock–scion relationship have revealed that vascular reconnection (compatibility) and graft-induced phenotypic variations are associated with the molecular signalling and sharing of genes and/or some related mobile factors including hormones. Since this technique has been widely used to improve resistance to pests and diseases in many fruit crops, it should be given preference to fix similar biotic stresses in guava. The wilt disease is a serious threat to guava cultivation all over the world. In many areas nematode infestation is also a grievous concern. Use of wild *Psidium* species' rootstocks as a source of resistance needs to be exploited. Unfortunately, no serious effort has been made in guava to realize modification in tree architecture, yield and fruit through the rootstocks.

The reasons for incompatibility reaction in guava with different rootstocks and methods of grafting have not yet been studied in detail at molecular level. Further research is required to understand whether specific rootstock effects on the scion are characteristics of the root system per se, or due to changes in hydraulic resistance and stem-sap flow across the union between the stock and the scion; or due to exchange of hormonal signals; or if the influences are due to genetic reasons, what is the extent and type of transfer of genetic elements between the scion and rootstock.

As regards micropropagation, plantlets of commercial cultivars of guava have been regenerated successfully through organogenesis and embryogenesis pathways but their protocols have not been standardized properly and are yet to be commercialized to harness their benefits. Limitations of *in vitro* propagation of guava are connected to occurrence of endophytic microorganisms, exudation of phenolic compounds and risk of somaclonal variation in the propagated materials. Nevertheless, the techniques hold very strong potential for fast, large-scale, disease-free clonal propagation of commercial cultivars as they are not weather-dependent and unlike other plant propagation techniques can be carried out in a laboratory setting year-round. As high performance of planting material is the cornerstone of a sustainable guava production system, a proactive role of researchers in further refinement of the plant propagation techniques, particularly lowering their per unit production cost, is envisaged.

References

Ahmad, I., Jaskani, M.J., Nafees, M., Ashraf, I. and Qureshi, R. (2016) Control of media browning in micropropagation of guava (*Psidium guajava* L.). *Pakistan Journal of Botany* 48(2), 713–716.

Akhtar, N. (2010) Evaluation of the efficiency of somatic embryogenesis in guava (*Psidium guajava* L.). *Journal of Horticultural Science and Biotechnology* 85, 556–562.

Akram, M.T., Qadri, R.W.K., Khan, I., Bashir, M., Jahangir, M.M. *et al.* (2017) Clonal multiplication of guava (*Psidium guajava*) through soft wood cuttings using IBA under low-plastic tunnel. *International Journal of Agriculture and Biology* 19, 417–442.

Ali, N., Mulwa, R.W.S., Norton, M.A. and Skirvin, R.M. (2007) Radical disinfection protocol eliminates *in vitro* contamination in guava (*Psidium guajava* L.) seeds. *Plant Cell, Tissue and Organ Culture* 91(3), 295–298.

Aloni, B., Cohen, R., Karni, L., Aktas, H. and Edelstein, M. (2010) Hormonal signaling in rootstock–scion interactions. *Scientia Horticulturae* 127, 119–126.

Alves, J.E. and Freitas, B.M. (2007) Pollination requirements of guava. *Ciência Rural* 37(5), 1281–1286.

Amin, M.N. (1987) *In vitro* clonal propagation of guava (*Psidium guajava* L.) and jackfruit (*Artocarpus heterophyllus* Lamk.). PhD thesis, Banaras Hindu University, Varanasi, India.

Amin, M.N. and Jaiswal, V.S. (1987) Rapid clonal propagation of guava through *in vitro* shoot proliferation on nodal explants of mature trees. *Plant Cell, Tissue and Organ Culture* 9, 235–243.

Anandhanambi, D., Arivazhagan, E. and Kandasamy, R. (2016) Influence of plant growth regulators and *Azospirillum* on rooting of air layers in guava (*Psidium guajava* L.). *The Asian Journal of Horticulture* 2(2), 261–268.

Babu, K.D., De, L.C., Patel, R.K. and Singh, A. (2009) Genotypic amenability of guava for patch budding. *Indian Journal of Horticulture* 66(2), 264–266.

Bajpai, A., Chandra, R., Mishra, M. and Tiwari, R.K. (2007) Regenerating *Psidium* sp. for screening wilt resistant rootstock under *in vitro* conditions. *Acta Horticulturae* 535, 145–154.

Bhagat, B.K., Jain, B.P., Singh, C. and Chowdhary, B.M. (1998) Propagation of guava (*Psidium guajava* L.) by ground layering. *Journal of Research, Birsa Agricultural University* 10(2), 209–210.

Bhandari, K.R. and Kologi, S.D. (1960) Studies on seasonal effects of growth regulators in combination on rooting of air-layers of guava (*P. guajava* L.) var. L-49. *The Lal Bag* 5, 12–19.

Bhandary, K.R. and Mukherjee, S.K. (1970) Effect of scion and ringing on veneer grafting of guava (*Psidium guajava* L.). *Indian Journal of Agricultural Sciences* 40, 495–498.

Bhatia, S. and Bera, T. (2015) Somatic embryogenesis and organogenesis. In: Bhatia, S., Sharma, K., Dahiya, R. and Bera, T. (eds) *Modern Applications of Plant Biotechnology in Pharmaceutical Sciences*. Academic Press, Amsterdam, pp. 209–230.

Bhatt, B.B., Tomar, Y.K. and Rawat, S.S. (2013) Effect of time and methods of budding in multiplication of guava cv. Allahabad Safeda under valley conditions of Garhwal Himalaya. *Progressive Horticulture* 45(2), 278–280.

Bhatt, R.K. and Chundawat, B.S. (1982) Studies on the effect of time of ringing and rooting media on rooting and survival of air-layers of guava (*Psidium guajava* L.). In: *Abstracts of National Seminar on Plant Propagation, Calcutta, India*. Society for Advancement of Horticulture, India, abstract no. 67.

Bhujbal, B.G. (1972) Effective concentration of IBA in the air-layering of guava. *Research Journal, Mahatma Phule Agriculture University* 3, 53–56.

Bisen, B.P. (2004) *In vitro* cloning of Brazilian guava (*Psidium guineense* Sw.). PhD thesis, G.B. Pant University of Agriculture and Technology, Pantnagar, India.

BPI (2011) *Production Guide for Guava*. Bureau of Plant Industry, Manila.

Butt, M., Usman, M. and Fatima, B. (2013) Enhanced seed germination and callogenesis under long days using leaf disc as explant in guava cultivars. *Biologia, Pakistan* 59(2), 293–298.

CABI (2020) *Psidium guajava* (guava). Datasheet updated by Rojas-Sandoval, J. and Acevedo-Rodríguez, P. In: *Invasive Species Compendium*. CAB International, Wallingford, UK. Available at: https://www.cabi.org/isc/datasheet/45141 (accessed 21 December 2020).

Casagrande, I.P., Oliveira, G.d.S., Leite, G.W.P and Teixeira, G.H.d.A. (2020) Feasibility of omega bench grafting in guava tree (*Psidium guajava* L.) propagated via herbaceous stem cutting. *Journal of Horticultural Science and Biotechnology* 95(2), 229–234.

Chacko, E.K. and Singh, R.N. (1968) Studies on the longevity of papaya, phalsa, guava and mango seeds. *15th International Seed Testing Congress, New Zealand* 7, 1–11.

Chandra, R. and Govind, S. (1990) Gibberellic acid, thiourea, ethrel and acid treatments in relation to seed germination and seedling growth in guava (*Psidium guajava* L.). *Progressive Horticulture* 22, 40–43.

Chandra, R. and Mishra, M. (2007) Biotechnological interventions for improvement of guava (*Psidium guajava* L.). *Acta Horticulturae* 735, 117–126.

Chandra, R., Mishra, M., Abida, M. and Singh, D.B. (2007) Triazole mediated somatic embryogenesis in guava (*P. guajava* L.) cv. Allahabad Safeda. *Acta Horticulturae* 735, 133–138.

Cheema, G.S. and Deshmukh, S.B. (1927) *Culture of Guava and Its Improvement by Selection in Western India*. Bulletin No. 148. Department of Agriculture, Bombay, India, p. 38.

Cronk, Q.C.B. and Fuller, J.L. (1995) *Plant Invaders: The Threat to Natural Ecosystems*. Chapman and Hall, London.

Debnath, G.C. and Maiti, S.C. (1990) Effect of growth regulators on rooting of soft-wood cuttings of guava (*Psidium guajava* L.) under mist. *Haryana Journal of Horticultural Science* 19(1–2), 79–85.

Dhua, R.S., Mitra, S.K., Sen, S.K. and Bose, T.K. (1982) Effect of ethyphon and IBA on regeneration of guava. *Science and Culture* 48, 444.

Dutta, P. and Mitra, S.K. (1991) Effect of etiolation on stooling of guava. *Indian Agriculturist* 35, 101–105.

Edward, J.C. and Shanker, G. (1964) Rootstocks trials for guava. *Allahabad Farmer* 38, 249–250.

El-Taweel, A.A., Osman, I.M.S. and Mikhail, E.G. (2015) Studies on the vegetative propagation of guava by grafting. *Egyptian Journal of Horticulture* 42(1), 87–100.

Essien, E.P. (2004) Breaking of seed coat dormancy in guava. *Tropical Science* 44(1), 40–42.

Fehér, A., Pasternak, T.P. and Dudits, D. (2003) Transition of somatic plant cells to an embryogenic state. *Plant Cell, Tissue and Organ Culture* 74(3), 201–228.

Fitchet-Purnell, M. (1990) Dimple guava established in tissue culture. *Inligtingsbulletin Navorsingsistituut vir Sitrus en Subtropiese Vrugte* 212, 5.

Fuentes, I., Stegemann, S., Golczyk, H., Karcher, D. and Bock, R. (2014) Horizontal genome transfer as an asexual path to the formation of new species. *Nature* 511, 232–235.

Gaffoor, A. and Alderson, P.G. (1994) Somatic embryogenesis in guava (*Psidium guajava* L.). In: Lumsden, P.J., Nicholas, J.R. and Davies, W.J. (eds) *Physiology, Growth and Development of Plants in Culture*. Springer, Dordrecht, the Netherlands, pp. 272–277.

Gamborg, O.L., Miller, R.A. and Ojima, K. (1968) Nutrient requirement of suspension culture of soybean root cells. *Experimental Cell Research* 50, 151–158.

Gautam, N.N., Singh, K., Singh, B., Seal, S., Goel, A. and Goal, V.L. (2010) Studies on clonal multiplication of guava (*Psidium guajava* L.) through cuttings under controlled conditions. *Australian Journal of Crop Science* 4(9), 666–669.

Gill, M.S. and Chahil, B.S. (2009) Response of rootstocks to scionic compatibility, wilt and fruit yield in guava. *Haryana Journal of Horticultural Science* 38(182), 16–19.

Gill, M.S. and Chahil, B.S. (2013) Performance of guava cv. Sardar on ten different rootstocks. *International Journal of Agricultural Science* 9(1), 317–319.

Gill, M.S., Chahil, B.S. and Singh, N. (2014) Effect of different rootstocks on scion relationship, tree growth and yield of guava (*Psidium guajava* L.). *Progressive Horticulture* 46(1), 34–40.

Goldschmidt, E.E. (2014) Plant grafting: new mechanisms, evolutionary implications. *Frontiers in Plant Science* 5, 727.

Gotur, M., Sharma, D.K., Chawla, S.L., Joshi, C.J. and Navya, K. (2017) Performance of wedge grafting in guava (*Psidium guajava* L.) under different growing conditions. *Plant Archives* 17(2), 1283–1287.

Harada, T. (2010) Grafting and RNA transport via phloem tissue in horticultural plants. *Scientia Horticulturae* 125, 545–550.

Hayes, W.B. (1974) *Fruit Growing in India*. Kitabistan, Allahabad, India.

Hirano, R.T. and Nakasone, H.Y. (1969) Chromosome numbers of ten species and clones in the genus *Psidium*. *Journal of the American Society for Horticultural Science* 94, 83–86.

Hooda, P.S. and Yamdagni, R. (1991) Salt tolerance of guava (*Psidium guajava* L.) and aonla (*Emblica officinalis*) at germination stage. *Research and Development Reporter* 8(1), 36–38.

Jaiswal, V.S. and Amin, M.N. (1987) *In vitro* propagation of guava from shoot cultures of mature trees. *Journal of Plant Physiology* 130, 7–12.

Joshi, M., Syamal, M.M. and Singh, S.P. (2014) Comparative efficacy of different propagation techniques in guava. *Indian Journal of Horticulture* 71(3), 315–320.

Kadam, S., Singh, P. and Patel, R.M. (2017) Rooting and acclimatization of *in vitro* raised plantlets of guava cv. Allhabad Safeda. *International Journal of Scientific and Research Publications* 7(8), 449–453.

Kalyani, M., Bharad, S.G. and Parameshwar, P. (2014) Effect of growth regulators on seed germination in guava. *International Journal on Biological Sciences* 5(2), 81–91.

Kamle, M. and Baek, K.-H. (2017) Somatic embryogenesis in guava (*Psidium guajava* L.): current status and future perspectives. *Biotech* 7, 203. https://doi.org/10.1007/s13205-017-0844-0

Kamle, M., Kumar, P., Bajpai, A., Kalim, S. and Chandra, R. (2013) Assessment of genetic fidelity of *in vitro* regenerated guava (*Psidium guajava* L.) plants using DNA based markers. *New Zealand Journal of Crop and Horticultural Science* 42, 1–9.

Kareem, A., Jaskani, M.J., Fatima, B. and Sadia, B. (2013) Clonal multiplication of guava through softwood cuttings under mist conditions. *Pakistan Journal of Agricultural Science* 50(1), 23–27.

Kaul, M.K., Mehta, P.K. and Bakshi, R.K. (1988) Note on effect of different salts on seed germination of *Psidium guajava* L. CV. L.-49 (Sardar). *Current Agriculture* 12, 83–85.

Kaundal, G.S., Gill, S.S. and Minhas, P.P. (1987) Budding techniques in clonal propagation of guava. *Punjab Horticulture Journal* 27, 208–211.

Kollmann, R. and Glockmann, C. (1985) Studies on graft unions. I. Plasmodesmata between cells of plants belonging to different unrelated taxa. *Protoplasma* 124, 224–235.

Kumari, B., Prakash, S. and Kumar, R. (2017) Effect of etiolation and plant growth substances on success, survival and growth behavior of air-layers of guava (*Psidium guajava* L.). *Environment and Ecology* 35(2), 712–716.

Lal, S., Tiwari, J.P., Awasthi, P. and Singh, G. (2007) Effect of IBA and NAA on rooting potential of stooled shoots of guava (*Psidium guajava* L.) cv. Sardar. *Acta Horticulturae* 735, 193–196.

Li, J.X., Wang, Y., Zhang, L.L., Liu, B., Cao, L.W. *et al.* (2013) Heritable variation and small RNAs in the progeny of chimeras of *Brassica juncea* and *Brassica oleracea*. *Journal of Experimental Botany* 64, 4851–4862.

Lin, M.-K., Belanger, H., Lee, Y.-J., Varkonyi-Gasic, E., Taoka, K.-I. *et al.* (2007) FLOWERING LOCUS T protein may act as the long-distance florigenic signal in the cucurbits. *The Plant Cell* 19, 1488–1506.

Liu, X. and Yang, G. (2011) Clonal propagation of guava (*Psidium guajava* L.) on nodal explants of mature elite cultivar. *International Journal of Plant Biology* 2, e2.

Liu, Y.S. (2006) Historical and modern genetics of plant graft hybridization. *Advances in Genetics* 56, 101–129.

Liu, Y.S., Wang, Q.L. and Li, B.Y. (2010) New insight into plant graft hybridization. *Heredity* 104, 1–2.

Lloyd, G. and McCown, B. (1981) Commercial feasible micropropagation of mountain laurel (*Kalmia latifolia*) by use of shoot tip culture. *Combined Proceedings, International Plant Propagators Society* 30, 421–427.

Majumder, P.K. and Mukherjee, S.K. (1968) Stooling in guava. *Indian Horticulture* 12, 11–13.

Manga, B. and Jholgiker, P. (2017) Studies on performance of softwood grafting in guava (*Psidium guajava* L.) cv. Sardar as influenced by different shade intensity. *International Journal of Current Microbiology and Applied Sciences* 6(6), 2792–2795.

Manga, B., Jholgiker, P., Swamy, G.S.K., Prabhuling, G. and Sandhyarani, N. (2017) Studies on effect of propagation environment for softwood grafting in guava (*Psidium guajava* L.) cv. Sardar. *International Journal of Current Microbiology and Applied Sciences* 6(6), 2779–2783.

Mangal, M., Sharma, D., Sharma, M., Kher, R. and Singh, A.K. (2008). *In vitro* plantlet regeneration in guava from nodal segments. *Phytomorphology* 58(1/2), 103–108.

Meghwal, P.R., Sharma, H.C. and Singh, S.K. (2010) Micropropagation studies on guava. *Indian Journal of Horticulture* 67, 55–58.

Melnyk, C.W. and Meyerowitz, E.M. (2015) Plant grafting. *Current Biology* 25, R183–R188.

Melnyk, C.W., Schuster, C., Leyser, O. and Meyerowitz, E.M. (2015) A developmental framework for graft formation and vascular reconnection in *Arabidopsis thaliana*. *Current Biology* 25, 1306–1318.

Mishra, D., Lal, B. and Pandey, D. (2007a) Clonal multiplication of *Psidium* species with mound layering. *Acta Horticulturae* 735, 339–342.

Mishra, M., Chandra, R., Pati, R. and Bajpai, A. (2007b) Micropropagation of guava (*Psidium guajava* L.). *Acta Horticulturae* 735, 155–158.

Mitra, S.K. (1996) Mist propagation of some tropical fruit crops by cuttings. In: *Proceedings of National Seminar on Plant Bio-regulators in Horticulture, BCKV, Kalyani, India*. Society for Advancement of Horticulture, India, pp. 70–75.

Mitra, S.K. and Bose, T.K. (1990) Guava. In: Bose, T.K. and Mitra, S.K. (eds) *Fruits: Tropical and Subtropical.* Naya Prokash, Kolkata, India, pp. 280–303.

Mitra, S.K. and Bose, T.K. (1996) Standardization of propagation techniques by cutting of some tropical fruit crops. *Scientific Horticulture* 5, 1–7.

Mitra, S.K. and Bose, T.K. (2001) Guava. In: Bose, T.K., Mitra, S.K. and Sanyal, D. (eds) *Fruits: Tropical and Subtropical.* Naya Udyog, Calcutta, India, pp. 609–653.

Mitra, S.K. and Pathak, P.K. (2018) Recent development in the propagation of tropical and subtropical fruit cops by cutting. *Acta Horticulturae* 1205, 721–725.

Mitra, S.K. and Sanyal, D. (2004) *Guava*. Indian Council of Agricultural Research, New Delhi.

Miyashima, S., Koi, S., Hashimoto, T. and Nakajima, K. (2011) Non-cell autonomous microRNA165 acts in a dose-dependent manner to regulate multiple differentiation status in the *Arabidopsis* root. *Development* 138, 2303–2313.

Mukherjee, S.K. and Singh, Y.M. (1965) Effect of season and nature of shoot on veneer grafting of guava (*Psidium guajava* L.). *Science and Culture* 31, 31–33.

Murashige, T. and Skoog, F. (1962) A revised medium for rapid growth and bioassay with tobacco tissue culture. *Physiologia Plantarum* 15, 473–497.

Nakasone, H.Y. and Paull, R.E. (eds) (1998) Guava. In: *Tropical Fruits.* CAB International, Wallingford, UK, pp. 149–172.

Nanditha, G.C., Patil, D.R., Patil, S.N., Venkateshalu and Gandolkar, K. (2017) Study on the effect of different growing conditions and varieties on graft-take, leaves and scion diameter in guava (*Psidium guajava* L.). *International Journal of Pure & Applied Bioscience* 5(4), 601–606.

Narayan, S. (2017) Effect of stratification duration and seed treatment with GA_3 on seed germination, transplanting success and seedling mortality in Chinese guava (*Psidium* sp. L.). *HortFlora Research Spectrum* 6(3), 192–195.

Pandey, D. and Singh, G. (2000) Effect of seed pre-treatment on promotion of germination in guava (*Psidium guajava* L.). *Annals of Agricultural Research* 21(2), 279–281.

Papadatou, P., Pontikis, C.A., Ephtimiadou, E. and Lydaki, M. (1990) Rapid multiplication of guava seedlings by *in vitro* shoot tip culture. *Scientia Horticulturae* 45(1–2), 99–103.

Patel, R.K., Bose, U.S. and Tripathi, S.K. (1996) Effect of growth regulators and wrappers on success and survival of air-layering in guava cv. Allahabad Safeda. *Crop Research* 12, 56–60.

Patel, R.M., Patel, R.B. and Patel, M.P. (1989) Effect of growth regulators and polythene wrappers on rooting of air layers of guava. *Bhartiya Krishi Anusandhana Patrika* 4(3), 145–148.

Patil, S.D., Desmukh, P.L., Swain, L. and Dadas, M.M. (2016) *In vivo* propagation through mound layering in guava (*Psidium guajava* L.). *Journal of Soils and Crops* 26(2), 281–284.

Paull, R.E. and Duarte, O. (2012) *Tropical Fruits,* Vol. 2. CAB International, Wallingford, UK.

Paultre, D.S.G., Gustin, M.P., Molnar, A. and Oparka, K.J. (2016) Lost in transit: long-distance trafficking and phloem unloading of protein signals in *Arabidopsis* homografts. *The Plant Cell* 28, 2016–2025.

Paxton, B., Saranah, J.U. and Chapman, K.R. (1980) *Guava Propagation by Cutting.* Biennial Report No. 1. Maroochy Horticultural Research Station, Queensland, Australia.

Pereira, F.M., Petrechen, E.d.H., Benincasa, M.M.P. and Banzatto, D.A. (1991) Effect of indolebutyric acid on rooting of guava (*Psidium guajava* L.) softwood cuttings of the cultivars Rica and Paluma under intermittent mist. *Cientifica Jaboticabal* 19(2), 199–206.

Pereira, F.M., Usman, M., Mayer, N.A., Nachtigal, J.C., Maphanga, O.R.M. and Willemse, S. (2016) Advances in guava propagation. *Revista Brasileira de Fruticultura* 39(4), 39–43.

Pirela-Fuenmayor, M.E. and Mogollon-Montero, N.J. (1997) *In vitro* clonal propagation of guava (*Psidium guajava* L.) from stem shoot of cv. Mara-7. *Acta Horticulturae* 452, 47–52.

Prakash, H. and Tiwari, J.P. (1996) Micropropagation of guava (*Psidium guajava* L.). *Journal of Applied Horticulture* 2(1–2), 98–101.

Prasad, J., Rabbani, A. and Ram, R.A. (1988). Rooting of hardwood cuttings of guava (*Psidium guajava* L.) through bottom heat. *Progressive Horticulture* 20(1), 20–23.

Qadri, R., Azam, M., Khan, S.B., Khan, I., Haq, I.U. *et al.* (2018) Growth performance of guava cuttings under different growing media and plant cutting taking height. *Bulgarian Journal of Agricultural Science* 24(2), 236–243.

Rahman, H.U., Khan, M.A., Khokhar, K.M. and Laghari, M.H. (1991) Effect of season on rooting ability of tip cuttings of guava (*Psidium guajava* L.) treated with paclobutrazol. *Indian Journal of Agricultural Sciences* 61(6), 404–406.

Rai, M.K., Akhtar, N. and Jaiswal, V.S. (2007) Somatic embryogenesis and plant regeneration in *Psidium guajava* L. cv. Banarasi Local. *Scientia Horticulturae* 113(2), 129–133.

Rai, M.K., Jaiswal, V.S. and Jaiswal, U. (2008) Effect of ABA and sucrose on germination of encapsulated somatic embryos of guava (*Psidium guajava* L.). *Scientia Horticulturae* 117(3), 302–305.

Rai, M.K., Jaiswal, V.S. and Jaiswal, U. (2009) Shoot multiplication and plant regeneration of guava (*Psidium guajava* L.) from nodal explants of *in vitro* raised plantlets. *Journal of Fruit and Ornamental Plant Research* 17(1), 29–38.

Rai, M.K., Asthana, P., Jaiswal, V.S. and Jaiswal, U. (2010) Biotechnological advances in guava (*Psidium guajava* L.): recent developments and prospects for further research. *Trees* 24(1), 1–12.

Rai, M.K., Phulwaria, M., Gupta, A.K., Shekhawat, N.S. and Jaiswal, U. (2012) Genetic homogeneity of guava plants derived from somatic embryogenesis using SSR and ISSR markers. *Plant Cell, Tissue and Organ Culture* 111(2), 259–264.

Rajan, S. (2013) Hi-tech nursery management. In: Singh, V.K. and Ravishankar, H. (eds) *Recent Advances in Subtropical Fruit Crop Production*. CISH, Lucknow, India, pp. 4–10.

Rani, S.P., Akash, A., Sharma, V.K., Wali, B.P. and Shahnawaz, A. (2015) Standardization of method and time of propagation in guava (*Psidium guajava*). *Indian Journal of Agricultural Sciences* 85(9), 1162–1169.

Rani, T.D., Srihari, D., Dorajeerao, A.V.D. and Subbaramamma, P. (2018) Effect of rooting media and IBA treatments on shoot production and survival of terminal cuttings in guava (*Psidium guajava* L.) cv. Taiwan Pink. *International Journal of Current Microbiology and Applied Sciences* 7(11), 231–242.

Reddy, Y.N. and Majumder, P.K. (1978) Synergism of phenols and flavonoids with IBA in regeneration of mango and guava cuttings. *Vatika* 1, 37–44.

Robaina, R.R., Campos, G.S., Marinho, C.S., Souza, R.M. and Bremenkamp, C.A. (2015) Grafting guava on cattley guava resistant to *Meloidogyne enterolobii*. *Ciência Rural, Santa Maria* 45(9), 1579–1584.

Rugini, E. (1984) *In vitro* propagation of olive (*Olea europaea sativa* L.) cultivars with different rootability and medium development using analytical data from developing shoots and embryos. *Scientia Horticulturae* 24, 123–134.

Rymbai, H., Sathyanarayana Reddy, G. and Reddy, K.C.S. (2012) Effect of cocopeat and sphagnum moss on guava air-layers and plantlets survival under open and polyhouse nursery. *Agriculture Science Digest* 32(3), 241–243.

Sardoei, A.S. (2014) Effect of different media of cuttings on rooting of guava (*Psidium guajava* L.). *European Journal of Experimental Biology* 4(2), 88–92.

Saroj, P.L. and Pathak, R.K. (1994) Propagation of *Psidium* species through stooling. *Indian Journal of Horticulture* 55, 183–189.

Saroj, P.L., Pathak, R.K. and Yunus, M. (1997) Anatomical indices for predicting vigour in clonal rootstock of guava. *Indian Journal of Horticulture* 54, 198–204.

Sharma, A. and Zheng, B. (2019) Molecular responses during plant grafting and its regulation by auxins, cytokinins, and gibberellins. *Biomolecules* 9(9), 397–403.

Sharma, K.K., Sandhu, A.S., Bajwa, M.S. and Dhillon, B.S. (1974) Effect of IBA and NAA on the rooting of the air-layers of guava (*P. guajava* L.). *Journal of Research, PAU* 12, 23–25.

Sharma, R.S., Sharma, T.R. and Sharma, R.C. (1991) Influence of growth regulators and time of operation on rooting of air layering in guava (*Psidium guajava* Linn.) cv. Allahabad Safeda. *Orissa Journal of Horticulture* 19(1–2), 41–45.

Sharma, Y.K., Goswami, A.M. and Sharma, R.R. (1992) Effect of dwarfing aneuploid guava rootstock in high density orcharding. *Indian Journal of Horticulture* 49, 31–36.

Singh, D.K. (1998) Regeneration of guava (*Psidium guajava* L.) cultivars by stooling with the aid of paclobutrazol. *Annals of Agricultural Research* 19(3), 317–320.

Singh, G., Gupta, S., Mishra, R. and Singh, G.P. (2005) *Wedge Grafting in Guava – A Novel Vegetative Propagation Technique*. Central Institute for Subtropical Horticulture, Lucknow, India.

Singh, G., Gupta, S., Mishra, R. and Singh, A. (2007) Technique for rapid multiplication of guava (*Psidium guajava* L.). *Acta Horticulturae* 735, 177–183.

Singh, G., Sahare, H. and Maninderdeep (2018) Recent trends in guava propagation – a review. *Biosciences and Biotechnology Research in Asia* 16(1), 143–154.

Singh, K. and Jain, B.P. (1996) Propagation of guava (*Psidium guajava* L.) by air layering. *Scientific Horticulture* 5, 49–50.

Singh, K.K. and Singh, S.P. (2018) A review: micropropagation of guava (*Psidium* spp.). *Horticulture International Journal* 2(6), 462–467.

Singh, M., Patel, Y.R. and Yadava, H.S. (1995) Success and survival of air-layering as influenced by growth regulators and wrappers in guava (*Psidium guajava* L.). *Advances in Plant Sciences* 8(1), 183–188.

Singh, S. and Soni, S.L. (1974) Effect of water and acid soaking periods on seed germination of guava. *Punjab Horticulture Journal* 14, 122–124.

Singh, S.K., Singh, S.P. and Sharma, H.C. (2001) *In vitro* clonal propagation of guava (*P. guajava* L.) from field grown mature plants. *Physiology and Molecular Biology of Plants* 7, 33–38.

Singh, U.R., Pandey, I.C., Upadhyay, N.P. and Prasad, R.S. (1976) Effect of different rootstocks on growth, yield and quality of guava. *Punjab Horticulture Journal* 16, 121–124.

Singh, V.A., Singh, J.N. and Surendra P. (2010) Standardization of wedge grafting in guava under North Indian plains. *Indian Journal of Horticulture* 67, 111–114.

Sinha, G.C., Reddy, Y.T.N. and Singh, G. (1993) Propagation and rootstocks of guava. In: Chadha, K.L. and Pareek, O.P. (eds) *Advances in Horticulture*, Volume 1. Malhotra Publishing House, New Delhi, pp. 551–562.

Sinha, M.M., Verma, J.P. and Koranga, D.S. (1973) Studies on seed germination of guava. I. Effect of scarification and plant growth regulator treatment. *Progressive Horticulture* 5, 37–40.

Srivastava, R.P. (1962a) Preliminary studies in budding of guava. *Science and Culture* 28(1), 28.

Srivastava, R.P. (1962b) Further studies in budding of guava. *Science and Culture* 28, 433–434.

Stegemann, S. and Bock, R. (2009) Exchange of genetic material between cells in plant tissue grafts. *Science* 324, 649–651.

Syamal, M.M., Katiya, R. and Mamta, J. (2012) Performance of wedge grafting in guava under different growing conditions. *Indian Journal of Horticulture* 69(3), 424–427.

Taller, J., Hirata, Y., Yagishita, N., Kita, M. and Ogata, S. (1998) Graft-induced genetic changes and the inheritance of several characteristics in pepper (*Capsicum annuum* L.). *Theoretical and Applied Genetics* 97, 705–713.

Tavares, M.S.W., Lucca Filho, O.A. and Kersten, E. (1995) Germination and vigour of guava (*Psidium guajava* L.) seeds submitted to different procedures for overcoming dormancy. *Ciência Rural* 25, 11–15.

Teaotia, S.S. and Phogat, K.P.S. (1971) Effect of rootstock on growth, yield and quality of guava (*Psidium guajava* L.). *Progressive Horticulture* 2, 37–45.

Tsaballa, A., Athanasiadis, C., Pasentsis, K., Ganopoulos, I., Nianiou-Obeidat, I. and Tsaftaris, A. (2013) Molecular studies of inheritable grafting induced changes in pepper (*Capsicum annuum*) fruit shape. *Scientia Horticulturae* 149, 2–8.

Usman, M., Butt, M. and Fatma, B. (2012) Enhanced *in vitro* multiple shoot induction in elite Pakistani guava cultivars for efficient clonal plant multiplication. *American Journal of Biotechnology* 11(44), 10182–10187.

Vanaja, L., Swami, D.V., Prasanna Kumar, B. and Subbaramamma, P. (2017) Effect of grafting time on growth and success rate of guava (*Psidium guajava* L.) wedge grafts grown under shade net and poly house conditions. *International Journal of Current Microbiology and Applied Sciences* 6(10), 771–779.

Vasconcelos, L.F.L. and Cardoso, A.A. (1997) Evaluation of guava cultivars as rootstocks, during nursery stage. *Acta Horticulturae* 452, 63–69.

Vijayakumar, A., Palanisamy, V., Jayaraj, T. and Arumugam, R. (1991) Studies on certain seed technological aspects in guava (*Psidium guajava* L.). *South Indian Horticulture* 39(5), 315–316.

Vilchez, P., Albany, V., Gomez-Kosky, R. and Garcia, L. (2002) Induction of somatic embryogenesis in *Psidium guajava* L. starting at the zygotic embryo stage. *Revista de la Facultad de Agronomia (LUZ)* 19, 284–293.
Wang, J., Jiang, L. and Wu, R. (2017) Plant grafting: how genetic exchange promotes vascular reconnection, *New Phytologist* 214, 56–65.
Xiaomei, L. and Guochen, Y. (2011) Clonal propagation of guava (*Psidium guajava* L.) on nodal explants of mature elite cultivar. *International Journal of Plant Biology* 2, 121–125.
Yang, G. and Lu, Z. (2007) *In vitro* callus initiation of guava. *Acta Horticulturae* 738, 501–506.
Youssef, M.A., El-Helw, M.R., Taghian, A.S. and El-Aref, H.M. (2010) Improvement of *Psidium guajava* L. using micropropagation. *Acta Horticulturae* 849, 223–230.
Zamir, R., Ali, N., Shah, S.T., Muhammad, T. and Shah, S.A. (2007) *In vitro* regeneration of guava (*Psidium guajava* L.) from shoot tips of mature trees. *Pakistan Journal of Botany* 39(7), 2395–2398.
Zhou, X. and Liu, Y. (2015) Hybridization by grafting: a new perspective? *HortScience* 50(4), 520–521.
Zimmerman, R.H. and Broome, O.C. (1980) Blueberry micropropagation. In: *Proceedings of the Conference on Nursery Production of Fruit Plants Through Tissue Culture: Application and Feasibility*. US Department of Agriculture, Science, Education and Administration, Agricultural Research (Northeastern Region), Beltsville, Maryland, ARR-RE-11, pp. 44–47.

5 Biotechnology

Maneesh Mishra*, Muthukumar, M. and Sandeep Kumar
ICAR–Central Institute for Subtropical Horticulture, Lucknow, Uttar Pradesh, India

5.1 Introduction

Guava is widely cultivated throughout the world encompassing tropical and subtropical areas (Morton, 1987). In the *Myrtaceae* family, which consists of 133 genera and more than 3800 species (Nwinyi *et al.*, 2008), about 150 species are enlisted in the genus *Psidium*; nevertheless, few of them are edible and of economic importance (Pommer and Murakami, 2009). Thus, there is a plethora of diversity among guava and *Psidium* species. There exists aneuploidy in *Psidium* species and these species seem to be potential resistance sources against several pests and diseases. This offers a lot of scope for interspecific hybridization and development of pre-breeding stocks for crop improvement programmes.

Guava scores the top place over other fruits in terms of adaptability and is less affected by extremes of temperature, hot winds, scanty rainfall, saline soils, waterlogging, etc. Unfortunately, the guava production system is beset with many agronomic and horticultural problems such as susceptibility to many pathogens, particularly guava wilt disease caused by *Fusarium oxysporum* f.sp. *psidii*.

Limitations and slow progress in breeding programmes in guava are due to the lack of genomic resources and whole-genome information. With the advent of next-generation sequencing approaches, the genome sequencing of the species and cultivars is being taken up by different research groups and the development of single nucleotide polymorphism (SNP) markers for high-throughput genotyping and high-resolution linkage/quantitative trait locus (QTL) mapping are underway. A great deal of research on utilization of available molecular markers such as simple sequence repeat (SSR), amplified fragment length polymorphism (AFLP) and sequence-related amplified polymorphism (SRAP) markers has been done in assessing genetic diversity, species discrimination, clonal fidelity and hybridity of guava genetic resources. Guava being a more amenable tree crop for crossing, means development of biparental mapping populations is easier. Such biparental segregating mapping populations have been utilized for linkage analysis and several QTLs have been mapped for horticulturally/agronomically important traits in different linkage groups. Improving the resolution of the QTL maps and their validation are essential for the technique's further utilization in molecular breeding programmes. Efforts are underway in this direction for utilization of genomics, transcriptomics, proteomics and metabolomics tools and their

*E-mail: maneeshmishra.cish@gmail.com

DOI: 10.1079/9781789247022.0005

integration with the QTL approach to expedite the breeding programmes (Fig. 5.1). In this context, accomplishments and technological advancements made using biotechnological interventions are discussed in different subsections in this chapter.

5.2 Euploidy in *Psidium* Species and Variation in Genome Size

The genus *Psidium* is reported to be composed of several species with varying ploidy levels, described as aneuploidy. Variations have been reported in the ploidy, which also contributes towards the variation in the genome size among different *Psidium* species (Table 5.1). Earlier, evolution and diversification of the *Psidium* species were documented based on SSR markers, karotyping and flow cytometry analysis, which also revealed euploidy (Costa *et al.*, 2008; Souza *et al.*, 2014). Polyploidy (diploid to octoploid) was evinced in seven *Psidium* species, besides the outcome of the whole-genome duplication about the nuclear DNA content, DNA sequence and distribution (Majumdar and Mukherjee, 1972; Tuler *et al.*, 2019). The common guava, *Psidium guajava*, is diploid in nature with $2C = 0.95$ pg and $2n = 2x = 22$ chromosomes (Marques *et al.*, 2016). But there are variations recorded in *Psidium* species where the chromosome number varies from $2n = 22$, $2n = 44$, $2n = 55$, $2n = 66$ to $2n = 88$ (diploid to octoploid) (Costa *et al.*, 2008; Souza *et al.*, 2014). For instance, *Psidium guineense* is a tetraploid species with $2C = 1.85$ pg, and the karyotype showed $2n = 4x = 44$ chromosomes. SSR markers were used for karyotypic refinements and precise prediction of the ploidy levels (Marques *et al.*, 2016; Tuler *et al.*, 2019). The direct relationship exists between genome size and *C* value as is evident in Table 5.1, wherein the genome size changes with *C* value and thereby is reflected in the ploidy levels of the *Psidium* species.

5.3 Genomic Resources

Genomic resources of guava are limited with only a few thousand SSR markers available in the public domain. Although the whole-genome sequence of common guava *P. guajava* L. cultivar 'Zhenshu' is available in the National Center for Biotechnology Information (NCBI) (PRJNA401640), genomic annotations and chromosome-wise distribution of sequences are not available. The genome size of cultivar 'Zhenshu' is reported to be 385 Mbp, deviating slightly from the predicted genome size 247.92 and 269.44 Mbp for white and red cultivars, respectively.

A total of ten genomic resources are available in NCBI to date (June 2020) which encompasses five transcriptomes, two whole genomes, three plastomes and two metagenomes (Table 5.2). Genes *mat*K, *psb*A and *rbc*L have been reported to be highly conserved genes that were used for developing barcodes in *Psidium* species, as well as these sequences being used for aligning and mapping chloroplast genomes of *Psidium galapageium*, *Psidium acidum* and *P. guajava* (Fig. 5.2) (Reatini *et al.*, 2018). The paucity of the genomic resources also limits the functional genomics and transcriptomics studies and hence only a few reports are available on these areas (which are discussed later in this chapter). This also limits the availability of markers for linkage and QTL mapping, eventually slowing the pace of the molecular breeding approach for crop improvement programmes in guava. To date, although a few thousand SSRs have been described by different groups (Risterucci *et al.*, 2005; Ritter *et al.*, 2010), no reports are available on SNP markers. Yet the genomic resources in the public domain give the overview of genomic information that could be utilized in guava improvement programmes.

5.4 Molecular Marker Systems and Their Applications in Guava Germplasm Characterization

Large genetic variation exists in *Psidium* species and cultivars contributing to their genetic diversity. Molecular markers are used for diversity analysis which helps to generate cultivar-specific DNA fingerprints not only as a prime source for accounting

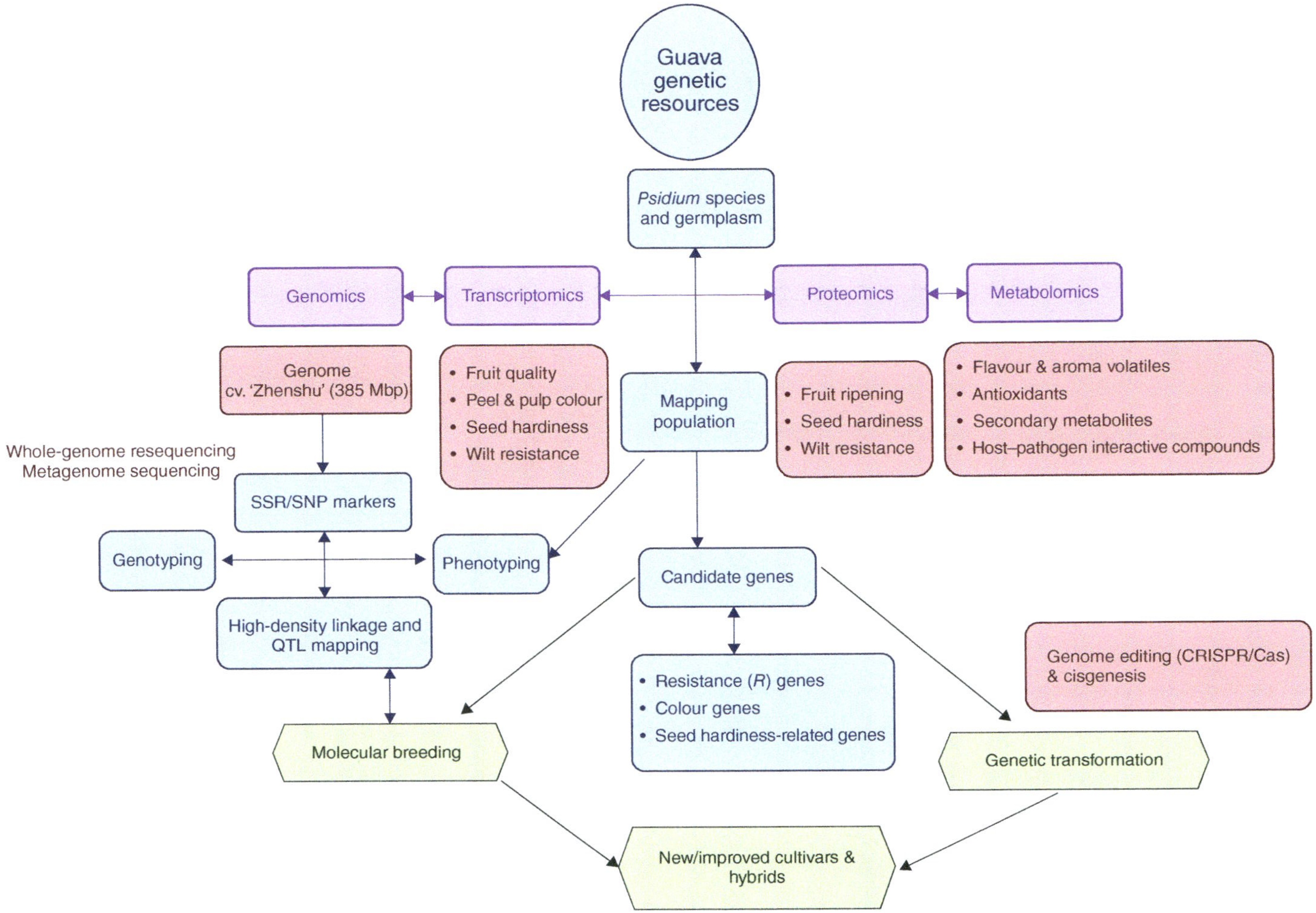

Fig. 5.1. Overview of biotechnological interventions in guava improvement programme. SSR, simple sequence repeat; SNP, single nucleotide polymorphism; QTL, quantitative trait locus; CRISPR/Cas, clustered regularly interspaced short palindromic repeats/CRISPR-associated.

Table 5.1. Ploidy and genome size variations among different *Psidium* species.

Species	Ploidy	2*n*	2*C* value (pg)	Genome size (Mpb)	Reference(s)
Psidium guajava (white cultivar)	2*x*	22	0.51	247.92	Costa and Forni-Martins (2006b); Costa *et al.* (2008)
P. guajava (red cultivar)	2*x*	22	0.55	269.44	Costa and Forni-Martins (2006b); Costa *et al.* (2008)
Psidium cattleyanum	4*x*	44	1.05	515.48	Costa and Forni-Martins (2006b); Costa *et al.* (2008)
Psidium acutangulum	4*x*	44	1.17	572.32	Costa and Forni-Martins (2006a); Costa *et al.* (2008)
Psidium guineense	4*x*	44	2.02	987.78	Souza *et al.* (2014); Marques *et al.* (2016)
Psidium australe	5*x*	55	2.97	1452.33	Souza *et al.* (2014)

variability, but also to document and conserve them. A wide array of marker systems such as random amplified polymorphic DNA (RAPD) and SSR markers have been extensively used in molecular characterization and diversity analysis in guava, as well as in *Psidium* species discrimination (Table 5.3). RAPD markers are highly efficient in guava diversity analysis, and thus have been documented extensively because of the high level of polymorphism detected in germplasm accessions of different geographical ecosystems (Bajpai *et al.*, 2008; Feria-Romero *et al.*, 2009; Pessanha *et al.*, 2011; Valera-Montero *et al.*, 2017), but have some inherent demerits such as low reproducibility, dominant nature and non-specificity. Another marker system, most commonly described as direct amplification of minisatellite DNA (DAMD), was used earlier for discriminating individuals in a half-sib population of *P. guajava* consisting of six half-sib progenies ('CISH-G-1', '-G-2', '-G-3', '-G-4', '-G-5' and '-G-6'), 'Allahabad Safeda' and two *Psidium* species (Saxena *et al.*, 2007). It was found that markers, namely HVR, HBV, M13 and 33.6b, discriminated the half-sib seedlings from 'Allahabad Safeda', *Psidium acutangulam* and *P. guineense*.

Genic and genomic SSR markers have yielded good results in genetic diversity analysis which is evident from the diversity indices, average heterozygosity levels and polymorphism information contents reported by several research groups (Valdés-Infante *et al.*, 2010; Kanupriya *et al.*, 2011; Costa and Santos, 2013). SSR markers, developed from a $(GA)_n$- and $(GT)_n$-enriched library, have been used to assess the diversity of *P. guajava* cultivars and were reported to be cross-transferable across taxa (Risterucci *et al.*, 2005). More markers were reported in guava in 2010 (Risterucci *et al.*, 2010). Kenyan guava cultivars could be clustered into two groups, namely white-fleshed and red-fleshed types, indicating shared ancestry and low diversity (Chiveu *et al.*, 2019). Sánchez-Teyer *et al.* (2010) used nine AFLP primer combinations and six SSR loci to assess the genetic diversity of 68 germplasm accessions from 12 different regions of Mexico. The microsatellite mPgCIR161 revealed the highest number of alleles. The AFLP- and SSR-allele based clustering divided guava accessions into four groups, with other diverse accessions failing to cluster. Most Mexican accessions were in groups B and C while Venezuelan cultivars were placed in group E, indicating enormous diversity in Mexican guava germplasm. Similarly, SSR markers were found to be efficient in discrimination of Venezuelan guava cultivars (Aranguren *et al.*, 2010; Briceño *et al.*, 2010). Youssef *et al.* (2015) used 88 amplicons generated from five SRAP markers for characterization of 49 guava germplasm accessions and identified cultivar-specific bands. Clonal uniformity and homogeneity of micropropagated guava plants were ascertained using molecular markers: inter-simple sequence repeat (ISSR) markers (Liu and Yang, 2012) and SSR markers (Rawls *et al.*, 2015). Chloroplast DNA (cpDNA) markers, especially maturase K (*mat*K) and

Table 5.2. Genomic resources available in the public domain (NCBI database as of June 2020).

Tissue	Bioproject	Genome/ transcriptome	Platform	Cultivar/species	Submitter	Year
Root and soil	PRJNA615656	Metagenome	Illumina MiSeq	–	University of Kansas	2020
Leaf	PRJNA401640	Whole genome	PacBio	'Zhenshu'	Hainan University	2018
Mature fruit red and green peel	PRJNA479714	Transcriptome	Illumina HiSeq 2500	'CISH-G5', 'Apple Colour'	Punjab Agricultural University	2018
Mature fruit with seeds	PRJNA479710	Transcriptome	Illumina HiSeq 2500	'Punjab Pink'	Punjab Agricultural University	2018
Fruits, leaves, buds	PRJNA472130	Transcriptome	Illumina HiSeq 2500	'Allahabad Safeda'	Punjab Agricultural University	2018
Leaf and shoot tip	PRJNA420020	Transcriptome	–	'Allahabad Safeda'	Punjab Agricultural University	2017
Leaf	PRJNA362375	Plastome	Illumina	*Psidium guajava*	Korea University	2017
Leaf	MH491846	Plastome	Illumina HiSeq 4000	*Psidium galapageium*	University of North Carolina	2018
Leaf	MH491847	Plastome	Illumina HiSeq 4000	*Psidium acidum*	University of North Carolina	2018
Young leaves	PRJNA392610	Whole-genome sequencing	Illumina HiSeq 2500	'Tikal' guava	King Abdulaziz University	2017
Fruit before and after *Bactrocera dorsalis* oviposition	PRJNA384497	Metagenome (bacteria)	Illumina HiSeq 2500	–	South China Agricultural University	2017
Leaf	PRJNA356080	*De novo* transcriptome	Illumina HiSeq 4000	–	Guangdong Pharmaceutical University	2016

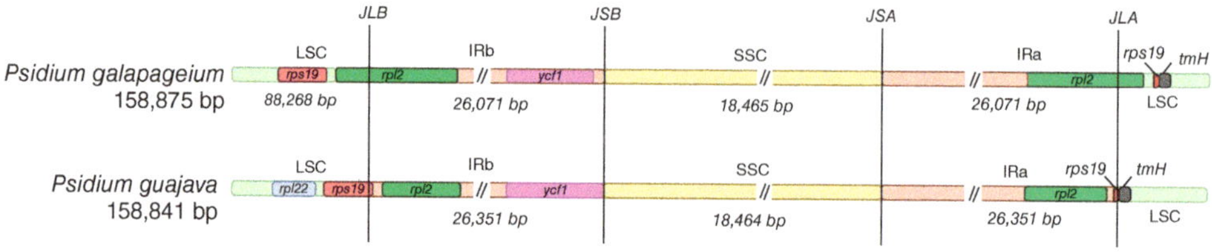

Fig. 5.2. Simplified version of large and small single copy (LSC, SSC) and inverted repeat regions (IRa, IRb) of chloroplast genomes of *Psidium guajava* L. in comparison with *Psidium galapageium*. Modified from Reatini *et al.* (2018).

transfer RNA genes (*trn*L, *trn*F), etc., have been well established for detection of species-level variation. SSR markers could discriminate species of *Psidium*, such as *Psidium friedrichsthalianum*, *Psidium chinensis* and *P. guineense* from *P. guajava* more efficiently than *mat*K, a cpDNA marker (Muthukumar *et al.*, 2019). Earlier, Kumari *et al.* (2017) confirmed that SSR primer mPgCIR256 was the most informative, with a high rate of polymorphism as well as number of alleles. Their study also confirmed the diversity in six species of *Psidium* and 28 genotypes of *P. guajava* using 39 SSR markers. Phenotype-based phylogenetic classification grouped all the *P. guajava* accessions into two major groups while genotype-based clustering grouped them into three clusters, with cluster A including only cultivars and cluster B including all species except *P. friedrichsthalianum*. Genetic diversity analysis among *Psidium* species showed that *P. guajava* had more similarity with *P. guineense* compared with other species. More recently, SNP markers of related species have been used in molecular characterization of *Psidium* species. Fifty-three *Psidium* accessions, including 47 *P. guajava*, were genotyped with EUCHIP60K, an SNP chip developed for *Eucalyptus grandis*. Phylogenetic analysis revealed five clusters with 44 guava accessions in one cluster, three *P. guineense* accessions in another cluster and three *Psidium cattleyanum* accessions in the third cluster, along with two minor clusters. The Bayesian analyses suggested seven subpopulations among these guava accessions (Costa and Santos, 2017).

5.5 Development of Mapping Populations, Linkage Analysis and QTL Mapping

Biparental mapping populations in different genetic backgrounds have been generated and used in linkage as well as QTL mapping studies in guava (Nimisha *et al.*, 2013). Naga Chaithanya *et al.* (2016) assessed 135 polymorphic SSR markers to identify pairs of genotypes contrasting for responses to infestation by the bark-eating caterpillar (*Indarbela tetraonis* Moore) to be used as putative parents to develop mapping populations for chromosomal localization of genomic regions controlling resistance to bark-eating caterpillar in guava. In their study, phylogenetic analysis was found to separate five morphologically resistant and three morphologically highly susceptible genotypes into two different clusters. Based on their assessment, four pairs of accessions, namely 'Superior Sour Lucidum' and 'Seedless', 'Portugal' and 'Seedless', 'Lalit' and 'Seedless', 'Spear Acid' and 'Seedless', which showed contrasting responses to bark-eating caterpillar and polymorphic SSR loci, were used as resistant donor parent and susceptible female recipient parent in developing mapping populations to be used in a resistance breeding programme against bark-eating caterpillar.

Molecular genetic linkage maps are particularly important and valuable tools for QTL mapping and marker-assisted selection (MAS) in guava with desirable traits. Linkage maps are genetic maps constructed using either genotypic or phenotypic data independently pertaining to a segregating mapping population. In guava, different mapping populations have been used to construct

Table 5.3. Overview of different marker systems used in assessing genetic diversity in guava.

Marker[a]	Geographical distribution	Diversity level[b]	Reference
RAPD	India	PP: 74.7%	Dahiya *et al.* (2002)
RAPD	India	PP: 54%–11%	Prakash *et al.* (2002)
AFLP	Cuba	–	Valdés-Infante *et al.* (2003)
RAPD	India	PIC: 0.916	Sharma *et al.* (2005)
SSR	Germany	Av. Ho: 0.42	Risterucci *et al.* (2005)
AFLP	Mexico	DI: 0.584	Hernández-Delgado *et al.* (2007)
RAPD	Taiwan	PP: 80%	Chen *et al.* (2007)
DAMD	India	PP: 87.5%	Saxena *et al.* (2007)
RAPD	Mexico	–	Feria-Romero *et al.* (2009)
SSR	Cuba	Av. Ho: 0.38	Valdés-Infante *et al.* (2010)
SSR	USA	Av. Ho: 0.61	Viji *et al.* (2010)
AFLP, SSR	Mexico	PP: 71.9% (AFLP), 64.4% (SSR)	Sánchez-Teyer *et al.* (2010)
SSR	Venezuela	–	Briceño *et al.* (2010)
SSR	Venezuela	Av. Ho: 0.7398 SII: 1.5853	Aranguren *et al.* (2010)
SSR	India	PIC: 0.749	Kanupriya *et al.* (2011)
RAPD	Bangladesh	PP: 33.19%	Ahmed *et al.* (2011)
RAPD, ISSR	India	PP: 77% (RAPD), 81.6% (ISSR)	Mani *et al.* (2011)
RAPD	Brazil	–	Pessanha *et al.* (2011)
ISSR	USA	PP: 9.67%	Liu and Yang (2012)
SSR	Brazil	–	Coser *et al.* (2012)
SSR	Brazil	PIC: 0.709	Costa and Santos (2013)
SSR	India	PIC: 0.563	Chaithanya *et al.* (2014)
SSR	USA	Av. Ho: 0.2	Sitther *et al.* (2014)
SSR, iPBS	Pakistan	SII: 0.130 (SSR) PIC: 0.1687–0.3522 (iPBS)	Mehmood *et al.* (2015)
SRAP	Egypt	PP: 65.9%	Youssef *et al.* (2015)
SRAP, ISSR	Egypt	PP: 25.35% (SRAP), 42.42% (ISSR)	Abouzaid *et al.* (2016)
SSR	India	Av. Ho: 0.25–0.50	Naga Chaithanya *et al.* (2016)
AFLP	Thailand and 15 other countries	PP: 54%	Thaipong *et al.* (2017)
SNP	Brazil	–	Costa and Santos (2017)
RAPD	New Delhi, India	PIC: 0.49–0.89	Shiva *et al.* (2017)
RAPD	Mexico	–	Valera-Montero *et al.* (2017)
SSR	Brazil	PIC: 0.54	Noia *et al.* (2017)
SSR	India	DI: 0.2766	Kumari *et al.* (2017)
SSR	India	–	Kherwar *et al.* (2018)
RAPD	Bangladesh	PP: 75.23%	Alam *et al.* (2018)
RAPD, DAMD	India	–	Bajpai *et al.* (2008)
SSR	Kenya	Av. Ho: 0.630	Chiveu *et al.* (2019)
SSR	Galapagos Islands	Av. Ho: 0.163	Urquía *et al.* (2019)
SSR	China	PIC: 0.60	Ma *et al.* (2020)
cpDNA and SSR	*Psidium* spp., India	Av. Ho: 0.48	Muthukumar *et al.* (2019)

[a]RAPD, random amplified polymorphic DNA; AFLP, amplified fragment length polymorphism; SSR, simple sequence repeat; DAMD, direct amplification of minisatellite DNA; ISSR, inter-simple sequence repeat; iPBS, inter-primer binding site; SRAP, sequence-related amplified polymorphism; SNP, single nucleotide polymorphism; cpDNA, chloroplast DNA.
[b]PP, percentage polymorphism; PIC, polymorphic information content; Av. Ho, average heterozygosity; DI, diversity index; SII, Shannon's information index.

linkage maps which were summarized in the review of Nimisha *et al.* (2013). The first molecular linkage map of guava was reported using AFLP markers by Valdés-Infante *et al.* (2003). This map was further refined by Rodríguez *et al.* (2007) saturating with 220 AFLP markers. Besides linkage groups, certain QTLs for several vegetative traits and fruit characteristics were also mapped on these linkage maps (Valdés-Infante *et al.*, 2003; Rodríguez *et al.*, 2007). Lepitre *et al.* (2010) reported an integrated parental linkage map of guava constructed for the mapping population MP1 derived from a cross between two heterozygous guava cultivars ('Enana' × 'N6') with 800 high-density AFLP and SSR markers. Padmakar *et al.* (2016) used RAPD, SSR and SRAP markers for construction of a parental linkage map and QTL detection in the F_1 progenies of a biparental mapping population derived from a cross between 'Kamsari' and 'Purple Local' (Fig. 5.3). A genomic cosmid (cos sites + plasmid = cosmid) library was constructed in guava from which resistance (RGL sequences) and homeotic genes (MADX-box and HOMEO-box genes) were characterized using heterologous probes (Ritter, 2012).

A QTL is defined as a region of the genome that is associated with an effect on a quantitative trait, which could be a single

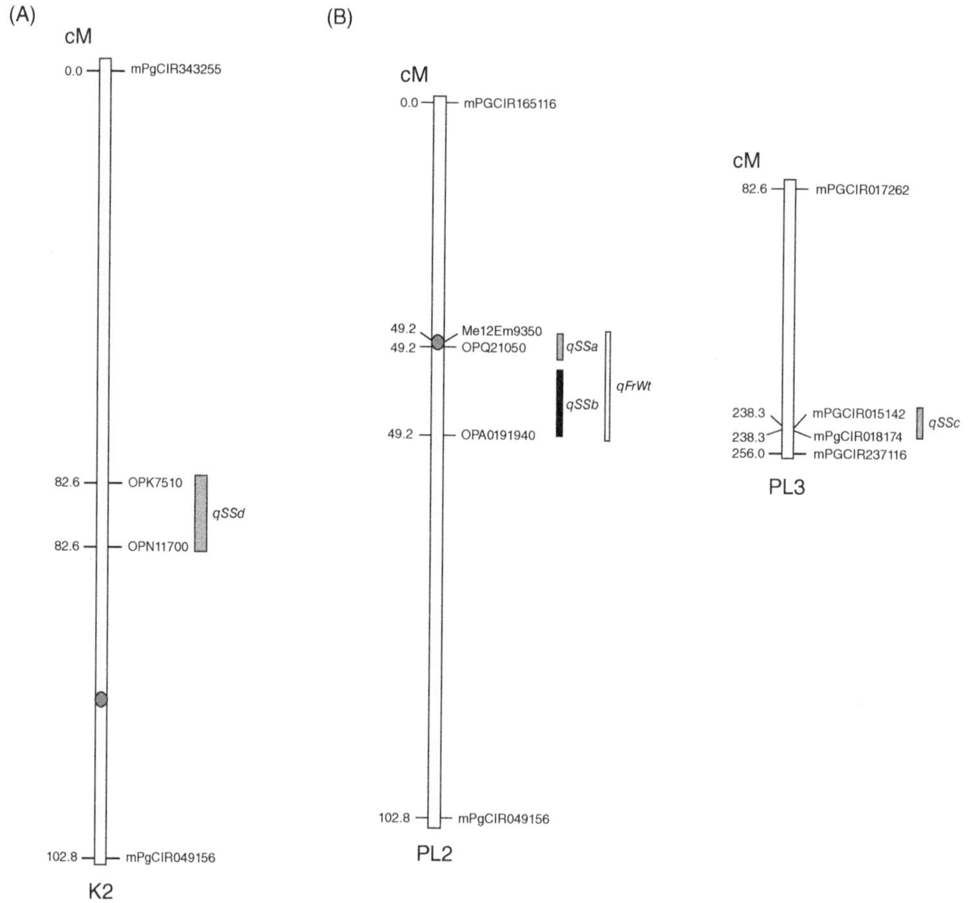

Fig. 5.3. A typical example of QTL regions of seed hardiness traits mapped on chromosome/linkage group of parental lines (A) 'Kamsari' and (B) 'Purple Local'. QTL, quantitative trait locus. Adapted and modified from Padmakar *et al.* (2016).

gene or a cluster. QTL mapping can enumerate genes contributing to a trait, indicate their chromosomal location and estimate their relative importance in phenotypic expression of the trait. Generation of segregating mapping populations forms the basis for linkage and QTL mapping. In the case of guava, developing half-sib populations and F_1 or F_2 progenies of a biparental population is relatively easier than in other fruit crops. The major constraint in the construction of QTL maps could be attributed to non-availability of adequate markers to cover the entire genome representing all chromosomes. Although a number of markers have been mapped into linkage groups, their utilization is limited due to their QTL validation and also the need for high-density markers in the QTL region for facilitating MAS. A few other QTLs have been identified in guava (Table 5.4) which are related to fruit quality traits, seed hardness, fruit width, fruit weight and pulp content (Rodríguez *et al.*, 2007; Padmakar *et al.*, 2015, 2016). Earlier, an integrated guava reference linkage map was established with a very high marker density of over 1700 markers that was used for QTL mapping in three mapping populations. Over 100 QTLs were detected which were associated with traits such as plant height, petiole length, leaf length and width, yield, fruit number and average fruit weight, fruit length and width, internal and external pulp thickness, seed number and weight, vitamin C content, acidity and total soluble solids. Individual QTLs explained up to 40% of the total variance and the total variance explained by the sum of all detected QTLs varied between 20% and over 50% between traits and populations (Grattapaglia *et al.*, 2012).

Pelea *et al.* (2012) observed vast genotype × environment (G × E) interactions for many quantitative traits that were evaluated in three Cuban guava populations for four consecutive years. It was also recorded that fruit length was the only trait with a high value of heritability in the narrow sense, but the other traits presented low and medium values. Such results have also been reported in different guava mapping populations. This warrants precise large-scale phenotyping in different environments and over seasons to be undertaken in varying mapping populations so that G × E interaction errors in phenotyping data could be minimized while detecting mega-effect QTLs harbouring major genes. However, developing and discovering more markers for QTL mapping also seems to be another prime focus area in guava genome-mapping programmes. This would not only enable large-scale genotyping but also facilitate fine mapping by saturation of the genome with an array of markers, thereby helping detection of significant QTLs and their linked markers. It is obvious that only with the detection of mega-effect QTLs with high degrees of linkage could marker-assisted breeding be successful in guava improvement programmes.

5.6 Use of Biotechnological Tools in Development and Characterization of Wilt-Resistant Rootstocks

Guava wilt disease is a serious pernicious disease reported in guava which affects the guava industry enormously (Mishra *et al.*, 2007). This disease is caused by either root-knot nematode, *Meloidogyne enterolobii*, invasion or a fungal incitant, i.e. *Fusarium oxysporum* f.sp. *psidii*, or a synergistic association of both the parasite and the pathogen. Generally wild relatives/species are good sources of gene pools for resistance. Wide hybridization is an efficient tool whereby, through distant hybridization involving interspecific crossing between two species, specific traits can be transferred. Thus, this wide hybridization technique has been used developing nematode- or *Fusarium* wilt-resistant rootstock lines. Sousa *et al.* (2017) characterized various *Psidium* species (*P. guajava* cultivar 'Paluma', *P. guineense*, *P. cattleyanum* and *P. friedrichsthalianum*) and histological analyses showed all *Psidium* species presented poorly developed feeding sites of nematode excepting *P. guajava*. Freitas *et al.* (2013) screened 51 accessions of *Psidium* spp. comprising *P. cattleyanum* (yellow guava), *P. friedrichsthalianum* (Costa Rican guava), *Acca sellowiana* (feijoa) and *Psidium rufum* (purple guava),

Table 5.4. Mapping of quantitative trait loci (QTLs) for major agronomically important horticultural and fruit quality traits in guava.

Mapping population	No. of linkage groups	Linked markers[a]	Trait	QTLs identified	Chromosome/linkage group position (distance in cM)	Reference
Cuban guava accessions	–	SSR	Vegetative characters	Different QTLs	Linkage map	Valdés-Infante *et al.* (2003)
'N3' (MP1), 'Suprema Roja' (MP2) and 'Belic L-207' (MP3)	11	SSR	Acidity	*qAC1*	VI (30)	Rodríguez *et al.* (2007)
				qAC2	VII (34.5)	
				qAC3	VIII (10.3)	
			TSS[b]	*qSW1*	V (30.5)	
				qSW2	VII (10.9)	
			Fruit weight	*qWF1*	I (1.6)	
				qWF2	III (12.3)	
				qWF3	X (12.3)	
			Fruit width	*qFW1*	I (1.6)	
				qFW2	III (12.3)	
				qFW3	X (12.3)	
			External pulp thickness	*qPE*	VII (34.6)	
			Internal pulp thickness	*qPI1*	VIII (3.8)	
			Seed number	*qSN1*	I (3.3)	
			Seed weight	*qWS1*	VIII (29.1)	
				qWS2	I (3.6)	
				qWS3	VII (20.8)	
				qWS4	VIII (23.2)	
				qWS5	IX (28.3)	
			Vitamin C content	*qVC1*	V (30.5)	
				qVC2	X (23.4)	
'N6' (MP1), 'Suprema Roja' (MP2) and 'Belic L-207' (MP3)	11	SSR	Not described	Different QTLs	–	Ritter *et al.* (2010)
Full sibs derived from 'Enana Roja Cubana' × 'N6' cross	11	AFLP (452) and SSR (126)	–	–	2179 cM linkage map	Lepitre *et al.* (2010)
'Kamsari' × 'Purple Local' mapping population	11	SRAP (126) and SSR (160)	Seed strength (hardness/softness), fruit quality, TSS	–	2551.3 cM ('Kamsari') 2113.0 cM ('Purple Local')	Padmakar *et al.* (2015)

F_1 progenies of two populations developed by pseudo-testcross ('Kamsari' × 'Purple Local', 'Purple Local' × 'Allahabad Safeda')	11	RAPD and SSR	Fruit weight Seed strength (hardness/softness)	*qFrWt* *qSSa* *qSSb* *qSSc* *qSSd*	PL2 (49.21) PL2 (51.21) PL2 (64.51) PL2 (234.81) K2 (61.61)	Padmakar *et al.* (2016)
'Allahabad Safeda' × purple guava, 'Allahabad Safeda' × 'CISH-G-1', 'Allahabad Safeda' × 'Seedless' and some other cross combinations	10	SSR	Fruit colour (anthocyanin gene)	MdMYB10	345.63 cM	Singh (2017)
F_1 progenies of 'Allahabad Safeda' × 'Arka Kiran'	–	SSR	Flesh colour	–	–	Jindal (2017)

[a]SSR, simple sequence repeat; AFLP, amplified fragment length polymorphism; SRAP, sequence-related amplified polymorphism; RAPD, random amplified polymorphic DNA.
[b]TSS, total soluble solids.

P. guajava (43), *P. guineense* (3, Brazilian guava), *Psidium acutangulum* (pear guava) and *P. guajava* cultivar 'Paluma' selected from the *Psidium* (Embrapa, Brasilía, Brazil) for assessing resistance against *M. enterolobii.* When used as rootstocks under greenhouse conditions, *P. cattleyanum* and *P. friedrichsthalianum* were compatible with cultivar 'Paluma'; however, under greenhouse and field conditions only 50% of both scions survived. No apparent hypersensitive response was seen in the resistant guava *P. cattleyanum* and *P. friedrichsthalianum.* Juveniles were able to develop normal feeding sites in resistant roots similar to those in susceptible roots but failed to reach maturity. In similar lines, an interspecific hybrid was developed with wilt resistance against *F. oxysporum* f.sp. *psidii* at ICAR–Central Institute for Subtropical Horticulture (CISH) (Lucknow, India) by crossing *P. guajava* with *Psidium molle* (Rajan *et al.*, 2005; Negi and Rajan, 2007). This interspecific hybrid is being used as a wilt-resistant rootstock as it showed enhanced tolerance to *Fusarium* wilt pathogen. Efforts are still underway at ICAR–CISH for characterization of the *Psidium* species as well as the interspecific hybrid for identification of resistance (*R*) genes which confer the resistance. An *in vitro* selection technique that involves selection of resistant lines from cell-free *Fusarium* culture filtrate selection medium was successfully developed (Bajpai *et al.*, 2007; Kamle *et al.*, 2012) and demonstrated field-level wilt resistance against *F. oxysporum* f.sp. *psidii.*

5.7 Structural Genomics

Structural genomics approaches of sequencing the genome and metagenome have been minimally attempted in guava. The genome of guava is only partially available in the public domain as described earlier and this limits reference mapping strategies. However, genome resequencing of two *Psidium* species, namely *P. guineense* (Species 7) and *P. friedrichsthalianum*, was performed using Illumina MiSeq sequencing technology. Besides this, an interspecific hybrid between *P. guajava* and *P. molle*, being used as a wilt-resistant rootstock, was also sequenced. Approximately 15 GB data each were generated for all three species and the data sets were assembled and annotated against the reference genome of cultivar 'Zhenshu'. Comparative analysis of the genomes with reference mapping strategy facilitated identification of native *R* genes. A large array of nucleotide-binding site–leucine-rich repeat (NBS-LRR) type *R* genes have been detected in the genome analysis. A wide range of SNPs have also been detected which need to be validated (ICAR–CISH, 2020, unpublished results). The complete chloroplast genome of guava has been sequenced (NCBI Accession No. KX364403) and reported to have a genome size of 158,841 bp consisting of a large single copy (LSC) of 87,675 bp and a small single copy (SSC) of 18,464 bp, separated by two inverted repeats of 26,351 bp (Jo *et al.*, 2016). The plastome contained 112 genes, of which 78 are protein-coding genes, 30 are tRNA genes and four are rRNA genes, and the chloroplast genome was found to be AT-rich. Earlier, using mapping of *trn*L, *trn*F and *atp*B–*rbc*L gene sequences of *Psidium* species on *Syzygium cumini*, a simplified version of the chloroplast genome map was constructed at ICAR–CISH which coincides with the reference chloroplast genome map developed by Jo *et al.* (2016).

5.8 Functional Genomics

Functional genomics is a field of genomics that involves isolation, characterization and expression analysis of a complete set of genes at transcriptional (mRNA) and translational levels (protein), thus referred to as transcriptomics and proteomics, respectively. Limited genomic information also restricts utilization of the functional genomics approach in guava molecular breeding. Although five transcriptomes have been deposited in NCBI, no significant genes associated with traits or novel genes have been characterized or reported that could be explored for genetic manipulation or gene-assisted trait improvement. More recently,

Mittal *et al.* (2020) used tissue-specific comparative gene expression of leaf, flower buds, peel and pulp tissues of guava cultivar 'Allahabad Safeda' by RNA sequencing. By means of comparative gene expression analysis in these different tissues of 'Allahabad Safeda' guava and its cross-comparison with peel and pulp colours of cultivars 'Apple' and 'Punjab Pink', respectively, the candidate genes governing colour development in guava were identified. Similarly, transcriptomics (RNA sequencing) and proteome approaches have been used at ICAR–CISH for elucidating the molecular mechanism of resistance against wilt disease caused by root-knot nematode *M. enterolobii* and fungal pathogen *F. oxysporum* f.sp. *psidii* in different *Psidium* species which could ease interspecific rootstock breeding programmes (ICAR–CISH, 2020, unpublished results).

5.9 Metabolomics in Guava Improvement Programmes

Metabolomics is an approach which involves complete or whole metabolite profiling of a cell or an organelle of a specific tissue, comprising both primary and secondary metabolites, under a given situation or developmental stage that defines its temporal, spatial and developmentally controlled production. Latest advancements in instrumentation technologies such as high-performance liquid chromatography (HPLC), gas chromatography (GC) and liquid chromatography (LC) coupled with mass spectrometry (MS) have opened up new scope for utilization of the metabolomics approach in characterization of guava cultivars and species. Most commonly, the functional compounds in fruits such as flavour and aromatic volatiles as well as nutritionally important compounds are characterized using metabolomics (Table 5.5). Some of the key metabolites that are characterized in guava are quercetin, myricetin, terpenes, gallic acid, chlorogenic acid, furanones, hexanal and ionones. Metabolite characterization of guava leaves has evinced presence of a high content of secondary metabolites such as antioxidants, polyphenols, antiviral compounds and anti-inflammatory compounds. Guava leaf is used as a medicine to cure cough, diarrhoea, oral ulcers, swollen gums and wounds. Quercetin is considered the most active antioxidant in guava leaves and is responsible for their spasmolytic activity (Naseer *et al.*, 2018). This metabolomics approach could also be utilized for selection of elite cultivars with specific flavour–aroma and antioxidants, which can be described as metabolomics-assisted selection.

5.10 Molecular Diagnostics of Pathogens Causing Diseases in Guava

Abiotic stresses do not affect guava much because of inherent resistance mechanisms for abiotic stresses. Mostly guava is categorized as a salinity-tolerant fruit species. However, biotic stresses such as pests and diseases pose serious problems in guava production systems. Among these biotic stresses, wilt disease caused by *F. oxysporum* f.sp. *psidii* is a major setback. Therefore, early detection of the pathogenic microorganism is highly essential for undertaking prophylactic and therapeutic measures for its management. Several molecular diagnostic tools have been developed for early detection of this wilt pathogen at ICAR–CISH (Pandey *et al.*, 2010; Muthukumar *et al.*, 2012, 2015). A colony polymerase chain reaction (PCR) assay was developed with 10 mg of fungal mycelium of *Fusarium* dissolved in 50 µl of $T_{10}E_1$ buffer using ITS1–ITS4 primer and species-specific primers, namely BKP1–BKP2 primer pair, for direct pathogen detection without a DNA extraction step (Muthukumar *et al.*, 2012) and another variant method of culture-independent PCR assay was developed for *Fusarium* detection (Mishra *et al.*, 2013). A multiplex PCR technique using a combination of three primer pairs, namely ITS1–4, chitin synthase and species-specific primers, was developed and validated for identification and detection of the wilt pathogen, *F. oxysporum* f.sp. *psidii* (Muthukumar *et al.*, 2012). Similarly, another molecular tool based on nested PCR assay with ITS1–ITS4 and ITS1F–R primer

Table 5.5. Characterization through metabolomics of key metabolites associated with flavour, aroma and nutritional value in guava.

Species (cultivar)	Technique used for metabolomics[a]	Key/novel metabolites identified	Pathway/functional properties	Reference
Psidium guajava ('Barbie Pink', 'Homestead', 'Sardina 1', 'Sardina 2', 'Yen 2', 'Sayla' and 'Thai Maroon')	HPLC–PDA and HR-ESI-MS	Delphinidin 3-*O*-glucoside (co-injection) Cyanidin-3-*O*-glucoside Myricetin-3-*O*-arabinoside Myricetin-3-*O*-xyloside Isorhamnetin-3-*O*-galactoside Abscisic acid Pinfaensin Turpinionosides A Pedunculoside Madecassic acid	Anthocyanins, flavonoids, triterpenes	Flores *et al.* (2015)
P. guajava L. (various pink- and white-pulped cultivars)	HPLC–UV/PDA	Gallic acid Chlorogenic acid Ellagic acid Catequin Rutin	Antioxidants	Santos *et al.* (2017)
P. guajava L. ('Regional Roja')	MS-EI	3-Sulfanyl-1-hexanol 3-Sulfanylhexyl acetate Hexanal Ethyl butanoate Acetaldehyde *Trans*-4,5-epoxy-(*E*)-2-decenal 4-Hydroxy-2,5-dimethyl-3(2*H*)-furanone Cinnamyl alcohol Methyl-(2*S*,3*S*)-2-hydroxy-3-methylpentanoate Cinnamyl acetate Methional 3-Hydroxy-4,5-dimethyl-2(5*H*)-furanone	Aroma	Steinhaus *et al.* (2009)

P. guajava L. (27 cultivars)	GC–MS	Hexanal (*E*)-Caryophyllene (*E*)-2-Hexenal (*Z*)-3-Hexenyl acetate Ethyl butanoate Ethyl octanoate Limonene (*E*)-Cadina-1,4-diene β-Ionone	Fruit aroma	Moon *et al.* (2018)

[a]HPLC, high-performance liquid chromatography; PDA, photodiode array; HR, high resolution; ESI, electrospray ionization; MS, mass spectrometry; EI, electron-impact ionization; UV, ultraviolet; GC, gas chromatography.

pair combinations was for rapid detection of *F. oxysporum* f.sp. *psidii* with a sensitivity level of 10^{-6} of PCR product to detect the pathogen at low titre (Muthukumar *et al.*, 2015).

5.11 Molecular Cloning and Gene Tagging for Specific Traits

Characterization and tagging genes associated with specific genes are conventionally done using cDNA cloning, transposon tagging and positional cloning approaches. Unfortunately, these approaches have not been much attempted in guava. HPL coding for 13-hydroperoxide lyase is an enzyme that has been patented in Germany (Patent No. DE69938343T2) and has been implicated in conferring frost resistance in guava. This is the first enzyme purified and characterized from guava fruits. The gene encoding this HPL enzyme was cloned and characterized to develop full-length cDNA using random amplified cDNA ends (RACE), via both the 3′-RACE and 5′-RACE approach (Tijet *et al.*, 2000). Sequence homology of HPL protein identified 60–70% identity to the known 13-hydroperoxide lyases orthologues. Since this enzyme shared homology with cytochrome P450, it was considered a cytochrome P45074B subfamily protein which was designated as CYP74B5. An alternative approach for tagging genes is characterization of transcript-derived fragments through cDNA-AFLP technique, which is also considered a technique of gene expression analysis. This technique was used for identification and isolation of genes associated with water stress and mechanical injury in guava (Furlan *et al.*, 2011). Four cDNA clones of guava encoding for a polygalacturonase (PG), an acid 1-aminocyclopropane-1-carboxylate (ACC) oxidase (ACCo) and two α-expansins (α-EXP) were characterized using real time-PCR (RT-PCR) and protein structural predictions (Morales *et al.*, 2013). Using RT-PCR, a partial cDNA fragment of 301 bp (*PgPG1*) from mature fruit, 320 bp (*PgACO1*) in overripe fruit and two fragments for α-expansins a 466 bp (*PgEXP2*) of overripe fruit and a 362 bp (*PgEXP3*) of peduncle were isolated and characterized. The amino-acid sequence of *PgPG1* indicated conserved regions of PGs in higher plants and was related to fruit ripening; while *PgACO1* was related to fruit maturation; *PgEXP2* and *PgEXP3* had two domains present in expansins; and they are phylogenetically grouped with α-expansins. Dot–blot analysis detected *PgPG1* expression in all stages of fruit ripening, with higher intensity at maturity. *PgACO1* expression was recorded in all five stages of fruit ripening and was highest during the transition stage. *PgEXP2* gene expression was detected in all tissues, with an increase from the green stage 2 to overripe stage 1; for *PgEXP3* the expression was visible in four stages of fruit ripening and at peduncle, with highest intensity at the mature stage. Four cDNA clones of ACC synthase (*PgACS1* and *PgACS2*) and ACC oxidase (*PgACO1* and *PgACO2*) involved in the ethylene biosynthetic pathway were isolated from guava pericarp and characterized for their expression patterns in the fruits of cultivar ‘Li-Tzy Bar’ (Chen *et al.*, 2007). Expressions of *PgACS1*, *PgACO1* and *PgACO2* increased with the advancement of fruit ripening and thus are responsible for the non-climacteric manner.

5.12 Genetic Transformation

Recombinant DNA technology can facilitate precision breeding in guava particularly for developing wilt-resistant scion varieties having enhanced shelf-life. Chandra and Mishra (2007) highlighted the significance of biotechnological interventions for improvement of guava. A robust *in vitro* regeneration system is prerequisite for developing a gene delivery system. Many workers have reported *in vitro* regeneration through enhanced axillary branching (Jaiswal and Amin, 1987; Amin and Jaiswal, 1988; Mishra *et al.*, 2007) or somatic embryogenesis (Chandra *et al.*, 2004, 2007). Mishra *et al.* (2014) reported a genetic transformation system in guava using *in vitro*-grown shoot-tip explant co-cultivated with *Agrobacterium tumefaciens* strain LBA4404 harbouring binary vector pIIHR-JBMch with *endochitinase* and *nptII* genes. The highest transformation efficiency was achieved by wounding explants with tungsten particles (0.5 µm) through a particle acceleration system, followed by infection for 45 min with *A. tumefaciens* grown overnight with 100 µM acetosyringone, corresponding to OD_{600} = 0.5, followed by co-cultivation for 72 h under dark condition on co-cultivation medium (MS (Murashige and Skoog) + acetosyringone 100 µM + L-cysteine 100 mg/l). Putative transformed explants' regenerated shoots on selection medium were stressed with kanamycin 200 mg/l for 12 weeks. Molecular analysis of putative transformants by PCR confirmed the integration of *endochitinase* and *nptII* genes in the plant nuclear genome. The histopathological analyses indicated the presence of mycelium in vascular bundles. However, notably, none of the plants showed symptoms of wilt disease during the investigation. *In vitro* pathogen inhibition assay and subsequently spore germination assay signified that the crude leaf extract of transformed plants inhibited the germination of fungal conidia. The leaf tissue studied for expression of *endochitinase* revealed that two individual transformation events manifested very high activity of *N*-acetyl-D-glucosamine (0.741 and 0.738 µM min^{-1} per µg protein, respectively) which clearly indicated that transgenic lines could not develop any symptoms of wilt disease due to overexpression of *endochitinase* (Mishra *et al.*, 2016).

5.13 Conclusion and Future Directions

Advancement in molecular breeding of guava has lagged behind owing to limited genomic resources in the public domain such as annotated whole-genome sequence information. With advancements in next-generation sequencing technologies, attempts are being made worldwide for augmenting the genomic resources with SSR and SNP markers that would enable QTL detection, validation and utilization in guava improvement programmes. The concern for management of guava wilt disease prompts the need for development of rapid RT-PCR based detection systems for early identification of the pathogen. Guava wild relatives, which are gene pools for the stress resistance, need to be explored for the *R* genes. Efforts made in characterization of the *Psidium* species and the interspecific hybrids using transcriptomics and proteomics approaches would help in better understanding of the mechanism and facilitate identification of candidate genes/QTLs along with their linked markers for both molecular breeding and genetic transformation systems using advanced genome editing tools such as clustered regularly interspaced short palindromic repeats (CRISPR)/CRISPR-associated (Cas) systems.

References

Abouzaid, E., El-Sayed, E.S.N., Mohamed, E.S.A. and Youssef, M. (2016) Molecular analysis of drought tolerance in guava based on *in vitro* PEG evaluation. *Tropical Plant Biology* 9(2), 73–81.

Ahmed, B., Mannan, M. and Hossain, S. (2011) Molecular characterization of guava (*Psidium guajava* L.) germplasm by RAPD analysis. *International Journal of Natural Sciences* 1(3), 62–67.

Alam, F., Islam, K.D. and Rahman, S.M. (2018) Variability among selective guava (*Psidium guajava* L.) varieties revealed by morphology and RAPD marker. *Jahangirnagar University Journal of Biological Sciences* 7(2), 89–98.

Amin, M.N. and Jaiswal, V.S. (1988) Micropropagation as an aid to rapid cloning of a guava cultivar. *Scientia Horticulturae* 36, 89–95.

Aranguren, Y., Briceño, A. and Fermin, G. (2010) Assessment of the variability of Venezuelan guava landraces by microsatellites. *Acta Horticulturae* 849, 147–154.

Bajpai, A., Chandra, R., Misra, M. and Tiwari, R. (2007) Regenerating *Psidium* spp. for screening wilt resistant rootstock under *in vitro* conditions. *Acta Horticulturae* 735, 145–153.

Bajpai, A., Chandra, R., Rajan, S. and Srivastava, N. (2008) RAPD and minisatellite markers for genetic diversity and relationship in guava varieties. *Indian Journal of Genetics and Plant Breeding* 68(4), 441–445.

Briceño, A., Aranguren, Y. and Fermin, G. (2010) Assessment of guava-derived SSR markers for the molecular characterization of Myrtaceae from different ecosystems in Venezuela. *Acta Horticulturae* 849, 139–146.

Chaithanya, M.V.N., Dinesh, M.R., Vasugi, C., Reddy, D.C.L., Sailaja, D. and Aswath, C. (2014) Assessment of genetic diversity in guava (*Psidium guajava*) germplasm using microsatellites. *Journal of Horticultural Sciences* 9(2), 117–125.

Chandra, R. and Mishra, M. (2007) Biotechnological interventions in guava (*Psidium guajava* L.). *Acta Horticulturae* 735, 117–126.

Chandra, R., Bajpai, A., Gupta, S. and Tiwari, R.K. (2004) Embryogenesis and plant regeneration from mesocarp of *Psidium guajava* L. (guava). *Indian Journal of Biotechnology* 3, 246–248.

Chandra, R., Mishra, M., Abida, M. and Singh, D.B. (2007) Triazole mediated somatic embryogenesis in guava (*Psidium guajava* L.). *Acta Horticulturae* 735, 133–138.

Chen, T.W., Ng, C.C., Wang, C.Y. and Shyu, Y.T. (2007) Molecular identification and analysis of *Psidium guajava* L. from indigenous tribes of Taiwan. *Journal of Food and Drug Analysis* 15(1), 82–88.

Chiveu, J.C., Mueller, M., Krutovsky, K.V., Kehlenbeck, K., Pawelzik, E. and Naumann, M. (2019) Genetic diversity of common guava in Kenya: an underutilized naturalized fruit species. *Fruits* 74(5), 236–248.

Coser, S.M., da Silva Ferreira, M.F., Ferreira, A., Mitre, L.K., Carvalho, C.R. and Clarindo, W.R. (2012) Assessment of genetic diversity in *Psidium guajava* L. using different approaches. *Scientia Horticulturae* 148, 223–229.

Costa, I.R. and Forni-Martins, E.R. (2006a) Chromosome studies in species of *Eugenia, Myrciaria* and *Plinia* (Myrtaceae) from southeastern Brazil. *Australian Journal of Botany* 54, 409–415.

Costa, I.R. and Forni-Martins, E.R. (2006b) Chromosome studies in Brazilian species of *Campomanesia* Ruiz et Pávon and *Psidium* L. (Myrtaceae Juss.). *Caryologia* 59, 7–13.

Costa, I.R., Dornelas, M.C. and Forni-Martins, E.R. (2008) Nuclear genome size variation in fleshy-fruited Neotropical Myrtaceae. *Plant Systematics and Evolution* 276(3–4), 209–217.

Costa, S.R. and Santos, C.A.F. (2013) Allelic database and divergence among *Psidium* accessions by using microsatellite markers. *Genetics and Molecular Research* 12(4), 6802–6812.

Costa, S.R. and Santos, C.A.F. (2017) Genetic divergence among *Psidium* accessions based on single nucleotide polymorphisms developed for *Eucalyptus. Genetics and Molecular Research.* 16(2), 1–9.

Dahiya, K.K., Archak, S. and Karihaloo, J.L. (2002) DNA fingerprinting of guava (*Psidium guajava* L.) cultivars using RAPD markers. *Indian Journal of Plant Genetic Resources* 15, 112–115.

Feria-Romero, I.A., Astudillo-de la Vega, H., Chavez-Soto, M.A., Rivera-Arce, E., Lopez, M. *et al.* (2009) RAPD markers associated with quercetin accumulation in *Psidium guajava. Biologia Plantarum* 53, 125–128.

Flores, G., Wu, S.B., Negrin, A. and Kennelly, E.J. (2015) Chemical composition and antioxidant activity of seven cultivars of guava (*Psidium guajava*) fruits. *Food Chemistry* 170, 327–335.

Freitas, V.M., Correa, V.R., Motta, F.C., Sousa, M.G., Gomes, A.C.M.M. *et al.* (2013) Resistant accessions of wild *Psidium* spp. to *Meloidogyne enterolobii* and histological characterization of resistance. *Plant Pathology* 63(4), 738–746.

Furlan, C., Zanotta, S. and Salatino, A. (2011) cDNA-AFLP analysis of *Psidium guajava* L. cultivars under water stress and mechanical injury: methodological implications. *Brazilian Journal of Plant Physiology* 24, 29–36.

Grattapaglia, D., René, E., Vaillancourt, R.E., Shepherd, M., Thumma, B.R. *et al.* (2012) Progress in Myrtaceae genetics and genomics: *Eucalyptus* as the pivotal genus. *Tree Genetics & Genomes* 8, 463–508.

Hernández-Delgado, S., Padilla-Ramirez, J.S., Nava Cedillo, A. and Mayek-Perez, N. (2007) Morphological and genetic diversity of Mexican guava germplasm. *Plant Genetic Resources* 5, 131–141.

Jaiswal, V.S. and Amin, M.N. (1987) *In vitro* propagation of guava from shoot culture of mature trees. *Journal of Plant Physiology* 130, 7–12.

Jindal, M. (2017) Linkage map construction in coloured guava using SSR markers. In: *Abstracts, 9th Global Summit on Agriculture and Horticulture, Beijing, 10–11 August 2017.*

Jo, S., Kim, H.-W., Kim, Y.-K., Cheon, S.-H. and Kim, K.-J. (2016) Complete plastome sequence of *Psidium guajava* L. (Myrtaceae). *Mitochondrial DNA Part B* 1(1), 612–614.

Kamle, M., Kalim, S., Bajpai, A., Chandra, R. and Kumar, R. (2012) *In vitro* selection for wilt resistance in guava (*Psidium guajava* L.) cv. Allahabad Safeda. *Biotechnology* 11, 163–171.

Kanupriya, Latha, P.M., Aswath, C., Reddy, L., Padmakar, B. *et al.* (2011) Cultivar identification and genetic fingerprinting of guava (*Psidium guajava*) using microsatellite markers. *International Journal of Fruit Science* 11(2), 184–196.

Kherwar, D., Usha, K., Mithra, S.V.A. and Singh, B. (2018) Microsatellite (SSR) marker assisted assessment of population structure and genetic diversity for morpho-physiological traits in guava (*Psidium guajava* L.). *Journal of Plant Biochemistry and Biotechnology* 27(3), 284–292.

Kumari, S., Arumugam, N., Singh, R., Srivastav, M., Banoth, S. *et al.* (2017) Diversity analysis of guava (*Psidium guajava*) germplasm collection. *Indian Journal of Agricultural Sciences* 88(3), 489–497.

Lepitre, V., Nansot, G., Grangeon, R., Pomies, V., Rivallan, R. *et al.* (2010) The microsatellite (SSR)/AFLP reference linkage map of guava. *Acta Horticulturae* 849, 183–192.

Liu, X. and Yang, G. (2012) Assessment of clonal fidelity of micro-propagated guava ('*Psidium guajava*') plants by ISSR markers. *Australian Journal of Crop Science* 6(2), 291–295.

Ma, Z., Liu, S., Liang, Z., Xu, S. and Hu, W. (2020) Analysis of genetic diversity of 45 guava germplasm evaluated using SSR markers. *International Journal of Fruit Science* 20(3), 385–393.

Majumdar, P.K. and Mukherjee, S.K. (1972) Aneuploidy in guava (*Psidium guajava* L.) I. Mechanism of variation in chromosome number. *Cytologia* 37, 541–548.

Mani, A., Mishra, R. and Thomas, G. (2011) Elucidation of diversity among *Psidium* species using morphological and SPAR methods. *Journal of Phytology* 3(8), 53–61.

Marques, A., Tuler, A.C., Carvalho, C.R., Carrijo, T., Ferreira, M.F. and Clarindo, W.R. (2016) Refinement of the karyological aspects of *Psidium guineense* (Swartz, 1788): a comparison with *Psidium guajava* (Linnaeus, 1753). *Comparative Cytogenetics* 10(1), 117–128. https://doi.org/10.3897/CompCytogen.v10i1.6462

Mehmood, A., Luo, S., Ahmad, N.M., Dong, C., Mahmood, T. *et al.* (2015) Molecular variability and phylogenetic relationships of guava (*Psidium guajava* L.) cultivars using inter-primer binding site (iPBS) and microsatellite (SSR) markers. *Genetic Resources and Crop Evolution* 63(8), 1345–1361.

Mishra, M., Chandra, R., Pati, R. and Bajpai, A. (2007) Micropropagation of guava (*Psidium guajava* L.). *Acta Horticulturae* 735, 155–160.

Mishra, M., Jalil, U., Sharma, N. and Hudedamani, U. (2014) An *Agrobacterium* mediated transformation system of guava (*Psidium guajava* L.) with *endochitinase* gene. *Crop Breeding and Applied Biotechnology* 14, 232–237.

Mishra, M., Jalil, S.U., Mishra, R.K., Kumari, S. and Pandey, B.K. (2016) *In vitro* screening of guava transformed with *endochitinase* gene against *Fusarium oxysporum psidii*. *Czech Journal of Plant Breeding* 52(1), 6–12.

Mishra, R.K., Pandey, B.K., Muthukumar, M., Pathak, N. and Zeeshan, M. (2013) Detection of *Fusarium* wilt pathogens of *Psidium guajava* L. in soil using culture independent PCR (ciPCR). *Saudi Journal of Biological Sciences* 20, 51–56.

Mittal, A., Yadav, I.S., Arora, N.K., Boora, R.B., Mittal, M. *et al.* (2020) RNA-sequencing based gene expression, landscape of guava *cv.* Allahabad Safeda and comparative analysis to colored cultivars. *BMC Genomics* 21, 484.

Moon, P., Fu, Y., Bai, J., Plotto, A., Crane, J. and Chambers, A. (2018) Assessment of fruit aroma for twenty-seven guava (*Psidium guajava*) accessions through three fruit developmental stages. *Scientia Horticulturae* 238, 375–383.

Morales, F., Reyes Silva, A., Palenius, H., Hernández-Guzmán, G., Alpuche-Solís, A. and Garcidueñas-Piña, C. (2013) Ripening-related cDNAs in guava fruit (*Psidium guajava* L.). characterization and expression analysis. *Revista Fitotecnia Mexicana* 36, 117–125.

Morton, J. (1987) Guava. In: Morton, J.F. (ed.) *Fruits of Warm Climates*. J.F. Morton, Miami, Florida, pp. 356–363.

Muthukumar, M., Pandey, B.K., Mishra, R.K. and Mishra, A.K. (2012) Molecular diagnostics for detection of fungal and viral pathogens in subtropical fruits. In: *Souvenir & Abstracts of National Conference on Managing Threatening Diseases of Horticultural, Medicinal, Aromatic and Field Crops in Relation to Changing Climatic Situation, IISR, Lucknow, India, 3–5 November 2012*, p. 167.

Muthukumar, M., Pandey, B.K., Chithrameenal, K. and Praghadeesh, M. (2015) Molecular diagnostics for detection of pathogens of subtropical fruits. In: *AGSC 2015 on Impact of Climate Risks on Agricultural and Horticultural Productivity, Tamil Nadu Agricultural University, Coimbatore, India, 13–14 May 2015*, pp. 49–50.

Muthukumar, M., Kumar, S., Veena, G.L., Hudedamani, U., Bajpai, A. and Rajan, S. (2019) Microsatellites presented high level of species level variation in *Psidium* species for utilization in rootstock breeding. In: *Abstracts, Progressive Horticulture Conclave–2019 on Futuristic Technologies in Horticulture, ISHRD, Uttarakhand, held at ICAR-IISR, Lucknow, India, 8–10 December 2019*, p. 98.

Naga Chaithanya, M.V., Ramesh, S., Dinesh, M.R., Sailaja, D. and Aswath, C. (2016) Developing mapping populations for identifying genomic regions controlling resistance to bark-eating caterpillar (*Indarbela tetraonis*) in guava. *Journal of Crop Improvement* 30(4), 371–377.

Naseer, S., Hussain, S., Naeem, N., Pervaiz, M. and Rahman, M. (2018) The phytochemistry and medicinal value of *Psidium guajava* (guava). *Clinical Phytoscience* 4, 32.

Negi, S.S. and Rajan, S. (2007) Improvement of guava through breeding. *Acta Horticulturae* 735, 31–37.

Nimisha, S., Kherwar, D., Ajay, K.M., Singh, B. and Usha, K. (2013) Molecular breeding to improve guava (*Psidium guajava* L.): current status and future prospective. *Scientia Horticulturae* 164, 578–588.

Noia, L.R., Tuler, A.C., Ferreira, A. and Ferreira, M.F.S. (2017) Relationship between *Psidium* species (Myrtaceae) by resistance gene analog markers: focus on nematode resistance. *Genetics and Molecular Research* 16(1), 1–13.

Nwinyi, O.C., Chinedu, N.S. and Ajani, O.O. (2008) Evaluation of antibacterial activity of *Psidium guajava* and *Gongronema latifolium*. *Journal of Medicinal Plants Research* 2(8), 189–192.

Padmakar, B., Kanupriya, Madhavi, L.P., Prashanta, K.S., Dinesh, M.R. *et al.* (2015) Development of SRAP and SSR marker-based genetic linkage maps of guava (*Psidium guajava* L.). *Scientia Horticulturae* 192, 158–165.

Padmakar, B., Kanupriya, C., Latha, P.M., Vasugi, C., Dinesh, M.R. *et al.* (2016) Enrichment of genetic linkage maps and mapping QTLs specific to seed strength – hardness/softness – in guava (*Psidium guajava* L.). *Journal of Horticultural Sciences* 11(1), 13–20.

Pandey, B.K., Mishra, R.K., Pandey, A., Kamle, M., Sareen, P. and Muthukumar, M. (2010) Molecular characterization of *Fusarium oxysporum* f.sp. *psidii*: a causal organism of wilt in guava. In: *Souvenir cum Abstracts of National Symposium on Molecular Approaches for Management of Fungal Diseases of Crop Plants, IIHR, Bangalore, India*, pp. 83–84 (Annexure-II-F).

Pelea, L.P., González, A.S., Fernández, E.B., Rodríguez Medina, N.N., Valdés-Infante Herrero, J. and Pommer, C.V. (2012) Heritability estimates of guava (*Psidium guajava* L.) agricultural important characters evaluated in three populations. *Acta Horticulturae* 959, 117–123.

Pessanha, P.G.d.O., Viana, A.P., Amaral Júnior, A.T.d., Souza, R.M.d., Teixeira, M.C. and Pereira, M.G. (2011) Avaliação da diversidade genética em acessos de *Psidum* spp. via marcadores RAPD. *Revista Brasileira de Fruticultura* 33(1), 129–136.

Pommer, C.V. and Murakami, K.R.N. (2009) Breeding guava (*Psidium guajava* L.). In: Jain, S.M and Priyadarshan, S. (eds) *Breeding Plantation Tree Crops*. Springer, New York, pp. 83–120.

Prakash, D.P., Narayanswami, P. and Sondur, S.N. (2002) Analysis of molecular diversity in guava using RAPD markers. *The Journal of Horticultural Science and Biotechnology* 77, 287–293.

Rajan, S., Yadava, L.P., Kumar, R. and Saxena, S.K. (2005) Selection possibilities for seed content – a determinant of fresh fruit quality in guava (*Psidium guajava* L.). *Journal of Applied Horticulture* 7, 52–54.

Rawls, B., Harris-Shultz, K.R., Dhekney, S. and Sitther, V. (2015) Analyzing clonal fidelity of micropropagated *Psidium guajava* L. plants using simple sequence repeat markers. *American Journal of Plant Sciences* 6, 2385–2392.

Reatini, B., Torres, M.d.L., Valdebenito, H. and Vision, T. (2018) Complete plastome sequences of two *Psidium* species from the Galápagos Islands. *F1000Research* 7, 1361.

Risterucci, A.M., Duval, M.F., Rohde, W. and Billotte, N. (2005) Isolation and characterization of microsatellite loci from *Psidium guajava* L. *Molecular Ecology Notes* 5(4), 745–748.

Risterucci, A.M., Nansot, G., Grangeon, R., Lepitre, V., de Reeper, A. *et al.* (2010) Development of guava microsatellite (SSR) markers using the SAT software. *Acta Horticulturae* 849, 113–119.

Ritter, E. (2012) Guava biotechnologies, genomic achievement and future needs. *Acta Horticulturae* 959, 131–140.

Ritter, E., Herran, A., Valdés-Infante, J., Rodríguez-Medina, N.N., Briceño, A. *et al.* (2010) Comparative linkage mapping in three guava mapping populations and construction of an integrated reference map in guava. *Acta Horticulturae* 849, 175–182.

Rodríguez, N., Valdés-Infante, J., Becker, D., Velázquez, B., González, G. *et al.* (2007) Characterization of guava accessions by SSR markers, extension of the molecular linkage map, and mapping of QTLs for vegetative and reproductive characters. *Acta Horticulturae* 735, 201–215.

Sánchez-Teyer, L.F., Barraza-Morales, A., Quiroz-Moreno, A., Ortiz-García, M.M., Becerril-Chi, K. *et al.* (2010) Genetic diversity of Mexican guava germplasm evaluated using AFLP and SSR markers. *Acta Horticulturae* 859, 255–260.

Santos, W.N.L.d., Sauthier M.C.d.S., Santos, A.M.P.d., Santana, D.d.A., Azevedo, R.S.A. and Caldas, J.d.C. (2017) Simultaneous determination of 13 phenolic bioactive compounds in guava (*Psidium guajava* L.) by HPLC-PAD with evaluation using PCA and neural network analysis (NNA). *Microchemical Journal* 133, 583–592.

Saxena, S.K., Rajan, S., Chandra, R., Srivastava, N. and Bajpai, A. (2007) Molecular characterization of closely related open pollinated seedling selections in guava. *Acta Horticulturae* 735, 49–55.

Sharma, A.S., Sehrawat, S.K., Singhrot, R.S. and Boora, K.S. (2005) Assessment of genetic diversity and relationship among *Psidium* spp. through RAPD analysis. *Acta Horticulturae* 735, 71–78.

Shiva, B., Nagaraja, A., Rakesh, S. and Manish, S. (2017) Genetic diversity of guava genotypes evaluated using RAPD molecular marker. *International Journal of Genetics* 9(5), 271–274.

Singh, H. (2017) Generation of linkage map and development of new hybrids for improved fruit colour in guava (*Psidium guajava* L.). PhD thesis, Punjab Agricultural University, Ludhiana, India.

Sitther, V., Zhang, D., Harris, D.L., Yadav, A.K., Zee, F.T. *et al.* (2014) Genetic characterization of guava (*Psidium guajava* L.) germplasm in the United States using microsatellite markers. *Genetic Resources and Crop Evolution* 61(4), 829–839.

Sousa, A.D., Pedrosa, E.M., Silva, C.U.C., Castro, J.M.C. and Riberiro, J.M. (2017) Penetration, development, and reproduction of *Meloidogyne enterolobii* on *Psidium* species and induced cellular responses in the roots. *Revista Brasileira de Fruticultura, Jaboticabal* 39(2), 453.

Souza, D.G., Resende, L.V.A., De Lima, I.P., Martins, L.S.S. and Techio, V.H. (2014) Chromosome number and nuclear DNA amount in *Psidium* spp. resistant and susceptible to *Meloidogyne enterolobii* and its relation with compatibility between rootstocks and commercial varieties of guava tree. *Plant Systematics and Evolution* 301(1), 231–237.

Steinhaus, M., Sinuco, D., Polster, J., Osorio, C. and Schieberle, P. (2009) Characterization of the key aroma compounds in pink guava (*Psidium guajava* L.) by means of aroma re-engineering experiments and omission tests. *Journal of Agricultural and Food Chemistry* 57(7), 2882–2888.

Thaipong, K., Promchot, S., Auvuchanon, A. and Boonprakob, U. (2017) Genetic analysis of guava germplasm using AFLP markers. *International Journal of Agricultural Technology* 13, 741–752.

Tijet, N., Waspi, U., Gaskin, D.J.H., Hunziker, P., Muller, B.L. *et al.* (2000) Purification, molecular cloning, and expression of the gene encoding fatty acid 13-hydroperoxide lyase from guava fruit (*Psidium guajava* L.). *Lipids* 35, 709–720.

Tuler, A.C., Carrijo, T.T., Peixoto, A.L., Garbin, M.L., Ferreira, M.F.d.S. *et al.* (2019) Diversification and geographical distribution of *Psidium* (Myrtaceae) species with distinct ploidy levels. *Trees* 33(4), 1101–1110.

Urquía, D., Gutierrez, B., Pozo, G., Pozo, M.J., Espín, A. and Torres, M.d.L. (2019) *Psidium guajava* in the Galapagos Islands: population genetics and history of an invasive species. *PLoS One* 14(3), e0203737.

Valdés-Infante, J., Sourd, D., Rodriguez, J., Becker, D., Rohde, W. and Ritter, E. (2003) Molecular characterization of Cuban accessions of guava (*Psidium guajava* L.), establishment of a first molecular linkage map and mapping of QTLs for vegetative characters. *Journal of Genetics and Breeding* 57, 349–357.

Valdés-Infante, J., Rodríguez, N.N., Velásquez, B., Rivero, D., Martínez, F. *et al.* (2010) Simple sequence repeats (SSRs) for diversity characterization of guava (*Psidium guajava* L.). *Acta Horticulturae* 849, 155–162.

Valera-Montero, L.L., Hernández-Dávila, A., Silos-Espino, H. and Flores-Benítez, S. (2017) Variación genética en guayaba mediante RAPDs y descriptores morfológicos en Calvillo, Aguascalientes. *Revista Mexicana de Ciencias Agrícolas* 8(1), 67. https://doi.org/10.29312/remexca.v8i1.72

Viji, G., Harris, D.L., Yadav, A.K. and Zee, F.T. (2010) Use of microsatellite markers to characterize genetic diversity of selected accessions of guava (*Psidium guajava*) in the United States. *Acta Horticulturae* 859, 169–176.

Youssef, M., Ibrahim, R.A. and Amein, K.A. (2015) Comparison of phenotypic and molecular assessment of genetic diversity in guava. *Acta Horticulturae* 1100, 115–120.

6 Cultivars and Plant Improvement

Sisir Mitra
Former Dean, Bidhan Chandra Krishi Viswavidyalaya, Mohanpur, West Bengal, India

6.1 Introduction

Guava is grown in more than 70 countries of the world. The major producers are India, Pakistan, Brazil, Mexico, Indonesia, Thailand, Malaysia, Taiwan and Cuba. The major problems facing the guava industry in different parts of the world are severity of wilt disease, root nematodes, high seed content of diploid commercial cultivars and poor yield with small, irregular and misshapen fruits of seedless cultivars. In the last two decades, guava improvement work carried out in different countries has resulted in the release of several superior selections and hybrids. Interspecific hybridization with wild *Psidium* species has yielded hybrids which are resistant or tolerant to wilt and root nematode and are grafting compatible.

6.2 *Psidium* Species Used in Breeding

The genus *Psidium* belongs to the family *Myrtaceae* and has a basic chromosome number $x = 11$. Most of the cultivated guava belongs to *Psidium guajava* L. The wild species of *Psidium* which have considerable importance in breeding programmes (Mitra, 2010; Mitra *et al.*, 2012) are the following.

6.2.1 *Psidium cattleyanum* Sabine

Common name: strawberry guava; other names: cattley guava, Chinese guava, purple guava, yellow strawberry guava, red strawberry guava, guayaba (Martin *et al.*, 1987).

Strawberry guava is an evergreen tree/shrub native to Brazil. Trees can grow from 9 to 12 m tall and have reddish, flaky bark. Leaves are opposite, oblong, glossy, entire and up to 7.5 cm long. Flowering occurs year-round. Flowers are white, 2.5 cm wide, with four or five petals and many stamens. Flowers can occur singly, or in groups of three, in the axils of the leaves. Fruits are dark red, edible, 2–4 cm in diameter, globose to obovoid with thin peel, yellow, red or purple and tipped with a protruding five-lobed calyx. The aromatic, white pulp is sweet to subacid, surrounding numerous, small, hard seeds.

The tree is grown as an ornamental plant or as a hedge. Fruits are eaten fresh or can be processed into various products such as jellies, preserves, desserts, drinks, sherbets, purée, jams, butter and paste. It can be

*E-mail: sisirm55@gmail.com

 Guava: Botany, Production and Uses (ed. S. Mitra)
DOI: 10.1079/9781789247022.0006

grown in low-lying degraded areas to increase the farming potential and generate additional income (Normand, 2002).

6.2.2 *Psidium guineense*

Common name: Brazilian guava; other names: guayabillo, guayaba agria; origin: southern Mexico, Argentina.

Small shrub or tree (1–9 m), hardy, successfully grown in a subtropical climate. Round or pear-shaped fruit, rind is yellow, enclosing a white acidic pulp with a guava–strawberry flavour. The fruit is a big berry (45–138 g), yellow-coloured when mature, round to ellipsoid-shaped of 6–8 cm diameter; it has a delicate pleasant flavour, and a peel like fine leather which has to be taken off since it is bitter (Manica, 2000).

Fruits are eaten fresh or mostly used to make jellies and preserves. Brazilian guava is known for a predominance of terpene compounds and volatile compounds of aroma and flavour (Franco and Shibamoto, 2000) and as a source of vitamin C, with a content of approximately 400 mg 100 g^{-1} fresh pulp (Andrade *et al.*, 2002).

6.2.3 *Psidium acutangulum* D.C.

The para guava or araca-pera, as it is commonly known as, originated in South America.

A shrub or small tree (7–15 m), more hardy than tropical guava, medium-sized yellow fruit with very tasty translucent white–yellow pulp similar to the guava but with a much more acidic flavour.

Fruit is eaten raw or used to flavour drink. It is combined with sugar or honey to make a lemonade-like drink. Peel has a high antioxidant content.

6.2.4 *Psidium friedrichsthalianum* Niedenzu

The species is believed to have originated somewhere between southern Mexico to North America, commonly known as cas guava or Costa Rican guava. Small tree (10–15 m), square branchlets, glossy leaves above and pubescent below, not as hardy as tropical guava, sulfur-yellow fruit, 2.5–3 cm in diameter and acidic. Flowers are near white or white.

Fruits are processed into jams, jellies, preserves, or used as flavouring agents for drinks. Contains 83.2 g water, 0.82 g protein, 0.44 g fat, 6.20 g carbohydrate and 22–50 mg vitamin C per 100 g fruit.

6.2.5 *Psidium angulatum*

Small tree, leaves are alternate and oblong. Fruit has a leathery peel and persistent calyx. Fruits are eaten raw, or used to flavour drink, or processed into ice cream, sorbet, gelatins and candies. Peel contains compounds with high antioxidant activity. Vitamin C content 389.3 mg 100 g^{-1} (Mitra *et al.*, 2012).

6.2.6 *Psidium littorale*

The species originated in Brazil, commonly known as lemon guava. Small bush or tree (6–9 m), frilly white flowers, fruits yellow, similar to the strawberry guava except fruits are often slightly larger (2.5–8 cm), flesh yellow, fragrant, lemon–guava like flavour. Fruits are eaten fresh, or used to flavour beverages, ice creams and desserts.

6.3 Breeding Objectives

The objectives of breeding programmes existing in different countries have been listed by researchers (Nakasone and Paull, 1998; Pereira and Nactigal, 2002; Ray, 2002; Negi and Rajan, 2007; Pommer and Murakami, 2009; Hsieh, 2011; Pommer, 2012; Nimisha *et al.*, 2013; Dinesh and Vasugi, 2015) and are as follows:

1. Dwarf plant habit suitable for high-density planting;
2. Large fruit size (200–340 g) with few soft seeds and thick pulp;
3. Ratio of pulp to total weight >70% and pericarp thickness >100 mm;

4. Absence of stone cells;
5. Green-yellow or yellow peel when ripe;
6. Productivity with a minimum yield of 30 t ha^{-1};
7. White pulp for desserts or dark pink pulp, particularly for processing;
8. More than 10% soluble solids content of fruit;
9. For processing, an acidity of 1.25–1.50% and for dessert guava, 0.2–0.6%;
10. Vitamin C content >300 mg 100 g^{-1} pulp;
11. Long shelf-life;
12. Resistance to fruit diseases and insects;
13. Resistance to *Fusarium* and *Myxosporium* wilt;
14. Resistant or tolerance to rust (*Puccinia psidii* Went.);
15. Resistance to nematodes (*Meloidogyne* spp.).

6.4 Introduction and Selection

Guava is predominantly a self-pollinated crop (80%), but cross-pollination does occur. This results in a heterozygous, open-pollinated seedling population, with adequate genetic variation for selection of desirable commercial types (Nakasone and Paull, 1998). The fruit from seedling guavas grown from unselected seeds may exhibit a wide variation in appearance, size, flavour, acidity, texture and colour. The shape may be round to oblong, ovate, globose or pyriform. The pulp colour may be white, yellow, pink, salmon or carmine (Ruehle, 1946). Selection can also be made from wild populations (Mitra and Sanyal, 2004). Several promising selections have been made in different guava-growing countries (Ruehle, 1946; Nakasone and Ito, 1978; Negi and Rajan, 2007; Padilla-Ramirez *et al.*, 2007; Milan, 2010; Dinesh and Vasugi, 2015; Thaipong *et al.*, 2017).

George D. Ruehle (1946) from the Florida Subtropical Experiment Station, USA selected 'Supreme' (in 1940) and 'Ruby' (in 1944) from open-pollinated seedlings. 'Supreme' was selected from open-pollinated seedling of 'PI 81849'. Another selection, 'Redland', was made by S.J. Lynch and H.S. Wolfe in 1938 from Florida, the seeds of which were obtained from the Atkins Institute, Arnold Arboretum, Cuba (Brooks and Olmo, 1952). 'Ka Hua Kula' (Golden fruit) was selected from over 1200 seedlings derived by open pollination of 'Beaumont' and originally designated as 'HAESO97', being a vigorous plant (Nakasone and Ito, 1978).

Guava production and consumption in Mexico are based on traditional varieties 'Media China' and 'Peruana'. They were selected by farmers using the Mexican germplasm available in the semi-arid region of Mexico (Jacobo *et al.*, 2009). However, studies have shown great variability among and within the orchards (Perales, 1993; Padilla-Ramirez *et al.*, 2002). Padilla-Ramirez *et al.* (2007) evaluated the characteristics of a group of selections from the guava *ex situ* collections of the Instituto Nacional de Investigaciones Forestales, Agrílas y Pecuarias (INIFAP) and selected 'S-106', 'S-11', 'S-10' and 'S-12' which showed a fruit yield from two to four times more than that obtained in the region of study. Padilla-Ramirez and Gonzalez-Gaona (2010) collected guava germplasm at 169 different sites in 11 Mexican states. The variability observed in the collected guava germplasm was very high in fruit characteristics like fruit size and shape, skin and pulp colour, pulp thickness and seed number. Mexican researchers also investigated phenotypic and genetic diversity in guava (Martinez de Lara *et al.*, 2004) and found large differences among genotypes unfavourable for production quality. Research was continued to develop pink-pulped guavas to diversify commercial production in Central Mexico (Mondragon-Jacobo *et al.*, 2010) and a number of genotypes were selected: '23-1', '26-8', '23-5', '25-5', '21-17', '22-11' and '25-9'.

The major cultivars of Brazil are 'Paluma', 'Rica' and 'Pedro Sato'. 'Paluma' is a seedling selection of 'Ruby' × 'Supreme', 'Rica' is a seedling selection of 'Supreme' and 'Pedro Sato' was probably selected from 'Red Ogawa #1' by a grower in Rio de Janeiro. Both 'Paluma' and 'Rica' were selected by Professor Fernando M. Pereira at Universidade Estadual Paulista (UNESP), Jaboticabal, São Paulo state. The ecoregions prospected in Amazonas and Roraima states are located in the upper northern part of South America, where, according to Risterucci *et al.* (2005), guava is native. Some ecoregions in the Brazilian north-eastern states are located in the least developed regions of the country, where very ancient *Psidium* accessions have been kept and propagated by slave descents or long-established rural communities (Fernandes-Santos *et al.*, 2010). 'Araca' is a general term used in Brazil to refer to wild *Psidium* species including *P. cattleyanum*, *Psidium incanescens* Martins, *Psidium grandifolium* Martins, *Psidium arboretum* Vell and *Psidium humile* Vell that are native to South America (Raseira and Raseira, 1996). Fernandes-Santos *et al.* (2010) sampled and characterized 119 guava and 40 araca accessions from 35 different Brazilian ecoregions. The large majority of araca accessions have widely spaced leaf veins compared with guava accessions, that have medium to close space between veins. Most fruit of araca accessions were classified as small, while most fruits of guava accessions were grouped into the class of medium size. The pulp colour of araca is mostly (91%) cream and white, while 58% of guava accessions have pale pink, pink or dark pink pulp. Canizares (1981) evaluated more than 3000 open-pollinated seedlings and selected about 60 promising genotypes. Five of these selected genotypes had fruits weighing more than 400 g each. The genotypes selected from seminal origin orchard (named 'Cortibel I' to 'Cortibel XIII'; 'CI' to 'CXIII') were compared with 'Paluma', 'Pedro Sato' and 'Roxa' (Coser *et al.*, 2014). There was divergence between 'Cortibel' selections. 'CI' was found as the most divergent between red-pulped genotypes and among all others, showing the best performance for fruit characteristics. 'CVIII' and 'CIV' genotypes showed the best performance among light-pulp genotypes. 'Cortibel' selections were as good as commercial cultivars in fruit quality, showing good genotypes for use as new cultivars or for hybridizations in breeding programmes.

The first introduced cultivar of South Africa was 'Madeira' that was introduced in the early 1700s by Jan Van Riebeeck (the founder of South Africa). These trees are still in a garden in Paarl (a town in the Western Cape). Later different guavas were introduced into the Cape. 'Malherbe', 'Rousseau', 'Du Preez' and 'vanZyl' are very old cultivars (1830–1835) that occurred through natural hybridization and were selected by farmers. Some of these cultivars still exist in orchards of the Western Cape (Mitra and Thingreingam Irenaeus, 2018). In 1938, 'Fan Ratief' was introduced to the Northern Province and planted widely, and until 1981 it was the most grown cultivar (Schoeman *et al.*, 2011).

More than 90 guava accessions have been collected from different countries since 1998 by the Department of Horticulture at Kasetsart University, Khamphaeng Sean campus in Nakhon Pathom, Thailand. These guava accessions vary widely in their fruit qualities (Thaipong and Boonprakob, 2006), plant types (Thaipong *et al.*, 2017) and salt tolerance (Thaipong and Boorprakob, 2019). Some of these genotypes are 'Klom Salee', 'Koha Um-porn', 'Na Suan', 'Paen Seethong', 'Daeng Siam', 'Pijit 13-30', 'KUHP38' and 'KUHP12'.

In Cuba, guava collection was established in the Scientific Technological Unit for Alquizer, adscript to the Institute for Research on Tropical Fruits. After evaluating 395 accessions, characterization was made for 18 genotypes (Rodriguez *et al.*, 2010). They are 'Belic L-123', 'Belic L-207', 'Belic L-213', 'BG 76-18', 'BG 76-10', 'BG 76-79', 'B 76-23', 'Cotorrea', 'EEA 1-23', 'EEA 18-40', 'EEA 384', 'Hemero No. 1', 'Ibarra', 'Indonesia Blanca', 'Micro-guayaba', 'N6', Selection Seychelles' and 'Supreme Roja'.

Guava is a priority fruit of Malaysia. Guava breeding at the Malaysian Agriculture and Research Institute (MARDI) was started by introducing planting materials from Thailand. These were evaluated and two varieties, 'JP1' and 'JP2', were developed for industry. However, both selections were found susceptible to nematode (Norlia, 1994). In the Horticulture Research Centre, MARDI, 198 accessions were collected to screen for resistance to root-knot nematode. Three accessions ('K-10', 'A-06' and 'J-16') were reported as resistant to nematodes (Milan, 2010). Cultivars 'GU-4', 'GU-5', 'GU-7', 'Hongkong Pink', 'Jambubiji',

'Taiwan', 'Laknaw' (Lucknow?), 'Vietnamese' and 'Kampuchia' are also available in Malaysia. The fruits of 'GU-4' (412 ± 8.3 g), 'GU-5' (520 ± 13.9 g) and 'GU-7' (718 ± 8.8 g) are higher in weight. The pulp of 'Jambubiji' and 'GU-5' is red in colour (Yusof, 1990). There are currently 14 guava clones ('GU-1' to 'GU-15') registered with the Department of Agriculture in Malaysia (Mitra and Thingreingam Irenaeus, 2018).

Taiwan collected guava seeds from Vietnam, Hawaii, Panama, Indonesia and other countries, raised the seedlings and selected several superior genotypes. Some of the oldest selected cultivars are 'Zhongshan Yueba', 'Lizaiba', 'White', 'Thailand', 'Pearl', 'Crystal, 'Yilan Red', Heart', 'Baiba', 'Emperor Pull', 'BaodaoYueba' and 'Diamond'. The Taiwan technical mission introduced three cultivars ('Taiwan Guava', 'Century Guava' and 'Pearl Guava'). Taiwanese guava cultivars produce fruits weighing 400–700 g. The other guava cultivars with high ascorbic acid content (210–392 mg 100 g^{-1}) available in Taiwan are 'Jen-Ju', 'Diwan' and 'Rainbow' (Hwang *et al.*, 2017). 'Yilan Red' is a local, pink-pulped cultivar that has a good storage life. In central Taiwan, cultivars 'Jen-Ju Bar', 'Sai Ji Bar' and 'Shui-Jing Bar' are grown; however, these cultivars are susceptible to nematodes.

Guava is an important crop of Pakistan. Some of the Indian cultivars like 'Safeda', 'Allahabad', 'Lucknow-49', 'Red Fleshed', 'Seedless' and 'Apple Colour' are popular in Pakistan (Mitra and Thingreingam Irenaeus, 2018). Cultivars 'Sufaida', 'Surahi', 'Surkha', 'Waikea', 'Beamount', 'Ruby' × 'Supreme', 'Hongkong' and 'Gola' are also available and have been evaluated by Adrees *et al.* (2010).

The Southern Horticultural Research Institute (SOFRI) at Vietnam introduced 27 cultivars/lines for breeding programmes (Van *et al.*, 2017). Among the cultivars, 'Le DaiLoan' (introduced from Taiwan) (a seedless cultivar), 'Nu Hoang' (a white-pulped cultivar with few seeds), 'Ruot do' (a red-pulped cultivar) and 'Se' (a dark-red-pulped cultivar suitable for juice making) have been recommended (Mitra and Thingreingam Irenaeus, 2018). SOFRI has developed cultivars 'LELD-TC 15', 'Nu Hoang', 'Tran Chau' and 'Tim'.

Guava has been in cultivation in Israel for more than 130 years. The only commercial cultivar 'Ben-Dov' was selected in the 1950s. Clones were introduced from Brazil, Mexico and Thailand and are being used to develop hybrids (Zipori *et al.*, 2007). Guava is not a popular crop of Turkey; seedling populations are available in Mersin and Antalya. Çelik (2019) evaluated about 550 genotypes and selected 'P2-S1-A3', 'AHR-A10', 'P1-S13-A1', 'P3-S7-A2' and 'P1-S12-A1' as superior genotypes having high fruit weight.

In Venezuela, some old guava cultivars like 'Cubana', 'Criolla Roja', 'Criolla Blanca' and 'Montalban' are grown mainly in Zulia state and 'Rio Chiquito' in Monagas state (Fermin, 2010). Aranguren *et al.* (2010) evaluated the genetic variability of 31 guava accessions from different regions of Venezuela and observed great genetic diversity of the natural population. The region with most guava landraces (abundance) corresponds to the immense area south to the Orinoco River basin and adjacent regions. However, despite the abundance of guava landraces in this region, it did not represent the place with the greatest variety of phenotypes (variability) compared with the remaining geographical zones of the country. In addition to the phenotypic variability with guava landraces in Venezuela, Aranguren *et al.* (2010) confirmed that the largest number of guava landraces as well as other *Myrtaceae* may be accounted for by the Andean and Orinoco regions.

In India, research on guava breeding is being carried out at the Central Institute of Subtropical Horticulture (CISH) in Lucknow, the Indian Institute of Horticulture Research (IIHR), other institutes, and central and state agricultural universities. The work on improvement was initiated during 1907 at Ganeshkhind Fruit Experimental Station, Pune, primarily with collection of seeds from different places. One selection was made from open-pollinated seedlings of 'Allahabad Safeda' collected from Lucknow and released as 'Lucknow-49'. This cultivar become very popular for its high productivity and good quality fruit (Cheema *et al.*, 1954) which was subsequently renamed as

'Sardar'. At CISH, 631 open-pollinated seedlings of the red-coloured guava collected from the guava belt of Allahabad were evaluated for various characteristics. Six selections, namely 'CISH-G-1', 'CISH-G-2', 'CISH-G-3', 'CISH-G-4', 'CISH-G-5' and 'CISH-G-6', were found promising. 'CISH-G-3' has been named as 'Lalit' and released for commercial cultivation. It gives 24% higher yield than 'Allahabad Safeda' (Negi and Rajan, 2007). Another selection named 'Swetha' was also released from CISH producing large fruit with good keeping quality. The 'CISH-G-5' was released as 'Lalima' in 2015. It has attractive, crimson-coloured fruits and good shelf-life (Anonymous, 2016). During 1995–2000, CISH surveyed for different variants of 'Allahabad Safeda' and developed a half-sib population of 267 seedlings out of which 'CISH-G-35' was selected for heavy bearing, large fruit size and high total soluble solids (TSS) content. The selection was evaluated for a decade, named as 'Dhawal' and released in 2015 (Anonymous, 2016). 'Vilas Pasand' was selected by a guava grower from Malihabad, Lucknow region. It is very productive year-round, having large fruits (>350 g), the fruit skin is pale yellow and the pulp is creamy with less seeds. At IIHR, Bangalore, from 200 open-pollinated seedlings of 'Allahabad Safeda', one seedling selection 'Selection 8' was promising and released as 'Arka Mridula' (Dinesh and Vasugi, 2015). In recent times, from a progeny population of more than 7000, three selections – namely 'Arka Kiran' and 'Arka Rashmi' having deep-pink pulp, with high lycopene and high vitamin C content, respectively, and 'Arka Poorna', big sized (200–230g) with white pulp – have been released, the first two are suitable for processing. Other important selections made in India are 'Allahabad Surkha', 'Pant Prabhat' and 'Dhareedar'. There are many more selections made in India and they were named according to shape of the fruit, skin or pulp colour, or after the place of origin. Some of these cultivars are 'Banarasi', 'Chittidar', 'Hafsi', 'Baruipur', 'Harijha', 'Nasik', 'Sind', 'Madhurima', 'Bangalora', 'Nagpur Seedless', 'Apple Colour', 'Khaza', 'Kerala Supreme' and 'Navalur' (Mitra and Bose, 1985).

6.4.1 Inheritance pattern

Heritability in the broad sense encompasses all types of gene action including dominance, additive and epistasis. Many researchers have studied the heritability pattern of guava. It has been observed that commercially important traits, such as yield, fruit size, certain types of disease resistance and fruit quality characteristics like vitamin C, acidity, pectin content, etc., are often in the low heritability category (Ray, 2002). Genetic studies indicated that red pulp colour is dominant over white (Soubihe Sobrinho and Gurgel, 1962; Subramanyam and Iyer, 1992) and this character is governed monogenically. A linkage was also found between pulp colour and seed size. Large and hard seeds were dominant over small and soft seeds. Red pulp type was linked with large and hard seeds, and white pulp type was linked to soft seeds (Subramanyam and Iyer, 1992). Raman *et al.* (1969, 1971) stated that triploidy and genetic factor(s) are responsible for female sterility and the variation among triploids is due to their independent origin from a distinct diploid cultivar. The seedless trait is related to many factors, of which self-incompatibility and chromosomal abnormalities were considered the major ones (Pommer and Murakami, 2009). Dinesh and Yadav (1998) carried out half-sib analysis in the progenies of the cultivar 'Apple Colour'. The F_1 progenies of 'Apple Colour' × 'Arka Mridula', 'Apple Colour' × 'Beaumont', 'Apple Colour' × 'Chittidar' and 'Apple Colour' × 'Allahabad Safeda' were used to study the variance, genotypic and phenotypic correlation, and heritability in a narrow sense. They observed that the genotypic variance was less than the phenotypic variance for all five characteristics studied (fruit weight, length, volume, width and TSS). The coefficient of variation also followed the same trend, implying greater manifestation of these characteristics. The low genotypic coefficient of variation indicated a low degree of genetic variability present in half-sib progenies. The higher phenotypic coefficients of variation imply the greater manifestation of these characteristics. The coefficients of variation indicated only the variability in different characteristics and did not indicate the heritable portion. Heritability in the narrow sense was observed to be moderately high in fruit length (44.45%) and TSS (42.88%). Heritability was low in fruit width (31.68%). Thus, selection can be practised to improve the yield characteristics since these traits are controlled by additive effects. Fruit weight had a positive correlation with fruit volume, fruit length and width. However, a negative correlation was observed between TSS and the other four characteristics. The genotypic correlation was higher than phenotypic correlation for all characteristics except TSS. This can be attributed to the relative stability of the genotypes. This occurs not only when genes governing the traits are similar but when environmental influence is also similar. Coheritability estimates were moderately high for most of the pairs of characteristics. TSS decreased with the selection of big-sized fruits while selection of medium-sized fruits would not decrease the TSS. Dinesh *et al.* (2017) raised 800 progenies using three white-pulped cultivars ('Allahabad Safeda', 'Sardar' and 'Apple Colour') and two pink-pulped cultivars ('Purple Local' and 'Thailand'). They observed significant variability within the progenies for most of the traits in all the half-sib families. The genotypic variance was greater than the phenotypic variance for all the characteristics except TSS, indicating non-additive gene action for TSS. Heritability was observed as high for all the characteristics except TSS, indicating that heterosis can be exploited for TSS. The genotypic correlation coefficient was observed to be higher than the phenotypic correlation coefficient between pairs of characteristics, indicating that strong intrinsic correlations were reduced at the phenotypic level. The association of characteristics was observed to be negative between fruit weight and seed hardness, both at the phenotypic and genotypic level, indicating that selecting medium-sized fruits would help in isolating progenies with moderately soft seeds. One of the important findings of these studies was the coheritability estimate, which was high for all pairs of characteristics, indicating that selection for one characteristic would help in heritability of the other. Validation of hybridity of the progenies was confirmed by simple sequence repeat (SSR) markers 220 and 185. High heritability coupled with high genetic advance as a percentage of the mean among nine genotypes of guava ('Allahabad Safeda', 'Sardar', 'Red Fleshed', 'Arka Amulya', 'Behat Coconut', 'Hisar Surkha', 'Hybrid Bahadurgarh', 'Bangalore Seedling' and 'Portugal') was recorded for fruit yield, plant height, fruit weight, acidity and number of seeds per fruit, indicating that these traits are under the control of additive gene action and phenotypic selection for their improvement will be effective. Significant positive association of fruit weight with fruit length (0.6134 cm) and fruit breadth (0.7622 cm) indicates the importance of these characteristics during selection for high-yielding genotypes in future breeding programmes (Gupta and Kour, 2019).

In South Africa, the variation was used as a means of selecting cultivars to improve the stability of an industry which is based only on one cultivar, 'Fan Retief'. From an evaluation of 8000 seedlings, five selections were made over a 3-year period. Differences were found among seedlings in fruit size, shape, pulp thickness, pulp colour, TSS, acidity and ascorbic acid. With the exception of ascorbic acid, all other characteristics were better in the selections than in 'Fan Retief'. The variability observed in these fruit traits indicates that they would be responsive to further controlled selection and breeding (Du Preez and Welgemoed, 1990).

Guava breeding focusing on the selection of elite genotypes via full-sib families has proven to be an efficient selection strategy, especially when low-heritability traits are involved (Sousa *et al.*, 2020). In this case, selection consists of ranking the individuals with high genotypic values within the full-sib families by using mixed models (Quintal *et al.*, 2017). However, it is essential to use methods that precisely estimate the variance components and allow for prediction of the individual values of candidates for selection

(Santos *et al.*, 2015). In mixed-model methodologies, variance components are estimated by the restricted maximum likelihood (REML) method and genetic values are predicted by the best linear unbiased prediction (BLUP) method (Resende, 2016). Sousa *et al.* (2020) evaluated 11 full-sib families. The parents used in the crosses showed a considerable degree of genetic divergence, since the plants were selected in orchards formed from seedlings. The seeds obtained from those crosses generated a segregating population of wide genetic variability that was subsequently evaluated and selected via REML/BLUP (Quintal *et al.*, 2017). The highest-yielding progenies were crossed again to give rise to the 11 full-sib families. After evaluating the 11 full-sib families, Sousa *et al.* (2020) concluded that: (i) the REML/BLUP statistical procedure was effective in selection of superior genotypes and prediction of genetic parameters in the population under study; (ii) the mean of the individuals selected exceeded the mean of their parents, confirming that the strategy of obtaining full-sib families was effective in generating gains in the guava breeding programme; (iii) families 1, 2, 3, 5, 8, 9, 10 and 11 were considered superior because they include many genotypes selected for fruit weight and thus they will be represented in possible new crosses for the formation of new full-sib families; and (iv) the most divergent guava genotypes, based on the UPGMA (unweighted pair group method with arithmetic mean) method, should be recommended for future crosses in order to obtain segregation populations and continue the guava breeding programme with the aim of obtaining superior genotypes. It was suggested to hybridize genetically distant individuals since that will generate segregating populations with greater genetic variability and could increase the possibility of obtaining superior genotypes.

6.5 Interspecific Hybridization

Interspecific hybridization has been attempted to develop rootstocks resistant to guava wilt. The *Psidium* species, *Psidium molle*, *P. guineense*, *P. friedrichsthalianum* and Philippine guava were found to be resistant to wilt (Edward and Shankar, 1964). The cross between *P. guajava* and *P. molle* was reported to be incompatible, but when *P. molle* was used as a female parent with *P. guajava*, the cross was compatible (Subramanyam and Iyer, 1982). At CISH, interspecific hybridization was done between *P. molle* and *P. guajava*. The interspecific hybrids have been found resistant to guava wilt and are graft compatible with commercial varieties of *P. guajava* (Anonymous, 2003–04). Even though some success has been obtained to develop interspecific progenies, still emphasis should be given to breed scion and rootstock separately for abiotic/biotic stress situations. Attempts were also made to use the wild species *P. guineense* and *Psidium chinensis* as potential gene donors with *P. guajava* cultivars 'Arka Mridula', 'Arka Rashmi' and 'Arka Kiran' using the standardized nutrient solution (sucrose 5% + H_3BO_3 100 ppm + $MgSO_4$ 200 ppm + $Ca(NO_3)_2$ 300 ppm + $K(NO_3)_2$ 100 ppm). Results indicated that 'Arka Rashmi' × *P. chinensis* caused highest fruit set (54.54%), followed by 'Arka Kiran' × *P. chinensis* (51.42%), as against the least fruit set (11.11%) in 'Arka Rashmi' × *P. guineense* (control). Thus wild species may be employed for the development of new varieties with wilt resistance (Alifa *et al.*, 2017). Phylogenic studies carried out in *Psidium* species by utilizing the differences in flavonoid patterns showed that there is close affinity between *P. guajava* and *P. molle*. *P. molle* and *P. guineense* were found similar morphologically with minute differences in chromatographic pattern. There was also close affinity between *P. guineense*, *Psidium pumilum* and *P. chinensis* (Dass and Prakash, 1981).

The red strawberry guava (*P. littorale* var. *longipes*) showed apparent tolerance to root-knot nematode (Abdul Karim *et al.*, 2002) and Costa Rican guava (*P. friedrichsthalianum*) was reported as highly resistant to *Meloidogyne* spp. (Cuadra and Quincosa, 1982). Interspecific hybrids between *P. guajava* accession 'GUA 161PE' and *P. guineense* accession 'ARA 138RR' were reported as highly tolerant to the nematode, the hybrids showed similar growth to that of the guava trees and have high compatibility with guava cultivar 'Paluma' (Costa *et al.*, 2012). The number of chromosomes in *P. guineense* was reported as $2n = 44$ (Chakraborti *et al.*, 2010). The 'GUA 161PE' guava tree was considered amphidiploid, that is, with $2n$ and $4n$ cells in the same individual, when analysed by flow cytometry, suggesting that success with *P. guineense* could be attributed to n gametes and $2n$ formation in 'GUA 161PE'.

6.6 Hybridization

6.6.1 Hybridization technique

In guava, flowers that are chosen for crossing are emasculated when at the 'calyx break stage', a day before flower opening (Dinesh and Vasugi, 2015). This stage can be easily identified by the change in bud colour from green to yellowish green and swelling of the bud (balloon stage). At this stage buds should be carefully emasculated, keeping the style and the stigma intact. The emasculated bud is enclosed in a paper bag and, 24 h later, pollen from the pollen parent is brought from an unopened flower, preferably at calyx break stage, and applied to the stigma with a small brush. Then the bud should be re-enclosed in a paper bag (Zipori *et al.*, 2007). Pollination during morning hours is suggested. The fruit should then be protected until its full development, harvested, and the seeds extracted and prepared for sowing; seedlings should be maintained in the nursery until they can be transplanted into the field in selection plots. Selection should be practised, and the best material has to be vegetatively propagated to go into evaluation plots for final screening.

6.6.2 Intervarietal hybridization

The majority of the commercial guava cultivars are diploid ($2n = 22$) while the 'Seedless' cultivar is triploid. 'Seedless' has been used as a parent to develop hybrids with few seeds. However, 'Seedless' is a shy bearing cultivar. To develop less-seeded hybrids with high yield potential, crosses were made between triploid ('Seedless') and diploid ('Allahabad Safeda') at the Indian Agricultural Research Institute, New Delhi (Majumder and Mukherjee, 1972; Mukherjee, 1977). Seventy-three F_1 hybrid seedlings were developed of which 26 were diploids ($2n$), nine were trisomics ($2n + 1$), five were double trisomic ($2n + 1 + 1$) and 14 tetrasomics ($2n + 2$). The trisomic plants had a dwarf growth habit and normal shape and size of fruits with few seeds.

In India, breeding work for guava improvement has been going on at several institutes and universities. Intervarietal hybridization carried out at the Fruit Research Station, Sangareddy, Andhra Pradesh resulted in the development of two superior hybrids, 'Safed Jam' ('Allahabad Safeda' × 'Kohir') and 'Kohir Safeda' ('Kohir' × 'Allahabad Safeda'), both having soft seeds and bigger fruit size than both the parents and being suitable for semi-arid climates (Mitra and Bose, 1985; Dinesh and Vasugi, 2015). The CCS Haryana Agricultural University, Hisar, Haryana released two guava hybrids, namely 'Hisar Safeda' ('Allahabad Safeda' × 'Seedless') and 'Hisar Surkha' ('Apple Colour' × 'Banarasi Surkha'). Fruits of both hybrids have high TSS (13.4–13.6°Brix) and low to medium seeds. At the Indian Institute of Horticultural Research, Bangalore, several F_1 hybrids were raised from various intervarietal crosses and three hybrids, namely 'Arka Amulya' ('Allahabad Safeda' × 'Triploid'), 'Arka Kiran' ('Kamsari' × 'Purple Local') and 'Arka Rashmi' ('Kamsari' × 'Purple Local'), were released. The fruits of hybrids 'Arka Kiran' and 'Arka Rashmi' are rich in lycopene and have soft to medium-soft seeds. Varietal crossing incompatibility has been reported when crosses were made between 'Behat Coconut' and 'Sardar', as well as 'Behat Coconut' and 'Apple Colour' (Dinesh and Vasugi, 2015).

In Israel, the local cultivar 'Ben-Dov' has a very strong flavour, not liked by many consumers, and suffers from a short postharvest life. To address these problems, breeding work was initiated in 1981 at the

Besor Experimental Farm in southern Israel for mass selection (Canizares, 1981) from the seedlings raised from open-pollinated Brazilian clones (1993–1997). In the second cycle, started in 1994, flowers of Mexican genotypes were pollinated with Thai genotypes, and flowers of Brazilian genotypes were pollinated with the Mexican types. In the third cycle, started in 1999, flowers of Thai types were hybridized with promising hybrids obtained in the second cycle. All parents of Mexican types used were characterized by small (30–50 g) fruit, sweet taste, white pulp, medium-size cavity and strong flavour; most parents of Brazilian types were characterized by medium (150–250 g) fruits, good taste, pink pulp, medium-size seed cavity and medium level of flavour; while all the parents of Thai types were characterized by large (300–350 g) fruits, white pulp, strong flavour and medium to poor taste, their fruits did not change colour upon ripening but remained green. Some of the hybrids, namely 61/1 (Thai × Mexican), 58/3 (Mexican × Brazilian), 68/4 (Mexican × Brazilian) (red pulp), 61/5 (Mexican × Brazilian) and 11/6 (Thai × Mexican), showed improvements in taste with mild flavour. These were selected for further evaluation (Zipori *et al.*, 2007). Another four red-pulped hybrids, 'Yuval' (2/4), 'Zohar' (11/51), 'Gili' (3/3) and 'Ido' (8/28), were subsequently developed (Fig. 6.1). All these hybrids are registered or in the process of registration, currently they are grown in Israel on a limited scale and are also in the process of testing in Spain. The long shelf-life of hybrids compared with 'Ben-Dov' is due to suppressed climacteric behaviour (Arnon Dag, Israel, 2020, personal communication).

In the Tropical Fruit Research Institute, Havana, Cuba, a team of researchers (Rodriguez *et al.*, 2010) evaluated the diversity of Cuban guava. A group of elite accessions was selected for commercialization and for breeding. Four cultivars, 'Enna Roja Cubana' ('EEA 18-40'), 'N6', 'Belic L-207' and 'Red Supreme', were selected to develop three crosses using the former as the female parent. From these crosses, 354 seedlings were planted and after evaluation 25 genotypes were selected for further evaluation. The cultivar 'XXI Century' ('Supreme-2' × 'Paluma') was released from Brazil, obtained from 219 plants originated from several crossings, after evaluation for 10 years. Its main characteristics are high productivity with a short cycle (130 days from flowering to harvest) and large fruit (236 g) with thick, rosy-red pulp and few small seeds, having appreciable flavour (Pereira *et al.*, 2003). The first hybrid guava from Taiwan is 'Tainung No. 1' ('Diwang Ba'), developed by crossing 'Pearl' and 'Crystal'. This has high TSS and a long shelf-life (Lee *et al.*, 2017).

6.7 Polyploidy

The basic chromosome number of *Myrtaceae* is $x = 11$ (Atchison, 1947). Usually, dysploid variations ($x = 5, 6, 7, 8, 9, 10, 11, 12$ and 14) occur in capsule-fruited taxa against polyploidy variations in fleshy fruited taxa (Rye, 1979) which evolved mainly by polyploidy, with various cytotypes (Costa and Forni-Martins, 2007). In *P. guajava* L., diploid, triploid, tetraploid and aneuploid forms have been reported (Naithani and Srivastava, 1966; Raman *et al.*, 1969, 1971; Srivastava, 1977). Considering that the genus *Psidium* shows polyploid species ($2n = 44 = 88$ chromosomes), the allo- and/or autopolyploidization in diploid species of this genus can be related to the occurrence of polyploidy (Marques *et al.*, 2016).

Seedlessness is an important commercial attribute and 'seeded' and 'seedless' types have been recognized. However, those commercially recognized as 'seedless' are not completely seedless and types ranging from 'seedless' to 'less-seeded' can be identified. The 'seedless' types are nevertheless found to be pollen fertile (Raman *et al.*, 1971). Cytological examination of the 'seeded', 'less-seeded' and 'seedless' types has shown that the difference in chromosome number as well as genetic dissimilarities contribute to this feature. The 'highly seeded' and 'less-seeded' types are diploids, while the 'highly seedless' types are triploids (Raman *et al.*, 1969). Chromosome counts were associated with $2C$ value estimated as 0.567 ± 0.019 pg

Fig. 6.1. Hybrids of guava developed in Israel: (A) 'King'; (B) 'Omri'; (C) 'Roni'; (D) 'Lior'; (E) 'Yuval'; (F) 'Zohar'; (G) 'Gili'; and (H) 'Ido'. Photograph courtesy of Dr Ron Polat.

DNA for white-pulped cultivars and 0.551 ± 0.02 pg DNA for red-pulped cultivars of *P. guajava*. The tetraploid species *Psidium acutangulum* and *P. cattleyanum* have higher 2*C* values of 1.167 and 1.053 pg DNA, respectively (Costa *et al*., 2008).

The possibility of occurrence of aneuploidy and its usefulness in guava have been explored (Majumder and Mukherjee, 1970; Mohammed, 1974; Sharma *et al.*, 1992). The tolerance of extra chromosomes in a diploid guava ($2n = 22$) was shown by D'Cruz and Babu Rao (1962). This was followed by the production of some aneuploid forms such as trisomics, tetrasomics and higher aneuploids (Majumder and Mukherjee, 1970). Mohammed (1974) identified aneuploids cytomorphologically in progenies from triploid and diploid–triploid crosses. Thirty trisomics, two double trisomics, one tetrasomic and two higher aneuploids were obtained. Some of the aneuploids were reported to have eight extra chromosomes. The plants with one extra chromosome occurred more frequently (67%) than other aneuploid types. Reduction in growth and size of leaf distinguished aneuploids from diploids. Aneuploids, particularly trisomics, had promising qualities and may be useful in developing plants with reduced seediness and possibly in providing dwarfing rootstocks (Negi and Rajan, 2007).

The occurrence of autotetraploids of guava in nature has been reported (Naithani and Srivastava, 1966; Raman *et al.*, 1971). Tetraploidy in guava has also been induced with colchicine (Janaki Ammal, 1951; Ram Kumar, 1975), oryzalin (Handayani *et al.*, 2017) and by chronic gamma irradiation via exposure to ^{60}Co (Das, 1971).

6.8 Mutation

The seedlessness in 'Seedless' guava was believed to be due to autoploidy (Dasarathy, 1951). Cheema and Desmukh (1927) stated that naturally occurring mutations are not rare in guava. Natural autotetraploid in guava has been reported (Naithani and Srivastava, 1966). Ram Kumar (1975) induced tetraploidy in cultivar 'Allahabad Safeda' by treating shoot tips with a 0.1% aqueous solution of colchicine. Brar and Bal (2003) investigated the effect of gamma rays (1, 2, 3, 4 and 5 kR) on buds of guava cultivar 'Sardar'. After the treatments, the buds were budded into the same 'Sardar' rootstock. Variability for plant height, internodal length and stem diameter was observed to be maximum in the 2 kR treatment; while for number of branches, number of leaves and breadth of leaves, the maximum variability was noted in 4, 1 and 3kR treatments, respective. However, the mutagenic treatments had no significant effect on stomata size. *In vitro* mutagenesis followed by micropropagation via axillary bud proliferation of shoot tips of guava was carried out by Zamir *et al.* (2003). Shoot tips were irradiated with gamma rays at 15–90 Gy and cultured in Murashige and Skoog's (MS) medium containing 3% sucrose, 6-benzylaminopurine (benzyl adenine) (BAP) and L-glutamine. Optimum shoot proliferation was recorded in the MS medium supplemented with BAP at 1.0 mg/l and L-glutamine at 250 mg/l. Rooting of cultured shoots was observed in half-strength MS medium supplemented with indole acetic acid and indole butyric acid. The LD_{50} was observed at 45 Gy. Rates of more than 75 Gy were found lethal to the explants.

6.9 Breeding for Disease and Nematode Resistance

Soil-borne vascular wilt pathogens cause among the most devastating plant diseases. Guava wilt disease (GWD) is a serious problem in most of the guava-growing countries of the world. The guava wilt pathogen, *Myxosporium psidii*, was identified from a diseased tree in the Fengshan area, Taiwan. The cultivar 'Peipa' was identified as a best source of resistance. The crosses between 'Peipa' and 'Clone R1' and 'Clone R4' have been selected for further study (Wan and Leu, 1999). In South Africa, the fungus causing GWD has been subjected to several name changes. The fungus was first identified as *Paecilomyces* (Manicom, 1980) and changed later to *Septofusidium* (Grech, 1982), *Gliocladium* sp. (Grech, 1984), *Acremonium diospyri* (Benade *et al.*, 1991), *Penicillium vermoesenii* (Schoeman *et al.*, 1997) and finally classified as *Nalanthamala psidii* (Schroers *et al.*, 2005). GWD caused by *N. psidii* resulted in the loss of more than half of the guava production area in Limpopo and Mpumalanga provinces of South Africa during the 1980s (Schoeman *et al.*, 2012; Severn-Ellis *et al.*, 2012). A tolerant/resistant guava selection 'TSG2' was developed by the Agricultural Research Council's Institute for Tropical and Subtropical Crops (ARC-ITSC). This cultivar was successfully used in replacing many of the production areas lost due to GWD during the late 1990s (Vos *et al.*, 2000). In 2009 a second outbreak of GWD occurred which in turn also affected the tolerant 'TSG2' cultivar, placing the guava industry under serious threat once again (Schoeman, 2011). Schoeman and Labuschagne (2014) made *in vitro* screening of 14 seedlings raised from escape trees in 'Fan Retief' and 'TSG2' orchards. A culture filtrate of *N. psidii* was used to screen guava seedlings *in vitro*. Promising selections were multiplied in tissue culture, hardened off and planted in bags before inoculation with GWD fungus in a shadehouse trial. Although none of the selections showed complete resistance, selection 'MS44' showed some tolerance against G2 isolate of the pathogen obtained from 'TSG2' trees, while selection 'MS70' showed some tolerance against G1 isolate obtained from diseased 'TSG1' trees. These selections were also resistant to the original 'Fan Retief' isolate of the pathogen.

More than 150 guava cultivars are available in India. Misra (1998–1999) evaluated field tolerance of 20 cultivars and reported that 'Apple Colour', 'Chittidar', 'Seedless', 'Super Acid', 'Superior Sour Lucidium', 'Red Flesh' and 'Smooth Green' are tolerant to GWD. The technique of expressing the *endochitinase* gene in the plant system to confer resistance against fungal disease has been successfully demonstrated (Bolar *et al.*, 2000). Mishra *et al.* (2014) developed transgenic guava overexpressing the *endochitinase* gene derived from *Trichoderma harzianum*. A total of 11 transgenic lines have been developed. These transgenic lines showed differential *endochitinase* expression. Two promising lines (T22 and T20) had high *endochitinase* expression in terms of *N*-acetyl-D-glucosamine release. The transgenic plantlets were screened *in vitro* for resistance against the wilt pathogen. *In vitro* pathogen inhibition assay and subsequent spore germination assays revealed that crude leaf extract of transformed plants inhibited the germination of fungal conidia. The leaf tissue studied for expression of *endochitinase* revealed that two transgenic plants, that showed no wilt symptoms, also showed very high activity of *N*-acetyl-D-glucosamine (0.741 and 0.738 μM min^{-1} per μg protein, respectively) which indicated that transgenic plants did not develop any symptoms of wilt disease due to overexpression of *endochitinase* (Mishra *et al.*, 2016).

The rust caused by *P. psidii* is another serious disease of guava. Vasconcelos *et al.* (1998) evaluated ten cultivars in Brazil for resistance to *P. psidii*. All ten cultivars were found susceptible. 'IAC-4', 'Campos' and 'Riverside' showed the lowest while 'Ouro' and 'Pirassununga Vermelha' showed the highest levels of rust infection. Ribeiro and Pommer (2004) studied half-sib progenies resulting from 22,950 seeds from fruits originated through open pollination of 306 accessions in Brazil. Seedlings were grouped into different numbers of accessions as: (i) 35 primary selections of white-pulped guava obtained in the breeding programme IAC (identification: White LxPy); (ii) 64 primary selections of red-pulped guava, obtained from the same programme (identification: Red LxPy); (iii) 118 commercial cultivars (some with two up to six accessions) of cultivars 'Supreme', 'Indiana', 'Weber', 'Australian', 'Patillo', 'Paluma', 'Rica', 'Ruby Supreme', 'IAC-4' and others; (iv) 55 advanced selections of IAC programme (with acronym MAS), of Conceição do Almeida, Bahia state (with acronym EEFT) and others named Sigla (II to XIII); and (v) 34 accessions not clearly identified or without identification. Selection was applied in the initial stages of the seedlings and after artificial inoculation with the fungus. The heritability of rust resistance was estimated in a broad sense, being h^2 = 0.275. The half-sib progenies showed variation in the proportion of plants without symptoms: 25% in Group 1 (IAC selections of white guava); 28% in Group 2 (IAC selections of red guava); 44% in Groups 3 (commercial cultivars) and 5 (others); and 64% of plants without symptoms in Group 4 (advanced selections of Monte Alegre do Sul, São Paulo state and of Conceição do Almeida). The analysis of variance showed that plants of Group 4 differed from the others. After 2 years, 105 individual plants were selected with absolutely no symptoms of the disease and are under selection for other traits, such as yield, fruit characteristics, colour and flavour.

In Brazil, the biggest challenge facing guava breeders is developing cultivars resistant to the 'guava decline'. First detected in 2001, this disease has decimated commercial orchards, where guava plants parasitized by the nematode *Meloidogyne enterolobii* become susceptible to the root rot caused by the *Fusarium solani* complex constituting the main disease affecting guava crops (Gomes *et al.*, 2012, 2014; Ribeiro *et al.*, 2019). Although genetic resistance to this nematode has not been reported in commercial guava genotypes, different studies have identified resistant species in the genus *Psidium*, such as *P. cattleyanum*, *P. friedrichsthalianum* and *P. guineense* (Martins *et al.*, 2013; Freitas *et al.*, 2014; Souza *et al.*, 2015). The search for the source of tolerance to *M. enterolobii* in wild *Psidium* species resulted in identifying the tolerance source, but it has presented limited or complete incompatibility when used as rootstock for commercial guava cultivars. From their studies to obtain interspecific hybrids between *P. guajava* and *P. guineense* for tolerance to *M. enterolobii*, Costa *et al.* (2012) concluded that: (i) the controlled crossing methodology can also be applied to interspecific crossing in the *Psidium* genus; (ii) there were barriers or need for additional care in crosses between *P. guajava* and *P. friedrichsthalianum* and *P. cattleyanum*; (iii) there was among and within variability in *P. guineense* accessions for *M. enterolobii*; (iv) interspecific hybrids between 'GUA 161PE' (*P. guajava*) × 'ARA 138RR' (*P. guineense*) were highly tolerant to nematodes; and (v) interspecific hybrids between 'GUA 161PE' × 'ARA 138RR' showed similar growth to that of 'Paluma' guava. Gomes *et al.* (2017) made interspecific crosses of *P. guineense* × *P. cattleyanum*, *P. guineense* × *P. guajava* and *P. cattleyanum* × *P. guajava*. These crosses resulted in hybrids immune, susceptible and resistant to *M. enterolobii*. The chi-square test rejected the hypothesis of monogenic inheritance with incomplete dominance, which corroborates that this trait has polygenic action.

Knowledge of the molecular relationship between *Psidium* species based on plant resistance gene analogues (RGAs) can be useful in the genetic breeding of guava for resistance to *M. enterolobii*. Noia *et al.* (2017) studied RGA markers from conserved domains and structural features of plant resistance (*R*) genes to characterize *Psidium* species and establish genetic proximity with a focus on nematode resistance. In ten evaluated *Psidium* species, high interspecific genetic variability was verified through RGA and SSR markers, with interspecific variation in *P. guajava* higher with SSR markers. Variability related to the number of species-specific amplicons detected by the RGA markers demonstrated the importance of using different species in breeding aiming at disease resistance, as *R* genes may be present in only one or few species. The genetic diversity of *Psidium* species observed by RGA markers represents a collection of random *R* genes conferring resistance to different pathogens in these species. Such characterization of the gene pool of wild relatives is useful for the management of genetic resources. Additionally species-specific amplicons can be explored, taking advantage of their potential as markers for use in the introgression of *R* genes in *P. guajava* by interspecific hybridization.

6.10 Molecular Characterization

Guava breeding through conventional approaches is a long-term and cumbersome process that relies on the arbitrary rearrangement of existing genes between two closely related parent plants. The combination of morpho-agronomic characteristics and DNA molecular markers constituted a novel tool of great utility to characterize guava germplasm, estimate diversity level and parentage relationships among accessions, and also to recommend genotypes with conservation and breeding potential (Nimisha *et al.*, 2013). Molecular markers can be gainfully employed to discriminate between species and cultivars of guava (Chandra and Mishra, 2007). Molecular markers are emerging as a prospective biotechnological tool for managing genetic variations, estimation of outcrossing rate, and percentage analysis and linkage map construction of guava. Different molecular markers such as random-amplified polymorphic DNA (RAPD) (Dahiya *et al.*, 2002; Prakash *et al.*, 2002; Chen *et al.*, 2007; Mani *et al.*, 2011; Pessanha *et al.*, 2011), amplified fragment length polymorphism (AFLP) (Valdes-Infante *et al.*, 2003; Sanchez-Teyer *et al.*, 2010) and SSR (Risterucci *et al.*, 2005; Valdes-Infante *et al.*, 2007; Aranguren *et al.*, 2010; Briceno *et al.*, 2010; Kanupriya *et al.*, 2011; Noia *et al.*, 2012a,b; Rai *et al.*, 2012, 2013; Padmakar *et al.*, 2015; Kareem *et al.*, 2018) have been used for guava germplasm analysis. RAPD markers were used to estimate molecular diversity of 41 genotypes of guava consisting of five *Psidium* species, 23 varieties, 12 selections and a hybrid. Analysis suggests that Indian guava can be rated as low to moderate diversity and also indicated that various triploid seedless cultivars of guava are not genetically identical and have independent origins (Prakash *et al.*, 2002). RAPD analysis for discriminating 13 North Indian cultivars of guava revealed that 'Hisar Safeda' and 'Allahabad Safeda' were the closest pair of cultivars with a distance of 0.051 on a scale of zero to one. Cultivars 'Pear Shaped' and 'Red Supreme' were most distantly placed in relation to each other with a distance of 0.423. Average similarity index among 13 cultivars was 0.064 and on average 81.85 bands were amplified per cultivar (Dahiya *et al.*, 2002).

A better understanding of the effectiveness of the different molecular markers is considered a priority step towards germplasm characterization and classification, as well as a prerequisite for effective breeding programmes (Belaj *et al.*, 2003). Valdes-Infante *et al.* (2010) compared the polymorphism level, the discriminating capacity and the informativeness of agro-morphologic traits and the polymerase chain reaction (PCR)-based molecular markers AFLP and SSR for genotype identification and genetic diversity analyses of 23 guava accessions maintained at the Research Institute on Tropical Fruit Crops, Havana, Cuba. The high level of polymorphic loci detected by the dominant AFLP marker highlights the discriminating capacity of this genetic marker. With a single primer combination all of the individuals were identified, while only few genotypes were differentiated with a single SSR combination or by morphological variables. The higher values of expected heterozygosity were, however, detected by SSR. This value doubled the one obtained with AFLP and reflects the high level of informativeness of the marker due to the multiallelic and co-dominant nature of SSR, which makes them suitable for diversity studies. The morphological diversity index provided a good estimate of diversity among guava accessions when phenotypic traits of high heritability were used, and it was comparable with the expected heterozygosity scored with DNA markers. SSR or microsatellite is a PCR-based molecular marker technique with many advantages such as abundance, high polymorphism and co-dominant mode of inheritance which permits easy transfer of markers between genetic maps of different crosses in contrast to dominant PCR marker types such as RAPD or AFLP (Risterucci *et al.*, 2010) and primer transferability (Padmakar *et al.*, 2015). SSR markers from guava were developed by Risterucci *et al.* (2005) and were applied in germplasm characterization and assessment of existing genetic variability (Risterucci *et al.*, 2005; Valdes-Infante *et al.*, 2007, Aranguren *et al.*, 2010; Viji *et al.*, 2010; Santos *et al.*, 2011; Jose *et al.*, 2012; Padmakar *et al.*, 2015; Kareem *et al.*, 2018). SSR markers were also used for cultivar identification (Kanupriya *et al.*, 2011), discrimination of wild guava species (Nogueira *et al.*, 2012) and assessing the genetic homogeneity of guava plants derived from somatic embryogenesis (Rai *et al.*, 2012; Kamle *et al.*, 2013).

Risterucci *et al.* (2005) developed and characterized 23 nuclear SSR loci in guava based on a $(GA)_n$ and $(GT)_n$ microsatellite-enriched library. The SSR markers amplified 4.5 alleles per locus in *P. guajava* and cross-species transferability was confirmed between *P. acutangulum*, *P. cattleyanum* var. *lucidum* and *P. friedrichsthalianum*, respectively. Phenotypically, *P. guajava* shares more similarity of molecular sequences with *P. cattleyanum* and *Psidium cujavillis* in comparison to others. While comparing *P. molle* with others, equidistance from all species was indicated, sharing maximum similarity with *P. guineense* and *P. cattleyanum*. *P. cujavillis* and *P. guineense* were the closest species with the coefficient of similarity being 0.804 (Bajpai *et al.*, 2011). Noia *et al.* (2012a,b) studied the genetic distance among guava genotypes that were collected from different altitudes. They stated that the wild genotypes have the potential to be used in breeding to increase the options of genotypes grown commercially. They demonstrated the cross-genera transferability of 23 SSR primer pairs developed for guava (*P. guajava*) to four new targets, two species of eucalyptus (*Eucalyptus citriodora* and *Eucalyptus camaldulensis*), bottlebrush (*Callistemon lanceolatus*) and clove (*Syzygium aromaticum*) belonging to family *Myrtaceae* and subfamily *Myrtoideae*. The high level of cross-genera transferability of guava SSRs may be applicable for the analysis of intra- and interspecific genetic diversity of target species, especially in *E. citriodora*, *C. lanceolatus* and *S. aromaticum*, for which no information about the expressed sequence tag (EST)-derived as well as the genomic SSR is available to date. Padmakar *et al.* (2015, 2016) carried out genotyping of guava full-sib population with 94 F_1 progenies (derived from a two-way, pseudo-testcross strategy) differing in fruit quality traits such as seed strength (hardness/softness), fruit weight and TSS, using SSR markers and sequence-related amplified polymorphic (SRAP) primer combinations. Two cultivars of guava 'Kamsari' (hard-seeded with pink pulp, good TSS, medium-sized fruits, green skin) and 'Purple Local' (soft-seeded with pink pulp, small-sized fruits, acidic, purple skin) were used as parental lines for developing the mapping population. In the case of SSRs, high-throughput genotyping, through a M13-tailed PCR principle using a set of 160 SSR primer pairs, revealed 64.3% parental polymorphism that generated 321 alleles during the mapping population survey. Twenty per cent of parental polymorphism was revealed with SRAPs that generated 126 scorable markers. Two linkage maps were constructed for parents using maternal-specific 143 testcross marker loci and parental-specific 127 testcross markers along with 60 intercross marker loci. At a minimum logarithm of the odds (LOD) score of 4.0 and a maximum map distance of 40 cM, the maternal 'Kamsari' map covered 2551.3 cM with an average marker interval distance of 13.21 cM having a total of 193 framework marker loci being ordered into 11 linkage groups. The paternal 'Purple Local' map covered 2113.0 cM with an average marker interval distance of 12.07 cM having 175 framework marker loci being ordered into 11 linkage groups. The estimated genome coverage was 87.32% in 'Kamsari' and 83.74% in 'Purple Local'. These genetic maps will help in identification of complex quantitative trait loci (QTLs) related to fruit quality (biotechnological interventions for improvement of guava are also dealt with in Chapter 5, this volume).

6.11 Cultivars Growing in Different Countries

6.11.1 Brazil

'Paluma'

'Paluma' is a seedling selection of 'Ruby Supreme'. Fruits are large (more than 200 g), pyriform with a short neck. Ripe fruit is smooth and yellow, pulp intense red (Fig. 6.2), firm and thick (1.3–2.0 cm) with few seeds, soluble solids around 10°Brix and balanced acidity. It is the most widely planted cultivar in Brazil.

Fig. 6.2. Cultivar 'Paluma'. Photograph courtesy of Valfrutas, Brazil.

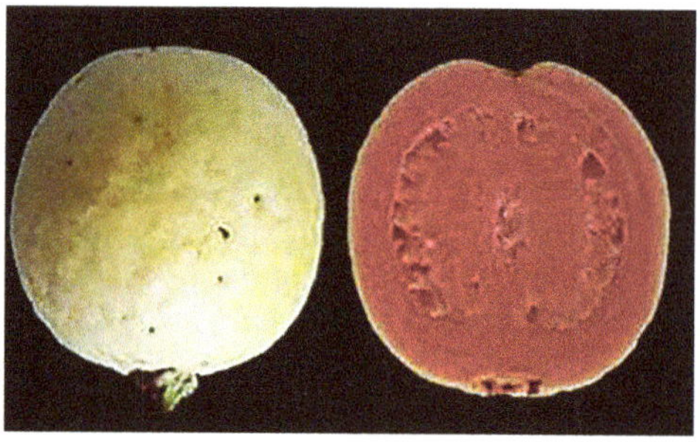

Fig. 6.3. Cultivar 'Rica'. Photograph courtesy of Dr Celso Pommer.

'Rica'

A seedling selection of 'Supreme'. Fruits are oval to slightly pyriform in shape with a short neck, average weight 100–250 g, yellow-green and slightly rough skin, red pulp, thick and firm, seeds are few and small (Fig. 6.3). Fruits have soluble solids around 10°Brix and balanced acidity. Mostly used in preparation of juices.

Fig. 6.4. Cultivar 'Pedro Sato'. Photograph courtesy of Valfrutas, Brazil.

'Pedro Sato'

Believed to be selected by a guava grower from 'Red Ogawa #1'. Fruits are 150 to 450 g, firm with pink pulp (Fig. 6.4), few seeds and pleasant taste. It is more productive and hardier than 'Paluma'.

'Cortibel' cultivars ('Cortibel RM' and 'Cortibel SLG')

The 'Cortibel' cultivar series was initially developed by Jose Corti and Isabel Corti guava growers in Espírito Santo state and was registered and commercialized by Frucafe. 'Cortibel RM' has fruits of medium size with rough skin, red pulp, very tasty, excellent postharvest life, very productive, resistance to several pests, rust and psyllid. 'Cortibel SLG' has red, semi-smooth pulp (Fig. 6.5), tasty, very productive, good postharvest life. This cultivar stands out for its easy flowering.

Fig. 6.5. Cultivar 'Cortibel SLG'. Photograph courtesy of Erli Ropke, Frucafe, Brazil.

'Seculo XXI'

A hybrid of 'Supreme-2' × 'Paluma'. Cultivar is more precocious than 'Paluma'. Fruit have rough, yellowish skin when ripe. Average weight around 250 g, pulp is pinkish red and has a very pleasant flavour, few small seeds. It has higher vitamin C content than both the parents.

6.11.2 China

'Xiguahong'

Fruit oblong, 7.6–10.6 cm in longitudinal diameter and 7.4–8.9 cm in transverse diameter. Mature fruit yellow-green with bright red pulp (Fig. 6.6). Contents of TSS, vitamin C, total sugar and titratable acid of the fruits are 8.3–11.5°Brix, 270–310 mg 100 g^{-1}, 7.2–8.4% and 0.26–0.37%, respectively (Wang, 2017).

'ZenZhu'

Fruit ovoid, average weight 200–400 g, ripe fruit pale green in colour, pulp white (Fig. 6.7), fine, crispy and sweet. Contents of TSS, vitamin C and total sugar of the fruits are 7–13°Brix, 190 mg 100 g^{-1} and 8.9%, respectively (Chen *et al.*, 2002; Liu *et al.*, 2008).

Fig. 6.6. Cultivar 'Xiguahong'. Photograph courtesy of Dr C. Haojun.

Fig. 6.7. Cultivar 'ZenZhu'. Photograph courtesy of Dr C. Haojun.

'Jindouxiang'

Fruit oval in shape, small in size (85 g), pericarp pale yellow, pulp white with strong aroma. Contents of TSS, vitamin C, total titratable acid and total sugar are 12°Brix, 216 mg 100 g^{-1}, 0.25% and 7.63%, respectively (Kuang *et al.*, 2018).

6.11.3 Egypt

'El-Sabahia'

Pear-shaped fruit (Fig. 6.8), small in size (100–110 g), medium sweet (8–9°Brix) and 0.30–0.32% acidity.

'El-Fakous'

Pear-shaped fruit, small to medium in size (110–150 g), pulp thick (1.2 cm), sweet (9.5–10.5°Brix) and 0.30–0.35% acidity.

'El-Mobaker'

Pear-shaped fruit, small in size (60–70 g), pulp yellowish white (Fig. 6.9), medium sweet (8.0–8.5°Brix).

'El-Banati'

Irregular-shaped fruit, seedless, medium in size (95–120 g), sweet (9–10°Brix) and medium in vitamin C content (100 mg 100 g^{-1} pulp).

Fig. 6.8. Cultivar 'El-Sabahia'. Photograph courtesy of Dr A.S. Elsoda.

'Malaysian Red'

Pear-shaped fruit, red pulp (Fig. 6.10), small in size (110 g), medium sweet (8.5–9.5°Brix) and high in vitamin C content (200 mg 100 g^{-1} pulp).

'Winter'

A late cultivar, pear-shaped fruit, big size (200–250 g), thick pulp (1.7–2.0 cm), sweet (9–11°Brix) and high in vitamin C content (200 mg 100 g^{-1} pulp).

6.11.4 India

'Allahabad Safeda'

Fruits are big in size (average weight 150–200 g), round, smooth skin, white pulp (Fig. 6.11), soft, firm, light yellow and on ripening develop very sweet taste. The °Brix, acidity and ascorbic acids contents are 11.0–12.6, 0.27–0.41% and 117–206 mg 100 g^{-1} pulp, respectively (Mitra *et al.*, 1983; Kundu *et al.*, 1995; Daulta *et al.*, 1998).

Fig. 6.9. Cultivar 'El-Mobaker'. Photograph courtesy of Dr A.S. Elsoda.

Fig. 6.10. Cultivar 'Malayasian Red'. Photograph courtesy of Dr A.S. Elsoda.

'Apple Colour'

Even though it is not a heavy bearer, this cultivar is grown because of its attractive colour and good-quality fruits. Fruits are 80–120 g in weight, spherical in shape, dawn-pink in colour with deep, minute dots on the fruit surface. The °Brix, acidity and ascorbic acid contents are 10–11.2, 0.17–0.38% and 205–225 mg 100 g^{-1} (Mitra *et al.*, 1983; Kundu *et al.*, 1995; Kumar, 1998).

'Allahabad Surkha'

Large fruit, soft deep pink pulp (Fig. 6.12), few seeds, strong flavour, very sweet, yield 120 kg per tree after 6 years (Ray, 2002).

Fig. 6.11. Cultivar 'Allahabad Safeda'. Photograph courtesy of Dr Kundan Kishore.

Fig. 6.12. Cultivar 'Allahabad Surkha'. Photograph courtesy of Dr Kundan Kishore.

'Apple Guava'

A selection from 'Apple Colour' seedlings, developed at Punjab Agricultural University, Ludhiana. Fruits medium in size and round in shape, dark red colour peel (Fig. 6.13) having creamy pulp with medium-sized seeds. Fruits have TSS of 11.83°Brix and 0.45% acidity.

'Harijiha'

Fruits are greenish yellow in colour, medium in size (80–135 g), sweet (8.8–10.0°Brix), with good keeping quality.

'Sardar' ('Lucknow-49')

Fruit is spherical to round in shape with primrose-yellow skin colour. Pulp is white with many seeds (Fig. 6.14). The °Brix, acidity and ascorbic acid contents are 9.2 to 11.6, 0.32 to 0.42% and 130 to 215 mg 100 g^{-1} pulp, respectively (Mitra *et al.*, 1983; Chandra and Govind, 1991; Kundu *et al.*, 1995).

Fig. 6.13. Cultivar 'Apple Guava'. Photograph courtesy of Dr K.B. Gill.

Fig. 6.14. Cultivar 'Sardar'. Photograph courtesy of Dr Kundan Kishore.

'Seedless'

Fruits are of irregular shape and yellow with thin skin, warty surface and swollen calyx end. Pulp white, good taste and aroma and has vitamin C content about 240 mg 100 g^{-1} pulp.

'Arka Mridula'

This is a selection from open-pollinated seedlings of 'Allahabad Safeda'. Fruits weigh about 180–200 g, white firm pulp (Fig. 6.15) with soft seeds, sweet (12°Brix) and good keeping quality (Bhalekar and Chalak, 2017).

'Lalit'

Seedling selection from a local, red-pulped type. Fruits are round, attractive saffron-yellow colour and medium in size (185 g), pink pulp (Fig. 6.16) with good sugar acid blend.

'Swetha'

Selected from open-pollinated seeds of a coloured guava variety. Fruits are subglobose in shape with few soft seeds (Fig. 6.17), very sweet (14°Brix), with good keeping quality.

Fig. 6.15. Cultivar 'Arka Mridula'. Photograph courtesy of Dr Kundan Kishore.

'Lalima'

A red-coloured selection. Attractive, crimson-coloured fruit (Fig. 6.18), high yielder, sweet (TSS of 13.7°Brix) and has a good shelf-life.

Fig. 6.16. Cultivar 'Lalit'. Photograph courtesy of Dr Kundan Kishore.

Fig. 6.17. Cultivar 'Swetha'. Photograph courtesy of Dr Kundan Kishore.

Fig. 6.18. Cultivar 'Lalima'. Photograph courtesy of Dr A. Bhattacherjee.

'Dhawal'

A half-sib selection from 'Allahabad Safeda'. High yielder, large attractive fruits (200–250 g), pulp is white (Fig. 6.19), sweet (TSS of 13.4°Brix) with muskiness, seeds are soft and moderate in number.

'Pant Prabhat'

Seedling selection from 'Allahabad Safeda'. Highly productive cultivar (100–125 kg per tree per year). Fruits are medium sized (110 g) with smooth skin, white pulp (Fig. 6.20) and medium-soft seeds. The °Brix, acidity and vitamin C contents are 11.82, 0.27%

Fig. 6.19. Cultivar 'Dhawal'. Photograph courtesy of Dr A. Bhattacherjee.

Fig. 6.20. Cultivar 'Pant Prabhat'. Photograph courtesy of Dr Kundan Kishore.

and 230.4 mg 100 g^{-1} pulp, respectively (Mehta *et al.*, 2016).

'Chittidar'

Fruits are small to medium, subglobose in shape, straw-yellow in colour with red spots of pin-head size on fruit skin. The °Brix, acidity and vitamin C contents are 10.1–13.07, 0.18–0.52% and 164–196 mg 100 g^{-1} pulp, respectively (Kumar, 1998; Pratibha and Lal, 2017).

'Safed Jam'

A hybrid of 'Allahabad Safeda' × 'Kohir' (a local collection from Hyderabad-Kanataka region). Fruits are big with few soft seeds and good keeping quality. The vitamin C content is higher than in both parents.

'Kohir Safeda'

This is a high-yielding hybrid developed by crossing of selected lines of 'Kohir' × 'Allahabad Safeda'. Tree is vigorous, fruits are larger with few soft seeds and white pulp.

'Arka Amulya'

A hybrid of 'Allahabad Safeda' and 'Triploid'. Plants are semi-vigorous and spreading (Bhalekar and Chalak, 2017). Fruits are medium sized, white pulp (Fig. 6.21), 12.5°Brix with good keeping quality (Satisha *et al.*, 2016).

'Arka Kiran'

This is from the cross of 'Kamsari' × 'Purple Local'. Pulp is pink in colour, fruit on average weighs 200–220 g, seed hardness is 4–6 kgf and TSS is 12°Brix with lycopene content of 7.45 mg 100 g^{-1}.

'Arka Rashmi'

This is also a hybrid of 'Kamsari' × 'Purple Local'. Medium-sized fruit (200 g), deep pink pulp (Fig. 6.22) with high vitamin C (200–220 mg 100 g^{-1}) and lycopene (4.5–6.0 mg 100 g^{-1}) contents. Fruits have TSS of 11–12°Brix and are low in oxalate content (28.3 mg 100 g^{-1}). The hybrid is suitable to use both as table fruit as well as for processing.

'Arka Poorna'

A hybrid of 'Purple Local' × 'Allahabad Safeda'. Medium-sized fruit (240 g), pulp white (Fig. 6.23) and firm with vitamin C content of 190–200 mg 100 g^{-1}. The hybrid is suitable to use both as table fruit as well as for processing.

'Hisar Safeda'

A hybrid of 'Allahabad Safeda' × 'Seedless'. Fruits are round, about 92 g in weight with creamy white pulp (Fig. 6.24) and few soft seeds. The °Brix and vitamin C contents are 13.4 and 185 mg 100 g^{-1} pulp, respectively (Daulta *et al.*, 1998).

Fig. 6.21. Cultivar 'Arka Amulya'. Photograph courtesy of Dr Kundan Kishore.

Fig. 6.22. Cultivar 'Arka Rashmi'. Photograph courtesy of Dr A. Bhattacherjee.

'Hisar Surkha'

A hybrid of 'Apple Colour' × 'Banarasi Surkha'. Round fruits weigh about 80 g, pulp is pink (Fig. 6.25) having 13.6°Brix, 0.48% acidity and vitamin C content of 169 mg 100 g^{-1} pulp (Daulta *et al.*, 1998).

Fig. 6.23. Cultivar 'Arka Poorna'. Photograph courtesy of Dr M.R. Dinesh.

Fig. 6.24. Cultivar 'Hisar Safeda'. Photograph courtesy of Dr Kundan Kishore.

Fig. 6.25. Cultivar 'Hisar Surkha'. Photograph courtesy of Dr Kundan Kishore.

'Punjab Pink'

A hybrid between ('Portugal' × 'L-49') × 'Apple Colour'. The tree is vigorous with drooping branches and a prolific bearer. Fruit is medium to large in size. The pulp is red (Fig. 6.26) having a pleasant flavour. TSS ranges from 10.5 to 12°Brix.

'Punjab Kiran'

This is a hybrid between 'Apple Guava' × '1716'. Fruits medium in size, round to oblong in shape, pink-coloured pulp (Fig. 6.27) with small and soft seeds and have a good shelf-life.

Fig. 6.26. Cultivar 'Punjab Pink'. Photograph courtesy of Dr K.B. Gill.

Fig. 6.27. Cultivar 'Punjab Kiran'. Photograph courtesy of Dr K.B. Gill.

'Punjab Safeda'

This is a hybrid of 'Shweta' × '1716'. Fruit medium to large in size, round with smooth creamy-white skin having white pulp and firm texture (Fig. 6.28). Fruits have TSS of 13.4°Brix and acidity of 0.62%. It can be stored for a long time (14 days at 7 ± 1°C and 85–90% relative humidity).

6.11.5 Israel

'Ben-Dov'

The only cultivar grown commercially in Israel, selected in the 1950s (Fig. 6.29). Pear-shaped fruit, white pulp, small seed cavity containing a relatively small number of seeds, and a sweet and good taste accompanied by a very strong and dominant flavour (Zipori *et al.*, 2007).

Fig. 6.28. Cultivar 'Punjab Safeda'. Photograph courtesy of Dr K.B. Gill.

6.11.6 Malaysia

'GU-8' ('Jambu Kampuchea')

A popular cultivar of Malaysia. Fruit roundish in shape, large size (450–750 g), sweet, pulp white, thick and crispy having many seeds.

'GU-10' ('Klom Sali')

Fruit round in shape, big fruits (300–600 g), greenish white skin, pulp white, sweet, thick and crispy.

'GU-12' ('Buah Hati Seronok')

This is a seedless cultivar. Fruit round in shape, large, pulp white, thick and crunchy. Sweet with some sour taste and good flavour.

'GU-13' ('Dam Rung')

It is believed to be introduced from Thailand. Fruit is round in shape, pulp white, crispy, sweet and seedless.

'GU-14' (TCG or 'Thai Cambodian Guava')

A seedless cultivar, round-shaped fruit, pulp white, sweet and coarse textured.

Fig. 6.29. Cultivar 'Ben-Dov'. Photograph courtesy of Dr Ron Port.

'GU-15' ('Jade Seedless')

Fruits are pear shaped, skin colour green, pulp white, crispy, sweet and seedless.

6.11.7 Mexico

'Media China'

Fruits are ovate to pear shaped, medium in size (60–100 g) with yellow skin at maturity, cream to light yellow pulp (Fig. 6.30), °Brix content of 12–14 and vitamin C content of 230 mg 100 g^{-1}. The fruit is well accepted because of its nice flavour.

'Peruana'

Pear-shaped fruit, average weight about 250 g, pulp white, sweet (13.7°Brix), 0.90% acidity and vitamin C content of 130 mg 100 g^{-1}.

6.11.8 Pakistan

'Gola'

Small- to medium-sized fruits, pulp white (Fig. 6.31). The fruits have 8.7°Brix, 1.67% acidity, 5.45% total sugar and vitamin C content of 165.1 mg 100 g^{-1} (Adrees *et al.*, 2010).

'Surahi'

Small- to medium-sized fruits, pulp white. The fruits have 8.23°Brix, 0.52% acidity, 6.24% total sugar and vitamin C content is 136.5 mg 100 g^{-1} (Adrees *et al.*, 2010).

6.11.9 South Africa

'Fan Retief'

Pear-shaped fruit, skin yellow, pulp pink/orange (Fig. 6.32), medium size (100–120 g). The °Brix, acidity and vitamin C contents are 11.2, 0.71% and 240 mg 100 g^{-1}, respectively. Susceptible to GWD (caused by *N. psidii*) (Du Preez and Welgemoed, 1990).

'TSG2'

Fruits are larger (130–140 g) than 'Fan Retief'. Oval-shaped fruit, skin greenish, pulp pink (Fig. 6.33). It is tolerant of the Malalane strain of GWD, but is susceptible to the other two strains, namely Nelspruit and Levubu.

Fig. 6.30. Cultivar 'Media China'. Photograph courtesy of Dr Rodrigo Lasa.

Fig. 6.31. Cultivar 'Gola'. Photograph courtesy of Dr A. Malik.

6.11.10 Taiwan

'Tainung No. 1' ('Diwang Ba')

This is a hybrid of 'Pearl' and 'Crystal'. Fruits are oval, big (440–700 g), pulp white (Fig. 6.34), crispy, sweet (TSS of 13°Brix), vitamin C content of 282 mg 100 g^{-1} and have a long shelf-life.

'Century'

Medium-sized fruit (250 g), peel yellow-green in colour (Fig. 6.35), pulp white and fruits have TSS of 8–9°Brix and vitamin C content of 186.5 mg 100 g^{-1}.

'Crystal'

Big-sized fruits (300–400 g), peel yellow-green in colour, pulp white (Fig. 6.36), hard (non-climacteric cultivar) and the fruits have TSS of 9–11.1°Brix and vitamin C content of 174.2 mg 100 g^{-1}.

'Jen-Ju' ('Pearl Guava')

Big-sized fruits (300–500 g), peel yellow-green in colour (Fig. 6.37), pulp white and

Fig. 6.32. Cultivar 'Fan Retief'. Photograph courtesy of S. Willemse.

Fig. 6.33. Cultivar 'TSG2'. Photograph courtesy of S. Willemse.

Fig. 6.35. Cultivar 'Century'. Photograph courtesy of Dr Huey-Ling Lin.

Fig. 6.34. Cultivar 'Tainung No. 1'. Photograph courtesy of Dr Huey-Ling Lin.

hard (non-climacteric cultivar), fruits have TSS of 9–10.6°Brix and vitamin C content of 234.8 mg 100 g^{-1}.

'Supreme'

Fruits are medium to large (170–280 g), pulp thick, flavour mild, small seed cavity and few seeds. The vitamin C and total sugar contents of the fruits are 247 mg 100 g^{-1} and 7–8%, respectively.

'Rainbow'

Medium-sized fruits (275–300 g), peel yellow-green in colour, pulp red (Fig. 6.38), hard, medium sweet (8–9.7°Brix) and high vitamin C content (300.6 mg 100 g^{-1}).

'Sweet Green'

Non-climacteric cultivar, big-sized fruits (450–500 g), peel green-yellow at maturity,

Fig. 6.36. Cultivar 'Crystal'. Photograph courtesy of Dr Huey-Ling Lin.

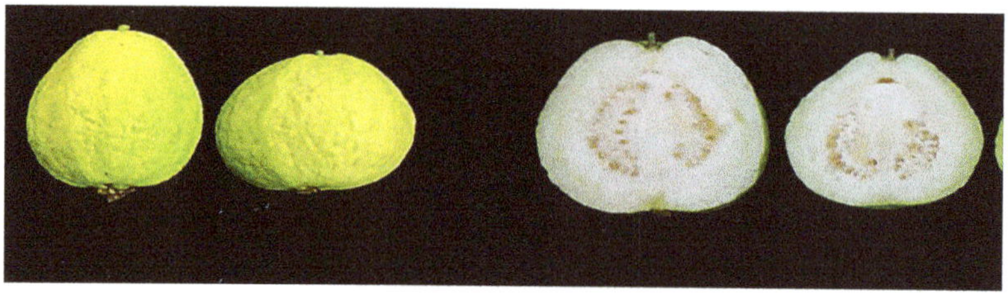

Fig. 6.37. Cultivar 'Jen-Ju'. Photograph courtesy of Dr Huey-Ling Lin.

Fig. 6.38. Cultivar 'Rainbow'. Photograph courtesy of Dr Huey-Ling Lin.

pulp white, TSS of 9.4–10°Brix and vitamin C content of 368.9 mg 100 g^{-1}.

6.11.11 Thailand

'Pan Si Thong'

Large fruits, approximately 300–600 g, peel light green in colour, white pulp (Fig. 6.39), seedy and medium sweet (8.5°Brix).

'Klom Sa Lee'

Large fruits (Fig. 6.40), approximately 500–600 g, light green peel, white pulp, seedy and medium sweet (8.5°Brix).

'Sam See Krob'

Large fruit, approximately 500–700 g, light green peel, white-pink pulp, crispy and sweet (9°Brix).

'Phet Pu Thon'

Medium-sized fruit, approximately 300–400 g, light green colour, white pulp, seedless, medium sweet (8.5°Brix).

Fig. 6.39. Cultivar 'Pan Si Thong'. Photograph courtesy of Dr T. Sangudon.

Fig. 6.40. Cultivar 'Klom Sa Lee'. Photograph courtesy of Dr T. Sangudon.

'Den Khun Wang'

Medium-sized fruit, approximately 300–350 g, light green colour, white pulp, seedless, crispy and sweet (9°Brix).

6.11.12 USA

'Blitch'

Fruits are oval in shape and medium in size. The skin is slightly rough and is yellow to greenish yellow in colour, pulp is light pink with a pleasant tart flavour. The aroma is relatively strong. Seeds are small and numerous. The ripe fruits contain 0.74% total acid and 120 mg vitamin C per 100 g (Campbell, 1984).

'Patillo'

Fruits are ovate to obovate in shape with smooth yellow skin. The pulp is pink in colour, seeds are small and moderate in number. The flavour is subacid with mild aroma. The ripe fruits contain 1.23% total acids and 170 mg vitamin C per 100 g (Campbell, 1984).

'Beaumont'

Medium to large, roundish fruits weighing up to 290 g. Pink pulp (Fig. 6.41), mildly acid and seedy. Excellent for processing. Somewhat susceptible to fruit rots (Pommer and Murakami, 2009).

'Ruby'

Fruits are large, pulp red, thick, sweet, mild flavour, few seeds, total sugar and vitamin C contents are 9–10% and 180 mg 100 g^{-1} pulp, respectively.

'Redland'

Fruits are large, pyriform, no musky guava odour, mild flavour, matures in winter, and a heavy cropper.

'Ka Hua Kula'

Fruits are oblong, larger in size (207–277 g) than those of 'Beaumont' with more intensely pink-coloured pulp (Fig. 6.42), fewer seeds and low in total acidity. It gives yields equal to those of 'Beaumont'.

'Ruby' × ('Ruby' × 'Supreme')

Small, roundish fruit. Skin greenish yellow. Flesh dark pinkish orange. Flavour delicious, seed cavity 33% of pulp.

6.11.13 Vietnam

'Leld-tc15'

Developed and released by SOFRI, it has large fruits (289–295 g), green-yellow colour skin, white pulp, soft seeds, sweet (10.18 to 11.46°Brix), 0.41 to 0.52% acidity and vitamin C content of 45.7 to 50.5 mg 100 g^{-1} pulp.

'Nu Hoang'

Medium-sized fruits (213.5 to 230.0 g), green-yellow colour skin (Fig. 6.43), white pulp, hard seeds, °Brix of 9.8 to 10.5, acidity of 0.19 to 32% and vitamin C content of 42.7 to 50.4 mg 100 g^{-1} pulp.

'Tran Chau'

Large fruits (230 to 290 g), greenish yellow colour fruit (Fig. 6.44), pulp thick (1.6 to 1.8 cm)

Fig. 6.41. Cultivar 'Beaumont'. Photograph courtesy of Dr T. Matsumoto.

Fig. 6.42. Cultivar 'Ka Hua Kula'. Photograph courtesy of Dr T. Matsumoto.

Fig. 6.43. Cultivar 'Nu Hoang'. Photograph courtesy of Dr N. Truong.

Fig. 6.44. Cultivar 'Tran Chau'. Photograph courtesy of Dr N. Truong.

and pink in colour, crunchy, low in TSS (7.0 to 8.07°Brix).

'Tim'

Fruits are small (132 to 148 g), have dark-purple colour skin as well as pulp (Fig. 6.45). Pulp thick (1.18 to 1.39 cm), hard seed, medium sweet (6.03 to 9.04°Brix).

Fig. 6.45. Cultivar 'Tim'. Photograph courtesy of Dr N. Truong.

6.12 Conclusion

Guava is grown in more than 70 countries of the world. However, guava is being considered as a commercial crop in about 15 countries of the world. Each of these countries has developed its own cultivars for commercial production. The major problems facing the guava industry are the severity of wilt disease, the high seed content of diploid cultivars and poor yield, with small, irregular and misshapen fruits of triploid seedless cultivars. The major breeding objectives are aimed at improving both plant and fruit characteristics such as to develop high-yielding, high-quality dwarf cultivars with fruits of uniform shape, good size, attractive skin and pulp colour, fewer and/or soft seeds, resistant to wilt, nematodes and long storage life. Selection of superior seedlings has resulted in the development of a number of cultivars in different countries. Several hybrids, particularly from India, have been developed through hybridization that are less seeded, soft-seeded or seedless, having high TSS, vitamin C and lycopene contents with desirable fruit aroma. Interspecific hybridization between wild *Psidium* species and *P. guajava* has led to the development of hybrid rootstocks that are resistant to wilt and nematodes and are graft compatible with *P. guajava*. Molecular characterization of germplasm is being carried out and intensified in the last decade to understand genetic distance so that good recombinations can be arrived at by crossing suitable parents.

Table 6.1. Classification based on fruit shape.

Globose	'Allahabad Safeda', 'Apple Colour', 'Arka Amulya', 'Arka Kiran', 'Benaras', 'Behat Coconut', 'Chittidar', 'Dharwad', 'EC 147037', 'Hafsi', 'Local 2', 'Mirzapur Seedling', 'Nasik', Philippine guava, 'Red Flesh', 'Sindh', 'Smooth Green', 'Surka Chitti', 'Dhareedar', 'Aneuploid-2', 'Lalit', 'Hisar Safeda', 'Hisar Surkha', 'Arka Rashmi'
Subglobose	'Swetha', '7-12 EC 147036', '9-35 EC 147036', 'EC 14089', 'EC 162904', 'Kamsari', 'Karela', 'Local 1', 'Sardar', 'Pati', 'Portugal', 'GR-1', 'Spear Acid'
Pyriform	'Redland', 'Paluma', 'Bangalore Local', 'G-6', 'White Flesh'
Ovate	'Florida Seedling', 'Surka Chitti Neptuani', 'Media China'
Oblong	'Aneuploid-1', '7-39 EC 147034', 'Nagpur Seedless', 'Seedless Triploid', 'Thailand 2', 'Ka Hua Kula', 'Xiguahong', 'Lucknow-42'
Oval	'Supreme', 'TSG2', 'Blitch', 'Jindouxiang'

Table 6.2. Classification based on fruit weight, pulp colour, total soluble solids (TSS) and vitamin C content of fruit.

Fruit weight (>150 g)	'Allahabad Safeda', 'Paluma', 'Pedro Sato', 'Dhawal', 'Beaumont', 'Supreme', 'Ka Hua Kula', 'Seculo XXI', 'Arka Mridula', 'Arka Rashmi', 'LELD-TC15', 'Nu Hoang', 'Tran Chau', 'ZenZhu', 'Pan Si Thong', 'Klom Sa Lee', 'Peruna', 'Sam See Krob', 'Phet Pu Thon', 'Den Khun Wang', 'Winter', 'GU-4', 'GU-5', 'GU-7', 'Emperor', 'Crystal', 'Jen-Ju', 'Tainung No. 1', 'Century', 'Rainbow'
White pulp colour	'Allahabad Safeda', 'Sardar', 'Arka Amulya', 'Arka Mridula', 'Hisar Safeda', 'Pant Prabhat', 'Swetha', 'ZenZhu', 'Nu Hoang', 'Le', 'Ben-Dov', 'Omri', 'Roni', 'Lior', 'Pan Si Thong', 'Kim Ju', 'Winter Late', 'El-Sabahia', 'Media China', 'Dhawal', 'Seedless', 'LELD-TC15', 'Klom Sa Lee', 'Peruna'
Pink pulp colour	'Homestead', 'Blanca', 'Blitch', 'Patillo', 'Beaumont', 'Ka Hua Kula', 'Lalit', 'Pink Supreme', 'Red Flesh', 'Arka Rashmi', 'H-724', 'Hisar Surkha', 'Lalima', 'Punjab Pink', 'Sassaoka', 'Xiguahong', 'Allahabad Surkha', 'Seculo XXI', 'Pedro Sato', 'Cortibel', 'Tran Chau', 'Gizy Red', 'Malaysian Red', 'Punjab Pink', 'Yilian Red'
Red pulp colour	'Kamsari', 'Purple Local', 'Paluma', 'Rica', 'Tim', 'Yuval', 'Zohar', 'Gili', 'Ido', 'Ruby'
TSS (°Brix) (>12)	'Swetha', 'Lalima', 'CISH-G-1', 'Dhawal', 'Dhareedar', 'Arka Mridula', 'Arka Amulya', 'Arka Kiran', 'Hisar Safeda', 'Hafsi', 'Allahabad Safeda', 'Behat Coconut'
Vitamin C (mg 100 g^{-1} pulp) (>150 mg)	'Cotorrera', 'Micro-guayaba', 'Dario 18-2', 'Mirzapur Seedling', 'Sardar', 'Seedless', 'Pant Prabhat', 'Xiguahong', 'Malaysian Red', 'Winter', 'Fan Retief', 'Ruby', 'Supreme', 'Jen-Ju', 'Diwan', 'Rainbow', 'Arka Rashmi', 'Arka Poorna', 'Sweet Green', 'Crystal', 'Tainung No. 1', 'Century'

Compared with other crops, genomic studies in guava have been neglected in the past and are still in their infancy. The advances of molecular biology and particularly in genomics allow the application of novel techniques and strategies for breeding (the progress in molecular breeding is also dealt with in Chapter 5, this volume, where future research needs have been suggested). The unfortunate scenario is that some of the major guava-growing countries have either discontinued guava breeding projects or minimized the activity.

Different cultivars are grouped based on fruit shape, large fruit weight, pulp colour, high TSS and vitamin C content of fruit which may be used as probable gene donors in breeding of guava (Tables 6.1 and 6.2).

References

Abdul Karim, S., Yuen, P.M. and Norila, Y. (2002) Challenges in the integrated management of the root-knot nematode on guava. In: *Proceedings Mesyuarat Teknikal Pusat HR, Port Dickson, Malaysia*, p. 27.

Adrees, M., Younis, M., Farooq, U. and Hussain, K. (2010) Nutritional quality evaluation of different guava varieties. *Pakistan Journal of Agricultural Science* 47(1), 1–4.

Alifa, M.A., Vasugi, C. and Dinesh, M.R. (2017) Interspecific hybridization – an approach for enhancing guava productivity. In: *National Seminar on Enhancing Productivity of Fruit Crops – Mitigating Major Challenges, Indian Institute of Horticultural Research, Bengaluru, India*, p. 113 (abstract no. S1A3).

Andrade, R.S.G., Diniz, M.C.T., Neves, E.A. and Nobrega, J.A. (2002) Determinacao e distribucao de acido ascorbico em tres frutos tropicais. *Ecletica Quimica* 27, 393–401.

Anonymous (2003–04) *Annual Report*. Central Institute for Subtropical Horticulture, Lucknow, India, pp. 26–27.

Anonymous (2016) *Annual Report*. Central Institute for Subtropical Horticulture, Lucknow, India, pp. 37–38.

Aranguren, Y., Briceno, A. and Fermin, G. (2010) Assessment of the variability of Venezuelan guava landraces by microsatellites. *Acta Horticulturae* 849, 77–86.

Atchison, E. (1947) Chromosome number in Myrtaceae. *American Journal of Botany* 34, 159–164.

Bajpai, A., Srivatava, N., Chandra, R. and Rajan, S. (2011) Guava. In: Singh, H.P., Parthasarathy, V.A. and Nirmal Babu, K. (eds) *Advances in Horticultural Biotechnology*, Vol. III. *Molecular Markers and Marker Assisted Selection: Fruit Crops, Plantation Crops and Spices*. Westville Publishing House, New Delhi, pp. 79–98.

Belaj, A., Zatovic, Z., Cipriani, G., Baldoni, L., Testolin, R. *et al.* (2003) Comparative study of the discriminating capacity of RAPD, AFLP and SSR markers and their effectiveness in establishing genetic relationship in olive. *Theoretical and Applied Genetics* 107, 736–744.

Benade, E., Kemp, C.H.J., Wingfield, M.J. and Kock, J.F.L. (1991) Comparison of *Acremonium diospyri* with guava wilt pathogen in South Africa. *Phytophylactica* 23, 98.

Bhalekar, S.G. and Chalak, S.U. (2017) Studies on performance of different guava cultivars under western Maharashtra conditions. *Electronic Journal of Plant Breeding* 8(2), 577–579.

Bolar, J.P., Norelli, J.L., Wong, K.W., Hayes, C.K., Harman, G.E. and Aldwinkle, H.S. (2000) Expression of *endochitinase* from *Trichoderma harzianum* in transgenic apple increases resistance to apple scab and reduces vigour. *Phytopathology* 90, 72–77.

Brar, H.S. and Bal, J.S. (2003) Studies on the use of gamma rays on the performance of guava budlings. *Annals of AgriBio Research* 8(2), 213–217.

Briceno, A., Aranguren, Y. and Fermin, C. (2010) Assessment of guava-derived SSR markers for the molecular characterization of Myrtaceae from different ecosystems in Venezuela. *Acta Horticulturae* 849, 139–146.

Brooks, R.M. and Olmo, H.P. (1952) *Register of the Fruit and Nut Varieties 1920–1950*. University of California Press, Berkley and Los Angeles, California.

Campbell, C.W. (1984) Guava: tropical fruits and nuts. In: Martin, F.W. (ed.) *CRC Handbook of Tropical Food Crops*. CRC Press, Boca Raton, Florida, pp. 254–256.

Canizares, Z.J. (1981) Breeding guava, *Psidium guajava* by mass selection. *Ciencia y Tecnica en la Agricultura, Citricos y Otros Frutales* 4, 7–21.

Chakraborti, S., Sinha, S. and Sinha, R.K. (2010) Chromosome number and karyotype analysis of wild guava *Psidium guineense* Sw. – a new report from Tripura, India. *Indian Journal of Science and Technology* 3, 925–927.

Chandra, R. and Govind, S. (1991) Studies on yield and quality of some guava cultivars. *Indian Journal of Hill Farming* 4, 15–18.

Chandra, R. and Mishra, M. (2007) Biotechnological interventions for improvement of guava (*Psidium guajava* L.). *Acta Horticulturae* 735, 117–126.

Cheema, G.S. and Desmukh, G.B. (1927) *Culture of Guava and Its Improvement by Selection in Western India*. Bulletin No. 148. Department of Agriculture, Bombay, India.

Cheema, G.S., Bhat, S.S. and Naik, K.C. (1954) *Commercial Fruits of India*. Macmillan, Calcutta, India.

Chen, J., Chen, K. and Ou, Y. (2002) Introduction and cultivation techniques for yield and quality of guava cv. Zhenzhu. *China Fruits* 1, 38–39.

Chen, T.W., Ng, C.C., Wang, C.Y. and Shyu, Y.T. (2007) Molecular identification and analysis of *Psidium guajava* L. from indigenous tribes of Taiwan. *Journal of Food and Drug Analysis* 15, 82–88.

Coser, S.M., Ferreira, M.F.d.S., Ferrera, A. and Saraiva, S.H. (2014) Genetic diversity in Cortibel guava selections. *Revista Brasileira de Fruticultura* 36(2), 391–399.

Costa, I.R. and Forni-Martins, E.R. (2007) Chromosome studies in *Gomidesia, Marlierea, Myrceugenia* and *Myrcia* (Myrtaceae, subtribe Myrciinae). *Kew Bulletin* 62, 113–118.

Costa, I.R., Dornelas, M.C. and Forni-Martins, E.R. (2008) Nuclear genome size variation in fleshy-fruited, *Neotropical* Myrtaceae. *Plant Systematics and Evolution* 276, 209–217.

Costa, S.R.d., Santos, C.A.F and Castro, J.M.C. (2012) Assessing *Psidium guajava* × *P. guineense* hybrids tolerance to *Meloidogyne enterolobii*. *Acta Horticulturae* 959, 59–65.

Cuadra, R. and Quincosa, A. (1982) Potential of different *Psidium guajava* species as source for resistance of guava to *Meloidogyne*. *Ciencias de la Agriculture* 13, 19–26.

Çelik, B. (2019) *Determination of the characteristic properties of some guava genotypes and investigation of hybridization possibilities*. Master's thesis, Akdeniz University, Turkey.

Dahiya, K.K., Archak, S. and Karihaloo, J.L. (2002) DNA fingerprinting of guava (*Psidium guajava* L.) cultivars using RAPD markers. *Indian Journal of Plant Genetic Resources* 15, 112–115.

Das, P.K. (1971) Induction of somatic mutations in some horticultural crops by chronic gamma irradiation. *Indian Journal of Horticulture* 28(1), 1–6.

Dasarathy, T.B. (1951) The guava. *Madras Agriculture Journal* 38, 521–527.

Dass, H.C. and Prakash, D. (1981) Phylogenic affinities in *Psidium* sp. as studied by flavonoid patterns. In: *Abstracts Book, National Symposium on Tropical and Subtropical Fruits Crops, Bangalore, India*, abstract no. 23.

Daulta, B.S., Godara, N.R., Bhatia, S.K., Singh, S.H.K. and and Bisla, S.S. (1998) New guava hybrids: 'Hisar Safeda' and 'Hisar Surkha'. *Indian Horticulture* 43, 22–23.

D'Cruz, R. and Babu Rao, G. (1962) Cytogenetic studies in two guava aneuploids. *Journal of the Indian Botanical Society* 41, 316–321.

Dinesh, M.R. and Vasugi, C. (2015) Guava breeding. In: Dinesh, M.R. (ed.) *Fruit Breeding*. New India Publishing Agency, New Delhi, pp. 277–288.

Dinesh, M.R. and Yadav, I.S. (1998) Half-sib analysis in guava (*Psidium guajava*). *Indian Journal of Horticulture* 55, 20–22.

Dinesh, M.R., Bharathi, K., Vasugi, C., Veena, G.L., Ravishankar, K.V. and Nischita, P. (2017) Inheritance study and validation of hybridity in guava (*Psidium guajava*). *Indian Journal of Agricultural Science* 87(1), 42-45.

Du Preez, R. and Welgemoed, C.P. (1990) Variability in fruit characteristics of five guava selections. *Acta Horticulturae* 275, 351–359.

Edward, J.C. and Shankar, G. (1964) Rootstock trial for guava (*Psidium guajava* L.). *Allahabad Farming* 38, 521–527.

Fermin, G. (2010) On the cultivation of guava in Venezuela. *Acta Horticulturae* 849, 77–86.

Fernandes-Santos, C.A., Cunha-Castro, J.M., Franca-Souza, F.d., Alcantara-Vilarinho, A., Ferreira, F.R.d. *et al.* (2010) Prospecting and morphological characterization of Brazilian *Psidium* germplasm. *Acta Horticulturae* 849, 63-68.

Franco, M.R. and Shibamoto, T. (2000) Volatile compound of some Brazilian fruits: umbu-caja (*Spondias citherea*), camu-camu (*Myrciaria dubia*), araca-boi (*Eugenia stipitata*) and cupuacu (*Theobroma grandiflorum*). *Journal of Agricultural and Food Chemistry* 48, 1263–1265.

Freitas, V.M., Correa, V.R., Motta, F.C., Sousa, M.G., Gomes, A.C.M.M. *et al.* (2014) Resistant accessions of wild *Psidium* spp. to *Meloidogyne enterolobii* and histological characterization of resistance. *Plant Pathology* 63, 738–746.

Gomes, V.M., Souza, R.M., Medorikawa, G., Miller, R. and Almeida, A.M. (2012) Guava decline: evidence of nationwide incidence in Brazil. *Nematropica* 42(1), 153–162.

Gomes, V.M., Souza, R.M., Almeida, A.M. and Dolinski, C. (2014) Relationships between *M. enterolobii* and *F. solani*: spatial and temporal dynamics in the occurrence of guava decline. *Nematoda* 1, e01014.

Gomes, V.M., Ribeiro, R.M., Viana, A.P., Souza, R.M.d., Santos, E.A. *et al.* (2017) Inheritance of resistance to *Meloidogyne enterolobii* and individual selection in segregating population of *Psidium* spp. *European Journal of Plant Pathology* 148(3), 699–708.

Grech, N.M. (1982) *Disease Management Progress Report.* CSFRI, Nelspruit, South Africa.

Grech, N.M (1984) *Disease Management Progress Report.* CSFRI, Nelspruit, South Africa.

Gupta, N. and Kour, A. (2019) Genetic parameters, character association and path analysis for fruit yield and its component characters in guava (*Psidium guajava* L.). *Electronic Journal of Plant Breeding* 10(1), 256–263.

Handayani, T., Witjaksono and Utami Nugraheni, K. (2017) *In vitro* induction of tetraploid in guava (*Psidium guajava* L.). *Journal Biologi Indonesia* 13(2), 271–278.

Hsieh, H.Y. (2011) Taiwan guava variety improvement and industrial development. In: *Proceedings of the Guava Cultivation Technology and Management Conference, Taichung District Agricultural Research and Extension Station Special Issue No.* 108, pp. 9–20.

Hwang, S.G., Li, Y.Y. and Lin, H.L. (2017) Excellent nutritional value in fruits of three guava cultivars in Taiwan. *Acta Horticulturae* 1166, 209–213.

Jacobo, C.M., Toriz Ahumada, L.M. and Maldonado, S.H.G. (2009) Characterization of guava selection for the Bajio Region of Guanajuato, Mexico. *Agricultura Tecnica en Mexico* 35(3), 315–322.

Janaki Ammal, E.K.J. (1951) Chromosomes and horticulture: tetraploids in guava. *Journal of the Royal Society of Horticulture* 76, 236–239.

Jose, M.C.C., Santos, C.A.F. and Flori, J.E. (2012) Reaction of *Psidium* accessions to the nematode *Meloidogyne enterolobii*. *Presented at 3rd International Symposium on Guava and Other Myrtacea, Petrolina, Brazil, 23–25 April 2012*, abstract no. 36.

Kamle, M., Kumar, P., Bajpai, A., Kalim, S. and Chandra, R. (2013) Assessment of genetic fidelity of *in vitro* regenerated guava (*Psidium guajava* L.) plants using DNA based markers. *New Zealand Journal of Crop and Horticulture Science* 42, 1–9.

Kanupriya, L.P.M., Aswath, C., Reddy, L., Padamkar, B., Vasugi, C. and Dinesh, M.R. (2011) Cultivar identification and genetic fingerprinting of guava (*Psidium guajava* L.) using microsatellite markers. *International Journal of Fruit Science* 11, 184–196.

Kareem, A., Jaskani, M.J., Mehmood, A., Khan, I.A., Awan, F.S. and Sajid, M.W. (2018) Morpho-genetic profiling and phylogenetic relationship of guava (*Psidium guajava* L.) as genetic resources in Pakistan. *Revista Brasileira de Fruticultura, Jaboticabal* 40(4), E-069. https://doi.org/10.1590/0100-29452018069

Kuang, S., Zhang, C. and Fang, J. (2018) Introduction of guava cv. ZhenZhu and integrated cultivation techniques. *Modern Agricultural Science and Technology* 14, 78–79.

Kumar, R. (1998) Performance of guava under rainfed conditions of Bihar. *Haryana Journal of Horticultural Science* 27, 145–148.

Kundu, S., Ghosh, S.N. and Mitra, S.K. (1995) Physico-chemical characters of twelve guava cultivars in the laterite tract of West Bengal. *Indian Food Packer* 49, 11–16.

Lee, W.-L., Huang, C.C., Kuan, C.S., Jiang, S.W. and Hsieh, H.Y. (2017) Development of GA3 tropical fruit varieties and cultivation techniques in Taiwan. *Acta Horticulturae* 11, 47–54.

Liu, Q., Ou, Z. and Gong, D. (2008) A report on the introduction experiment on Pearl variety of *Psidium guajava*. *Guizhou Agricultural Sciences* 36(4), 21–22.

Majumder, P.K. and Mukherjee, S.K. (1970) Isolation of trisomics and tetrasomics in guava (*Psidium guajava* L.). *Current Science* 39, 409–410.

Mani, A., Mishra, R. and Thomas, G. (2011) Elucidation of diversity among *Psidium* species using morphological and SPAR methods. *Journal of Physiological Research* 3, 53–61.

Manica, I. (2000) *Frutas Nativas, Silvestres e Exoticas I: Tecnicas de Producao e Mercado: abiu, amora-preta, araca, bacuri, biriba, carambola, cereja-do-rio-grande, jabuticaba.* Cinco Continentes, Porto Alegre, Brazil.

Manicom, B.Q. (1980) *Disease Management Progress Report.* CSFRI, Nelspruit, South Africa.

Marques, A.M., Tuler, A.C., Carvalho, C.R., Carrijo, T.T., Ferreira, M.F.d.S. and Clarindo, W.R. (2016) Refinement of the karyological aspects of *Psidium guineense* (Swartz, 1988): a comparison with *Psidium guajava* (Linnaeus, 1753). *Comparative Cytogenetics* 10(1), 117–128.

Martin, F.W., Campbell, C.W. and Ruberte, R.M. (1987) *Perennial Edible Fruit of the Tropics: An Inventory. Agriculture Handbook No. 642.* US Department of Agriculture, Agricultural Research Service, Washington, DC, p. 42.

Martinez de Lara, J., Barrientos Lara, M.C., Reyes de Anda, A.C., Hernandez Delgado, S., Padilla Ramirez, J.S. and Mayek Perez, N. (2004) Diversidad fenotipica y genetic en huertas del guayabo (*Psidium guajava* L.) de Calvillo, Aguascalientes. *Revista Fitotecnia Mexicana* 27, 243–249.

Martins, L.S.S., Musser, R.S., Souza, A.G., Resende, L.V. and Mulaf, W.R. (2013) Parasitismo de *Meloidogyne enterolobii* em species de Myrtaceae. *Revista Brasileira de Fruiticultura* 35, 477–484.

Mehta, S.K., Singh, K.K., Jat, D.K. and Rana, D.K. (2016) Comparative studies of physico-chemical characteristics of various cultivars of guava (*Psidium guajava* L.) under sub-tropical valley condition of Garhwal Himalaya (Uttarakhand), India. *Plant Archives* 16(1), 361–364.

Milan, A.R. (2010) Collection and evaluation of guava (*Psidium guajava* L.) for nematode resistance in Malaysia. *Acta Horticulturae* 849, 359–361.

Mishra, M., Jalil, S.U., Sharma, N. and Hudedamani, U. (2014) An *Agrobacterium* mediated transformation system of guava (*Psidium guajava* L.) with *endochitinase* gene. *Crop Breeding and Applied Biotechnology* 14, 232–237.

Mishra, M., Jalil, S.U., Mishra, R.K., Kumari, S. and Pandey, B.K. (2016) *In vitro* screening of guava plantlets transformed *with endochitinase* gene against *Fusarium oxysporum* f.*sp. psidii*. *Czech Journal of Genetics and Plant Breeding* 52(10), 6–13.

Misra, A.K. (1998–1999) Screening against wilt. In: *Annual Report*. CIHNP, Lucknow, India, p. 10.

Mitra, S.K. (2010) Important Myrtaceae fruit crops. *Acta Horticulturae* 849, 33–38.

Mitra, S.K. and Bose, T.K. (1985) Guava. In: Bose, T.K. (ed.) *Fruits of India: Tropical and Subtropical*. Naya Prakash, Calcutta, India, pp. 277–297.

Mitra, S.K. and Sanyal, D. (2004) *Guava*. Indian Council of Agricultural Research, New Delhi.

Mitra, S.K. and Thingreingam Irenaeus, K.S. (2018) Guava cultivars of the world. *Acta Horticulturae* 1205, 905–910.

Mitra, S.K., Maiti, S.C., Sen, S.K. and Bose, T.K. (1983) Physico-chemical characters of some guava varieties of West Bengal. *South Indian Horticulture* 31, 62–65.

Mitra, S.K., Irenaeus, T.K.S., Gurung, M.R. and Pathak, P.K. (2012) Taxonomy and importance of Myrtaceae. *Acta Horticulturae* 959, 23–34.

Mohammed, S. (1974) Aneuploidy in guava. *Biologia Plantarum* 16(5), 382–388.

Mondragon-Jacobo, C., Toriz-Ahumada, L.M. and Guzman-Maldonado, H. (2010) Generation of pink-fleshed guava to diversify commercial production in central Mexico. *Acta Horticulturae* 849, 333–339.

Mukherjee, S.K. (1997) Improvement of mango, grapes and guava. In: Nijjar, G.S. (ed.) *Fruit Breeding in India*. Oxford and IBH, New Delhi, pp. 15–20.

Naithani, S.P. and Srivastava, H.C. (1966) Autotetraploidy in *Psidium guajava* L. *Naturwissenchaften* 8, 205–206.

Nakasone, H.Y. and Ito, P.J. (1978) 'Ka Hua Kula' guava. *HortScience* 13(2), 197.

Nakasone, H.Y. and Paull, R.E. (eds) (1998) Guava. In: *Tropical Fruits*. CAB International, Wallingford, UK, pp. 149–172.

Negi, S.S. and Rajan, S. (2007) Improvement of guava through breeding. *Acta Horticulturae* 735, 31–37.

Nimisha, S., Kherwar, D., Ajay, K.M., Singh, B. and Usha, K. (2013) Molecular breeding to improve guava (*Psidium guajava* L.): current status and future prospects. *Scientia Horticulturae* 164, 578–588.

Nogueira, A.M., Ferreira, M.F.S. and Mangaravite, E. (2012) Preliminary study of wild guava from Espirito Santo and Minas Gerais by continuous descriptors. *Acta Horticulturae* 959, 35–40.

Noia, L.R., Coser, S.M., Guilhen, J.H.S., Ferreira, A. and Ferreira, M.F.S. (2012a) Genetic distance of guava genotypes from different altitudes by microsatellites. *Presented at 3rd International Symposium on Guava and Other Myrtacea, Petrolina, Brazil, 23–25 April 2012*, abstract no. 67.

Noia, L.R., Nogueira, A.M., Mangaravite, E., Ferreira, A. and Ferreira, M.F.S. (2012b) Genetic diversity of wild guava from southern Espirito Santo and commercial cultivars by microsatellites. *Presented at 3rd International Symposium on Guava and Other Myrtacea, Petrolina, Brazil, 23–25 April 2012*, abstract no. 68.

Noia, L.R., Tuler, A.C., Ferreira, A. and Ferreira, M.F. (2017) Relationship between *Psidium* species (Myrtaceae) by resistance gene analog markers: focus on nematode resistance. *Genetics and Molecular Research* 16(1), gmr16019441. https://doi.org/10.4238/gmr16019441

Norlia, Y. (1994) Klon-klonjambu batu yang berpotensi. *Tecknologi Buah-buahan* 10, 9–12.

Normand, F. (2002) The strawberry guava: a new fruit species for humid areas in Reunion Island. *Acta Horticulturae* 575, 245–251.

Padilla-Ramirez, J.S. and Gonzalez-Gaona, E. (2010) Collection and characterization of Mexican guava (*Psidium guajava* L.) germplasm. *Acta Horticulturae* 849, 49–54.

Padilla-Ramirez, J.S., Gonzalez-Gaona, E., Valdez-Marin, C., Meracado-Silva, E., Hernandez-Zelgado, S. and Mayek-Perez, N. (2002) Caracterizacion de germplasma sobersaliente de guayabo de la region Calvillo-Canones, Mexico. *Revista Fitotecnica Mexicana* 25, 393–399.

Padilla-Ramirez, J.S., Gonzalez-Gaona, E., Cruz, M.A.P.d., Guitierrez-Acosta, F. and Mayek-Perez, N. (2007) Fruit yield and quality of twelve outstanding selections of guava (*Psidium guajava*) from the Calvillo-Canones region, Mexico. *Acta Horticulturae* 735, 25–30.

Padmakar, B., Kanupriya, C., Madhavi Latha, P., Prashant, K.S. and Dinesh, M.R. (2015) Development of SRAP and SSR marker-based genetic linkage maps of guava (*Psidium guajava* L.). *Scientia Horticulturae* 192, 158–165.

Padmakar, B., Kanupriya, C., Madhavi Latha, P., Vasugi, C., Dinesh, M.R. *et al.* (2016) Enrichment of genetic linkage maps and mapping QTLs specific to seed strength – hardness/softness – in guava (*Psidium guajava* L.). *Journal of Horticultural Sciences* 11(1), 13–20.

Perales, M.A. (1993) Evaluacion del rendimiento y calidad de fruta de colectas de guayaba tipo criollo en la region Calvillo-Canones. In: *V Congreso Nacional de Horticultura, Veracruz, Mexico*, p. 74.

Pereira, F.M. and Nachtigal, J.C. (2002) Goiabeira. In: Bruckner, C.H. (ed.) *Melhoramento de Fruteiras Tropicias*. UFV, Vicosa, Brazil, pp. 267–289.

Pereira, F.M., Carvalho, C.A. and Nachtigal, J.C. (2003) Seculo XXI: nova cultivar de goiabeira de dupla finalidade. *Revista Brasileiria de Fruticultura* 25, 498–500.

Pessanha, P.G.D.O., Viana, A.P., Junior, A.t.D.A., Souza, R.M.D., Teixeira, M.C. and Pereira, M.G. (2011) Assessment of genetic diversity in access to *Psidium* spp. via RAPD markers. *Revista Brasileiria de Fruticultura* 33, 129–136.

Pommer, C.V. (2012) Guava worldwide breeding: major techniques and cultivars and future challenges. *Acta Horticulturae* 959, 81–88.

Pommer, C.V. and Murakami, K.R.N. (2009) Breeding of guava (*Psidium guajava* L.). In: Mohan Jain, S. and Priyadarshan, P.M. (eds) *Breeding Plantation Tree Crops: Tropical Species*. Springer, New York, pp. 83–120.

Prakash, D.P., Narayanaswamy, P. and Sondur, S.N. (2002) Analysis of molecular diversity in guava using RAPD markers. *Journal of Horticultural Science and Biotechnology* 77, 287–293.

Pratibha and Lal, S. (2017) Evaluation of guava varieties for growth, yield and quality attributes in Tarai conditions of Uttarakhand. *International Journal of Tropical Agriculture* 35(3), 699–706.

Quintal, S.S.R., Viana, A.P., Campos, B.M., Vivas, M. and Amral Junior, A.T. (2017) Selection via mixed models in segregating guava families based on yield and quality traits. *Revista Brasileira de Fruticultura* 39, 1–8.

Rai, K.M., Phulwaria, M., Gupta, K.M., Shekhawat, N.S. and Jaiswal, U. (2012) Genetic homogeneity of guava plants derived from somatic embryogenesis using SSR and ISSR markers. *Plant Cell, Tissue and Organ Culture* 111, 259–264.

Rai, K.M., Phulwaria, M. and Shekhawat, N.S. (2013) Transferability of simple sequence repeat (SSR) markers developed in guava (*Psidium guajava* L.) to four Myrtaceae species. *Molecular Biology Reports* 40, 5067–5071.

Raman, W.M., Manimekalai, G. and Ramalingam, R.S. (1969) Observation on seedlessness, fruit development and cytology of varieties of guava. *Madras Agriculture Journal* 56, 255–261.

Raman, V.S., Sri Rangaswamy, S. and Manimekalai, F. (1971) Triploidy and seedlessness in guava (*Psidium guajava* L.). *Cytologia* 36, 392–399.

Ram Kumar (1975) Inducing polyploidy and cytological studies in guava. *Indian Journal of Horticulture* 32(3–4), 128–130.

Raseira, M.d.C.B and Raseira, B.M. (1996) *Contribuicao ao Estudo do Araca*. Embrapa–CPACT, Pelotas, Brazil.

Ray, P.K. (2002) Guava. In: *Breeding Tropical and Subtropical Fruits*. Narosa Publishing House, New Delhi, pp. 143–155.

Resende, M.D.V. (2016) Software Selegen-REML/BLUP: a useful tool for plant breeding. *Crop Breeding and Applied Biotechnology* 16, 330–339.

Ribeiro, I.J.A. and Pommer, C.V. (2004) Breeding guava (*Psidium guajava*) for resistance to rust caused by *Puccinia psidii*. *Acta Horticulturae* 632, 75–78.

Ribeiro, R.M., Gomes, V.M., Viana, A.P., da Souza, R.M. and dos Santos, P.R. (2019) Selection of interspecific *Psidium* spp. hybrids resistant to *Meloidogyne enterolobii*. *Acta Scientiarum Agronomy* 41, e42702.

Risterucci, A.M., Duval, M.F., Rohde, W. and Billote, N. (2005) Isolation and characterization of microsatellite loci from *Psidium guajava* L. *Molecular Ecology Notes* 5, 745–748.

Risterucci, A.M., Nansot, G., Grangeon, R., Lepitre, V., de Reeper, A. *et al.* (2010) Development of guava microsatellite (SSR) markers using the SAT software. *Acta Horticulturae* 849, 113–120.

Rodriguez, M.N., Valdes, J., Rodriguez, J.A., Velasquez, J.B., Rivero, D. *et al.* (2010) Genetic resources and breeding of guava (*Psidium guajava* L.) in Cuba. *Biotechnology Application* 27(3), 238–241.

Ruehle, G.W. (1946) Promising new guava varieties. *Proceedings of the Florida State Horticultural Society* 59, 127–131.

Rye, B. (1979) Chromosome number variation in the Myrtaceae and its taxonomic implications. *Australian Journal of Botany* 27, 547–573.

Sanchez-Teyer, L.F., Barraza-Morales, A., Quiroz-Moreno, A., Ortiz-Garcia, M.M., Becerril-Chi, K. *et al.* (2010) Genetic diversity of Mexican guava germplasm using DNA molecular markers. *Acta Horticulturae* 849, 133–138.

Santos, C.A.F., Correa, L.C. and Costa, S.R. (2011) Genetic divergence among *Psidium* accession based on biochemical and agronomic variables. *Crop Breeding and Applied Biotechnology* 11, 149–156.

Santos, E.A., Viana, A.P., Freitas, J.C.O., Rodrigues, D.L., Tavares, R.F. *et al.* (2015) Genotype selection by REML/BLUP methodology in a segregating population from an interspecific *Passiflora* spp. crossing. *Euphytica* 204, 1–11.

Satisha, J., Kurian, R.M. and Dinesh, M.R. (2016) *Production Technology of Tropical Fruits – A Hand Book.* Indian Institute of Horticultural Research, Bengaluru, India, pp. 31–44.

Schoeman, M.H. (2011) The current status of guava wilt disease in South Africa. *South African Fruit Journal* 2011(Aug–Sep), 46–49.

Schoeman, M.H. and Labuschagne, N. (2014) Preliminary evaluation of guava selections for guava wilt disease resistance in South Africa. *South African Journal of Plant and Soil* 31(2), 109–112.

Schoeman, M.H., Benade, E. and Wingfield, M.J. (1997) The symptoms and causes of guava wilt in South Africa. *Journal of Phytopathology* 145, 37–41.

Schoeman, M.H., Botha, F.A. and Manicom, B.Q. (2012) Guava wilt disease – the South African perspective. *Acta Horticulturae* 959, 67–72.

Schroers, H.J., Geldenhuis, M.M., Wingfield, M.J., Schoeman, M.H., Yen, Y.F. *et al.* (2005) Classification of guava wilt fungus *Myxosporium psidii,* the palm pathogen *Gliocladium vermoesenii* and the persimmon wilt fungus *Acremonium diospyri* in *Nalanthamala*. *Mycologia* 97, 373–395.

Severn-Ellis, A., Schoeman, M.H., Willemse, S., Sippel, A., Rees, J. and Castro, M.d. (2012) Towards guava wilt resistance in South Africa. *Acta Horticulturae* 959, 73–79.

Sharma, Y.K., Goswami, A.M. and Sharma, R.R. (1992) Effect of dwarfing aneuploid guava rootstock in high density orcharding. *Indian Journal of Horticulture* 49, 31–36.

Soubihe Sobrinho, J. and Gurgel, J.T.A. (1962) Taxa da panmixia na goiabera (*Psidium guajava* L.). *Bragantia* 21, 15–20.

Sousa, C.M.B., Ribeiro, R.M., Viana, A.P., Cavalcante, N.R., Silva, F.A.d. *et al.* (2020) Guava breeding via full-sib family selection: conducting selection cycle and divergence between parents and families. *Crop Breeding and Applied Biotechnology* 20(1), e256520112.

Souza, A.D.G., Resende, L.V., Lima, I.P., Martins, L.S.S. and Techio, V.H. (2015) Chromosome number and DNA nuclear amount in *Psidium* spp. resistant and susceptible to *Meloidogyne enterolobii* and its relation with compatibility between rootstock and commercial varieties of guava tree. *Plant Systematics and Evolution* 301, 231–237.

Srivastava, H.C. (1977) Chromosome behaviour of a spontaneous autotetraploid guava (*Psidium guajava* L.). *Cytologia* 42, 389–394.

Subramanyam, M.D. and Iyer, C.P.A. (1982) *Report.* Fruits Research Workshop, Nagpur, India, pp. 117–118.

Subramanyam, M.D. and Iyer, C.P.A. (1992) Studies on inheritance in guava (*Psidium guajava* L.). *Acta Horticulturae* 317, 255–258.

Thaipong, K. and Boonprakob, U. (2006) Repeatability, optimal sample size of measurement and phenotypic correlations of quantitative traits in guava. *Kasetsart Journal (Natural Science)* 40, 11–19.

Thaipong, K. and Boonprakob, U. (2019) Salt tolerance evaluation in guava germplasm. *International Journal of Agricultural Technology* 15, 791–796.

Thaipong, K., Promchot, S., Auvuchanon, A. and Boonprakob, U. (2017) Genetic analysis of guava germplasm using AFLP markers. *International Journal of Agricultural Technology* 13, 741–752.

Valdes-Infante, J., Becker, D., Rodriguez, N., Velazquez, B., Gonzalez, G. *et al.* (2003) Molecular characterization of Cuban accession of guava (*Psidium guajava* L.), establishment of a first molecular linkage map and mapping of QTLs for vegetative characters. *Journal of Genetics and Breeding* 57, 349–358.

Valdes-Infante, J., Rodriguez, N., Becker, D., Velazquez, B., Sourd, D. *et al.* (2007) Microsatellite characterization of guava (*Psidium guajava* L.) germplasm collections in Cuba. *Cultivos Tropicales* 28, 61–67.

Valdes-Infante, J., Rodriguez, N., Velazquez, B., Rivero, D., Martinez, F. *et al.* (2010) Comparison of the polymorphism level, discriminating capacity and informativeness of morpho-agronomic traits and molecular markers in guava (*Psidium guajava* L.). *Acta Horticulturae* 849, 121–131.

Van, H.N., Le Quoc, D., Uyen, D.T.K. and Hang, N.T.N. (2017) Overview on guava, pineapple, wax apple and sugar apple research and production in Vietnam. *Acta Horticulturae* 1166, 15–24.

Vasconcelos, L.F.L., Alfenas, A.C. and Maffia, L.A. (1998) Resistance of guava cultivars to rust caused by *Puccinia psidii*. *Fitopatologia Brasileira* 23(4), 492–494.

Viji, G., Harris, D.L., Yadav, A.K. and Zee, F.T. (2010) Use of microsatellite markers to characterize genetic diversity of selected accessions of guava (*Psidium guajava* L.) in the United States. *Acta Horticulturae* 859, 169–176.

Vos, J.E., Schoeman, M.H., Berjak, P., Watt, M.P. and Toerien, A.J. (2000) *In vitro* selection and commercial release of guava wilt resistant rootstocks. *Acta Horticulturae* 513, 69–79.

Wan, W. and Leu, L. (1999) Breeding guava resistant lines against *Myxosporium* wilt. *Plant Protection Bulletin Taipei* 41(2), 149–154.

Wang, A. (2017) A report on introduction of four red guava varieties in Zhanzhou of Fujan Province. *South China Fruits* 46(3), 103–105.

Yusof, S. (1990) Physico-chemical characteristics of some guava varieties in Malaysia. *Acta Horticulturae* 269, 301–306.

Zamir, R., Khattak, G.S.S., Mohammad, T., Shah, S.A., Khan, A.J. and Ali, N. (2003) *In vitro* mutagenesis in guava (*Psidium guajava* L.). *Pakistan Journal of Botany* 35(5), 825–828.

Zipori, I., Shuker, S., Dag, A. and Tomer, E. (2007) Guava breeding in Israel. *Acta Horticulturae* 735, 39–47.

7 Plant Nutrition and Irrigation

Sisir Mitra*
Former Dean, Bidhan Chandra Krishi Viswavidyalaya, Mohanpur, West Bengal, India

7.1 Introduction

Among the various factors that affect the productivity of guava, adequate nutrients and water assume greatest significance. It is essential that these two inputs are managed in a manner that provides maximum output. Inadequacy of either nutrients or water at critical stages of fruit development can drastically reduce yields and adversely affect the quality of produce. It is essential that an optimum quantity of nutrients be applied at the appropriate time to achieve targeted production (Singh and Singh, 2007). Similarly, availability of adequate, timely and assured irrigation is critical for obtaining optimum growth, yield and quality of guava fruits.

7.2 Soil

Guava trees are adaptable to a wide range of soils including sands, loams, rock-based soils and muck. However, the ideal soil for guava is a deep, well-drained, friable and fertile loam soil. A soil pH of 5 to 7 is ideal but plants do well in high-pH soils (7–9) if supplied with chelated iron. The main guava-producing areas of Brazil are in São Paulo state where the orchard soil was classified as Ultisol, or Red-Yellow Argisol with a pH of 5.3 and organic matter content of 11 g dm^{-3} (Souza *et al.*, 2012). Although guava may be tolerant of rather poor soil conditions, for example the trees are fairly resistant to salt and drought, it responds well to good soils, climate and both organic and chemical fertilizers (Hamilton and Seagrave-Smith, 1959).

In India, the major guava-growing states are Maharashtra, Bihar, Madhya Pradesh, Uttar Pradesh, West Bengal, Andhra Pradesh and Chhattisgarh. The Allahabad region of Uttar Pradesh has the reputation of producing high-quality guavas. The soil in this region is mostly sandy loam with a pH between 6.61 and 7.24 (Das *et al.*, 2018). Guava is reported to be very (Ogden *et al.*, 1981) to moderately (Schaffer *et al.*, 1992) flood-tolerant and can thrive on poorly drained land where most other fruit trees do not grow (Morton, 1987).

7.3 Salinity

Guava is regarded as moderately salt-tolerant (Mass, 1993). It bears successfully up to an electrical conductivity (EC) of soil saturation extract of 8 to 9 dS m^{-1} (Mehta *et al.*, 1988).

*E-mail: sisirm55@gmail.com

DOI: 10.1079/9781789247022.0007

However, Patil *et al.* (1984) reported a threshold EC of 4.7 dS m^{-1} and a decrease in growth of almost 10% per dS m^{-1} above this threshold. Varying degrees of susceptibility of different cultivars to salt stress are characterized by leaf burn which appears from young to mature and from the margin to the centre of leaves (Thaipong and Boonprakob, 2019). Increasing salinity level (EC values from 4.5 to 12 mmhos cm^{-1}) caused a decrease in total N, P, K and Na concentrations in leaves and total chlorophyll content in seedlings of guava while the concentrations of Ca, Mg and Cl increased (Singh and Pathak, 1994). However, Dhankar *et al.* (1981) and Haggag and Maksoud (1996) reported a substantial increase in leaf Na and Cl contents with increasing salinity, resulting in leaf injury and reduced shoot growth, which might be attributed to Na and Cl toxicity and the imbalance of other nutrients at higher levels of salinity (Marler, 1994). Thaipang and Boonprakob (2019) evaluated 20 genotypes for their salt tolerance in sand culture. Based on leaf burn severity, the 20 genotypes were separated into four groups by level of tolerance. One genotype ('KUHP38') was highly tolerant, three genotypes ('Paen Seethong', 'Na Suan' and 'KUHP12') were moderately tolerant, eight genotypes were sensitive, and another eight genotypes were highly sensitive to salt stress. Based on the results obtained, 'Paen Seethong' and 'Na Suan' were recommended for desert-type commercial plantings in the saline soil regions of Thailand while 'KUHP38' and 'KUHP12' may be used as rootstocks for salt tolerance.

Guava seedlings are sensitive to salt and if irrigated with water having an EC above 1.5 dS m^{-1} do not produce quality planting materials suitable for planting (Cavalcante *et al.*, 2007). Under saline conditions, shoot growth and leaf chlorophyll content were reduced. However, moderate N supply improved these parameters. Salinity due to sodium chloride (NaCl) reduced net photosynthesis rate (Pn) and transpiration (T) but moderate N supply (10 mM in Sl pots) enhanced Pn and T (Ali-Dinar *et al.*, 1999). Guava cultivar 'Paluma' leaves became chlorotic, and subsequently necrotic, even with the use of marginally saline water (2dS m^{-1}). High-salinity (8 dS m^{-1}) water caused the complete cessation of growth due to accumulation of Na^+ and Cl^- ions to toxic levels (Cavalcante *et al.*, 2009). These observations point to the fact that guava tolerance to salinity varies with the level and duration of exposure to salinity, the growing conditions and the genotype (Singh *et al.*, 2016). Singh *et al.* (2016) reported negligible Na^+ accumulation in leaves up to 1.4 dS m^{-1} salinity, but exposure to elevated salinity (2 dS m^{-1}) significantly increased leaf Na^+ (0.16% of dry weight). The salinity tolerance threshold of guava cultivar 'Allahabad Safeda' was about 1.5 dS m^{-1}.

7.4 Nutrient Uptake

Guava removed 6.0 kg N, 2.5 kg P_2O_5 and 7.5 kg K_2O per tonne of fruit and the removal ratio of K_2O and P_2O_5 relative to N (100) is 125 and 42, respectively (Pathak *et al.*, 2002). With the exception of Cu and Zn, the estimated quantities of macro- and micronutrients exported per tonne of fresh guava fruit were greater for 'Rica' than for 'Paluma' (Table 7.1). The data indicated that plant nutrient requirements depend on cultivar and fertilizer programme even in the same growing situation.

Table 7.1. Quantities of macro- and micronutrients exported per tonne of fresh guava fruit for two Brazilian cultivars. Adapted from Natale *et al.* (2002).

	g t^{-1}	
Nutrient	Cultivar 'Rica'	Cultivar 'Paluma'
N	1325	1179
P	166	121
K	2180	1554
Ca	110	94
Mg	110	107
S	152	107
B	0.83	0.67
Cu	0.83	1.34
Fe	2.21	1.88
Mn	2.90	1.88
Zn	1.52	1.88

7.5 Role of Nutrients

7.5.1 Nitrogen

Deficiency of nitrogen (N) caused stunted growth and reduced the size of leaves and new growth. Leaves had purple-coloured patches on both sides of the midrib and reduced fruiting (Mitra and Sanyal, 2004; Singh and Singh, 2007).

Guava bears fruits mostly on newly emerged shoots. Application of nutrients, especially N, encourages vegetative growth. Increasing the rate of N application was reported to increase growth, yield and improve fruit quality (Mitra, 1987; Kaj *et al.*, 1989; Natale *et al.*, 1994a; Kumar *et al.*, 1996a). Kumar *et al.* (1996a) suggested application of 300 g N per tree per year for 5- to 6-year-old trees and 600 g N per tree per year for 7- to 8-year-old trees of cultivar 'Allahabad Safeda'. Mitra and Bose (1985) reported higher yield from a 3-year-old tree of cultivar 'Sardar' with 260 g N per tree per year, while Wagh and Mahajan (1985) recorded highest yield with 600 g N per tree per year for 5-year-old 'Sardar' guava trees. Natale *et al.* (1994) observed that yield increased by increasing the quantity of N applied to 59.34 kg ha^{-1} in the third year, equivalent to 422 g N per plant. Interpolation from the results showed that 90% of the maximum yield was obtained when the leaf N content was 2.35–2.55%, which was associated with 184, 262 and 422 g N per plant in years 1, 2 and 3, respectively.

Foliar sprays of urea (1.0–3.0%) in January and July have been reported to be effective in improving fruit set, fruit size and yield (Arora and Singh, 1970a; Doraipandian and Shanmugavelu, 1972; Pandey *et al.*, 1988; Kundu *et al.*, 2007). However, spraying >2% urea sometimes caused marginal leaf burning and reduced fruit set (Ahmad *et al.*, 1998).

7.5.2 Phosphorus

Deficiency of phosphorus (P) caused stunted growth, reduced number and size of leaves and premature defoliation. Deficiency of P also resulted in the development of typical foliage symptoms of leaf bronzing and premature leaf drop under acute deficiency (Mitra and Sanyal, 2004; Singh and Singh, 2007).

The critical level of leaf P was estimated as 0.03% on a dry weight basis under sand culture (Tiwari and Tiwari, 1993). However, from a pot culture study, Kadam and Patil (1993) observed increased growth and dry matter when leaf P content was 0.42%. Radioactive P uptake and redistribution in guava trees where the leaves were treated with monoammonium phosphate (2%) was studied by Natale *et al.* (1999). They observed that about 12% of the applied P was taken up by plants after 20 days and about 20% of the P taken up by leaves was redistributed mainly to the tenderest parts of the tree. Using ^{32}P-labelled superphosphate, Murthy and Kotur (1998) studied the effect of time of application and method of placement on the absorption of P by 10-year-old guava cultivar 'Arka Mridula' plants. The results indicated that application during July (south-west monsoon season) resulted in very high absorption efficiency of P compared with October (north-east monsoon season) and April (summer). Placing the fertilizer in a circular band of 60 cm at a radial distance of 100–160 cm or 160–220 cm was superior to placing the fertilizer nearer the plant. To derive maximum absorption efficiency, it was suggested that the entire dose of superphosphate be applied during July in 60 cm circular bands within a radial distance of 100–160 cm from the tree.

The quantity of P applied depended on soil nutrient status, tree age and varied among the cultivars. Reports suggest application of P at 270 to 900 g per tree per year (Mitra and Bose, 1985; Chelvan and Singh, 1986; Kotur *et al.*, 1997) in different guava-growing regions of India.

7.5.3 Potassium

The deficiency of potassium (K) reduces plant growth; leaves show necrotic patches which are mainly concentrated more towards the margin, tip and base. Brown spots appear on the leaves, which later increase to form strips on the small leaves (Mitra and Sanyal, 2004; Singh and Singh, 2007). Critical leaf K nutrient limits (on a dry weight basis) were suggested as 1.0–1.60% from various locations (Du Plessis *et al.*, 1973; Singh and Rajput, 1976; Sanyal and Mitra, 1990).

Tree growth, fruit weight and yield increased significantly with K application at 400 g per tree per year (Natale *et al.*, 1996). The role of K on quality improvement of fruit has been well documented (Rajput *et al.*, 1978; Mitra, 1987; Kumar *et al.*, 1996b; Kundu *et al.*, 2007).

7.5.4 Calcium

Deficiency of calcium (Ca) caused necrosis of leaf margins and severe leaf drop (Singh and Singh, 2007). Singh and Chauhan (1982) reported that foliar application of 2% calcium nitrate on 'Sardar' guava 3 weeks before harvest reduced loss of ascorbic acid and sugar content of fruit compared with control at ambient storage (27°C) for 8 days. Foliar spraying of 0.6% Ca as calcium chloride (or 1.0% calcium nitrate) at 80–85 days after fruit set increased fruit weight and improved fruit quality of 'Sardar' guava at harvest. The treatments also reduced the rate of ripening and physiological weight loss during storage (Raychaudhuri *et al.*, 1992). Post-harvest dipping of fruits in 1% calcium nitrate solution reduced weight loss, respiration rate and maintained edible quality of fruit for more than 6 days in storage (Singh, B.P. *et al.*, 1981).

7.5.5 Zinc

Zinc (Zn) deficiency symptoms result in smaller leaves, many leaves in a cluster, sparse interveinal chlorosis on leaves and branch dieback. The deficiency caused reduced plant growth, flower numbers and immature/mature fruits develop cracking (Singh and Singh, 2007).

A foliar spray of 0.2 to 0.4% zinc sulfate on Zn-deficient 'Allahabad Safeda' guava plants significantly improved elongation of the terminal shoot, number of leaves, leaf area and leaf chlorophyll. The treatment also increased yield and improved fruit contents of total soluble solids (TSS), vitamin C and pectic substances (Arora and Singh, 1970b; Rajput and Chand, 1975). Singh (1985) suggested spraying twice with 1 lb (0.454 kg) of zinc sulfate and 0.7 lb (0.318 kg) of slaked lime dissolved in 16 gallons (60.57 litres) of water in soil deficient in Zn, which improved the health of the plant and increased fruit size and yield. Spraying of Zn at 0.2 to 0.4%, twice before flowering and again at full bloom, significantly increased leaf N and K contents but no marked increase in P was observed (Sharma and Bhattacharyya, 1989). Abd-Elhamied and Fouda (2018) noted higher P, K and Zn uptake by spraying 0.6% Zn four times at monthly intervals from June to September. Kundu and Mitra (1999) reported increased TSS, total sugar and ascorbic acid contents and a decrease in fruit acidity by spraying of Zn at 0.3% in May and September. A similar increase in TSS and vitamin C contents of fruit and a decrease in fruit acidity were recorded by Arshad and Ali (2016) by spraying 0.5% Zn three times (time of spraying not mentioned) in a Zn-deficient soil.

7.5.6 Boron

Symptoms of boron (B) deficiency of guava include leaf fall, reduced fruit size, internal necrosis and hardening of affected portions of fruit, and fruit cracking in extreme cases (Singh and Singh, 2007). Application of B in the form of boric acid (17% B), borax (11% B) or Solubor (22% B) has been used to overcome the B deficiency. Rajput and Chand (1975) and Kundu and Mitra (1999) recorded significant improvement in fruit weight, yield and fruit quality by foliar spraying of boric acid at 0.1 to 0.4%, two or three times in a year. Preharvest spray of borax (0.6–1%) twice in October increased fruit size, weight, TSS and ascorbic acid contents of fruits, while the 0.2 or 0.4%

borax spraying improved the storage life of 'Allahabad Safeda' guava (Raghava and Tiwari, 1998). Spraying of borax at 0.4% increased fruit set (78.37%) and fruit retention (56.93%) of 'Sardar' guava compared with untreated plants (70.92 and 43.47%, respectively) (Hada *et al.*, 2014).

7.6 Fertilizer Rates and Time of Application

The amounts of N, P and K or other nutrients applied to guava to achieve high yields and good-quality fruit differ with age of the tree, soil condition, cultivar and agro-climatic situation of the area where the crop is growing. The aim of early management is to obtain the greatest tree size and bearing surface. This is achieved by regular application of nutrients and water along with appropriate pruning. The fertilizer programme for young trees begins when the plants begin to flush after planting. The quantity of N, P and K applied increase each year.

In bearing orchards, the aim, through annual leaf and soil tests, is to maintain nutrient concentrations within a specified optimum range. Soil analysis is used to monitor and adjust pH and the quantities (mg kg^{-1}) of Ca, Mg, P and B. The quantities of macronutrients required for 'Allahabad Safeda' and 'Sardar' (g per tree per year), when grown in different states of India (Table 7.2), ranged from 360 to 1000 for N, from 300 to 1000 for P and from 300 to 1000 for K (Singh and Singh, 2007). In an experiment conducted for 5 years at Ranchi, India, separately for N, P and K, the quantities (g per tree per year) were in the range of 0–800 for N, 0–349 for P and 0–482 for K in soil having a pH of 6.2 and organic carbon of 0.65%, and availability of N, P and K were 40, 3 and 170 kg ha^{-1}, respectively, and exchangeable Ca and Mg were 4.5 and 0.45 meq 100 g^{-1}, respectively. Critical examination of the results indicated that 583 g N, 175 g P and 400 g K per plant were optimum in that soil (Kotur *et al.*, 1997; Singh and Singh, 2007).

In most of the experiments conducted in India, inorganic sources of fertilizer, superphosphate and potassium sulfate, have been used. Potassium sulfate is preferred over potassium chloride. Calcium ammonium sulfate is the best source in acidic soils. Fertilizers are usually applied at the onset and end of the monsoon. However, considering the need for regulation of flowering, time of application of nutrients may be adjusted. Fertigation using the fertilizer and water together would be one of the best methods to provide the required nutrients in the root zone to achieve a targeted yield.

In sandy loam soil in southern Taiwan, 4-year-old guava ('Thai-Kou-Bar') produced the maximum yield, fruit weight and quality with N, P and K at 200, 100 and 400 g per tree per year, respectively (LihShang and Weider, 1997). However, when working with different amounts of N, P and K (each at 0–300 g per tree per year) in the areas of Jaboticabal ('Rica') and São Paulo ('Paluma'), Brazil, Natale *et al.* (1995) did not find any effect of grams of N, P and K on °Brix values of fruits, which ranged between 8.0 and 10.8 for 'Rica' fruits and between 8.4 and 9.65 for 'Paluma' fruits.

Experiments were conducted south-east of Gharo Sindh, Pakistan on 'Allahabad Safeda' in the soil (30 cm depth) having a pH of 9.84 and EC of 0.192 dS m^{-1}, and available nutrients (in kg ha^{-1}) were N, 83.3; P, 30.75; and K, 355.5; with different rates of N, P and K (Arshad, 2015). The results suggested application of gypsum at 3 kg per tree, farmyard manure (FYM) at 40 kg per tree and N, P and K each at 1000 g per tree per year, which caused maximum fruit weight (110.45 g), number of fruits (375.9 per tree), contents of TSS (9.28%) and vitamin C (44.70 mg 100 ml^{-1} juice), fruit firmness (5.91 kg cm^{-2}) and reduced fruit acidity (0.48%).

7.7 Foliar Application of Nutrients

Foliar application of nutrients in tropical fruits has gained much importance because more fertilizer is needed for soil application as some of the applied nutrients leach below the root zone or become unavailable due to complex chemical reactions and soil pH. Spraying of 1 or 2% urea, 2% triple superphosphate and 1% potassium sulfate during July (Arora

Table 7.2. Quantities of nitrogen, phosphorus and potassium applied to bearing guava orchards in different states of India.

Place/location	Cultivar	Nutrients (g per tree) N	P	K	Month applied	Reference
Varanasi (Uttar Pradesh)	'Allahabad Safeda'	500	900	300	May & Jul	Rajput and Singh (1976)
Lucknow (Uttar Pradesh)	'Sardar'	800 g neem cake, urea	600 SSP	600 MOP	Jul & Sep	Ram and Rajput (1998)
Pantnagar (Uttarakhand)	'Sardar'	450	400	300	N in 2 split doses: Jul & Nov–Dec	Tiwari and Lal (2000)
Punjab	'Allahabad Safeda'				Jun & Sep	Gupta and Nijjar (1978)
	Rainy	120	200	120		
	Winter	40	100	40		
Kalyani (West Bengal)	'Sardar'	260	320	260	Jan &Aug	Mitra *et al.* (2008)
Coimbatore (Tamil Nadu)	'Sardar'	1000	1000	1000	Jun–Jul & Dec	Azhakiamanavalan and Khader (1981)
Rahuri (Maharashtra)	'Sardar'	600	300	300	Jun & Aug	Wagh and Mahajan (1985)
Sabour (Bihar)	'Allahabad Safeda'	490	560	375	Jun & Oct	Chaudhury *et al.* (1975)
Ranchi (Jharkhand)	'Allahabad Safeda'	580	270	400	Jun & Aug	Kotur *et al.* (1997)
Bangalore (Karnataka)	'Sardar'	900	600	600	Jun & Sep	Satisha *et al.* (2016)
Bhubaneswar (Orissa)	'Allahabad Safeda'	600	300	600	Jun & Aug	Kumar *et al.* (2009)

SSP, single superphosphate; MOP, muriate of potash.

and Singh, 1970a) or 3% urea along with 1% calcium phosphate and 1% muriate of potash (Chhonkar and Singh, 1981) increased vegetative growth and yield. Spraying of 2–4% superphosphate in July and January increased the number of flower buds, yield and size of fruit (Singh and Rajput, 1977). Combined application of potassium nitrate, calcium chloride and calcium nitrate, each at 1%, increased fruit weight (Singh, H.K. *et al.*, 1981). Foliar spraying of 4–6% urea in January and July significantly increased plant growth, flowering, yield and improved fruit quality of 'Allahabad Safeda' guava (Singh, 1985). Yield and quality attributes improved significantly by spraying Ca and K compounds and calcium nitrate (1.5%) was found more effective (Brahmachari *et al.*, 1997). Kundu *et al.* (2007) obtained highest yield (48.09 kg per plant) from 8-year-old 'Sardar' guava by combined spraying with 3% urea and 2% each of calcium phosphate and muriate of potash, twice in May and September. However, combined spraying of 1% urea, 1% calcium phosphate and 2% muriate of potash caused maximum TSS (12.03°Brix) and ascorbic acid (252.4 mg 100 g^{-1} pulp) of winter-harvested fruit.

Foliar spraying of micronutrients is effective in regulating growth, yield and fruit quality of guava. Combined spraying of Cu (0.3%), B (0.1%) and Zn (0.3%) in May and September increased the yield (56.6 kg per tree), average fruit weight (156 g) and ascorbic acid content (256 mg 100 g^{-1} pulp) of fruits of winter-season 'Sardar' guava (Kundu and Mitra, 1999). Ripening could be advanced by spraying 2% urea and 0.4% zinc sulfate before flowering followed by a spray 3 weeks after fruit setting (Pandey *et al.*, 1988). Foliar spraying of 1% zinc sulfate and 1% borax in August and September increased fruit set (65.43% versus 46.08% in control), fruit weight (134.63 g versus 102.42 g in control) and yield (167.99 quintal ha^{-1} versus 124.09 quintal ha^{-1}) (Yadav, A. *et al.*, 2017).

7.8 Integrated Nutrient Management

There is an increasing awareness worldwide about alternative agriculture systems known as integrated plant nutrient management. The term implies the maintenance or adjustment of soil fertility and of plant nutrient supply to optimum levels for sustaining desired crop productivity through optimization of benefits from all possible sources of plant nutrients in an integrated manner. The appropriate combination of mineral fertilizers, organic sources and biofertilizers according to land use and social and economic conditions is the objective (Ram *et al.*, 2007). Application of a combination of manure with mineral fertilizer was reported to be economically viable in guava (Pereira and Mitra, 1999; Corrales *et al.*, 2000).

Guava cultivation in subtropical India is mainly confined to nutrient-poor marginal lands under rainfed conditions, resulting in low productivity and poor fruit quality. Restoration of the soil quality of such guava orchards with low soil fertility requires an integrated approach of using organic manures and biofertilizers along with inorganic nutrients (Adak *et al.*, 2014). Application of 10 kg of vermicompost along with 500 g N, 200 g P, 500 g K and 20 g phosphate-solubilizing bacteria per tree resulted in maximum fruit set (59.66%), fruit retention (58.63%) and yield (24.0 t ha^{-1}) of 'Sardar' guava (Singh *et al.*, 2007). Pereira and Mitra (1999) suggested application of 75 g N, 100 g P and 75 g K along with 1.5 kg neem cake for 3-year-old 'Sardar' guava. However, 7-year-old 'Sardar' guava trees fertilized with 250 g N, 100 g P, 250 g K, 10 kg FYM and 250 g *Azotobacter* produced maximum fruit yield of 150.25 kg per tree with a TSS content of 13.5°Brix in the fruit (Ram *et al.*, 2007). Trivedi *et al.* (2012) obtained highest fruit yield of 11,851.85 kg ha^{-1} (at a density of 278 trees ha^{-1}) and 42.6 kg per tree with the application of biofertilizers (*Azotobacter* and phosphate-solubilizing bacteria, each at 100 g per tree) and FYM at 40 kg per tree from 11- to 13-year-old guava cultivar 'Sardar', compared with 8236.11 kg ha^{-1} by application of N, P and K at 500, 250 and 250 g per tree per year and FYM at 40 kg per tree. They concluded that application of organic manures in guava favoured vegetative growth, encouraged nutrient uptake and enhanced soil nutrient availability, whereas addition of biofertilizers with manures proved effective in increasing fruit yield. Effects of organic and inorganic substrates were studied through field experimentation for 5 years by Adak *et al.* (2014). Different combinations of organic (FYM, vermicompost, mulching, *Azotobacter*, phosphate-solubilizing microbes and *Trichoderma harzianum*) and inorganic (N, P, K) substrates were applied each year within the tree basin under subtropical environmental conditions. At the end of the study, soil fertilized with vermicompost (10 kg per tree), biofertilizers and organic mulching showed improvement in yield compared with N, P, K (120, 60, 120 g per tree per year of age) and FYM (10 kg per tree) application. Higher soil moisture retention, soil fertility and organic carbon build-up and better soil temperature regulation were observed under mulching treatment. Microbial population was significantly higher in plots treated with vermicompost, microbial inoculants and mulching compared with NPK + FYM. Barne *et al.* (2011) reported maximum yield (39.45 kg) and average fruit weight (289.4 g) by application of N, P, K at 487.5 g + 243.75 g + 281.25 g, respectively, along with 50 kg FYM and 250 g each of *Azotobacter* and phosphate-solubilizing bacteria per tree per year, compared with 12.74 kg yield and 131.80 g average fruit weight under control conditions.

7.9 Tissue Analysis

In perennial fruit trees such as guava, leaf analysis is better than soil analysis to ensure the nutritional status of the tree (Chadha *et al.*, 1973), not only by determining the response to different nutrients but also to assess techniques for the recommendation of fertilizers (Natale *et al.*, 2002). Guava-leaf nutrient element composition varies with different factors such as age of the leaf (Sanyal and Mitra, 1990; Chetri *et al.*, 1996), direction of the shoot in the tree and position of the leaf on the shoot (Sanyal and Mitra, 1990; Chetri *et al.*, 1999), fruiting or non-fruiting shoots (Rodriguez *et al.*, 1984), cultivar (Ravi-Kumar *et al.*, 1990; Chetri *et al.*, 1999) and time of sampling (Arora and Singh, 1972; Natale *et al.*, 1994b; Chetri *et al.*, 1999). The effects are large enough to influence nutrient diagnosis in many orchards, suggesting that the protocol for sampling leaf nutrient concentrations needs to be standardized across different growing regions.

Most researchers have suggested sampling fully developed leaves, corresponding to the third pair of leaves with petiole from the tip of the branch at the beginning/during flowering (Sanyal and Mitra, 1990; Dahiya and Joon, 1995; Natale *et al.*, 2002; Maia *et al.*, 2007). It was suggested to sample 30 to 60 leaves of the same age covering all sides of the tree at a height of about 1.5 m from the soil for estimation of elements (Arora and Singh, 1972; Chadha *et al.*, 1973; Sanyal and Mitra, 1990; Dahiya and Joon, 1995).

Research has resulted in the development of leaf nutrient standards for guava in India, Brazil, Venezuela and Hawaii, USA. The suggested optimum nutrient concentrations show that the N content in leaf varies between 1.57 and 2.26%, while the K content between 1.25 and 2.26%. The variation in P content of leaves was not very wide. These elements were below the concentrations of N (1.25 to 1.40%) and K (1.15 to 1.50%) for the conditions of South Africa (Koen and Hobbs, 1990) that were sampled in March when the leaves were 5 months old after pruning (Koen, 1987). These variations may be due to several reasons such as differences in experimental conditions, cultivar studied, fruit load on the tree, season and orchard management, etc. (Marchal, 1984).

Plant nutrient contents for N and K are greater than for P in guava (Queiroz *et al.*, 1986; Mitra and Dhaliwal, 2009) and higher leaf nutrient concentrations are found during flowering than at the time of fruit set (Natale *et al.*, 2002) and fruit ripening (Chadha *et al.*, 1973). Critical leaf nutrient contents developed in India, Brazil, Venezuela and Hawaii are presented in Table 7.3.

The usual methods for interpretation of tissue analysis in leaves is to compare the analysis with reference values such as critical level (CL) or sufficiency ranges (SR), characterized by the independence of nutrients (Souza *et al.*, 2015). Such values can be

Table 7.3. Suggested optimum leaf nutrient concentrations for guava on a dry weight basis.

Nutrient	India (Ravi-Kumar *et al.*, 1990)		Brazil (Natale *et al.*, 2002)		Venezuela (Guerra and Bautista, 2002)	
	Cultivar 'Allahabad Safeda'	Cultivar 'Sardar'	Cultivar 'Rica'	Cultivar 'Paluma'	Cultivars 'Mara 4', 'Mara 6' and 'Mara 8'	Hawaii, USA (Shigeura and Bullock, 1983)
N (%)	1.10	1.27	2.2–2.6	2.0–2.3	1.57	1.70
P (%)	0.17	0.19	0.15–0.19	0.14–0.18	0.18	0.25
K (%)	1.46	1.20	1.7–2.0	1.4–1.7	1.25	1.50
Ca (%)	1.08	1.68	1.1–1.5	0.7–1.1	1.64	1.25
Mg (%)	–	–	0.25–0.35	0.34–0.40	0.30	0.25
S (%)	–	–	0.30–0.35	0.25–0.35	–	–
Zn (mg kg^{-1})	26	29	25–35	25–35	31.3	20
Fe (mg kg^{-1})	282	184	50–150	60–90	120.3	–
Cu (mg kg^{-1})	4	9	10–40	26–40	11.1	8
Mn (mg kg^{-1})	53	45	180–250	40–80	88.8	60
B (mg kg^{-1})	–	–	20–25	20–25	–	20
Time of sampling	–	–	Recent fully developed leaves, corresponding to the third pair with petiole	–	Third and fourth nodal position from the base of non-fruiting shoots	Fourth leaf in a whorl of leaves of an actively growing terminal

established in field calibration experiments where genetic and environmental characteristics and their interactions with nutrients are controlled (Bhargava and Chadha, 1988). The reference values are, however, updated regularly because new genetic material and management or cultivation techniques are being introduced and environmental conditions vary (Rozane *et al.*, 2016). Alternatively, surveys of commercial stands provide data from a broad variation of environmental conditions to obtain the reference values. The Diagnosis and Recommendation Integrated System (DRIS) and Compositional Nutrient Diagnosis norms (CND) can track nutrient sufficiency ranges and help in assessing the most limiting nutrient for tree management.

Anjaneyulu and Raghupathi (2009) conducted a study to develop diagnostic leaf nutrient norms to identify yield-limiting nutrients in guava cultivar 'Allahabad Safeda'. By using DRIS, the whole population was divided into two subgroups and 45 nutrient expressions were identified as diagnostic norms from the data collected on nutrient concentration in leaves and yield. Among the nutrients selected from diagnostic parameters, P/N (0.105), N/K (1.375), P/Mg (0.485), P/Zn (0.006), K/Zn (0.042), Mg/K (0.302), Mg/S (1.147) and Fe/Zn (4.302), among others, had shown higher variance and lower coefficient of variation that were found to have greater diagnostic precision. The overall imbalance of the nutrients was assessed based on the sum of the indices irrespective of sign, which was referred to as nutritional balance index. The higher the sum value, the larger will be the plant nutritional imbalance and vice versa. The diagnosis of nutrient balance through DRIS indices indicated that the most common yield-limiting nutrients were Zn and K. In addition, five nutrient ranges were derived using mean and standard deviation as low, deficient, optimum, high or excess for each nutrient, to serve as a guide for diagnostic purposes. The results indicated that the optimum ranges for N, P and K were 1.69–2.19, 0.168–0.236 and 1.20–1.67%, respectively. Similarly the optimum ranges for Ca, Mg and S were 0.60–1.27, 0.35–0.50 and 0.29–0.47%, respectively. The optimum Fe ranged from 114 to 178 ppm, Mn from 34 to 77 ppm, Zn from 29 to 41 ppm and Cu from 6 to 12 ppm. Hundal *et al.* (2007) employed DRIS for interpreting nutrient analysis of leaf tissue of guava cultivated in Punjab, India. They stated that DRIS evaluation and the sufficiency range approach were equally effective and in agreement for diagnosing deficiencies of nutrients. The results also showed that the position of leaf tissue sampled does not have a major effect on the DRIS diagnosis. Leaf nutrient sufficiency ranges derived from DRIS norms (%) for N, P, K, Ca, Mg and S were 1.41–1.65, 0.10–0.17, 0.51–0.97, 1.16–2.12, 0.31–0.51 and 0.18–0.28, respectively, and norms (mg kg^{-1}) for Fe, Mn, Zn and Cu were 105–153, 58–110, 15–29 and 6–16, respectively. According to this, sufficiency ranges of 35, 62, 51, 75, 70 and 68% of the samples were sufficient and 4, 29, 36, 9, 10 and 22% of the samples were low in N, P, K, Ca, Mg and S, respectively. For micronutrients, 15, 61 and 7% of the samples were found to be low in Mn, Zn and Cu, respectively.

Souza *et al.* (2015) proposed critical levels and nutrient sufficiency ranges in the leaves of guava plants in nursery stage by using DRIS. They observed that the limiting nutrients by deficiency, in descending order, were: N > Cu > P = K > Mn > Fe = Zn > S > B = Mg > Ca; and the limiting ones by excess, in descending order, were: B > Ca > Fe > Mn > S > Mg > Cu > P > Zn > N = K.

The DRIS techniques proved to be geometrically unsuitable and thus were replaced by CND norms (Parent and Dafir, 1992), which is another approach. CND is a multivariate norm that gives due weightage to all the elements, including unmeasured factors, and therefore has higher diagnostic sensitivity. Multivariate nutrient diagnostic norms were developed for guava using CND through leaf nutrient concentrations versus yield databank (Anjaneyulu *et al.*, 2008). During July–August, Anjaneyulu *et al.* (2008) conducted a survey in guava orchards (cultivar 'Allahabad Safeda') in and around Bangalore and Kolar (mostly Alfisols) in Karnataka, India. Samples were collected from 70 orchards (around 15 years of age) by selecting the third pair of leaves from the apex, which provides the index leaf (recently matured leaf) in guava. The mean N concentration was 1.91% and ranged from 1.33 to 2.48%. The mean P concentration was 0.20% and varied from 0.14 to 0.26%. The mean K concentration was 1.35% and varied widely between 0.90 and 1.85%. A higher range of Ca concentration (1.50 to 2.60%) was observed, whereas a majority of the samples were in the optimum range with regard to Mg (0.30 to 0.75%). The mean S concentration was 0.38% and varied from 0.21 to 0.53%. Concentration of Fe and Mn ranged from 104 to 197 ppm and from 25 to 98 ppm, respectively. The mean Zn concentration was 31.66 ppm and the mean Cu concentration was 8.37 ppm. Interactions among different nutrients was explained by principal component analysis (PCA) conducted on log-transformed data which produced four significant principal components (PCs), explaining about 73.66% of the variance. The four eigenvalues added up to 8.1, denoting the four significant PCs. The first PC (PC1) was positively correlated with P, Zn and R (residue; a reflection of dry matter accumulation in the plant) and negatively correlated with Ca, Mg, S and Fe, indicating that P and Zn behaved in one direction and the other elements in the opposite direction. In PC2, an antagonistic effect of N and Fe with P and Cu was evident. In PC3, P and Mg were negatively correlated with Mn and Cu. In PC4, N and S showed behaviour in the same direction. Multi-nutrient diagnosis developed through CND and nutrient interactions elucidated through PCA were found to have higher precision in diagnosing nutrient imbalance in guava and are thus helpful in evolving nutrient management strategies (Anjaneyulu *et al.*, 2008).

The CND-Goiaba 1.0 software for nutritional diagnosis of guava has been developed in Brazil. A database was created based on 205 leaf samples collected in commercial plots (sampling units) of cultivated 'Paluma' guava trees with ages between 5 and 20 years. The production data were normally distributed according to the Shapiro–Wilk test ($W = 0.988$; $P = 0.11$). The software made it possible to diagnose that 63% of the orchards evaluated needed to improve the nutritional status of the trees. The CND method showed severe nutritional imbalances in Mg and Zn in these orchards (Rozane *et al.*, 2012).

7.10 Organic Production

Organic farming in fruit crops can be defined as an approach where the aim is to create integrated, human, environmental and economically sustainable agricultural production systems. Maximum reliance is placed on self-regulating agroecosystems, locally or farm-derived renewable resources and the management of ecological and biological processes and interactions. Reliance on external inputs, whether chemical or organic, is reduced as far as possible (Ram *et al.*, 2007). Guava responds well to application of manures and fertilizers, which indicates the feasibility of using organic sources of nutrition in guava. Organic materials such as FYM, cakes of plant origin, vermicompost and microbial biofertilizers are used for organic production.

Application of native mycorrhizal fungi and beneficial strains of bacteria and fungi ensures plants' better adaptability and survival under prevailing environmental conditions, which is an extremely important factor in their long-term effects on plants (Regvar *et al.*, 2003). Enriching fertilizers with beneficial strains of bacteria and fungi can increase their effectiveness in crop production (Chen, 2006) by enhancing the physiology of plants, stimulating their growth and yield, as well as by increasing their resistance to environmental and biotic stresses (Corte *et al.*, 2014).

Azotobacter spp. are capable of converting nitrogen to ammonia (Bishop *et al.*, 1980) which in turn is taken up by the plants. They have the ability to produce different growth hormones, vitamins and siderophores (Narula *et al.*, 1981; Neito and Frankenberger, 1989; Tindale *et al.*, 2000). Vesicular arbuscular mycorrhizae (VAM) are beneficial fungi that penetrate and colonize the roots of the plant, then send out filaments (hyphae) into the surrounding soil. The hyphae literally form a bridge that connects the plant root with large areas of soil and serves as a pipeline to funnel nutrients back to the plant (Prajapati *et al.*, 2008).

There are many reports on the effect of biofertilizers alone or in combination with organic manures on growth of the plant, productivity and quality of fruit of guava (Pereira and Mitra, 1999; Yang and Hung, 2001; Dey *et al.*, 2005; Naik and Hari Babu, 2007; Chakraborty *et al.*, 2008; Devi *et al.*, 2012; Das *et al.*, 2017); however, in most of the investigations, the effect was not compared with inorganic sources.

Haggag *et al.* (1994) reported that application of VAM improved fruit quality by increasing the TSS (10.1°Brix) content of fruit. Application of P-solubilizing bacteria increased P uptake from soil more than when calcium superphosphate was used alone (Haggag *et al.*, 1994). Das *et al.* (2017) recommended application of biofertilizer, *Azospirillum brasilense* and arbuscular mycorrhizal fungi for 6-year-old 'Sardar' guava, along with FYM at 10 kg ha^{-1}, in January and August. The treatment improved yield (41.3 kg per tree), TSS (10.30°Brix) and ascorbic acid (153.44 mg 100 g^{-1} pulp) contents of fruit. Combined application of poultry manure and biofertilizers (*Azotobacter* and *Azospirillum*) increased the leaf N (1.73%), P (0.24%), K (1.23%), Ca (1.96%) and Mg (0.80%) contents (Sharma *et al.*, 2011).

Batista *et al.* (2015) concluded that organic inputs use for growing guava provide high yield and fruit quality compatible with market demands for soluble solids, pH, titratable acidity, pulp firmness and soluble solids/titratable acidity ratio. Chavez and Torres (2012) compared organic and conventional production systems partially replaced by biofertilizers application through fertigation and reported that biofertilizer was better than other treatments. Fertigation with bovine biofertilizer at 5.66% along with recommended N fertilizer has been recommended for guava cultivation in semi-arid region of Brazil by Santana *et al.* (2017). The treatment increased fruit firmness, vitamin C and pH and reduced titratable acidity content of fruit. In 4-year-old guava cultivar 'Sardar', Devi *et al.* (2012) recommended application of FYM (26 kg per tree per year) + *Azotobacter* (100 g per tree) + P solubilizers (100 g per tree) + potash mobilizers (100 g per tree) in January and August. The result was 114 kg fruit yield per plant and fruit with a TSS content of 9.25°Brix and vitamin C content of 154.6 mg 100 g^{-1} pulp.

7.11 Water Management

Guava can withstand a long period of drought (Shigeura and Bullock, 1983). The hardy nature of guava allows it to survive in adverse climatic and soil conditions. Optimum annual rainfall for guava is reported to be 100–200 cm (Mitra and Bose, 1990). In India, the major guava-growing states are Uttar Pradesh, Bihar, Maharashtra, Madhya Pradesh, Rajasthan and West Bengal where the average annual rainfall is between 80 and 140 cm. Other than in high-input, managed commercial orchards, guavas are seldom irrigated in India. In Hawaii, guava trees are found growing in the 500 cm rainfall belt, with continuous free-standing water, as well as in desert-like areas found in Kawaihae and Ka'u where annual rainfall is less than 25 cm. In these areas, the trees are not too productive, demonstrating the ability to withstand extreme conditions in water supply (Shigeura and Bullock, 1983). Brazilian north-east states (Piauí, Ceará, Rio Grande do Norte, Paraíba, Pernambuco, Alagoas, Sergipe and Bahia) are considered favourable for guava cultivation. The semi-arid Brazil region is the wettest semi-arid region of the world, with an average rainfall of about 80 cm concentrated between February and May, and reference evapotranspiration exceeds 200 cm $year^{-1}$ (Marengo *et al.*, 2011; Tinco *et al.*, 2018). The dry season takes place between August and October. The region presents 70% of water deficit annually (Marengo *et al.*, 2011). In this region, orchards are exposed to low rates of precipitation and high atmospheric demands, with irrigation being necessary for guava cultivation (Teixeira and Hernandez, 2012). About 67% of guava cultivation in Mexico is in the Calvillo-Tube region. The orchards are irrigated by flooding of 'cajete' (cajete is a hole (width is larger than depth) which is created around the trunk of the tree to increase water retention) using water volume of 551 m^3 ha^{-1} irrigation (Luna-Jimenez *et al.*, 2010). However, there is not enough water for guava cultivation in this region. Ortega (1971) found that with year-round irrigation, guava trees flowered throughout the year in this region. Temporary suspension of irrigation (short water stress) can cause the total loss of leaves and modify the flowering period in Calvillo, Aguascalientes (Mata and Rodriguez, 1989).

Although guava is sufficiently hardy to survive some moisture stress, vegetative growth and fruit development are considered to be the most affected where soil moisture is limited (Singh and Singh, 2007). Flowering is sensitive to water availability and heavy flowering has been reported in a dry tropical belt following the onset of the rainy period (Nakasone and Paull, 1998). Moisture stress during the fruit setting period may cause severe fruit drop (Mitra and Bose, 1990). The amount of water required depends largely upon climatic conditions, soil type and phenological stage of the plant. In the vegetative stage, until the plant starts bearing, it is essential that optimum soil moisture be maintained for continuous growth. However, some stress is beneficial before flowering to encourage reproductive development. Soil moisture should be at optimum from fruit set to maturity to minimize fruit drop and for proper fruit growth and development (Singh and Singh, 2007).

Conventionally guava is irrigated by flooding the field, furrows (small channels which carry water down the land slope between the tree rows; water infiltrates into the soil as it moves along the slope while the trees are usually grown on ridges between the furrows) and ring basins (for each tree a separate basin is made which is usually circular in shape) (Fig. 7.1). Water to the basins is supplied through small field channels. The conventional irrigation systems have several disadvantages in areas where water is a limiting resource, rains are uneven, drought is a recurring factor and topography is undulating; moreover, this requires a high volume of water for each irrigation. Most of the high-input commercial guava orchards around the world are now using drip irrigation/fertigation systems due to their higher water-use efficiency (Sharma *et al.*, 2012). Drip irrigation (Fig. 7.2) can save 40–60% water compared with a conventional basin irrigation system. Drip irrigation is superior to flood and ring basin methods of irrigation in terms of water saving

Fig. 7.1. Ring basin irrigation system. Photograph courtesy of Dr A.K. Bhattacherjee.

Fig. 7.2. Guava orchard with drip irrigation system. Photograph courtesy of Dr A.K. Bhattacherjee.

and water-use efficiency (Hanson *et al.*, 1997; Mandal *et al.*, 2007). Drip irrigation along with polythene mulch resulted in the highest yield of guava (37.7 t ha^{-1}) and 164% greater yield compared with ring basin irrigation (Sharma *et al.*, 2012).

Irrigation in guava orchards may be managed either by monitoring soil water data with a neutron probe or by Class A pan evaporation. The crop coefficient (K_c) is one of the most important variables in the estimation of irrigation water needs for a specific crop. K_c integrates crop properties, related to the plant phenological stage, plus the effect of soil evaporation in one value (Kisekka *et al.*, 2019). Due to variations in phenological stages (flowering, fruit set, fruit development, fruit maturity) of perennial crops with time, the water requirement of crops varies during growing periods (Yadav, D. *et al.*, 2017; Kisekka *et al.*, 2019) and K_c varies on a monthly basis depending on the phenological stage of the plant and the percentage of the ground shaded by the tree canopy (Allen *et al.*, 1998; Singh and Singh, 2007).

Teixeira *et al.* (2003) estimated the crop evapotranspiration (ET_c) and the crop coefficient (K_c) of guava cultivar 'Paluma' irrigated with a microsprinkler for 2 years and 3 months after planting and over the growing season at Petrolina, Pernambuco state, Brazil. The ET_c was estimated by the Bowen ratio and the reference evapotranspiration (ET_0) was estimated using data collected by conventional and automatic agrometeorological stations. The accumulated ET_c from pruning (June) to harvest (December) was 906 mm in 200 days, which corresponds to an average of 4.53 ± 0.68 mm day^{-1}. The K_c varied from 0.75 to 0.93 when the conventional station was used and from 0.61 to 0.84 when using an automatic station. Under Hawaiian conditions (latitude 21°N and longitude 157°W), the K_c varied from 0.80 and 1.00, the higher values (1.00) were in the months from June to September and the lower values (0.80) from November to March (Fares, 2008).

In Madhya Pradesh state, India (21°6′N latitude and 74°9′E to 82°48′E longitude), where the rainfall is distributed unevenly both temporally and spatially, accurate assessment of crop water requirement for precise irrigation water application was considered necessary (Yadav, D. *et al.*, 2017). Using the K_c values of different months suggested by Fares (2008) and long-term weather data of the state, Yadav,

D. *et al.* (2017) observed an increase in water requirement of guava continuously from the 1st Standard Meteorological Week (SMW) (about 20 l day^{-1}) to the 23rd SMW (about 75 l day^{-1}) during the fruit growth stage. The water requirement of guava at Kharagpur, West Bengal state, India (22°19′N latitude and 87°19′E longitude) was estimated (Singh *et al.*, 2007). The daily crop water requirement for a drip-irrigated guava tree ranges from 13.76 l day^{-1} per tree in December to 39.10 l day^{-1} per tree in May and operation time varies accordingly (Table 7.4). The water requirement and evapotranspiration decrease during the monsoon months of July to September and the average annual net water requirement was estimated at 206 mm ha^{-1}.

The crop water requirement of guava can be estimated by the following equation:

$$V = (ET_0 \times K_c \times A_p) - (A_p \times R_c)$$

where:

V = net irrigation water requirement (l day^{-1});
ET_0 = reference evapotranspiration (mm day^{-1});
K_c = crop coefficient;
A = area allocated to each plant (spacing between plants and rows) (m^2);
$A_p = A \times W_p$ = effective area for irrigation;
W_p = wetting fraction; and
R_c = effective rainfall (mm day^{-1}).

Guava production and guava water requirement (GWR) are two closely linked processes (Teixeira and Hernandez, 2012). The guava water productivity (GWP) can be considered as the rate of yield to GWR. The economic water productivity is the value derived per unit of water used and the economic indicators (GWPs) (Teixeira and Bassoi, 2009). The approach of modelling actual evapotranspiration (ET_a) through the crop coefficient, suggested by the Food and Agriculture Organization of the United Nations, is a viable alternative for use together with a Geographic Information System (GIS) where the spatial variations in soil hydrology and weather conditions make parameterization of hydrological models a difficult task (Teixeira and Bassoi, 2009).

The intensification of agricultural crops in the Brazilian north-east results in a change of natural vegetation, making the quantification and evaluation of additional water use important. Teixeira and Hernandez (2012) estimated GWP by using GIS on a large scale. Long-term weather data were used together with regression models involving the crop coefficient (K_c), reference evapotranspiration (ET_0) and accumulated degree-days to quantify GWR in the Brazilian north-east producer states, considering an average growing season of 6 months and the cultivar 'Paluma' as a reference. By coupling GWR data with total precipitation for a growing season, it was possible to quantify the guava water deficit (GWD), giving an estimate of irrigation needs. Considering the whole region, the variation of the averaged GWD values ranged from 50 mm for pruning dates in January to 520 mm with pruning done in May. Associating the average GWR values with yield data for 2010 from the Brazilian Geographical and Statistical Institute, the average biophysical and economic values of GWP were estimated for each guava producer state. The biophysical values were between 0.86 and 4.95 kg m^{-3} for pruning dates in July and January in Rio Grande do Norte and Pernambuco states, respectively, while the economic ones ranged from 0.40 to 3.18 Brazilian real m^{-3} for the same pruning periods. The states of Pernambuco, Bahia and Piauí showed the largest biophysical and economic GWP values.

The growth of guava trees was positively correlated with increasing soil moisture (El-Khoreiby and Salem, 1989) and trees that suffered from water stress showed a significant reduction in fruit quality (El-Khoreiby and Salem, 1989; Lal, 1996; Pereira *et al.*, 2000). LihShang (1997) reported that fruit yield, fruit soluble solids, firmness, vitamin C and juice pH were higher when the trees were irrigated at 20–40 cb than at 40–60 cb. Guava trees planted at 5 m × 5 m spacing had higher irrigation efficiency (6.79 kg m^{-3}) than trees planted at a 6 m × 6 m (4.70 kg m^{-3}) spacing (Mandal *et al.*, 2007).

7.11.1 Fertigation

The application of soluble fertilizer through the irrigation water has many advantages

Table 7.4. Estimation of water requirement for guava in West Bengal, India. Adapted from Singh *et al.* (2007).

Month	ET_0 (mm day^{-1})	K_c	ET_a (mm day^{-1})	Wetted area (30%) per tree (m^2)	Volume of water (l day^{-1} per tree)	Duration of system operation (min)	Total water requirement (l ha^{-1})
Jan	2.89	0.65	1.88	8.25	15.50	116	174,701
Feb	3.94	0.60	2.36	8.25	19.50	146	198,576
Mar	5.66	0.60	3.40	8.25	28.02	210	315,828
Apr	6.24	0.75	4.68	8.25	38.61	290	421,200
May	6.77	0.70	4.74	8.25	39.10	293	440,727
Jun	5.25	0.70	3.68	8.25	30.32	227	330,750
Jul	4.29	0.55	2.36	8.25	19.47	146	219,434
Aug	3.97	0.55	2.18	8.25	18.01	135	203,066
Sep	3.86	0.55	2.12	8.25	17.51	131	191,070
Oct	3.37	0.55	1.85	8.25	15.29	115	172,376
Nov	3.35	0.60	2.01	8.25	16.58	124	180,900
Dec	2.78	0.60	1.67	8.25	13.76	103	155,124
Total water requirement (l ha^{-1})							3,003,750
Total water requirement (mm ha^{-1})							300
Water requirement during monsoon months (mm ha^{-1})							94
Net water requirement during full season (mm ha^{-1})							206

ET_0, reference evapotranspiration; K_c, crop coefficient; ET_a, actual evapotranspiration.

over broadcasting solid fertilizers (Menzel *et al.*, 2002). Furthermore, fertigation ensures savings in fertilizer usage of about 30–50% of N and K requirements compared with standard under-tree fertilizing. Similar to frequent application of water, optimum split applications of fertilizer improve quality and quantity of crop yield compared with the conventional practice (Ramniwas *et al.*, 2013).

The fertilizer is dissolved in water in a drum or tank and sucked or injected through the watering system via a venturi injection pump or a pressure differential system. Fertilizer must be highly soluble to avoid damaging the pump and lines; and must also be compatible to avoid precipitates that block the dripper nozzles or sprinklers. Liquid fertilizers are most suited for fertigation; however, conventional fertilizers such as urea, monoammonium phosphate and potassium chloride can be used in fertigation (Rao *et al.*, 2017). High concentration of Fe can promote the growth of certain bacteria. To address these problems, a full water test is advised at the start.

Field studies conducted at Bhopal, India in Vertisols indicated that irrigation scheduling at 80% of ET_c and 75% of the recommended dose of fertilizer (RDF) can save 45% of irrigations and 25% of fertilizers in guava cultivar 'Sardar' by using fertigation over the basin irrigation (15,843 litres per plant required between October and June) and under-tree fertilization. The number of fruits produced per plant and average yield increased by 28.5 and 32.7%, respectively, compared with control (Singh *et al.*, 2012). In north-west India (30°56′N and 75°52′E) where annual rainfall is only about 44 cm concentrated between June and September, it was recommended to control soil metric potential (SMP) between –40 and –45 kPa at 0.2 m depth immediately under the drip emitter and to use of 80% of RDF. The maximum yield (68.66 kg per plant) was recorded in guava cultivar 'Allahabad Safeda' with irrigation at –40 kPa and 80% of RDF (Khan *et al.*, 2013).

7.12 Conclusion

Guava exceeds the majority of tropical and subtropical fruit trees in adaptability, productivity and tolerance to adverse mild cold and night frosts, besides soil pH (4.5 to 8.2). Among the various factors that affect the production and productivity of guava, nutrient and water assume great significance. These two inputs are essentially required to be managed in a manner that provides maximum output. Guava orchards in most countries are planted in low-fertility soil, semi-arid or arid climates and under rainfed situations. The low precipitation along with impaired nutrient management systems are the main limiting factor for orchard productivity.

Response to particular fertilizer regimes is normally site-specific. Most orchards require regular applications of N, K, Mg and Zn. Integrated nutrient management using inorganic fertilizer along with organic substrates and biofertilizers has shown marked response particularly in low-fertility and marginal soils. Optimum leaf nutrient concentrations for high productivity have been established and calibrations between yield and leaf N, P and K calculated. Acceptable yields and fruit quality are generally achieved over a wide range of leaf nutrient concentrations.

Most of the guava orchards in India, Brazil and Mexico depend on regular rainfall, with irrigation being either too expensive or not available. Where water is available from a nearby reservoir or canal or there is a facility to lift groundwater, irrigation is provided during establishment and at some important phenological stages like fruit set and fruit development by conventional methods (flood, furrow or ring basin) of irrigation. However, in the last two decades, many orchards have been developed in different countries by planting productive cultivars/hybrids, high-density planting, canopy management and micro-irrigation systems (most commonly drip irrigation) with capability of applying fertilizer (fertigation).

References

Abd-Elhamied, A.S. and Fouda, K.F. (2018) Effect of foliar application of zinc and soil phosphorus rates on guava seedlings growth and nutrient uptake. *Journal of Soil Science and Agriculture Engineering, Mansoura University, Egypt* 9(3), 149–154.

Adak, T., Kumar, K., Singh, A., Shukla, S.K. and Singh, V.K. (2014) Assessing soil characteristics and guava orchard productivity as influenced by organic and inorganic substrates. *The Journal of Animal and Plant Sciences* 24(4), 1157–1165.

Ahmad, F.M.D., Shankar, G. and Sharma, R.R. (1998) Effect of different concentrations of urea on reproductive and physico-chemical parameters of guava. *Annals of Agricultural Research* 19(2), 199–201.

Ali-Dinar, H.M., Ebert, G. and Ludders, P. (1999) Growth, chlorophyll content, photosynthesis and water relations in guava (*Psidium guajava* L.) under salinity and different nitrogen supply. *Gartenbauwissenshaft* 64(2), 55–59.

Allen, R.G., Pereira, L.S., Raes, D. and Smith, M. (1998) *Crop Evapotranspiration – Guidelines for Computing Crop Water Requirements*. FAO Irrigation and Drainage Paper 56. Food and Agricultural Organization of the United Nations, Rome.

Anjaneyulu, K. and Raghupathi, H.B. (2009) Identification of yield-limiting nutrients through DRIS leaf nutrient norms and indices in guava (*Psidium guajava* L.). *Indian Journal of Agricultural Science* 79(6), 418–421.

Anjaneyulu, K., Raghupathi, H.B. and Chandraprakash, M.K. (2008) Composition nutrient diagnosis norms (CND) for guava (*Psidium guajava* L.). *Journal of Horticultural Science* 3(2), 132–135.

Arora, J.S. and Singh, J.R. (1970a) Effect of nitrogen, phosphorus and potassium sprays on guava (*Psidium guajava* L.). *Journal of Japanese Society for Horticultural Science* 39, 55–62.

Arora, J.S. and Singh, J.R. (1970b) Some effects of foliar spray of zinc sulphate on growth, yield and fruit quality of guava (*P. guajava* L.). *Journal of Japanese Society for Horticultural Science* 39, 207–211.

Arora, J.S. and Singh, J.R. (1972) Response of guava (*P. guajava* L.) to boron spray. *Journal of Japanese Society for Horticultural Science* 41, 239–244.

Arshad, I. (2015) Integrated application of NPK fertilizers on the growth and yield of guava (*Psidium guajava* L.) in arid region of lower Sindh, Pakistan. *International Research Journal of Plant and Crop Science* 2(1), 19–23.

Arshad, I. and Ali, W. (2016) Effect of foliar application of zinc on growth and yield of guava (*Psidium guajava* L.). *Advances in Science Technology and Engineering Systems Journal* 1(1), 19–22.

Azhakiamanavalan, R.S. and Khader, J.B.Md.A. (1981) Effect of certain fertilizer treatments on tree vigour, fruit yield and quality of Lucknow-49. In: *Abstracts of National Symposium on Tropical and Subtropical Fruit Crops, Bangalore, India, 21–24 January 1981*, p. 6.

Barne, V.G., Bhard, S.G., Dod, V.N. and Baviskar, M.N. (2011) Effect of integrated nutrient management on yield and quality of guava. *The Asian Journal of Horticulture* 6(2), 546–548.

Batista, P.F., Lima, M.A.C., Trindade, D.C.G. and Alves, R.E. (2015) Quality of different tropical fruit cultivars produced in the lower basin of the Sao Francisco Valley. *Revista Ciencia Agronomia* 46, 176–184.

Bhargava, B.S. and Chadha, K.L. (1988) Leaf nutrient guide for fruit and plantation crops. *Fertilizers News* 33(7), 21–29.

Bishop, P.E., Jarlenski, D.M.L. and Hetherington, D.R. (1980) Evidence for an alternative nitrogen fixation system in *Azotobacter vinelandii*. *Proceedings of the National Academy of Sciences, USA* 77, 7342–7346.

Brahmachari, V.S., Kumar, N. and Kumar, R. (1997) Effect of foliar feeding of calcium, potassium and growth substance on yield and fruit quality of guava (*Psidium guajava* L.). *Haryana Journal of Horticultural Science* 26, 169–176.

Cavalcante, I.H.L., Cavalcante, L.F., Hu, Y. and Cavalcante, M.Z.B. (2007) Water salinity and initial development of four guava (*Psidium guajava* L.) cultivars in north-eastern Brazil. *Journal of Fruit and Ornamental Plant Research* 15, 71–80.

Cavalcante, L.F., Silva, G.F., Gheyi, H.R., Dias, T.J., Alves, J.C. and Casta, A.P.M. (2009) Crescimento de mudas de maracujazeiro amarelo em solo salono com estercoborino liquid fermentado. *Revista Brasileira de Ciencias Agrarias, Recife* 4(4), 414–420.

Chadha, K.L., Arora, J.S., Ravel, P. and Shikhamany, S.D. (1973) Variation in the mineral composition of the leaves of guava (*Psidium guajava* L.) as affected by leaf position, season and sample size. *Indian Journal of Agricultural Science* 43(6), 555–561.

Chakroborty, B., Tiwari, J.P., Lal, S. and Kumar, R. (2008) Effect of organic manure and mulching on growth, yield and quality of winter season crop of guava (*Psidium guajava* L.) cv. Pant Prabhat. *Pantnagar Journal of Research* 6(2), 239–242.

Chaudhury, D.N., Shyamal, M.M. and Maurya, K.R. (1975) Influence of inorganic and organic manures alone and in combination on growth, yield and chemical quality of guava (*Psidium guajava* L.). *Indian Food Packer* 29, 24–26.

Chavez, J.C.L. and Torres, A.I.Z. (2012) Conventional guava in Zitacuaro's Region, Michoacan, Mexico. *Sustainable Agricultural Research* 1, 19–25.

Chelvan, R.C. and Singh, G. (1986) *Annual Report*. Indian Institute of Horticultural Research, Bangalore, India, pp. 23–25.

Chen, J. (2006) The combined use of chemical and organic fertilizers and/or biofertilizer for crop growth and soil fertility. In: *International Workshop on Sustained Management of the Soil–Rhizosphere System for Efficient Crop Production and Fertilizer Use, Bangkok, 16 20 October 2006*. Land Development Department, Bangkok, pp. 1–11.

Chetri, K., Sanyal, D. and Kar, P.L. (1996) Variation in foliar micronutrient composition of guava due to month, cultivar, direction of shoot and zone of sampling. *The Horticulture Journal* 9, 21–26.

Chetri, K., Sanyal, D. and Kar, P.L. (1999) Changes in nutrient element composition of guava leaves in relation to season, cultivar, direction of shoot, and zone of leaf sampling. *Communications in Soil Science and Plant Analysis* 30(1/2), 121–128.

Chhonkar, V.S. and Singh, P.N. (1981) Effect of nitrogen, phosphorus and potassium as foliar spray on growth, flowering and fruiting of guava (*Psidium guajava* L.) cv. Allahabad Safeda. In: *Abstracts of National Symposium on Tropical and Subtropical Fruit Crops, Bangalore, India, 21–24 January 1981*, p. 53.

Corrales, G.I., Lopaz, L.P. and Gomez, G.A. (2000) Use of poultry manure and mineral fertilizer on guava (*Psidium guajava* L.). *Centro Agricola* 27(4), 47–57.

Corte, L., Dell'Abate, M.T., Magini, A., Migliore, M., Felici, B. *et al.* (2014) Assessment of safety and efficiency of nitrogen organic fertilizers from animal-based protein hydrolysates – a laboratory multidisciplinary approach. *Journal of the Science of Food and Agriculture* 94, 235–245.

Dahiya, S.S. and Joon, M.S. (1995) Variation in mineral composition of leaves of guava cultivar L-49 as affected by sample size. *Crop Research* 9, 121–122.

Das, A., David, A.A., Swaroop, N., Thomas, T., Rao, S. and Hasan, A. (2018) Assessment of physic-chemical properties of river bank soil of Yamuna in Allahabad city, Uttar Pradesh. *International Journal of Chemical Studies* 6(3), 2412–2417.

Das, K., Sau, S., Datta, P. and Sengupta, D. (2017) Influence of bio-fertilizer on guava (*Psidium guajava* L.) cultivation in Gangetic alluvial plain of West Bengal, India. *Journal of Experimental Biology and Agricultural Science* 5(4), 476–482.

Devi, H.S., Mitra, S.K. and Poi, S.C. (2012) Effect of different organic and biofertilizer sources on guava (*Psidium guajava* L.) 'Sardar'. *Acta Horticulturae* 959, 201–208.

Dey, P., Raii, M., Kumar, S., Nath, V., Das, B. and Reddy, N.M. (2005) Effect of biofertilizers on physico-chemical characteristics of guava (*Psidium guajava* L.). *Indian Journal of Agricultural Science* 75(2), 95–96.

Dhankar, O.P., Makhija, M., Chauhan, K.S. and Singrot, R.S. (1981) Foliar mineral composition of guava and phalsa grown on saline soils. *Transactions of Indian Society of Desert Technology, University Centre for Desert Studies* 6, 33.

Doraipandian, A. and Shanmugavelu, K.G. (1972) Effect of foliar spray of urea on the yield of guava (*Psidium guajava* L.). *South Indian Horticulture* 20, 80–81.

Du Plessis, S.K., Smart, G. and Koen, T.J. (1973) A few aspects of fertilizing guavas. *Citrus Growers and Sub-Tropical Fruit Journal* 478, 18–19.

El-Khoreiby, A.M. and Salem, A.T. (1989) Effect of different irrigation regimes on growth, fruiting and fruit quality of seedy guava trees. *Annals of Agricultural Science, Faculty of Agriculture, Ain Shams University, Egypt* 34, 313–321

Fares, A. (2008) *Water Management Software to Estimate Crop Irrigation Requirements or Consumptive Use Permitting in Hawaii. Department of Land and Natural Resources*, Honolulu, Hawaii. Available at: http://hawaii.gov/dlnr/cwrm/publishedreports/PR200808.pdf (accessed 7 January 2021).

Guerra, E. and Bautista, D. (2002) Leaf mineral status of three guava (*Psidium guajava* L) cultivars during high growth activity. *Bioagro* 14(2), 99–104.

Gupta, M.R. and Nijjar, G.S. (1978) Crop regulation in guava. *Indian Journal of Horticulture* 35, 23–27.

Hada, T.S., Singh, B.K., Veer, K. and Singh, S.P. (2014) Effect of different levels of boron and zinc on flowering, fruiting and growth parameter of winter season guava (*Psidium guajava* L.) cv. L-49. *The Asian Journal of Horticulture* 9(1), 53–56.

Haggag, L.F. and Maksoud, M.A. (1996) Effect of salinity of irrigation water on some properties of guava fruits. *Egyptian Journal of Horticulture* 23, 203–209.

Haggag, L.F., Azzazy, M.A. and Maksoud, M.A. (1994) Effect of biofertilizer 'phosphorine' on phosphorous content and dry matter of guava seedlings growing in sandy soil conditioned with composted town refuse. *Annals of Agricultural Science Cairo* 39(1), 345–353.

Hamilton, R.A. and Seagrave-Smith, H. (1959) *Growing Guava for Processing*. Extension Bulletin No. 63. University of Hawaii, Honolulu, Hawaii.

Hanson, B.R., Schwanki, L.J., Schulbach, K.F. and Pettygove, G.S. (1997) A comparison of furrow, surface drip and sub-surface drip irrigation on lettuce yield and applied water. *Agricultural Water Management* 33, 139–157.

Hundal, H.S., Singh, D. and Singh, K. (2007) Monitoring nutrient status of guava fruit trees in Punjab, northwest India through the diagnostic and recommendation integrated system approach. *Communications in Soil Science and Plant Analysis* 38(15/16), 2117–2130.

Kadam, A.S. and Patil, V.K. (1993) Phosphorus nutrition studies in Sardar guava. *Annals of Plant Physiology* 7(2), 150–152.

Kaj, T., Tewari, J.P. and Lal, S. (1989) Effect of different levels of nitrogen on growth, yield and quality of guava (*Psidium guajava* L.) cv. Sardar. *Progressive Horticulture* 20, 213–217.

Khan, J.N., Jain, A.K., Sharda, R., Singh, N.P., Gill, P.P.S. and Kaur, S. (2013) Growth, yield and nutrient uptake of guava (*Psidium guajava* L.) affected by soil matric potential, fertigation and mulching under drip irrigation. *Agricultural Engineering Institute CIGR Journal* 15(3), 17–28.

Kisekka, I., Migliaccio, K.W., Dukes, M.D., Crane, J.H., Schaffer, B. *et al.* (2019) *Evapotranspiration-based Irrigation for Agriculture: Crop Coefficients of Some Commercial Crops in Florida*. Document No. AE-456. UF/IFAS Extension, University of Florida, Gainesville, Florida.

Koen, T. (1987) *Manurial Requirements and Leaf Norms of Guavas*. Information Bulletin No. 184. Citrus and Subtropical Fruit Research Institute, Nelspruit, South Africa.

Koen, T. and Hobbs, A. (1990) *Guava. Leaf and Soil Analysis Service*. Information Bulletin No. 210. Tropical Fruit Research Institute, Nelspruit, South Africa, pp. 7–8.

Kotur, S.C., Kumar, R. and Singh, H.P. (1997) Influence of nitrogen, phosphorus and potassium on composition of leaf and its relationship with fruit yield in Allahabad Safeda guava (*Psidium guajava*) on a Alfisol. *Indian Journal of Agricultural Science* 67, 568–570.

Kumar, D., Pandey, V., Anjaneyulu, K. and Nath, V. (2009) Optimization of major nutrients for guava and quality under east coastal conditions. *Indian Journal of Horticulture* 66(1), 18–21.

Kumar, R., Kotur, S.C. and Singh, H.P. (1996a) Effect of nitrogen on growth, fruit yield and quality of guava under rainfed conditions of Bihar plateau. *Indian Journal of Horticulture* 53, 109–113.

Kumar, R., Kotur, S.C. and Singh, H.P. (1996b) Response of guava (*Psidium guajava* L.) cv. Allahabad Safeda to potassium on a sandy loam soil. *Journal of Potassium Research* 12, 59–64.

Kundu, S. and Mitra, S.K. (1999) Response of guava to foliar spray of copper, boron and zinc. *India Agriculturist* 43, 49–54.

Kundu, S., Ghosh, B., Mitra, S.K. and Mazumdar, D. (2007) Effect of foliar spraying of nitrogen phosphorus and potassium on yield and fruit quality of guava (*Psidium guajava* L.). *Acta Horticulturae* 735, 433–440.

Lal, G. (1996) Scheduling and depth of irrigation on growth and yield of guava (*Psidium guajava* L.) variety Sardar. *Annals of Biology* 12, 238–241.

LihShang, K. (1997) Effects of different soil moistures on fruit yield and quality of guava (*Psidium guajava* L.). *Special Publication Taichung District Agricultural Improvement Station* 38, 231–237.

LihShang, K. and Weider, W. (1997) Effect of nitrogen, phosphorus and potassium fertilizer levels on yield and quality of guava (*Psidium guajava* L.). *Special Publication Taichung District Agricultural Improvement Station* 38, 239–250.

Luna-Jimenez, A.d., Meraz-Jimnez, A.d.J., Ponce-Montoya, A., Luna-Ruiz, J.d. and Lara, J.M.d. (2010) Coberturas de Cajete, manejo del reigo y rendimiento de guayaba. *Acta Horticulturae* 849, 381–386.

Maia, P.L.T., Bassoi, L.H., Silva, D., Lima, M.A.C., Assis, J.S. and Morais, P.L.D. (2007) Assessment of nutrient levels in the aerial biomass of irrigated guava in San Francisco valley, Brazil. *Revista Brasileira de Fruticultura, Jaboticabal* 39(3), 705–709.

Mandal, G., Kumar, S., Kumar, R. and Singh, R. (2007) Effect of drip irrigation and plant spacing on yield, quality and economic return of guava (*Psidium guajava* L.) grown in saline soil. *Acta Horticulturae* 735, 427–432.

Marchal, J. (1984) Frutiers tropicaux divers. In: Martin-Prevel, P., Gagnard, J. and Gautier, P. (coords) *L'analyse vegetale dans le controle de l'alimentation des plantes temprees et tropicales*. Tec & Doc Lavoisier, Paris, pp. 496–510.

Marengo, J.A., Alves, L.M., Beserra, E.A. and Lacerda, F.F. (2011) Variabilidade e mudancas climaticas no semiardo brasileiro. In: *Recarsos Hídricos em Regioes do Aridas e Semiáridas*. Instituto Nacional do Semiarido, Campina Grande, Brazil, pp. 384–422.

Marler, T.K. (1994) Guava. In: Schaffer, S. and Anderson, P.C. (eds) *Handbook of Environmental Physiology of Fruit Crops*, Vol. II. *Sub-tropical and Tropical Crops*. CRC Press, Boca Raton, Florida, pp. 213–216.

Mass, E.V. (1993) Testing crops for salinity tolerance. In: *Proceedings of the Workshop on Adaptation of Plants to Soil Stresses, Lincoln, Nebraska, 1–4 August 1993*, pp. 234–247.

Mata, B.I. and Rodriguez, M. (1989) *El Guayabo: Aspectos de su Cultivo y Produccion. Universidad Autonoma Champing Agraira Antonio Narro*, Coahuila, Mexico.

Mehta, P.K., Kachroo, A., Kaul, M.K. and Yamdagni, R. (1988) Salt tolerance in fruit crops – a review. *Agriculture Reviews* 9, 57–68.

Menzel, C.M., Bagshaw, J., Campbell, T., Greer, N., Noller, J. *et al.* (2002) *Lychee Information Kit*. Queensland Department of Primary Industries, Nambur, Australia.

Mitra, S.K. (1987) Studies on guava nutrition with special reference to potassium and nitrogen. *Journal of Potassium Research* 3(4), 199–201.

Mitra, S.K. and Bose, T.K. (1985) Effect of varying levels of nitrogen, phosphorus and potassium fertilization on the growth, yield and quality of guava (*Psidium guajava* L.) var. L-49. *South Indian Horticulture* 33, 286–292.

Mitra, S.K. and Bose, T.K. (1990) Guava. In: Bose, T.K. and Mitra, S.K. (eds) *Fruits: Tropical and Subtropical*. Naya Prokash, Kolkata, India, pp. 280–303.

Mitra, S.K. and Dhaliwal, S.S. (2009) Effect of potassium on fruit quality and their storage life. In: Brar, M.S. and Mukhopadhyay, S.K. (eds) *Proceedings of the IPI-OUAT-IPNI International Symposium on 'Potassium Role and Benefits in Improving Nutrient Management for Food Production, Quality and Reduced Environmental Damages', Bhubaneswar, India, 5–7 November 2009*. International Potash Institute, Zug, Switzerland, pp. 327–342.

Mitra, S.K. and Sanyal, D. (2004) *Guava*. Indian Council of Agricultural Research, New Delhi.

Mitra, S.K., Gurung, M.R. and Pathak, P.K. (2008) Sustainable guava production in West Bengal, India. *Acta Horticulturae* 773, 179–182.

Morton, J.F. (1987) *Fruits of Warm Climates*. Creative Resource Systems, Inc., Winterville, North Carolina.

Murthy, K.S.V. and Kotur, S.C. (1998) Phosphorus absorption by guava (*Psidium guajava*) as influenced by time of application and placement method in typic Haplusfalf. *Journal of Nuclear Agriculture and Biology* 27(2), 117–121.

Naik, M.H. and Hari Babu, R. (2007) Feasibility of organic farming in guava (*Psidium guajava* L.). *Acta Horticulturae* 735, 365–372.

Nakasone, H.Y. and Paull, R.E. (eds) (1998) Guava. In: *Tropical Fruits*. CAB International, Wallingford, UK, pp. 149–172.

Narula, N., Lakshminarayana, K.L. and Tauro, P. (1981) Ammonia excretion by *Azotobacter chroococcum*. *Biotechnology and Bioengineering* 23, 467–470.

Natale, W., Coutinho, E.L.M., Boaretto, A.E. and Pereira, F.M. (1994a) Nitrogen fertilization of guava. *Fruits, Paris* 49(3), 205–210.

Natale, W., Coutinho, E.L.M., Boaretto, A.E. and Banzatto, D.A. (1994b) Effect of sampling time on leaf chemical composition in guava (*Psidium guajava* L.). *Revista de Agricultura Piracicaba* 63(3), 247–265.

Natale, W., Coutinho, E.L.M., Pereira, F.M., Martinez Junior, M. and Mortins, M.C. (1995) Effect of N, P and K fertilization on total soluble solids of guava (*Psidium guajava* L.) fruits. *Alimentos e Nutricao* 6, 69–75.

Natale, W., Coutinho, E.L.M., Boaretto, A.E. and Pereira, F.M. (1996) Effect of potassium fertilization in 'Rica' guava (*Psidium guajava*) cultivation. *Indian Journal of Agricultural Sciences* 66(4), 201–207.

Natale, W., Boaretto, A.E. and Maraoka, T. (1999) Radioactive phosphorus uptake and redistribution in guava trees when directly applied to the leaves. *Fruits, Paris* 54(1), 23–29.

Natale, W., Coutinho, E.L.M., Pereira, F.M. and Boaretto, A.E. (2002) Nutrients foliar content for high productivity cultivars of guava in Brazil. *Acta Horticulturae* 594, 383–386.

Neito, K.F. and Frankenberger, W.T. (1989) Biosynthesis of cytokinins by *Azotobacter chroococcum*. *Soil Biology and Biochemistry* 21, 967–972.

Ogden, M.A.H., Jackson, L.K. and Campbell, C.W. (1981) Florida tropical fruit culture via master gardener. *Proceedings of the Florida State Horticulture Society* 94, 222–225.

Ortega, O. (1971) Relacion entre la fecha de inicode riego y fecha de floracion y fructificacion de la guayaba in Colvillo, Ags., Mexico. *Revista Fitotecnia Mexicana* 2, 34–35.

Pandey, D.K., Pathak, R.A. and Pathak, R.K. (1988) Studies on the foliar application of nitrogen and plant growth regulators in Sarder guava. *I. Effect on yield and quality. Indian Journal of Horticulture* 45, 197–202.

Parent, L.E. and Dafir, M. (1992) A theoretical concept of composition nutrient diagnosis. *Journal of the American Society for Horticultural Science* 117(2), 239–242.

Pathak, R.K., Ram, R.A. and Mishra, M. (2002) Nutrient management in fruit crops of Uttar Pradesh and Uttaranchal: an overview. In: Pathak, R.K., Tandan, D.K., Singh, V.K. and Tiwari, K.N. (eds) *Proceedings of the Workshop on Nutrient Status, Needs and Recommendations for Major Fruit Crops of Uttar Pradesh and Uttaranchal, 10–11 December 2002*, pp. 1–4.

Patil, P.K., Patil, V.K. and Ghonsikar, C.P. (1984) Effect of soil salinity on growth and nutritional status of guava (*Psidium guajava* L.). *International Journal of Agriculture* 2, 337–344.

Pereira, L.S. and Mitra, S.K. (1999) Studies on organic along with inorganic nutrition in guava. *Indian Agriculturist* 43(3/4), 155–160.

Pereira, W.E., Couto, F.A., Siqueira, D.A., Bruckner, C.H. and Barros, R.S. (2000) Yield and some physico-chemical characteristics of fruit of six guava cultivars under water stress. *Revista Ceres* 47, 349–362.

Prajapati, K., Yami, K.D. and Singh, A. (2008) Plant growth promotional effect of *Azotobacter chroococcum, Piriformospora indica* and vermicompost on rice plant. *Nepal Journal of Science and Technology* 9, 85–90.

Queiroz, E.F., Kliemann, H.J., Vieira, A., Rodrigues, A.P.M. and Guilherme, M.R. (1986) Nutricado mineral e adubacado da goiabeira (*Psidium guajava* L.). In: Haag, H.P. (ed.) *Nuticado Mineral Adubacado de Fruteiras Tropicais no Brazil.* Fundacao Cargil, Campinas, Brazil, pp. 165–187.

Raghava, M. and Tiwari, J.P. (1998) Effect of boron on growth, quality and shelf life of fruits of guava (*Psidium guajava* L.) cv. Sardar. *Progressive Horticulture* 30(1/2), 68–72.

Rajput, C.B.S. and Chand, S. (1975) Significance of boron and zinc in guava (*Psidium guajava* L.). *Bangladesh Horticulture* 3(2), 22–27.

Rajput, C.B.S. and Singh, N.P. (1976) Leaf analysis and phosphorus fertilization of guava (*Psidium guajava* L.). *Journal of Japanese Society for Horticultural Science* 44, 555–559.

Rajput, C.B.S., Singh, N.P. and Tiwari, J.P. (1978) Effect of potash on yield attributes of guava. *Indian Journal of Horticulture* 35, 19–22.

Ram, R.A. and Rajput, M.S. (1998) Effect of slow release nitrogen fertilizers on yield and quality of guava cv. Sardar. In: *National Seminar on New Horizons in Production and Postharvest Management of Tropical and Subtropical Fruits, 8–9 December 1998.* Indian Agricultural Research Institute, New Delhi, p. 29.

Ram, R.A., Bhriguvanshi, S.R. and Pathak, R.K. (2007) Integrated plant nutrient management in guava (*Psidium guajava* L.) cv. Sardar. *Acta Horticulturae* 735, 345–350.

Ramniwas, R., Kaushik, A., Pareek, S., Sarolia, D.K. and Singh, V. (2013) Effect of drip fertigation scheduling on fertilizer use efficiency, leaf nutrient status, yield and quality of 'Shweta' guava (*Psidium guajava* L.) under meadow orcharding. *National Academy of Science Letters, India* 36(5), 483–488.

Rao, K.V.R., Gangwar, S., Bajpai, S., Chourasiya, L. and Soni, K. (2017) Influence of growth, yield and quality of guava (*Psidium guajava* L.) by drip irrigation and fertigation. *Journal of Applied and Natural Science* 9(1), 642–655.

Ravi-Kumar, Ahlawt, V.P., Chauhan, K.S. and Singh, H.K. (1990) Assessment of some guava (*Psidium guajava* L.) cultivars for their leaf mineral composition. *Haryana Journal of Horticultural Science* 19(1–2), 101–105.

Raychaudhuri, R., Kabir, J., Ray, S.K.D. and Dhua, R.S. (1992) Influence of pre-harvest spray of calcium salts in the improvement of fruit quality in guava cv. L. 49. *Advances in Horticulture and Forestry* 2, 70–76.

Regvar, M., Vogel-Mikus, K. and Severkar, T. (2003) Effect of AMF inoculums from field isolates on the yield of green pepper, parsley, carrot and tomato. *Folia Geobotanica* 38, 223–224.

Rodriguez, H., Perera, B. and Carrera, N. (1984) Influence of fruiting and leaf position along the branch on the mineral content of guava (*Psidium guajava* L.) cv. E.E.A., 18–40 leaves. *Ciencia y Tecnica en la Agricultura Citricos y Otros Fruitales* 7, 83–97.

Rozane, D.E., Natale, W., Parent, L.E. and dos Santos, E.M.H. (2012) The CN-Goiaba 1.0 software for nutrimental diagnosis of guava (*Psidium guajava* L.) 'Paluma' in Brazil. *Acta Horticulturae* 959, 161–166.

Rozane, D.E., Parent, L.E. and Natale, W. (2016) Evolution of the predictive criteria for the tropical fruit tree nutritional status. *Cientifica, Jaboticabal* 44(1), 102–112.

Santana, E.A., Cavalcante, I.H.L., Brito, D.d.S., Carmo, R.M. and de Sousa, K.d.S.M. (2017) Fruit production and quality of guava as a function of biofertilizer and nitrogen fertigation in Brazilian semiarid. *Emirates Journal of Food and Agriculture* 29(4), 242–249.

Sanyal, D. and Mitra, S.K. (1990) Standardization of leaf sampling technique for mineral composition of guava (*P. guajava* L.). *Indian Journal of Horticulture* 55, 183–189.

Satisha, J., Kurian, R.M. and Dinesh, M.R. (2016) *Production Technology of Tropical Fruits – A Hand Book.* Indian Institute of Horticulture Research, Bengaluru, India, pp. 31–44.

Schaffer, B., Anderson, P.C. and Plotez, R.C. (1992) Response of fruit crops to flooding. *Horticulture Reviews* 13, 257–269.

Sharma, A., Wali, V.K., Bakshi, P. and Jamwal, M. (2011) Effect of organic manures and biofertilizers on leaf and fruit nutrient status in guava (*Psidium guajava* L.) cv. Sardar. *Journal of Horticulture Science* 6(2), 169–171.

Sharma, R. and Bhattacharyya, R.K. (1989) Effect of foliar nutrition of zinc on the nutrient concentration of guava leaves. *South Indian Horticulture* 37(6), 323–325.

Sharma, S., Halder, A., Patra, S.K. and Ray, R. (2012) Effect of drip irrigation and nitrogen fertigation on water use efficiency (WUE) and cost economics of guava cv. *Khaja. Progressive Horticulture* 44(1), 136–141.

Shigeura, G.T. and Bullock, R.M. (1983) *Guava (Psidium guajava L.) in Hawaii – History and Production.* Hawaii Institute of Tropical Agriculture and Human Resources, Research Station Series No. 035. University of Hawaii, Honolulu, Hawaii.

Singh, A., Kumar, A., Yadav, R.K., Dutta, A. and Sharma, D.K. (2016) Growth and mineral nutrition in salt stressed guava (*Psidium guajava* L.) cv. Allahabad Safeda. *Journal of AgriSearch* 3(1), 21–25.

Singh, A.K. and Pathak, R.K. (1994) Effect of salinity levels on nutrient status and chlorophyll content in guava leaves. *Orissa Journal of Horticulture* 22(1/2), 31–35.

Singh, B.K., Tiwari, K.N., Chourasia, S.K. and Mandal, S. (2007) Crop water requirement of guava (*Psidium guajava* L.) cv. KG/Kaji. *Acta Horticulturae* 735, 399–405.

Singh, B.P., Singh, H.K. and Chauhan, K.S. (1981) Effect of post-harvest calcium treatment on the storage life of guava fruit. *Indian Journal of Agricultural Science* 15, 44–47.

Singh, H.K., Singh, B.P. and Chauhan, B.P. (1981) Effect of foliar feeding of various chemicals on physio-chemical quality of guava fruit. *Haryana Agriculture Journal of Research* 11, 411–414.

Singh, H.P. and Singh, G. (2007) Nutrient and water management in guava. *Acta Horticulturae* 735, 389–397.

Singh, K. and Chauhan, K.S. (1982) Effect of pre-harvest application of calcium, potassium and alar on fruit quality and storage life of guava fruits. *Haryana University Journal of Research* 12, 649–654.

Singh, N.P. and Rajput, C.B.S. (1976) Leaf analysis and potassium fertilization in guava. *Indian Journal of Horticulture* 33, 152–153.

Singh, N.P. and Rajput, C.B.S. (1977) Effect of phosphorus on yield attributes and quality of guava (*P. guajava* L.). *Indian Journal of Horticulture* 34, 120–125.

Singh, R.M., Bhandarkar, D.M., Singh, D.K., Reddy, K.S., Rao, K.V.R. and Mathankar, S.K. (2012) Techno-economic feasibility of drip fertigation in guava (*Psidium guajava* L). *Environment and Ecology* 30(2), 271–274.

Singh, V. (1985) Effect of foliar spray of urea on growth, yield and quality of guava (*P. guajava*) cv. Safeda. *Udyanika* 5, 11–16.

Souza, H.A., Rozane, D.E., Amorim, D.A., Dias, M.J.T., Modesto, V.C. and Natale, W. (2015) Assessment of nutritional status of guava seedlings using preliminary DRIS norms and sufficiency ranges. *Journal of Plant Nutrition* 38, 1611–1618.

Souza, W.J.O., Souza, H.A., Rozane, D.E. and Natale, W. (2012) Evaluation of soil management and use in an Utisol in a guava orchard in comparison with a sugarcane field and native forest area. *Acta Horticulturae* 959, 173–177.

Teixeira, A.H.d.C. and Bassoi, L.H. (2009) Crop water productivity in semi-arid regions: from field to large scales. *Annals of Arid Zone* 48, 1–13.

Teixeira, A.H.d.C. and Hernandez, F.B.T. (2012) Delimitation of guava water productivity in the Brazilian Northeast. *Acta Horticulturae* 959, 179–186.

Teixeira, A.H.d.C., Bassoi, L.H., Reis, V.C.d.S., Silva, T.G.F.d., Ferreira, M.d.N.L. and Maia, J.L.T. (2003) Evaluation of water consumption of guava crop by using automatic and conventional agrometeorological station. *Revista Brasileira de Fruticultura* 25, 457–460.

Thaipong, K. and Boonprakob, U. (2019) Salt tolerance evaluation in guava germplasm. *International Journal of Agricultural Technology* 15(5), 791–796.

Tinco, I.C.M, Bezerra, B.G., Lucio, P.S. and Barbosa, L.d.M. (2018) Characterization of rainfall patterns in the semiarid Brazil. *Anuario do Instituto de Geociencias* 41 (2), 397–409.

Tindale, A.E., Mehrotra, M., Ottem, D. and Page, W.J. (2000) Dual regulation of catecholate siderophore biosynthesis an *Azotobacter vinelandii* by iron and oxidative stress. *Microbiology* 146, 1617–1626.

Tiwari, J.P. and Lal, S. (2000) *Amrood Ki Kheti evem bagh prabandh*. Directorate of Publication, G.B. Pant University of Agriculture and Technology, Pantnagar, India, pp. 19–23.

Tiwari, R.B. and Tiwari, J.P. (1993) Studies on the problem of leaf bronzing in guava – induction of phosphorus deficiency symptoms under sand culture. *Indian Journal of Horticulture* 50, 53–56.

Trivedi, Y.V., Patel, N.L., Ahlawat, T.R., Gaikwad, S.S. and Bhalerao, P.P. (2012) Impact of organic manures and inorganic fertilizers on growth, yield, nutrient uptake and soil nutrient status in guava. *Indian Journal of Horticulture* 69(4), 501–506.

Wagh, A.N. and Mahajan, P.R. (1985) Effect of nitrogen, phosphorus and potassium on growth and yield of guava cv. Sardar. *Current Research Report* 1, 124–126.

Yadav, A., Verma, R.S., Ram, R.B., Kumar, V. and Yadav, R.K. (2017) Effect of foliar application of micronutrients on physical parameters of winter season guava (*Psidium guajava*) cv. *Lalit*. *Plant Archives* 17(2), 1457–1459.

Yadav, D., Awasthi, M.K. and Nema, R.K. (2017) Estimation of crop water requirement of micro-irrigated orchard crops for different agro-climatic conditions of Madhya Pradesh. *International Archive of Applied Science and Technology* 8(3), 18–24.

Yang, S. and Hung, H.Y. (2001) Effect of different fertilization on the fruit quality of guava (*Psidium guajava* L.). *Journal of Agricultural Research of China* 50(4), 23–28.

8 Orchard Management

Sisir Mitra[1]* and P.K. Pathak[2]

[1]*Former Dean, Bidhan Chandra Krishi Viswavidyalaya, Mohanpur, West Bengal, India;* [2]*KVK Murshidabad, West Bengal University of Animal and Fishery Sciences, Murshidabad, West Bengal, India*

8.1 Introduction

Commercial production from a guava orchard begins on the third year after planting and cropping may continue for 40 years or more. The performance of an orchard depends on its management, which includes water and nutrient management, selection of the correct cultivars, planting technique, canopy management for flowering and fruiting, and improved light efficiency through pruning to optimize the quality of young and bearing trees (Singh, 2007). Under normal planting distance (6–7 m × 6–7 m), the interspace may be utilized by growing short-duration intercrops to make the orchard productive in the early years (second to fourth) of planting. Guava is grown mainly as a rainfed crop in most places. In such situation, mulching is an important and useful practice. Mulches are important in conserving soil moisture, maintaining root-zone temperature, controlling weeds and mitigating soil erosion.

8.2 Planting System

Guava is usually planted in square or rectangular systems of planting. However, the recent trend is in developing new orchards with high plant densities by planting at a closer spacing for obtaining higher yield per unit area. Several new systems of planting such as hedgerow system, double-hedgerow system, pair planting, meadow orchard, etc. (Fig. 8.1) have been developed and evaluated with different cultivars of guava. A long-term experiment was conducted at Central Horticultural Experiment Station, Ranchi, India with four planting systems and different plant densities. The treatments included square system (5 m × 5 m, 400 plants ha^{-1}), hedgerow system (2.5 m × 5 m × 5 m, 800 plants ha^{-1}), double-hedgerow system (5 m × 2.5 m × 5 m, 530 plants ha^{-1}) and paired planting (2.5 m × 2.5 m × 2.5 m, 1060 plants ha^{-1}). The results revealed that plant growth characteristics were significantly influenced by planting systems-cum-densities. The girth and volume of plants decreased with increasing plant density while tree height increased with increasing plant density. The average fruit weight and yield per plant were higher with wider spaced plants (Table 8.1) while total yield per unit area was maximum in the paired system of planting accommodating 1060 plants ha^{-1} (Kumar and Singh, 2000). Evaluating 5-year-old 'Sardar' guava under

*E-mail: sisirm55@gmail.com

DOI: 10.1079/9781789247022.0008

Fig. 8.1. Systems of planting: (A) square system; (B) hedgerow system; (C) cluster planting system; and (D) paired planting system. Photographs courtesy Dr A. Guha Choudhury (A); Dr Pratibha (B–D).

Table 8.1. Effect of planting systems-cum-densities on growth and yield of guava 'Allahabad Safeda' (mean of 5 years, 1988–1992). Adapted from Kumar and Singh (2000).

Planting density (trees per ha)	Tree girth (cm)	Tree height (m)	Tree volume (m^2)	Fruit weight (g)	Yield (kg per tree)	Yield (kg ha^{-1})
400 (square system)	37.8	3.44	20.65	130.3	20.54	8,212
800 (hedgerow system)	35.3	3.74	16.36	108.3	14.87	11,868
530 (double-hedgerow system)	36.6	3.62	18.48	118.4	17.52	9,065
1060 (paired planting)	34.4	4.15	15.09	104.2	12.62	13,697
SEM (±)	0.42	0.06	0.67	2.24	0.46	2.56
CD (P = 0.05)	1.28	0.19	2.03	6.74	1.38	7.78

SEM, standard error of the mean; CD, critical difference.

different planting systems, Lal *et al.* (2007) reported maximum yield of 64.88 kg per plant in the square system of planting compared with 57.90 kg per plant in the double-hedgerow system of planting. The yield per hectare was, however, recorded maximum in the double-hedgerow system (26.23 t ha^{-1} at 453 plants ha^{-1}) compared with 13.20 t ha^{-1} (204 plants ha^{-1}) in the square system. Irrespective, the planting systems showed no significant variation in average fruit weight, °Brix, total sugar and vitamin C contents of fruit.

8.3 Planting Density

Many variables must be included in making a decision about plant spacing. Tree vigour and growth habit as influenced by variety and rootstock (where it could be used) are important. Site quality in terms of climate, soil characteristics and water availability should be considered. If water and nutrients are provided in adequate amounts, the interception and utilization of sunlight becomes the next most important consideration in orchard design. The ultimate limit on productivity of any crop is the amount of photosynthetically active radiation intercepted. Much of the planning that goes into the design of high-density planting (HDP) is actually based on average light interception over the life of the planting. This involves both minimizing the amount of light which strikes the ground and providing canopy structures in which the largest amount of canopy receives the optimum light intensity. In designing an orchard, the relative amounts of bearing and non-bearing areas can be planned. A tree with a large portion of its volume devoted to bearing has a greater potential for productivity than one with most of its volume devoted to tree support, middles or other functions (Mitra *et al.*, 2018).

Tree shape and planting density influence light interception and its distribution through the canopy. Various models of canopy photosynthesis and dry matter production have been developed in apple, and optimum tree shapes and planting densities have been determined for different environments. In recent years, the tropical fruit industries, including guava, have moved to closer plantings, although these developments have occurred in the absence of sound tree physiology. Preliminary observations in guava revealed that trees spaced at 6 m × 6 m intercepted higher radiation than close-spaced trees and the maximum interception occurred on the upper part of the tree, irrespective of planting distance (Singh and Dhaliwal, 2007). Data collected by Brar *et al.* (2013) indicated a significant increase in radiation with increase in plant spacing from 6 m × 2 m (60.3%) to 6 m × 4 m (68.91%). The reduced solar radiation in close-spaced trees was due to vertical orientation of auxiliary shoots and leaves leading to less absorption of solar radiation. The plants at 6 m × 4 m spacing intercepted higher radiation owing to horizontal orientation of shoots and leaves with higher foliage mass.

Guava trees tend to have a spreading growth habit and are relatively fast-growing in nature. In a commercial 15-year-old orchard in Hawaii requiring some pruning, trees are spaced at 5.2 m × 7.6 m, with 225 trees ha^{-1}, height being controlled at about 3.7 m (Nakasone and Paull, 1998). In Venezuela, planting density of 5 m × 5 m is recommended for cultivar 'Criolla Roja' trees and 5 m × 3.5 m when 'Cuban' type trees are grafted on 'Criolla Roja' (Araujo *et al.*, 1999). In Mexico, the cultivar 'Meridia China' is usually planted at 1.2 m × 1.2 m and the same spacing is also used for cultivars 'P.R-6-65' and 'P.R-7-65' growing in Puerto Rico. In Australia, the recommended spacing is 4 m × 6 m, giving a tree density of 416 trees ha^{-1} (Nakasone and Paull, 1998). In India, traditionally guava was planted at 8 m × 8 m or 6 m × 6 m (Bose *et al.*, 1992).

8.3.1 High-density planting

Recent trends in guava cultivation are planting at a closer spacing for obtaining higher yield per unit area. HDP is a highly efficient and advanced production system of fruit cultivation (Pal and Lal, 2015). In most of the temperate tree and nut crops it is now commercially followed all over the world because of the availability of suitable dwarfing rootstocks to control tree vigour. Important tropical fruit crops are mostly propagated by suckers (banana, pineapple), seed (papaya) or by layering (guava) and grafting (mango). The use of dwarfing rootstocks, as in many temperate tree fruits, is not possible/available in most of the commercial tropical fruits for high-density orcharding.

HDP is one of the novel methods to achieve high productivity per unit area both in short-duration and perennial horticultural crops. High yield and high fruit quality can be achieved with a high-density orchard when the orchard has good light distribution throughout the tree canopy and there is a balance between vegetative growth and cropping. Planting density is one of the most important factors which determine the yield of an orchard. After the first few years, the fertilization regime should be maintained with a balance between fruiting and cropping. Excess fertility often results in excessive vegetative growth, delayed cropping, and soft and poorly coloured unmarketable fruit. The benefit of HDP is to get the trees into cropping as soon as possible from a limited scope. This is best accomplished by proper canopy management (Goswami *et al.*, 2014). In terms of disadvantages of HDP, some observations indicate a reduction in yield and quality of fruit due to overcrowding or shading. Such reductions may result from delays in pruning and topping for tree size control (in tree crops) or use of excessive plant density causing shade and inhibiting light and air movement. In many countries, HDP has proved useful in many tropical and subtropical fruit crops such as mango, citrus, pineapple, banana, guava and papaya.

However, regular canopy management by pruning and thinning is considered a prerequisite for HDP in guava (Mitra and Sanyal, 2004); without canopy management, high density caused decreased fruit size and quality (Mitra *et al.*, 1984; Kundu *et al.*, 1993). Trials conducted at Bidhan Chandra Krishi Viswavidyalaya, West Bengal, India showed that HDP in guava increased yield per unit area but decreased fruit weight and fruit quality (Mitra *et al.*, 1984; Table 8.2).

Lal *et al.* (1996) evaluated four plant spacings (2 m × 2 m, 4 m × 4 m, 6 m × 6 m and 8 m × 8 m) with 'Sardar' guava. They observed lowest yield per plant at 2 m × 2 m spacing but maximum yield per unit area. Kumar and Singh (2000) also observed highest fruit yield per hectare in cultivar 'Allahabad Safeda' at 1060 trees ha^{-1}, compared with 400, 530 and 800 trees ha^{-1}. With proper canopy management from the first year of planting, planting density of 3 m × 6 m (555 trees ha^{-1}) has been found most suitable and highly productive for cultivar 'Allahabad Safeda' (Singh, 2005). Effect of pruning levels (one leaf pair pruning and unpruned control) and planting systems (square, hedgerow, double-hedgerow, paired and cluster planting) on 'Sardar' guava was studied by Pratibha (2008) and Pratibha *et al.* (2013). They observed that regardless of planting system, one leaf pair pruning significantly decreased the annual increment in tree height, tree spread, trunk diameter and tree volume. Number of flower buds was significantly increased with one leaf pair pruning during winter-season crop and the maximum number of flower buds and yield was noted in the treatment combination of one leaf pair pruning and square system of planting. They concluded that one leaf pair pruning in guava cultivar 'Sardar' planted in the square system is useful to maximize yield in the winter season.

One of the characteristics of guava is flowering on newly emerging lateral shoots, irrespective of time of year. Consequently, the occurrence of blooming and fruiting in the course of the year may be erratic or seasonal, depending on how the environment affects shoot growth (Singh, 2005). The excessive vegetative growth delayed cropping

Table 8.2. Effect of plant density on growth, yield and fruit quality of guava cultivar 'L-49'. Adapted from Mitra *et al.* (1984).

		Yield at 6th year		
Plant density per ha	Spacing (m × m)	Per plant (kg)	Per ha (t)	TSS/acid ratio of fruit
278	6.0 × 6.0	36.2	10.06	81.3
625	4.0 × 4.0	28.8	17.38	29.5
1111	3.0 × 3.0	25.0	27.76	27.9
1600	2.5 × 2.5	19.6	31.36	27.1

TSS, total soluble solids.

and produced soft and poorly coloured unmarketable fruit. Proper canopy management induces early flowering by allowing more light to penetrate inside the tree canopy and availability of more stored carbohydrates to cause flowering (Goswami *et al.*, 2014). Regular pruning after fruit harvest encourages development of lateral shoots from which flowering occurs. Some researchers suggested spraying of potassium nitrate (4%), ethephon (2-chloroethylphosphonic acid) (600 ppm) or Dormex (1.5%) after moderate pruning for induction of flowers and higher yield (Shaban and Haseeb, 2009). Serrano *et al.* (2008a) studied the effect of pruning time and intensity of pruning on 'Paluma' guava in Brazil. They reported that the period between pruning time and beginning of fruit ripening varied from 189 days (pruning in November and December) to 203 days (pruning in February). Light pruning caused more flowering, fruit set and yield, while heavy pruning produced fewer number of fruits with higher fruit weight. Pruning in the month of February was suggested to get the maximum yield.

Under HDP systems, although guava gives high yield during the initial stage, as the tree advances in age, light interception and aeration decrease inside the canopy due to shading and overcrowding, resulting in low yield and poor-quality fruits. In order to develop efficient canopy architecture for HDP, espalier technology has been developed (Srivastava *et al.*, 2018; Fig. 8.2). In this system, iron angle posts having 3 m length are fixed at 6–7 m intervals on which five or six wires (tiers) are erected. The first wire is erected at 45 cm from ground level and the remaining ones at 45–50 cm intervals. After erecting the structure, well-feathered grafting plants are planted at 1.5 m × 3.0 m (row-to-row × plant-to-plant) distance. In this system, 2222 trees in a hectare are accommodated as compared with 204 trees ha^{-1} in traditional systems. First scaffold branches are trained at 45 cm from ground level and the rest of the scaffolds at 45–50 cm intervals. Thus, eight to ten main branches are allowed on four or five tiers, on which fruiting shoots are promoted. In order to promote side branches, notching and girdling are done as required. In this training system, the topmost scaffolds are 260–270 cm in height. The main trunk is terminated leaving 10–15 cm from the last scaffolds, thus final

Fig. 8.2. Guava trees trained in espalier architecture. Photograph courtesy of Dr K.K. Srivastava.

tree height is 260–270 cm in espalier architecture. Total spread of the tree canopy is around 45–60 cm on both sides (between rows).

Since branch bending is practised on the wire erected at 90° angles, it alters the apical dominance, which results in better fruit bud formation at the basal portion of the tree scaffolds. The tree trained on espalier architecture is compact and its periphery is close to the main trunk, hence effective cropping area is more, leading to higher production. Guava in this architecture bears 35–40 fruits (7–8 kg per tree) and about 15 t ha^{-1} from 1-year-old trees (Srivastava *et al.*, 2018).

8.3.2 Meadow orchard

The Central Institute for Subtropical Horticulture, Lucknow India has developed the meadow orchard system for guava which accommodates 5000 plants ha^{-1} (1 m × 2 m) (Fig. 8.3). This system of planting requires regular topping and hedging of plants. In this system, trees are trained to dwarf structure during the initial stage of planting. To provide dwarf tree stature, plants are topped to a uniform height at 2 months after planting (October) for emergence of new shoots below the cut end. After the appearance of flowers on the new shoots, 50% of the semi-hardwood portions of these are pruned in May the following year. Renewal growth is initiated, flowers differentiate, and original size of the well-feathered tree is recovered by the end of September. The entire tree is re-pruned in October for initiation of new shoots. These new shoots, which resulted from the May pruning, produce fruits in the following season. Thus, annual cropping is obtained with a rhythmic change in

Fig. 8.3. Meadow orchard. Photograph courtesy of Dr A. Bhattacherjee.

tree size between two constant leaves (May and October). An average yield of 12.5 t ha^{-1} is obtained in the first year and it reaches 45 t ha^{-1} after 4 years (Singh, 2005).

The advantages of growing guavas in the meadow system compared with the traditional system are: (i) bearing starts from the first year of planting; (ii) easy to manage due to small tree size; (iii) lower cost of production and high returns; (iv) small canopy that facilitates better air and sunlight penetration; (v) minimum incidence of diseases; and (vi) high-quality fruits with good colour development. Moreover, the average yield in the meadow system was reported to be about 40–60 t ha^{-1} compared with 12–20 t ha^{-1} under the traditional system (Singh, 2007).

8.4 Training and Pruning

Training of guava trees after planting is necessary to improve yield as well as fruit quality. The main objective of training is to provide a strong framework and scaffold branches suitable for bearing a heavy crop without damaging the branches. The system usually followed is open centre, in which the plants are headed back and four primary shoots are retained for the initial framework which is subsequently pruned by cutting one-third to one-half of their length after 3 months. After making the initial framework, the two side shoots are permitted to grow initially and after 3–4 months subsequent doubling of selected branches is continued.

The other system is delayed open centre, in which two tiers of framework are developed and the centre is kept open at a height of 1.4 to 1.5 m. Four side shoots are retained in this system for the initial framework, but only one-third of their growth is pruned. For obtaining the second tier, buds growing just below the cut of the central leader are retained. It was observed that training reduced the tree areas up to 11.1% in open centre and 27.6% in delayed open centre (Mitra and Bose, 1990), thus providing an opportunity to increase the number of trees in a unit area and subsequently the productivity.

The objective of pruning is to open the canopy, as more sunlight leads to more shoots and higher yield. Pruning begins at an early stage of plant growth to develop single-trunk trees with well-spaced scaffold branches to form the framework. A single-trunk tree with no interfering branches 0.5–1 m from the ground is desirable for hand-harvesting and shaker/catcher type of mechanical harvesting (Nakasone and Paull, 1998). A light annual pruning after fruit harvest is considered necessary to encourage new shoots that will flower in the following season. All dead, diseased, crowded growths and suckers coming out from rootstock and sides of the framework should be pruned back annually. The intensity of pruning has a profound influence on flowering and fruiting of guava which depends on cultivar, plant density, climatic situation and cropping pattern.

8.4.1 Canopy management

Canopy management is the manipulation of tree canopies to optimize the production of quality fruits. The canopy management, particularly its components like tree training and pruning, affects the quality of sunlight intercepted by trees, as tree shape determines the presentation of leaf area to incoming radiation. Differential light interception within tree canopies by canopy management influences vegetative growth, photosynthetic efficiency, flower initiation, fruit set, fruit colour, fruit size and fruit quality (Syvertsen, 1984). Closely spaced guava orchards are highly productive with proper canopy management. However, without proper pruning, dense canopy development characterized by intermingling, overcrowding and insect- and disease-infested branches with more wood mass occurs and a canopy of unhealthy thin shoots develops. This reduces the productivity of the orchard. The quality of fruits deteriorates as the tree becomes crowded and few good-quality fruits are available only at the upper canopy. Asrey *et al.* (2007) evaluated fruit quality of 10-, 15- and 20-year-old 'Allahabad Safeda' guava. They recorded higher total soluble solids (11.85°Brix) and total sugar (7.5%) contents in fruits from the upper canopy of 15-year-old trees compared with 9.5°Brix and 4.75% total sugar in fruits from the lower canopy of 20-year-old trees.

8.4.2 Time and intensity of pruning

Rodrigues da Silva *et al.* (2016) suggested pruning of 'Paluma' guava in August in Brazil (20°55′S; 48°26′W). They also suggested pruning of one-third of the original length of branch, without considering the branch diameter. Pruning on 27 August produced 64.98 t ha^{-1} yield compared with 63.30, 38.68 and 27.89 t ha^{-1} when pruning was done on 11 September, 26 September and 11 October, respectively. Hojo *et al.* (2007) characterized 'Pedro Sato' guava tree phenology at four pruning times (September and December of 2003 and March and June of 2004) in Minas Gerais, Brazil. Four-year-old plants were used for the investigation. They established the different phenophases and reported that the duration between pruning and beginning of the first sprout was from 30.8 to 39.2 days; between pruning and flowering was from 68.6 to 133 days; between opening of the flower (complete flowering) and fruit ripening was from 118.3 to 148.4 days; and the pruning to complete crop cycle on average was 214.2, 211.4, 247.8 and 237.3 days for the pruning carried out in September, December, March and June, respectively. In another similar study Serrano *et al.* (2008b) evaluated the effects of different pruning times and intensities on 'Paluma' guava in two cultivation systems (with and without irrigation) at Pedro Canario, Espírito Santo, Brazil. The pruning times were 10 November and 9 December of 2005 and 13 January and 10 February of 2006 at three different intensities (heavy, medium and light). They observed that the time between pruning and beginning of fruit ripening was from 182 days (pruning in November and December) to 203 days (pruning in February). Fruit drop continued for 56 days after completion of flowering. Number of bud sprouts and formation of new shoots were higher in irrigated plants that were lightly pruned. The highest number of fruits per branch was recorded in plants that were lightly pruned in February. Irrespective of the pruning time, plants subjected to heavy pruning produced the least number of fruit set and fruits per branch. The largest fruit size was noted in irrigated plants that were pruned in December and January.

Depending upon spacing, cultivar and agroclimatic conditions of the growing area, high-density orchards become overcrowded after 5–7 years of planting in the absence of sufficient exposure to light. The leaves inside deeper layers of the canopy become photosynthetically inactive and act as an unproductive sink. This leads to disturbance of the source–sink relationship and as a result a major portion of the plant canopy becomes unfruitful. Moreover, restricted distribution of solar radiation and circulation of air inside the dense canopy builds up the microclimate congenial for pests and diseases. All these contribute towards low fruit yield and poor quality of produce (Samant and Kishore, 2019). Hence, success of HDP essentially requires manipulation of the canopy; that is, judicious pruning to facilitate better light penetration inside the canopy to keep the orchard productive.

Pruning of terminal shoots at 45 cm length during summer (May) caused maximum yield (14.7 t ha^{-1} versus 6.35 t ha^{-1} in control) and improved fruit quality (13.17°Brix versus 9.20°Brix in control) in 10-year-old 'Lalit' guava (Meena *et al.*, 2017). In an experiment with 6-year-old 'Pant Prabhat' guava, Thakre *et al.* (2016) suggested pruning of the upper portions of all fruited shoots and keeping only one leaf pair at the base of the shoot in February. The treatment reduced the rainy-season crop and caused maximum winter yield (19.19 t ha^{-1} versus 2.18 t ha^{-1} in control). Susanta *et al.* (2019) stated that pruning (pruning was done after the shoot leaves grew fully) had a significant effect in increasing the total number of shoots as well as generative shoots of 'Crystal' guava. Pruned plants produced more flowers and fruits than unpruned ones. They suggested pruning by leaving four pairs of leaves that subsequently produced higher number of flowers and fruits. However, they did not find any response of pruning on average fruit weight and its quality compared with control. Response to three levels of pruning (30, 50 and 70% of shoots) in winter (December) and summer (May) was studied on 8-year-old 'Sardar' guava spaced at 2.5 m × 1.25 m by Samant and Kishore (2019) in the hot and humid climate situation of Odisha, India. They reported that pruning encouraged shoot emergence, irrespective of time and intensity; however, winter pruning resulted in the production of more shoots or laterals (27.02 ± 3.85 shoots per metre of branch) compared with summer pruning (14.55 ± 3.3). Shoot emergence increased with the severity of pruning. Shoot pruning during winter was found effective in enhancing flowering intensity (38.58 ± 4.25%), fruit set (73.68 ± 0.48%) and fruit yield (6.12 ± 1.13 kg per plant), whereas summer pruning did not show significant influence on these parameters. Shoot pruning by 70% in winter showed the maximum values for flowering intensity (42.83%), fruit set (74.15%) and yield (7.25 kg per plant or 23.2 t ha^{-1}). All the pruning treatments caused improvement of fruit quality (°Brix and vitamin C contents).

8.4.3 Rejuvenation of old and senile orchards

Rejuvenation technology for old and unproductive guava orchards has been developed (Singh *et al.*, 2007). Older trees of cultivars 'Allahabad Safeda' and 'Sardar' were headed back up to 1.0–2.5 m height above ground level depending on the structure of the individual tree in the month of May. The newly emerged shoots as a result of rejuvenation pruning were allowed to grow up to a length of 40–50 cm. These shoots were further pruned to about 50% of their total length in the month of October for emergence of multiple shoots below the pruning point. This was mainly done to modify the tree structure and maintained canopy size. Emerging shoots of the inner canopy were also pruned to promote branching. This operation (pruning of shoots) was continued every year in the months of May and October. Pruning at different periodicity was performed for attainment of healthy and vigorous canopy and development of short tree architecture conducive for canopy management, plant protection measures, cultural practices and harvesting operations. As a result of topping and hedging, the height was reduced by 34 to 43% over control trees in both cultivars. Similarly, May and October pruning restricted tree canopy dimension by 29.89 to 50.03% without affecting yield. An increased yield of 82.39 kg per plant in 'Sardar' compared with unpruned trees after the first year of topping and hedging was recorded. Further, yield enhancement of 104–112 kg per plant in 'Allahabad Safeda' and 74.90–79.20 kg per plant compared with unpruned trees was noted in the second year. The fruit quality from rejuvenated trees was superior to that from unpruned trees. The unpruned trees produced a significantly higher number of small fruits. Even with proper heading back, management of shoots within the tree canopy is important for maintenance of fruit quality and production (Campbell and Wasielewaski, 2000). The success of this technology depends upon the proper management of shoots through precise and timely pruning and cropping. Singh and Singh (2007) also studied the photosynthetic efficiency and canopy microclimate of the rejuvenated orchards. They observed that the level of net photosynthesis was more pronounced in flowering branches of trees which were topped from 1.5 m height compared with the non-flowering ones in both cultivars. Light penetration was greater in trees pruned to 1.5 and 2.0 m height. Canopy temperature was maximum in rejuvenated orchards and lowest in controls.

8.5 Intercropping

Traditionally guava is planted at 6–8 m × 6–8 m distance. The tree requires 4–5 years to eliminate sunlight reaching the ground by canopy cover. During this period, the space between rows can be utilized for growing different intercrops (Fig. 8.4). The intercrops should be grown in the inter-row spaces, leaving 1.0 m radial space under-tree basin in the centre. The recommended package practice should be followed separately for both the guava and the intercrop. Intercrops inside guava orchards with pineapple, vegetables and tuber crops (Hugar *et al.*, 1991; Singh *et al.*, 2016; Ghosh *et al.*, 2017), legumes (Shweta *et al.*, 2015; Raut, 2018) and fodders (Gill, 1998) were reported as productive in the early years of orchard establishment.

Growing of pulses as an intercrop in non-bearing guava orchards improved the growth (height and canopy volume) of young plants, and growers can get some income from the orchard during the pre-bearing period (Shweta *et al.*, 2015). Intercropping with vegetables like *Solanum melongena* and *Trichosanthes dioica* was suggested by Ghosh *et al.* (2017). Growing of leguminous crops like black gram (*Vigna mungo*) and pigeon pea (*Cajanus cajan*) not only generates extra income but also improves soil health (Raut, 2018). Tuber crops such as *Amorphophallus campanulatus*, *Colocasia esculenta* var. *antiquerum*, *C. esculenta* var. *esculenta*, *Curcuma domestica* and *Zingiber officinale* are well suited as intercrops even under old guava plantations due to their shade-loving nature (Singh *et al.*, 2016). In rainfed situations, an arable crop like rainfed groundnut can be grown in the interspace of an orchard (Rajanikanth *et al.*, 2016).

8.6 Weed Control

Weed control is essential during the initial phase of orchard establishment. The shade provided by the canopy, particularly under high-density orchards, checks weed growth. Weeds compete with the plant at all stages of development for soil moisture, soil nutrients and light, besides harbouring insects, pests and diseases. In tropical and subtropical climates, high rainfall and humidity provide

Fig. 8.4. Intercropping in a guava orchard with: (A) *Brassica oleracea* var. *botrytis*; (B) *Solanum melongena*; (C) *Trichosanthes dioica*; and (D) *Ananas comosus*. Photographs courtesy of Dr S.K. Mitra (A–C) and Dr B.R. Jana (D).

a congenial environment for the luxuriant growth of a wide range of weeds.

8.6.1 Weed control in the nursery

Control of weeds in the nursery could be done by various methods. Manual removal of weeds is uneconomical and mechanical weeding in nurseries may affect the plant root. Weed control in the nursery is difficult due to the sensitivity of young guava seedlings to herbicides when used at the rate recommended for orchards. Pathak *et al.* (2007) suggested that spraying of paraquat at 3.0 kg ha^{-1} or glyphosate at 7.5 ml l^{-1} after 20 days of seed germination in a nursery controls the weeds, with the effect lasting for the next 6 months. Pendemethlin at 4.5 l ha^{-1} as a pre-emergence application was also found effective to control weeds in guava nurseries (Kour *et al.*, 2019).

8.6.2 Weed control in the orchard

Weed control is essential during the initial phase of orchard establishment. The plant height of young guava seedlings is checked by 50% due to weeds like *Cynodon dactylon* (Maurya and Shankar, 1982). In established orchards, growing of a leguminous cover in the tree basin, surface mulching, mowing and complete weed control with herbicides are the options available to control the weeds.

8.6.3 Cover crops

Planted cover crops are characterized by the type of plant used: grasses, legumes, or a mixture of both. Grasses are often low-maintenance cover crops that are very effective in binding the surface soil. Grass–legume mixture is desirable because of the nitrogen contribution of the legume. Leguminous cover crops, which act as a living cover on the orchard floor, suppress the weed growth as well as fix atmospheric nitrogen in the soil (Jayasinghe, 2008). Mung bean (*Vigna radiata*) and black gram (*V. mungo*) were reported to be effective leguminous cover crops for guava orchards (Bhattacharjee and Debnath, 2019). In subtropical red soils of South China, cover cropping with *Paspalum natatu* along with arbuscular mycorrhizal fungal inoculation was suggested for guava orchards (Hang *et al.*, 2015). This combination was found effective for hydrolysis of organic phosphorus in the soils where immobilization of phosphorus by iron and aluminium caused phosphorus deficiency.

8.6.4 Surface mulching

Mulching at the base of the trees can be done very inexpensively using black polythene sheet or with organic materials such as wood shavings, sawdust, rice husks, rice straw, banana bio-mat, water hyacinth (*Eichhernia crassipes*), etc. (Evans *et al.*, 1988; Borthakur and Bhattacharyya, 1999; Patra *et al.*, 2003; Brar *et al.*, 2013; Bhattacharjee and Debnath, 2019; Kour *et al.*, 2019). Borthakur and Bhattacharyya (1999) reported suppression of weed growth and increased yield in 'Allahabad Safeda', 'Banarasi' and 'Seedless' guavas by using rice straw as surface mulch. Use of banana bio-mat (webbed leaf sheath) was suggested by Bhattacharjee and Debnath (2019) that not only suppressed weed growth but also increased nutrient (nitrogen, phosphorus and potassium) contents of guava leaf and soil nutrient status.

8.6.5 Mowing

Mowing throughout the orchard with an offset tow mower is a good method by which tall woody plants or grasses are eliminated (Shigeura and Bullock, 1983). However, to control weeds in the tree basin, growing of cover crops, use of mulch or use of herbicides should be practised.

8.6.6 Use of herbicides

The use of chemical weed control may appear expensive initially, but when properly applied, it can become a good economical method to achieve gradual elimination of weed seeds and vegetative propagules (Shigeura and Bullock, 1983). Herbicides are not generally recommended in young plantations as there may be severe damage by spray drift or direct contact (Nakasone and Paull, 1998). Pre-emergence use of diuron (1.6 kg ha^{-1}), oryzalin (1.67 l ha^{-1}), simazine (1.6 kg ha^{-1}) or atrazine (1.6 kg ha^{-1}) showed good control of weeds in guava orchards. The best weed control was reported with a combination of diuron + oryzalin and diuron alone (Martinez and Pereira, 1984). During trials in Hawaii, Nishimoto and Yee (1980) obtained good control of weeds by use of paraquat. Oxyfluorofen at 0.50 kg of active ingredient per hectare (kg a.i. ha^{-1}) and atrazine at 5.0 kg a.i. ha^{-1} were found very effective to control weeds in a guava orchard in a hot and humid climate (Kundu *et al.*, 1997). Maji *et al.* (2008) suggested use of a combination of glyphosate (0.5 kg a.i. ha^{-1}) and 2,4-D (2,4-dichlorophenoxyacetic acid) sodium salt (1 kg a.i. ha^{-1}) to control weeds in guava orchards.

8.7 Conclusion

Orchard profitability and sustainability are largely dependent on the proportion of crop that can be marketed as first quality. The visual component of quality (i.e. colour, size and skin finish) predominantly determine whether a premium price is achieved. The design of most modern-day guava orchards has shifted to medium- or high-density planting where topography, cultivar and pruning/training systems become key issues in determining tree spacing and orchard layout. It has been demonstrated by several studies that HDP in guava is profitable with proper canopy management. Both the degree and timing of pruning can affect crop load, fruit size and fruit quality.

References

Araujo, F.J., Urdanrta, T., Salazar, N. and Simancas, R. (1999) Effect of planting density on guava (*Psidium guajava* L.) yield in the Maracaibo plant, Venezuela. *Revista de la Facultad de Agronomia, Universidad del Zulia* 16, 13–16.

Asrey, R., Pal, R.K., Sagar, V.R. and Patel, V.P. (2007) Impact of tree age and canopy position on fruit quality of guava. *Acta Horticulturae* 735, 259–262.

Bhattacharjee, A. and Debnath, S. (2019) Evaluation of leguminous cover crop and banana bio-mat mulching for weed suppression and conservation of soil moisture and nutrient in guava orchard. *Journal of Crop and Weed* 15(1), 170–177.

Borthakur, P.K. and Bhattacharyya, R.K. (1999) Response of mulching on yield and mineral composition in guava (*Psidium guajava* L.). *Haryana Journal of Horticultural Science* 28, 35–37.

Bose, T.K., Mitra, S.K. and Chattopadhyay, P.K. (1992) Optimum plant density for some tropical fruit crops. *Acta Horticulturae* 296, 171–176.

Brar, J.S., Dhaliwal, H.S., Bal, J.S. and Singh, S.P. (2013) Influence of inter-plant spacing on microclimatic and horticultural parameters of guava (*Psidium guajava* L.) orchard. *Research on Crops* 14(1), 205–212.

Campbell, R.J. and Wasielewaski, J. (2000) Mango tree training techniques for the hot tropics. *Acta Horticulturae* 509, 641–651.

Evans, D.O., Joy, J.R. and Chia, C.L. (1988) *Cover Crops for Orchards in Hawaii.* Research Extension Series No. 094. College of Tropical Agriculture and Human Resources, University of Hawaii, Honolulu, Hawaii.

Ghosh, S., Sarkar, S., Sau, S., Karmakar, S. and Brahmachari, K. (2017) Influence of guava (*Psidium guajava* L.) based intercropping systems on soil health and productivity in alluvial soil of West Bengal, India. *International Journal of Current Microbiology and Applied Sciences* 6(11), 241–251.

Gill, A.S. (1998) Grow stylo in the inter space of fruit trees. *Indian Farming* 38, 33–34.

Goswami, A.K., Prakash, J. and Singh, A.K. (2014) High density planting system in tropical fruits. *HortFlora Research Spectrum* 3(3), 298–300.

Hang, C., Yang, Z., Zhenhong, G., Honghui, Z., Shenlei, F. and Qing, Y. (2015) The combined effects of cover crops and symbiotic microbes on phosphatase gene and organic phosphorus hydrolysis in subtropical orchard soil. *Soil Biology and Biochemistry* 82, 119–126.

Hojo, R.H., Chalfun, N.N.J., Hojo, E.T.D., Souza, H.A.d., Paglis, C.M. and Sao Jose, A.R. (2007) 'Pedro Sato' guava tree phenological characterization in different pruning times. *Revista Brasileira de Fruticultura* 29(1), 20–24.

Hugar, L.B., Murthy, P.S.S., Umesh, K.B. and Reddy, B.S. (1991) Economic feasibility of cultivation of guava cultivation under scientific management an empirical evidence. *Agriculture Situation in India* 46, 211–214.

Jayasinghe, C.K. (2008) The role of leguminous cover crops in soil improvement with special reference to nitrogen economy of tropical rubber crops. *Bulletin of Rubber Research Institute, Sri Lanka* 28, 22–25.

Kour, A., Gupta, N. and Brar, S.K. (2019) Integrated weed management practices in guava nursery. *Journal of Pharmacognosy and Phytochemistry* 8(2), 982–985.

Kumar, R. and Singh, H.P. (2000) Effect of planting system cum densities on growth fruit size and yield of guava cv. Allahabad Safeda under rainfed conditions. *Annals of Agriculture Research* 21(1), 152–153.

Kundu, S., Ghosh, S.N. and Mitra, S.K. (1993) Yield and fruit quality of guava cv. L-49 under different plant densities. *Indian Agriculturist* 37, 157–162.

Kundu, S., Ghosh, S.N. and Mitra, S.K. (1997) Chemical weed control in guava. *The Horticulture Journal* 10, 49–58.

Lal, S., Tiwari, J.P. and Misra, K.K. (1996) Effect of plant spacing and pruning intensity on flowering and fruiting of guava. *Annals of Agriculture Research* 17(1), 83–89.

Lal, S., Tiwari, J.P. and Mahajan, A.R. (2007) Studies on planting systems in guava (*Psidium guajava* L.) cv. Sardar. *Acta Horticulturae* 735, 263–266.

Maji, S., Das, B.C. and Bandyopadhyaya, P. (2008) Studies on weed management practices in guava cv. L-49. *Journal of Crop and Weed* 4(2), 52–56.

Martinez, J.M. and Pereira, F.M. (1984) Effect of different pre-emergence herbicides on the control of weeds in a guava (*Psidium guajava* L.) orchard. *Empresa Catarinese de Pesquisa Agropecuaria, S.A.* 2, 472–476.

Maurya, K.R. and Shankar, G. (1982) Effect of weed competition on growth of young seedling of guava (*Psidium guajava* L.), Karna katta (*Citrus karna*) and Lagzi lime (*C. aurantifolia*). In: *Abstracts Book, Annual Conference of the Indian Society of Weed Science, Allahabad, India*, pp. 32–33.

Meena, K.R., Maji, S. and Meena, S.C. (2017) Use of shoot pruning for crop regulation and quality fruit production of guava (*Psidium guajava* L.). *International Journal of Agricultural Science* 13(2), 184–191.

Mitra, S.K. and Bose, T.K. (1990) Guava. In: Bose, T.K. and Mitra, S.K. (eds) *Fruits: Tropical and Subtropical*. Naya Prokash, Kolkata, India, pp. 280–303.

Mitra, S.K. and Sanyal, D. (2004) *Guava*. Indian Council of Agricultural Research, New Delhi.

Mitra, S.K., Sen, S.K., Maiti, S.C. and Bose, T.K. (1984) Effect of plant density on growth yield and fruit quality in guava. *Bangladesh Horticulture* 12, 7–9.

Mitra, S.K., Ghosh, B. and Pathak, P.K. (2018) High density orcharding and canopy management in guava. *Acta Horticulturae* 1205, 955–958.

Nakasone, H.Y. and Paull, R.E. (eds) (1998) Guava. In: *Tropical Fruits*. CAB International, Wallingford, UK, pp. 149–172.

Nishimoto, R.K. and Yee, W.Y.J. (1980) *A Guide to Chemical Weed Control in Tropical and Subtropical Fruit and Nut Crops in Hawaii*. Cooperative Extension Service Circular No. 423. University of Hawaii, Honolulu, Hawaii.

Pal, M. and Lal, S. (2015) Effect of different high density planting on growth and yield of guava (*Psidium guajava* L) cv. Pant Prabhat. *International Journal of Basic and Applied Agricultural Research* 13(3), 420–425.

Pathak, S.M., Singh, D.K. and Singh, V.K. (2007) Effect of herbicides on growth, weed infestation and physico-biochemical changes in guava (*Psidium guajava* L.) seedlings. *Acta Horticulturae* 735, 315–320.

Patra, R.K., Das, B.C. and Hasan, M.A. (2003) Flowering behaviour and fruit yield of guava cv. Sardar as influenced by different soil covers. *Research on Crops* 4(3), 383–387.

Pratibha, S. (2008) Response of planting system and pruning on growth, yield and quality of guava (*Psidium guajava* L.) cv. Sardar. PhD thesis, G.B. Pant University of Agriculture & Technology, Pantnagar, India.

Pratibha, S., Lal, S. and Goswami, A.K. (2013) Effect of pruning and planting system on growth, flowering, fruiting and yield of guava cv. Sardar. *Indian Journal of Horticulture* 70, 496–500.

Rajanikanth, E., Manjulatha, G., Anjaiah, T. and Mallaiah, B. (2016) Influence of integrated nutrient management practices on soil properties under intercropping of groundnut with guava based agri-horti system. *International Journal of Tropical Agriculture* 34(6), 1567–1573.

Raut, R.L. (2018) Evaluation of various intercrops in guava orchard. *Plant Archives* 18(1), 466–488.

Rodrigues da Silva, M.J., Tecchio, M.A., Domiciano, S., Leonel, S. and Balestrero, R.I. (2016) Phenology, yield and fruit quality of 'Paluma' guava tree at different pruning times. *Ciencia e Agrotechnologia* 40(3), 317–325.

Samant, D. and Kishore, D. (2019) standardization of pruning for high density 'Sardar' guava orchards under hot and humid climate of Eastern India. *Indian Journal of Horticulture* 76(1), 70–74.

Serrano, L.A.L., Martins, M.V.V., Lima, I.d.M., Marinho, C.S. and Tardin, F.D. (2008a) Effect of pruning time and intensity on 'Paluma' guava trees in Pinheiros, ES, Brazil. *Revista Brasileira de Fruticultura* 30, 994–1000.

Serrano, L.A.L., Marinho, C.S., Lima, I.d.M., Martins, M.V.V., Ronchi, C.P. and Tardin, F.D. (2008b) Phenology of 'Paluma' guava trees under different cultivation systems, times and intensities of pruning. *Bragantia* 67(3), 701–712.

Shaban, A.E.A. and Haseeb, G.M. (2009) Effect of pruning severity and spraying some chemical substances on growth and fruiting of guava trees. *American–Eurasian Journal of Agricultural & Environmental Sciences* 5, 825–831.

Shigeura, G.T. and Bullock, R.M. (1983) *Guava (*Psidium guajava *L.) in Hawaii – History and Production*. Hawaii Institute of Tropical Agriculture and Human Resources, Research Station Series No. 035. University of Hawaii, Honolulu, Hawaii.

Shweta, Baloda, S., Bhatia, S.K. and Sharma, J.R. (2015) Intercropping studies in guava orchard. *International Journal of Tropical Agriculture* 33(3), 2189–2192.

Singh, A. and Dhaliwal, G.S. (2007) Solar radiation interception and its effect on physical characteristic of fruits of cv. Sardar. *Acta Horticulturae* 735, 297–302.

Singh, G. (2005) Strategies for improved production of guava. In: *Souvenir, 1st International Guava Symposium, Lucknow, India*, pp. 26–39.

Singh, G. (2007) Recent development in production of guava. *Acta Horticulturae* 735, 161–173.

Singh, G., Mishra, R. and Gupta, S. (2007) Modifying existing guava tree canopies for increased production efficiency. *Acta Horticulturae* 735, 243–248.

Singh, S.K., Sharma, M. and Singh, P.K. (2016) Intercropping – an approach to reduce fruit drop and improve fruit quality in guava. *Journal of Chemical and Pharmaceutical Sciences* 9(4), 3182–3187.

Singh, V.K. and Singh, G. (2007) Photosynthetic efficiency, canopy micro climate and yield of rejuvenated guava trees. *Acta Horticulturae* 735, 249–257.

Srivastava, K.K., Rajan, S., Kumar, D., Pandey, G., Verma, A. and Pandey, S. (2018) Espalier architecture for revolutionizing high quality guava production. *Indian Horticulture* 63(6), 18–20.

Susanta, S., Melati, M. and Aziz, S.A. (2019) Pruning to improve flowering and fruiting of 'Crystal' guava. *AGROVITA Journal of Agricultural Science* 41(1), 48–54.

Syvertsen, J.P. (1984) Light acclimatization in citrus leaves. II. CO_2 assimilation and light, water, and nitrogen use efficiency. *Journal of the American Society for Horticultural Science* 109, 812–817.

Thakre, M., Lal, S., Uniyal, S., Goswami, A.K. and Prakash, P. (2016) Pruning for crop regulation in high density guava plantation. *Spanish Journal of Agricultural Research* 14(2), e0905. http://dx.doi.org/10.5424/sjar/2016142-7846

9 Flowering

Shu-Yen Lin[1] and Po-An Chen[2]*
[1]*National Taiwan University, Taiwan;* [2]*Agricultural Technology Research Institute, Taiwan*

9.1 Introduction

Guava is an evergreen fruit species native to tropical Central America from southern Mexico to northern South America (Salazar *et al.*, 2006). Guava trees are grown normally in tropical and subtropical regions. Therefore, it can be successfully introduced to more than 60 countries of the world.

The best temperature for guava cultivation ranges from 15 to 30°C, with an annual average temperature of 18°C (Salazar *et al.*, 2006). The flowering season of guava varies from region to region. In most producing areas, guava bears fruit twice a year. However, the fruit yield was maximum during the rainy season while fruit quality characteristics were better during winter as compared with the rainy season in north Indian conditions (Rathore, 1976; Aulakh, 2004). Since improvement of fruit quality is an important issue, guava trees are forced to produce their fruits in winter season through different horticultural practices such as pruning, bending, fertilization and defoliation by the application of some safe compounds like urea. Practically, guava bears fruit all year round.

Knowledge of floral morphology, blossom biology and mode of pollination is essential for fruit crop improvement programmes. This chapter covers the various factors affecting the transformation of vegetative growth to reproductive phase of guava trees, and the phenology, anthesis and pollination in guava. The agricultural practices for regulating crop production are discussed as well.

9.2 Phenology and Shoot Development

Flowering physiology in fruit trees is highly related to environmental factors. Generally, the progress of formation of flower buds to flowering shows two different patterns in temperate and subtropical/tropical regions. Flowering progress of temperate fruit trees should pass through a long cold period to satisfy the chilling requirement for budbreak (Fig. 9.1). Breaking dormancy is an essential step for temperate fruit trees. The flowering progress from flower bud to flowering is discontinuous and takes a long time. The obvious characteristic of subtropical and tropical fruit trees is continuous growth year-round. The induction of reproductive progress starts when the vegetative growing strength is weak and shoots cease to grow. This short period of growth cessation is a vegetative dormancy for subtropical and

*E-mail: chenpoan@mail.atri.org.tw

 Guava: Botany, Production and Uses (ed. S. Mitra)
DOI: 10.1079/9781789247022.0009

tropical trees (Fig. 9.2). During this period, the environmental signals may evoke bud transformation to become a flower bud.

A conceptual flowering model for subtropical and tropical fruit trees has been described as a consequence resulting from physiological and environmental interactions. Individual shoots of these fruit trees are in quiescent state most of the time. Growth occurs as periodic flushes of shoots emerging from apical or lateral resting buds by breaking signals such as warm temperature, sufficient nutrients or water availability. Most often, these flushes are vegetative. Individual shoots typically produce only one reproductive flush per year in those fruit trees originating from the higher-latitude tropics and the subtropics, such as mango (Davenport and Nuñez-Elisea, 1997), lychee (Menzel, 1984) or citrus (Davenport, 1990). In tropical fruit trees like guava or durian, flowers are produced once from lateral buds of each vegetative shoot and there may be several vegetative flushes emerging per year (Ketsa *et al.*, 2019). Reproductive flushes or floral buds generally occur after extended periods of shoots in the rest mode (or vegetative dormancy) with cool temperature or non-lethal water stress. Before buds burst, conditions suitable for floral induction may be present lasting to the bud initiation (Shigeura and Bullock, 1976;

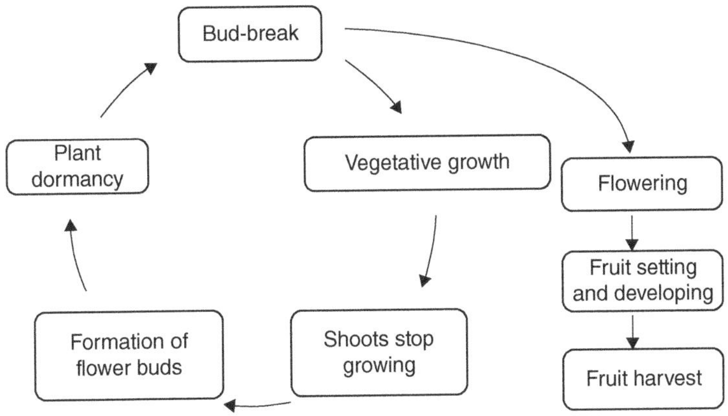

Fig. 9.1. Schematic diagram showing the growth cycle of temperate fruit trees.

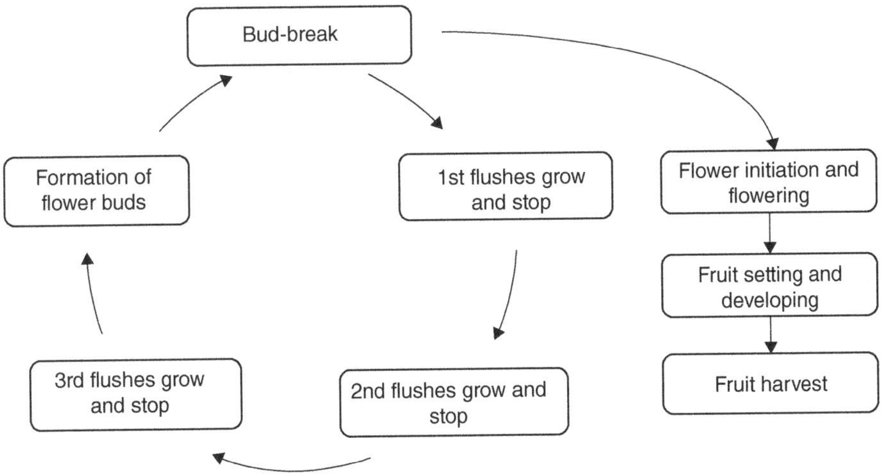

Fig. 9.2. Schematic diagram showing the growth cycle of subtropical and tropical fruit trees.

Batten and McConchie 1995; Davenport, 2000; Ketsa *et al.*, 2019).

Guava is native to the tropics and has high environmental adaptability. Given the contrasting conditions (temperature–elevation) at which guava trees are grown, it is possible to find guava orchards over a wide altitudinal range (Padilla-Ramirez *et al.*, 2007). It is possible to find guava trees growing from sea level up to 2000 m above sea level (Yadava, 1996). These broad altitudinal variations in the areas where guava trees are grown have an important influence on their phenological behaviour mainly due to temperature difference (Padilla-Ramirez *et al.*, 2012).

Flower buds appear soon after the first pair of leaves mature, but there is no direct association between leaf appearance and flower production (Menzel and Paxton, 1986). Flowers occur either singly or in cymes of two or three at leaf axils of current (Braganza, 1990) and preceding growth (Nakasone and Paull, 1998). The bearing twigs normally grow a few centimetres, putting forth four or five pairs of leaves, and thereafter either flower buds start developing or twigs cease to grow until the next season. Guava requires about 30 days from flower-bud differentiation to complete development up to calyx-cracking stage (Prakash, 1976). The flower buds when fully developed have two distinct parts, namely an ovoid proximal or adnate part and a distal free part which is ovoid or round and slightly pointed at the apex (Sehgal and Singh, 1967).

Both terminal and lateral flowerings have been reported, occasionally mixed type flowering also occurs. The blooming period ranges from 28 to 45 days depending on cultivar, season and climatic conditions during flowering (Mitra and Bose, 1990). The flower consists of a superior calyx with five lobes and the corolla consists of six to ten petals arranged in one or two whorls. The androecium consists of 160–400 thin filaments carrying bi-lobed anthers, packed closely together. The gynoecium consists of an inferior ovary syncarpus with axil placentation and subulate style. The style is smooth and bearded at the summit (Kahlon *et al.*, 1987).

There are three distinct flowering seasons of guava in South Asia that occur in spring (February), rainy season (June) and autumn (October), with their corresponding harvesting periods being July–August, November–December and March–April (Rathore and Singh 1974; Singh *et al.*, 2001; Sahoo *et al.*, 2017) (Fig. 9.3). In subtropical regions, guava flowers mainly from April to July and harvest is from June to September. The rainy-season crop of guava is poor in quality and fruits are affected by many biotic and abiotic stresses compared with the winter-season crop. The winter-season crops are superior in quality, free from diseases and pests, and fetch higher incomes.

9.3 Anthesis

Longitudinal splitting of the calyx occurred a day before the opening of the flowers. The buds swelled apparently by the evening and the flowers opened next morning (Kundu, 1992). The petals took about 95 to 120 min to stretch out completely, opening in the order of their aestivation (Sehgal and Singh, 1967). At first the outermost petal started straightening up and was followed by the next in succession. Gradually they pushed

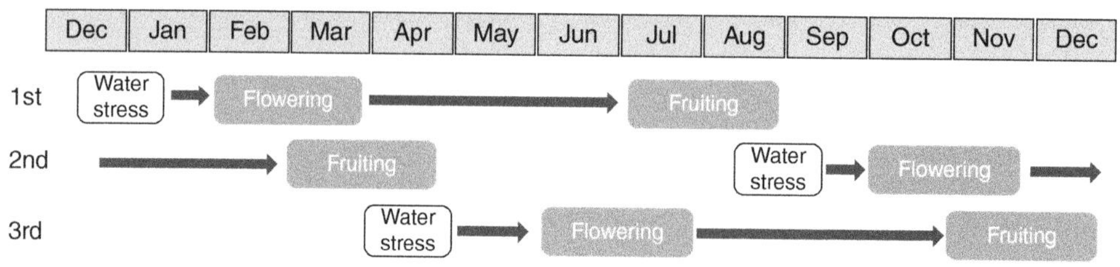

Fig. 9.3. The three flowering seasons and harvesting periods of guava production in South Asia.

backwards exposing the stamens and pistil. Flower opening is usually in the morning hours (4 a.m. to 8 a.m.) and depends on cultivar to show the peak of flower opening (Sehgal and Singh, 1967; Srivastava, 1974; Ojha *et al.*, 1986; Kundu and Mitra, 1994; Caraballo, 2001). Kahlon *et al.* (1987) reported that opening of flowers started at 5 a.m., increased with rise of temperature until 7 a.m., followed by a sharp decline and continued at a diminishing rate up to 11 a.m. in cultivars 'Allahabad Safeda' and 'Lucknow-49' ('Sardar'). In cultivars 'Hisar Safeda', 'Hisar Surkha', 'Lucknow-49' and 'Allahabad Safeda', peak anthesis was observed between 5.30 a.m. and 6.30 a.m. under north Indian conditions (Sharma *et al.*, 2013). In cultivar 'Seedless', the petals sometimes failed to unfold but were bodily thrown off as a cap. Sometimes, the petal caps turned brown and the stamens inside withered completely (Kundu, 1992).

9.3.1 Dehiscence

The anthers commenced dehiscing 15–60 min prior to opening of the petals and reached peak dehiscence in about 45–120 min after the opening of flowers. The period differs from cultivar to cultivar. The stigma became receptive as the flowers opened and remained receptive for about 48 h. However, the stigma was non-receptive before 24 h prior to anthesis (Sharma *et al.*, 2013). The anthers burst longitudinally and required 5–8 min for bursting completely (Sehgal and Singh, 1967; Srivastava, 1974; Ojha *et al.*, 1986; Kahlon *et al.*, 1987; Kundu and Mitra, 1994). The anthers, after dehiscence, turned yellowish brown and faded away. The dehiscence started from the peripheral anthers towards the centre and many of the filaments dropped away by the evening or next day (Kundu, 1992).

9.3.2 Pollen

Pollen grains of *Psidium* are isopolar, oblate, peroblate or oblate-spheroidal, 3-syncolporate or 4-syncolporate. The sexine ornamentation is regulate, granulate or spinulose-granulate and differs between the mesocolpium and apocolpium (Tuler *et al.*, 2017). Viability of the freshly collected pollen varies from 42 to 95% depending upon the variety. Seedless cultivars, in general, have less than 50% pollen viability at the time of dehiscence (Kundu, 1992; Ray, 2002). Some of the Egyptian guava cultivars ('Fakous', 'Banaty', 'Gize Yellow', 'Montakhab-Elsabaheya' and 'Mobakker') showed 92.16 to 97.86% pollen viability (El-Halwagi *et al.*, 2007). Pollen grains remain viable for 1 day under field conditions but are viable for 90 to 135 days at low temperature (0–4.5°C) and low (0–25%) relative humidity (RH) (Seth, 1960; Mandloi, 1973). Tiwari (1969) stated that pollen grains of cultivar 'Chittidar' could be stored for about 5 months at 0°C with 25% RH. Chezhiyan (1989) succeeded in storing pollen of cultivar 'Seedless' for 7 months in a refrigerator at 5°C, but the pollen lost its viability fully in 1 month at room temperature.

9.4 Pollination

In guava, self-pollination is conspicuous. However, the distribution of cross-pollination by insects is about 35% (Menzel and Paxton, 1986). Some researchers (Purseglove, 1968; Hedström, 1988; da Silva *et al.*, 2017), however, consider cross-pollination is the most frequent form of pollination in guava. The reduction in yield via self-pollination has been reported, which has been attributed to self-incompatibility phenomena (Alves and Freitas, 2007). Usman *et al.* (2013) reported metaxenia effect of pollen donor genotype on the physicochemical traits of guava. Honeybees are the chief pollinators of guava, which contributes 25.70–41.30% of cross-pollination (Vinod and Sattagi, 2018a). Several *Apis* species such as *Apis dorsata*, *Apis cerana*, *Apis mellifera* and *Apis florea*, the stingless bees *Melipona subnitida*, *Partamona cupria* and *Trigona spinipes*, and the carpenter bee *Xylocopa frontalis*, were reported to be involved in pollinating guava (Rajagopal and Eswarappa, 2005; Alves and Freitas, 2006; Siqueira *et al.*, 2012). Among the honeybees, *A. dorsata* and *A. mellifera* are the major floral visitors (Rajagopal and Eswarappa, 2005) and *A. mellifera* produced more fruits after a single visit to the flowers (Alves and Freitas, 2006).

The stigmas, anthers and nectaries are exposed in guava flowers and are available to insects. The flowers are highly attractive to honeybees as they produce large amounts of pollen and nectar, have a sweet odour and a low depth of flowers (Alves and Freitas, 2006). A large number of species have been found on flowers in India (Rajagopal and Eswarappa, 2005; Vinod and Sattagi, 2018b), Brazil (Alves and Freitas, 2006, 2007; Guimaraes *et al.*, 2009; Siqueira *et al.*, 2012) and Venezuela (Caraballo, 2001).

Guava sets fruit when pollinated by wind or biotic agents, but flowers protected from flower visitors set only 15.09%. Cross-pollination caused higher fruit set than self-pollination (Alves and Freitas, 2007).

9.5 Flowering

Flower induction in guava requires both internal and external factors. The competence of nutrition and the balance of hormones are considered the two main internal factors (Balakrishnan, 2000; Lal *et al.*, 2013). Temperature and water availability are the primary environmental factors. Moderately low water potentials delay shoot growth by reduced turgor, thus contributing to improve the maturity of stems and reducing the levels of floral inhibitor in the leaves (Núñez-Elisea and Davenport, 1994; Boora *et al.*, 2016).

9.5.1 Effect of temperature on flowering

Temperature affects growth of guava trees in physiological development. Among the seasonal changes, guava trees show vegetative dormancy when cool temperature comes and begin active growth when temperatures rise (Salazar *et al.*, 2006). Guava trees undergo several physiological changes including shoot growth, bud swelling and break, and the increment of trunks, and all these changes vary with temperature. The changes in reproductive phases such as flower initiation, fruit setting, and ripening are also temperature dependent.

Guava produces flowers on new as well as on mature vegetative shoots (Fig. 9.4). In subtropical regions, guava flowers two to three times a year, and guava flowers continuously throughout the year if the soil water availability is enough in tropical regions (Mitra *et al.*, 2008; Abbas *et al.*, 2014). The phenological stages of the guava tree from a winter bud (cease bud or vegetative dormancy bud) included bud swelling, bud growing, first leaf sprouting, leaves unfolding and complete development in sequence. When the shoots are mature, flower buds appear at axils, calyx becomes visible and the internode ceases to grow. Flowers develop with the elongating petals. When sepals are fully extended, the petals can open. After 50% of flowers are open, the first petals will fall (Salazar *et al.*, 2006).

The ambient temperature affects the number of guava flowers. The higher the temperature (between 22 and 28°C), the more the number of flowers. There are fewer flowers when the temperature is lower than 20°C and the flowers drop easily when the temperature is over 30°C (Huang, 1961). The cambial growth and leaf initiation of guava in Taiwan were found accelerated when temperature was above 15°C up to a maximum of about 28°C; however, there was no association between floral initiation and cambial growth (Chou *et al.*, 1973). Similarly, Rathore and Singh (1974), Mitra and Bose (1990) and Kundu (1992) reported that flower production was greater in India during summer when the temperature was high.

The accumulations of the temperature variations between each critical growth and phenological phase indicate heat units or temperature requirements. Bittenbender and Kobayashi (1990) investigated the effect of temperature on 'Beaumont' guava at two different sites by calculating growing degree-days (GDD) using the standard formula with a base temperature of 15°C:

$$\mathrm{GDD} = \left[(\max + \min)/2\right] - 15$$

Fig. 9.4. Phenological stages of the guava. (A) New and young flush. (B) Flower buds ready to bloom. (C) Flowers opened. (D) Petal fall. (E) Fruit set and developed. Bar represents 2 cm.

where max is the maximum daily temperature (°C) and min is the minimum daily temperature (°C). GDD was calculated from historical weather records in 30-day intervals starting at the last day of cycling and summed at 60, 90, 120, 150, 180 and 210 days after cycling. Correlating these GDD sums with days to harvest, they found that the sum of GDD for the 150 days after cycling (150GDD) had the highest correlation. Correlating with photoperiod at cycling and the 150GDD, it was found that they were highly correlated (r = 0.99 and r = 0.98 at two sites).

Ruiz *et al.* (1992) estimated the accumulated heat units (HU) or GDD of four phenological phases of guava as well the threshold temperature for each phenological phase. Threshold temperatures were estimated as 9.2, 13.8, 10.0 and 8.4°C for the periods from post-dormancy irrigation to first buds, from first buds to first blossom, from first blossom to anthesis and from anthesis to beginning of harvest, respectively. To complete each stage, 193, 244, 428 and 2076 GDD were needed, respectively.

Phenological development of guava trees and their relationships with ambient temperature at three locations in Mexico having different climatic conditions were investigated by Padilla-Ramirez *et al.* (2012). The three locations were: (i) Santiago Ixcuintla, Nayarit (hot–subhumid); (ii) Huanusco, Zacatecas (semi-hot–semi-dry); and (iii) Temascaltepec, Mexico (temperate to semi-cool–subhumid). Annual mean temperature varies from 25 to 18°C at the hottest and coolest sites, respectively. They observed that phenological development was greatly influenced by temperature at each location. Duration from pruning to flowering (P–F) varied from 60 to 115 days, while the period from flowering to beginning of harvest (F–BH) varied between 100 and 180 days at Santiago Ixcuintla and Temascaltepec, respectively. Phenology of the trees at Huanusco was found intermediate between the other two locations. Although great variations were observed in the number of days for the different phenological developments among the locations, the accumulated HUs (using a threshold temperature of 9°C) were similar for all sites, being 800–850 HU for stage P–F and 1950–2000 HU for F–BH stage.

Chen *et al.* (2017) studied the phenology of 'Diwang Ba' guava in Taiwan. Three thermal models: (i) anthesis (pruning to 50% of the branches showing anthesis); (ii) fruit growth (from 50% branch anthesis to 25% fruit harvest); and (iii) full growth period (from pruning to 25% fruit harvest), were estimated. The models were established by coefficient of variation (CV) in an accumulated thermal unit method and a regression method. The average number of days for each period were 70.83 (CV = 28.06%), 113.50 (CV = 22.54%) and 184.33 (CV = 10.85%), respectively. The thermal requirements (degree-hour) of three models were estimated as 25,754.39 (CV = 9.59%), 40,273.18 (CV = 6.08%) and 85,704.53 (CV = 4.31%), respectively. These three thermal requirement models could estimate the separate growth periods with more accuracy than those predicted by growth days.

9.5.2 Nutritional effects on flowering

In perennial fruit crops which have wood framework (nutrients locked therein), their extended physiological stages of growth, differential root distribution pattern and phenological stages make them nutritionally more efficient than annual crops (Scholberg and Morgan, 2012). In the tropics and subtropics, flowers in guava are always borne on newly growing vegetative shoots, irrespective of the time of year. Where the nutrition level is high and where abundant moisture and hot temperatures prevail, one vegetative flush is succeeded by a succession of vegetative flushes.

Nitrogen is the most important nutrient affecting vegetative flushing. Increased concentration of nitrogen in tissues of leaves and shoots influences the post-monsoon vegetative growth and, as guava bears flowers and fruits on the current season's growth (Shigeura *et al.*, 1975; Shigeura and Bullock, 1983; Sanyal, 1991; Jat and Kacha, 2014), the percentage of flowering shoots increases. Spraying of urea (10–15%) used for de-blossoming increases the development of new shoots and the majority of them also flower subsequently. However, spraying with high rates of urea (25%) sometimes causes excessive vegetative growth at the expense of flowering (Shigeura and Bullock, 1983).

Nitrogen appears to have a special importance in flowering and generally the number of flowers formed on a tree is dependent on the level of nitrogen in the tree as it translocates from leaves to flowers at flowering (Smith, 1970). Spraying with 10% urea for de-blossoming of summer crops in the month of April showed a leaf carbohydrate:nitrogen (C:N) ratio from 6.6 to 10.0 between June and October, compared with 5.9 to 7.2 in control. This high C:N ratio caused 45–55% increased flowering and 20–25% increased fruit set in the following winter compared with control (Sanyal, 1991).

Spraying with urea rapidly increased the urea-N concentration in the leaves which rose to a maximum concentration within 4 days. Urea-N metabolite in the tissue transformed into ammonium-N in the first 4 days and thereafter to nitrate-N within 12 days after urea application (Singh *et al.*, 2002). Urea spray suppressed shoot growth, producing severe leaf fall followed by the initiation of new shoots on which flower buds formed in the following season (this aspect is also dealt with in Chapter 10, this volume).

9.6 Regulation of Flowering

Guava crop usually bears twice a year, during the summer (such as in Taiwan) or the rainy season (such as in India) and the winter season, in most production areas (Wang and Pech, 2011; Mitra, 2017). Usually the production is higher during summer or the rainy season, while fruit quality is always superior during winter as compared with the rainy season (Rathore, 1976; Aulakh, 2004). The rainy-season crop is mostly infested by fruit fly which causes heavy losses (Rathore, 1976; Singh and Joon, 1984; Singh and Singh, 2000; Adhikari and Kandel, 2015), while the winter-harvested fruits are superior in size, quality and taste, less attacked by pests and diseases, and have better storage life and thus can be transported to a destination offering remunerative prices (Abbas *et al.*, 2014; Maji *et al.*, 2015; Nautiyal *et al.*, 2016).

A cropping cycle of guava is started from pruning or defoliation to harvest. The cropping cycle varies in different locations and seasons (Singh *et al.*, 1996; Chang and Lin, 1998). In Hawaii, it needed 160 days in summer and 220 days in winter (Bittenbender and Kobayashi, 1990; Nakasone and Paull, 1998). In Spain, it took almost 1 year from bud emergence in winter to fruit harvest (Salazar *et al.*, 2006). In Australia, time between floral bud emergence and fruit set was 30–38 weeks later by defoliation and fruit was harvested in mid-June (Menzel and Paxton, 1986). In Mexico, the duration from pruning to flowering varied even by elevation from 60 to 115 days, and the period from flowering to harvest was from 100 to 180 days (Padilla-Ramirez *et al.*, 2012). In India, it required an average of 30–40 days to develop flowers and more than 124–138 days to harvest fruits (Sahoo *et al.*, 2017). In Taiwan, the cycling started in September lasted 133 days, while when the growth started in December, it needed 194–205 days to finish the cropping cycle (Wang, 1988, 1989).

Cycling of crop to avoid rainy-season harvest and increased winter yield is widely accepted and followed by guava growers (Mitra, 2017). Coordination of the fruiting cycle can help in maintaining fruit supplies during most months (Lopez and Perez, 1977; Manica *et al.*, 1982; Shatat, 1993).

Crop regulation can be achieved by shifting the flowering time. The principle of crop cycling was developed by Shigeura *et al.* (1979) in Hawaii to harness the natural flowering and fruiting tendencies of guava and to contribute to increase yield and profitability. The crop cycling procedure disrupts the normal production tendencies of guavas obtained under natural circumstances. To force the tree into increased production by satisfying its other needs, such as fertilizer, supplemental pruning and defoliation, is good farming practice since the cycling procedure extends the production potential of guava beyond its natural tendency (Shigeura and Bullock, 1983). However, regulating flowering time causes nutrition redistribution, may affect the development of growing fruits, and sometimes causes the competition or the interaction of vegetative and reproductive growth. The different methods used for crop cycling are: (i) water management to induce stress; (ii) pruning of shoots at different levels and at different times; (iii) training (bending) of shoots; and (iv) manual or chemical thinning of flowers.

9.6.1 Water management to induce stress

Withholding irrigation from December to June (in the southern hemisphere) or until the beginning of monsoon depending on prevailing conditions at a particular location has been recommended in southern peninsular India (Kaul, 1974). Withholding of irrigation and removing the soil from around the upper roots by 10 June, and again covering it with soil and manure mixture, has been suggested. Two light irrigations were also suggested before the normal heavy one if rain did not start. The treated plants shed all their leaves in May and the regrowth starts with emerging new shoots in July–August that flower in September and fruits mature in November–December as a winter crop.

9.6.2 Pruning of shoots

Pruning in guava has a rejuvenating effect on aged trees and re-establishes the leaf architecture, which becomes more compact. Annual pruning is used as a cheap and effective cultural technique for regulating cropping pattern (Jadhav *et al.*, 2002; Bhagawati *et al.*, 2015; Singh *et al.*, 2018). Regular pruning is practised to make better light interception inside the tree canopy, leading to better photosynthetic rate and better nutrient and water supply with reduced canopy and shortening the distance from the source to sink for higher productivity and better-quality fruits (Bhagawati *et al.*, 2015; Sah *et al.*, 2017). A precise level of pruning is required to regulate the balance between vegetative and reproductive phases.

Appropriate pruning methods with different intensity improve growth and production of guava by increasing the number of flower buds and fruits on plants (Mohammed *et al.*, 2006; Lakpathi *et al.*, 2013). Pruning favoured the production of more flowers in the July–August flush and thereby more fruits in the winter season (Gaur, 1996). Pruning the current-season growth of the spring flush to avoid rainy-season crop has been advocated in northern parts of India (Tiwari and Lal, 1984). Pruning of 25–50% of shoots at any of the dates of 20 April, 10 May or 30 May was found to evade flowering in the rainy season and encouraged winter-season flowering of 'Sardar' guava (Dhaliwal *et al.*, 1998). Shoot pruning by 50% in April and July has a positive effect towards vegetative growth, emergence of new shoots and increased yield in winter (Sah *et al.*, 2017). Pruning by leaving four pairs of leaves caused higher number of flowers and fruits than by leaving eight pairs of leaves (Susanto *et al.*, 2019). Pinching is another useful way to improve flowering from axillary buds (Adhikari and Kandel, 2015).

Crop regulation in guava was successfully achieved in Taiwan by pruning (Fig. 9.5). The period of the growth cycle is dependent on temperature and is different from the pruning seasons. Temperature is the important factor in the effect of the pruning date on fruit yield (Chandra and Govind, 1995; Dhaliwal and Kaur, 2003; Sarkar *et al.*, 2005; Shaban and Haseeb, 2009). Using temperature models of each critical stage of growth and vegetative phases, it was possible to predict the harvesting time in guava. Three thermal requirement models of: (i) anthesis (from pruning to 50% anthesis); (ii) fruit growth (from 50% anthesis to 25% fruit harvest); and (iii) full growth period (from pruning to 25% fruit harvest), could be estimated and the thermal requirement models were established by the 'coefficient of variation in accumulated thermal unit method' and 'regression method' (Chen *et al.*, 2017). Suitable regression formulas and accurate base temperatures for growth are the two essential factors in a prediction model. The thermal requirement models could be estimated more precisely by growth hours than by growth days (Bittenbender and Kobayashi, 1990; Salazar *et al.*, 2006; Chen *et al.*, 2017). A well-predicted temperature model can significantly improve the correctness of the crop season estimation.

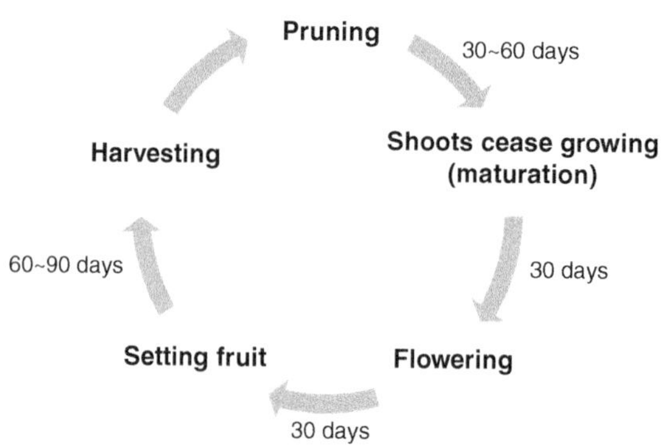

Fig. 9.5. Flowering regulation by pruning and its crop cycle in Taiwan.

9.6.3 Training (bending) of shoots

This method of crop cycling is effective when the plants are between 2 and 5 years of age. The plants are trained in such a manner that only four shoots will grow in all the four directions at 30–45 cm trunk. Certain cultivars ('Allahabad Safeda' and 'Bengal Safeda') were found more responsive to the treatment (Mitra, 2017). The shoot bending should be done 15–30 days before flowering. Leaves, small developing shoots and flowers (if any) should be removed before training (bending) of shoots (Fig. 9.6). The bent shoots are tied with ropes to pegs which should be removed after emergence of new shoots. Trees should be fertilized 15 days before treatment (bending of shoots), followed by irrigation. During summer (April–June), the new shoots emerge within 8–10 days of treatment, while in autumn (September–October) the new shoots emerge after 20–25 days. Flowering occurs in the new emerging shoots at the four-pair to five-pair leaf stage, after 45–50 days of summer and 60–65 days of autumn treatment (Sarkar *et al.*, 2005) (Fig. 9.6). A second dose of fertilization is required at the pea stage of fruit growth.

This technology is commercially used in South Asia for the regulation of flowering in the off season (autumn to early summer) (Mamun *et al.*, 2012). Nandi *et al.* (2017) reported that shoot bending treatment is more effective in increasing flowering shoots and fruit yield when carried out in the month of October. Crop regulation by shoot bending increased winter harvest and improved fruit quality in terms of fruit size, total soluble solids (TSS) and vitamin C content of fruit (Sarkar *et al.*, 2005). Bending of shoots increased the total free amino acid contents in leaf and bark compared with control (Bagchi *et al.*, 2008). This may be on account of the inefficient utilization of amino acids in protein synthesis or due to proteolysis. Higher lipids in bark at the initial stage and in leaves at the later stage signify the tendency of plants to overcome the shock effects of bending and pruning. In leaves and bark, total soluble sugars and reducing sugars were also higher in the bending/pruning treatments; for starch, however, it followed the reverse trend. As proline biosynthesis is stimulated under stress, profuse flower bud initiation occurs in the stressed plant (by bending). A positive relationship between proline concentration in the leaves and yield was observed by Bagchi *et al.* (2008). Definite biochemical changes in guava shoots under an episode of stress may produce off-season flowering in guava, leading to higher fruit production.

9.6.4 Flower and fruit thinning by chemicals

Manual de-blossoming of flowers for crop regulation, although very effective, is not used in practice because it is very cumbersome, laborious and uneconomical (Mitra *et al.*, 1982, 1995; Singh *et al.*, 2018). Growth regulators and certain chemicals have been found very effective in crop cycling. Flower thinning by using naphthalene acetic acid (NAA), naphthalene acetamide (NAD), 2,4-dichlorophenoxyacetic acid (2,4-D), potassium iodide (KI), 2-chloroethylphosphonic acid (ethephon), 4,6-dintro-*o*-cresol (DNOC) and urea have been tried with varying degrees of success. This variation may be due to cultivars, tree condition, soil types and environment. Most workers are of the opinion that chemical thinning is economical and a suitable method for crop regulation.

The most used de-blossoming chemical in India is NAA to prevent flowering and cropping during the rainy season (Rathore, 1976; Kumar and Hoda 1977; Parvez *et al.*, 1999; Abbas *et al.*, 2014; Maji *et al.*, 2015; Singh *et al.*, 2018). Spraying of NAA, NAD and 2,4-D caused blossom drop in guava; the most effective treatment was reported as NAD at 50 ppm which thinned 90% of flowers, followed by 2,4-D at 30 ppm (80%). Both treatments markedly increased fruit set in the following winter (Mitra *et al.*, 1982). Spraying urea at a high concentration is also an effective treatment for the removal of summer-season flowers (Dwivedi *et al.*, 1990). Two sprays of 10 and 15% urea at 10-day intervals during summer flowering season was found effective for crop regulation (Gorakh *et al.*, 2000;

Fig. 9.6. Crop cycling in guava by training of shoots. (A) Bending of shoots. (B) Emergence of new shoots. (C) Emergence of new flower buds. (D) Flowering. (E) Developing fruits. (F) Reaching maturity. Photographs courtesy of Professor S. Mitra.

Dhaliwal *et al.*, 2002). Thinning of rainy-season flowers not only increased the winter yield but also markedly improved fruit quality in terms of TSS, total sugar and ascorbic acid contents (Singh *et al.*, 1992; Mitra *et al.*, 1982, 1995; Singh and Singh, 2000; Lal *et al.*, 2013). The applications of deblossoming chemicals and their effects on fruit quality in crop regulation in India are presented in Tables 9.1 and 9.2.

Table 9.1. Crop regulation in guava by using chemicals in India. Adapted from Mitra *et al.* (2008).

Location	Cultivar	Chemical	Time of application	Reference
Lucknow, Uttar Pradesh	'Allahabad Safeda', 'Lucknow- 49' ('Sardar')	Urea 10% KI 0.05%	30 April and 10 May (2 sprays)	Singh and Singh (2000)
Ludhiana, Punjab	'Allahabad Safeda'	Urea 10%	3rd week of May	Singh *et al.* (1992)
Patiala, Punjab	'Allahabad Safeda'	Urea 15%	1st week of April or 1st week of May and 10–15 days of first spray (2 sprays)	Singh *et al.* (1994)
Sabour, Bihar	'Allahabad Safeda' 'Lucknow-49'	NAD 50 ppm NAA 250 ppm	4th week of April and 2nd week of May (2 sprays)	Singh *et al.* (1993) Choudhary *et al.* (1997)
Midnapur, West Bengal	'Lucknow-49' ('Sardar')	DNOC 10 ppm	4th week of May	Kundu and Mitra (1997)
Nadia, West Bengal	'Lucknow-49' ('Sardar')	NAD 50 ppm	2nd week of April	Mitra *et al.* (1982)
	'Harijha'	Urea 10%	2nd week of April	Mitra *et al.* (1995)
Sabour, Bihar	'Allahabad Safeda'	Carbaryl 300	4th week of April	Singh (1986)

KI, potassium iodide; NAD, naphthalene acetamide; NAA, naphthalene acetic acid; DNOC, 4,6-dintro-*o*-cresol.

Table 9.2. Fruit quality (at the season of harvesting: rainy or winter) due to crop regulation in guava. Adapted from Mitra (2017).

	Fruit weight (g)		TSS/acid ratio		Ascorbic acid (mg 100 g^{-1})	
Treatment	Rainy	Winter	Rainy	Winter	Rainy	Winter
Control	95	113	19.3	41.6	122	187
Manual	–	130	–	30.2	–	222
NAA, 100 ppm	140	157	19.1	30.7	115	253
2,4-D, 100 ppm	163	161	20.2	34.8	167	220
DNOC, 10 mg/l	97	160	22.4	33.6	138	247
Urea, 10%	154	176	25.9	37.8	172	230
Critical difference (P = 0.05)	20.2	25.0	3.8	NS	26.2	NS

TSS, total soluble solids; NAA, naphthalene acetic acid; DNOC, 4,6-dintro-*o*-cresol; 2,4-D, 2,4-dichlorophenoxyacetic acid; NS, not significant.

Defoliation is another function of the plant growth regulators that are applied to induce new shoots and flowers. Defoliation with ethephon at 600–1800 ppm induced lateral shoot production (Kobayashi, 1987). Leaf drop increased with increasing concentration of NAA (200–600 ppm) and urea (10–20%) (Singh *et al.*, 1989; Singh *et al.*, 1992).

9.7 Conclusion

Guava flowers are always borne on newly growing vegetative shoots, irrespective of the time throughout the year, in the tropics and subtropics. Floral buds generally occur after shoot growth ceases for a period with cool temperature or non-lethal water stress. Flowers occur either singly or in cymes of two or three at leaf axils of current and preceding growth. It requires 30 days from flower-bud differentiation to calyx-cracking stage. The blooming period ranges from 28 to 45 days. In guava, both self-pollination and cross-pollination occur; however, the yield decreases via self-pollination due to self-incompatibility phenomena in some cultivars. The nutrition level, especially nitrogen, affects vegetative flushing and influences the vegetative growth and flowering in guava. Methods for regulating flowering time are practical for crop cycling. A precise level of pruning with a well-predicted temperature model can significantly improve the correctness of the crop season estimation. Shoot bending, spraying chemicals such as NAA or 2,4-D for thinning flowers, or defoliation with ethephon may cause the flowering time shift to achieve the crop regulation.

References

Abbas, M.M., Ahmad, S. and Javaid, M.A. (2014) Effect of NAA on flower and fruit thinning of guava. *Journal of Agricultural Research* 52(1), 111–116.

Adhikari, S. and Kandel, T.P. (2015) Effect of time and level of pruning on vegetative growth, flowering, yield, and quality of guava. *International Journal of Fruit Science* 15(3), 290–301.

Alves, J.E. and Freitas, B.M. (2006) Foraging behavior and pollination efficiency of five bee species on guava (*Psidium guajava* L.) flowers. *Revista Ciencia Agronomica* 37(2), 216–220.

Alves, J.E. and Freitas, B.M. (2007) Pollination requirements of guava. *Ciencia Rural* 37(5), 1281–1286.

Aulakh, A.K. (2004) Fruit yield and quality of six outstanding selections of guava (*Psidium guajava*) from the Calvillo-Canones region, Mexico. *Acta Horticulturae* 735, 25–30.

Bagchi, T.B., Sukul, P. and Ghosh, B. (2008) Biochemical changes during off-season flowering in guava (*Psidium guajava* L.) induced by bending and pruning. *Journal of Tropical Agriculture* 46, 64–66.

Balakrishnan, K. (2000) Foliar spray of zinc, iron, boron and magnesium on vegetative growth, yield and quality of guava. *Annals of Plant Physiology* 14(2), 151–153.

Batten, D.J. and McConchie, C.A. (1995) Floral induction in growing buds of lychee (*Litchi chinensis*) and mango (*Mangifera indica*). *Australian Journal of Plant Physiology* 22, 783–791.

Bhagawati, R., Bhagawati, K., Choudhary, V.K., Rajkhowa, D.J. and Sharma, R.J. (2015) Effect of pruning intensities on the performance of fruit plants under mid-hill condition of eastern Himalayas: case study on guava. *International Journal of Natural Sciences* 46, 46–51.

Bittenbender, H.C. and Kobayashi, K.D. (1990) Predicting the harvest of cycled 'Beaumont' guava. *Acta Horticulturae* 269, 197–204.

Boora, R.S., Dhaliwal, H.S. and Arora, N.K. (2016) Crop regulation in guava – a review. *Agricultural Reviews* 37(1), 1–9.

Braganza, M.A. (1990) Floral biology studies and varietal evaluation in genus *Psidium*. MSc (Agric.) thesis, University of Agricultural Sciences, Bangalore, India.

Caraballo, H.B.M. (2001) Floral biology of guava (*Psidium guajava* L.) in Maracaibo Plateau, Zulia, Venezuela. *Revista de Facultad de Agronomia, Universidad del Zulia* 18(1), 41–55.

Chezhiyan, N. (1989) Palynological studies in guava and its related species. *Madras Agricultural Journal* 76(12), 671–675.

Chandra, R. and Govind, S. (1995) Influence of time and intensity of pruning on growth, yield, and fruit quality of guava under high-density planting. *Tropical Agriculture* 72(2), 110–113.

Chang, J.C. and Lin, T.S. (1998) Current status and suggestions on improvements of guava production in Taiwan. *Journal of the Chinese Society for Horticultural Science* 44(2), 116–124.

Chen, P.A., Huang, M.Y., Lin, S.Y., Roan, S.F. and Chen, I.Z. (2017) Temperature growth models of guava (*Psidium guajava* L.). *Acta Horticulturae* 1166, 157–160.

Chou, T., Hwa, S. and Chiang, T. (1973) Seasonal changes in cambial activity in the young branch of *Psidium guajava* Linn. *Taiwania* 18, 35–41.

Choudhary, R., Singh, U.P. and Sharma, R.K. (1997) Crop regulation in guava cv Lucknow-49. *Orissa Journal of Horticulture* 25(1), 10–13.

da Silva, S.N., Silva, M.A., Marçal, T.d.S., Ferreira, A., Fontes, M.M.P. and Ferreira, M.F.d.S. (2017) Genetic parameters of pollen viability in guava (*Psidium guajava* L.). *Australian Journal of Crop Science* 11(1), 1–8.

Davenport, T.L. (1990) Citrus flowering. *Horticultural Reviews* 12, 349–408.

Davenport, T.L. (2000) Processes influencing floral initiation and bloom: the role of phytohormones in a conceptual flowering model. *HortTechnology* 10(4), 733–739.

Davenport, T.L. and Nuñez-Elisea, R. (1997) Reproductive physiology. In: Litz, R.E. (ed.) *The Mango: Botany, Production and Uses*. CAB International, Wallingford, UK, pp. 69–146.

Dhaliwal, G.S. and Kaur, R. (2003) Effect of time and pruning intensity on age of bearing shoot and fruit quality of Sardar guava. *Haryana Journal of Horticultural Sciences* 32, 21–23.

Dhaliwal, G.S., Gill, H.S. and Rattanpal, H.S. (1998) Effect of time and severity of pruning on shoot growth and flowering in guava. *Haryana Journal of Horticultural Sciences* 27, 223–229.

Dhaliwal, G.S., Nanra, N.K. and Rattanpal, H.S. (2002) Effect of chemicals on flower drop, fruit set and yield on rainy and winter season crops of guava. *Indian Journal of Horticulture* 59(1), 31–33.

Dwivedi, R., Pathak, R.K., and Pandey, S.D. (1990) Effect of various concentrations of urea on crop regulation in guava (*Psidium guajava* L.) cv. Sardar. *Progressive Horticulture* 22(1–4), 134–139.

El-Halwagi, A., Khalaf, R.M., Sayed, H.A., Tawfik, A.A. and Khalifa, M.A. (2007) Morphological, physico-chemical and pollen grain description of some guava varieties of Egypt. *Egyptian Journal of Horticulture* 34(1), 27–42.

Gaur, G.S. (1996) Studies on crop regulation in guava. *Recent Horticulture* 8(1), 21–23.

Gorakh, S., Singh, A.K. and Ajay, V. (2000) Economic evaluation of crop regulation treatments in guava (*Psidium guajava*). *Indian Journal of Agricultural Sciences* 70(4), 226–230.

Guimaraes, R.A., Perez-Maluf, R. and Castellani, M.A. (2009) Bee diversity in a commercial guava orchard in Salinas, Minas Gerais state, Brazil. *Bragantia* 68(1), 23–27.

Hedström, I. (1988) Pollen carriers and fruit development of *Psidium guajava* L. (Myrtaceae) in the Neotropical region. *Revista de Biologia Tropical* 36(2B), 551–553.

Huang, P.C. (1961) The investigations on the flowering and fruiting habits of guava tree. *Journal of the Chinese Society for Horticultural Science* 7(3), 27–36.

Jadhav, B.J., Mahurkar, V.K. and Kale, V.S. (2002) Effect of time and severity of pruning on growth and yield of guava (*Psidium guajava* L.) cv. Sardar. *The Orissa Journal of Horticulture* 30(2), 83–86.

Jat, G. and Kacha, H. (2014) Response of guava to foliar application of urea and zinc on fruit set, yield and quality. *Journal of AgriSearch* 1(2), 86–91.

Kahlon, P.S., Sharma, P.K. and Rambadi, J.L. (1987) Studies on floral biology of guava (*Psidium guajava* L.) cultivar Allahabad Safeda and Lucknow-49. *Haryana Journal of Horticultural Sciences* 16(1&2), 65–73.

Kaul, G.L. (1974) Guava cultivation – the right way – cropping and its regulation. *Indian Farmers Digest* 7(5), 33.

Ketsa, S., Wisutiamonkul, A., Palapol, Y. and Paull, R.E. (2019) The durian: botany, horticulture, and utilization. *Horticultural Reviews* 47, 125–211.

Kobayashi, K.D. (1987) Defoliation and vegetative growth of *Psidium guajava* with ethephon and gibberellic acid. *Acta Horticulturae* 201, 145–148.

Kumar, R. and Hoda, M.N. (1977) Crop regulation studies in Allahabad Safeda guava. *Indian Journal of Horticulture* 34(1), 13–14.

Kundu, S. (1992) Investigation on the improvement of guava production in the laterite tracts of West Bengal. PhD thesis, Bidhan Chandra Krishi Viswavidyalaya, Mohanpur, India.

Kundu, S. and Mitra, S.K. (1994) Studies on floral biology of different guava cultivars. *Crop Research* 8(1), 80–85.

Kundu, S. and Mitra, S.K. (1997) Regulation of cropping in guava. *Indian Journal of Horticulture* 54, 139–145.

Lakpathi, G., Rajkumar, M. and Chandrasekhar, R. (2013) Effect of pruning intensities and fruit load on growth, yield and quality of guava (*Psidium guajava* L.) cv. Allahabad Safeda under high density planting. *International Journal of Current Research* 7(12), 4083–4090.

Lal, N., Das, R.P. and Verma, L.R. (2013) Effect of plant growth regulators on flowering and fruit growth of guava (*Psidium guajava* L.) cv. Allhabad Safeda. *The Asian Journal of Horticulture* 8(1), 54–56.

Lopez, G.J. and Perez, R.P. (1977) Effect of pruning and harvesting methods on guava fields. *Journal of Agriculture of the University of Puerto Rico* 61, 143–151.

Maji, S., Das, B.C. and Sarkar, S.K. (2015) Efficiency of some chemicals on crop regulation of Sardar guava. *Scientia Horticulturae* 188, 66–70.

Mamun, A.A., Rahman, M.H. and Rahim, M.A. (2012) Effect of shoot bending and fruit thinning on productivity of guava. *Journal of Environmental Science and Natural Resources* 5(2), 167–172.

Mandloi, K.C. (1973) Longevity of guava pollen. *JNKVV Research Journal, India* 7(2), 104.

Manica, I., Passos, L.P., Mundstock, E.C., Chaves, J.B. and Stringheta, P.C. (1982) Effect of four periods of pruning on the production of two guava varieties in Minas Gerais, Brazil. *Proceedings of the Tropical Region, American Society for Horticultural Science Annual Meeting* 25, 259–261.

Menzel, C.M. (1984) The pattern and control of reproductive development in lychee: a review. *Scientia Horticulturae* 22, 333–345.

Menzel, C.M. and Paxton, B.F. (1986) The pattern of growth, flowering and fruiting of guava varieties in subtropical Queensland. *Australian Journal of Experimental Agriculture* 26, 123–128.

Mitra, S.K. (2017) Crop regulation for round-the-year harvesting of guava. *Acta Horticulturae* 1166, 35–39.

Mitra, S.K. and Bose, T.K. (1990) Guava. In: Bose, T.K. and Mitra, S.K. (eds) *Fruits: Tropical and Subtropical*. Naya Prokash, Kolkata, India, pp. 280–303.

Mitra, S.K., Sen, S.K., Maiti, S.C. and Bose, T.K. (1982) Effect of growth substances on deblossoming, regulation of cropping and fruit quality in guava. *The Horticultural Journal* 1, 81–88.
Mitra, S.K., Sanyal, D. and Bose, T.K. (1995) Quality improvement of guava by flower regulation. In: *Proceedings of the National Seminar on Advances in Research and Development in Horticultural Crops for Export, Haryana Agricultural University, Hisar, India*, pp. 81–82.
Mitra, S.K., Gurung, M.R. and Pathak, P.K. (2008) Guava production and improvement in India: an overview. *Acta Horticulturae* 787, 59–66.
Mohammed, S., Sharma, J.R., Kumar, R., Gupta, R.B. and Singh, S. (2006) Effect of pruning on growth and cropping pattern in guava cv. Lucknow-49. *Haryana Journal of Horticultural Sciences* 35(3/4), 211–212.
Nakasone, H.Y. and Paull, R.E. (eds) (1998) Guava. In: *Tropical Fruits*. CAB International, Wallingford, UK, pp. 149–172.
Nandi, P., Roy, D., Ghosh, B. and Kundu, S. (2017) Effect of bending of shoots on flowering, yield and quality of guava cv. Khaja. *Journal of Applied and Natural Science* 9(3), 1365–1368.
Nautiyal, P., Lal, S. and Singh, C.P. (2016) Effect of shoot pruning severity and plant spacing on leaf nutrient status and yield of guava cv. Pant Prabhat. *International Journal of Basic and Applied Agricultural Research* 14(3), 288.
Núñez-Elisea, R. and Davenport, T.L. (1994) Flowering of mango trees in containers as influenced by seasonal temperature and water stress. *Scientia Horticulturae* 58(1–2), 57–66.
Ojha, A.P., Tiwari, J.P. and Misra, K.K. (1986) Studies on floral biology of guava (*Psidium guajava* L) cultivars under Tarai conditions of Uttar Pradesh. *Progressive Horticulture* 18(3&4), 308–311.
Padilla-Ramirez, J.S., Gonzalez-Gaona, E., Perales, d.C.M.A., Gutierrez, A.F. and Meyek, P.N. (2007) Fruit yield and quality of twelve outstanding selections of guava (*Psidium guajava*) from Calvillo-Canones region, Mexico. *Acta Horticulturae* 735, 25-30.
Padilla-Ramirez, J.S., Gonzalez-Gaona, E., Perez-Barraza, M.H., Osuna-Garcia, J.A., Espindola-Barquera, M.d.I.C. and Reyes-Aleman, J.C. (2012) Phenological behavior of guava trees (*Psidium guajava* L.) under different climatic conditions of Mexico. *Acta Horticulturae* 959, 97–102.
Parvez, M.A., Muhammad, F. and Ahmad, M. (1999) Scientific approach to enhance the income from guava orchards. *Pakistan Journal of Biological Sciences* 2(4), 1637–1638.
Prakash, N.A. (1976) Studies on growth and fruiting of Sardar guava (*Psidium guajava* L.). MSc (Agric.) thesis, University of Agricultural Science, Dharwad, India.
Purseglove, J.W. (1968) Tropical crops – dicotyledons 1 and 2. In: *Tropical Crops: Dicotyledons*. Longman, London, pp. 346–381.
Rajagopal, D. and Eswarappa, G. (2005) Pollination potentiality of honey bees in increasing productivity in Karnataka. In: Raju, A.J.S. (ed.) *Changing Trends in Pollen Spore Research*. Today & Tomorrow Printers and Publishers, New Delhi, pp. 131–141.
Rathore, D.S. (1976) Effect of season on the growth and chemical composition of guava (*Psidium guajava* L.) fruits. *Journal of Horticultural Science* 51, 41–44.
Rathore, D.S. and Singh, R.N. (1974) Flowering and fruiting in the three cropping patterns of guava. *Indian Journal of Horticulture* 31(4), 331–336.
Ray, P.K. (2002) Guava. In: Ray, P.K. (ed.) *Breeding Tropical and Subtropical Fruits*. Narosa Publishing House, New Delhi, pp. 143–155.
Ruiz, C.J.A., Ortiz, C., Aceves, L. and Becerril, E. (1992) Phenological characterization of guava. *Agroscience Series Water–Soil–Climate* 3(2), 95–114.
Sah, H., Lal, S. and Negi, S.S. (2017) Effect of pruning on growth, flowering and yield in high density planting of guava. *International Journal of Pure & Applied Bioscience* 5(1), 285–292.
Sahoo, J., Tarai, R.K., Sethy, B.K., Sahoo, A.K., Swain, S.C. and Dash, D. (2017) Flowering and fruiting behavior of some guava genotypes in India. *International Journal of Current Microbiology and Applied Sciences* 6(11), 3902–3911.
Salazar, D.M., Melgarejo, P., Martínez, R., Martínez, J.J., Hernández, F. and Burguera, M. (2006) Phenological stages of the guava tree (*Psidium guajava* L.). *Scientia Horticulturae* 108, 157–161.
Sanyal, D. (1991) Physiological studies on flowering of mango, litchi and guava. PhD thesis, Bidhan Chandra Krishi Viswavidyalaya, Mohanpur, India.
Sarkar, A., Ghosh, B., Kundu, S. and Sukul, P. (2005) Effect of shoot pruning and bending on yield and fruit quality in guava cv. L-49. *Journal of Ecology and Environment* 235, 621–623.
Scholberg, J. and Morgan, K.T. (2012) Nutrient use efficiency in citrus. In: Srivastava, A.K. (ed.) *Advances in Citrus Nutrition*. Springer, Dordrecht, The Netherlands, pp. 205–229.
Sehgal, O.P. and Singh, R. (1967) Studies on blossom biology of guava (*Psidium guajava* L.). I. Flowering season, flowering habit, floral bud development, anthesis and dehiscence. *Indian Journal of Horticulture* 24, 118–126.
Seth, J.N. (1960) Varietal cross incompatibility in *Psidium* species. *Horticulture Advance* 4, 161–164.

Shaban, A.E.A. and Haseeb, G.M.M. (2009) Effect of pruning severity and spraying some chemical substances on growth and fruiting of guava trees. *American–Eurasian Journal of Agricultural & Environmental Sciences* 5(6), 825–831.

Shatat, F.A. (1993) Pruning guava (*Psidium guajava* L.): effect of pruning date on yield and quality. *Dirasat* 1993, 153–159.

Shigeura, G.T. and Bullock, R.M. (1976) Flowers induction and fruit production of guava. *Acta Horticulturae* 57, 247–251.

Shigeura, G.T. and Bullock, R.M. (1983) *Guava (*Psidium guajava *L.) in Hawaii – History and Production*. Hawaii Institute of Tropical Agriculture and Human Resources, Research Station Series No. 035. University of Hawaii, Honolulu, Hawaii.

Shigeura, G.T., Bullock, R.M. and Silva, J.A. (1979) Defoliation and fruit set in guava. *HortScience* 10(6), 509.

Singh, A.K., Singh, G., Pandey, D. and Rajan, S. (1996) Effect of cropping pattern on quality attributes of guava fruit. *Physiologia Plantarum* 71, 187–192.

Singh, B., Sharma, D.P. and Kashyap, R. (1992) Effect of defoliation on survival and growth of guava (*Psidium guajava* L.) layers cv. Seedless. *Advances in Plant Sciences* 5, 176–179.

Singh, G. and Singh, A.K. (2000) Regulation of summer flowering in guava (*Psidium guajava* L.) and subsequent impact on yield in winter. *The Horticulture Journal* 13, 1–16.

Singh, G., Rao, O.P. and Mishra, J.S. (1989) Chemical defoliation of guava with urea and NAA. *Progressive Horticulture* 21, 49–50.

Singh, G., Singh, A.K. and Rajan, S. (2001) Influence of pruning date on fruit yield of guava (*Psidium guajava* L.) under subtropics. *Journal of Applied Horticulture* 3(1), 37–40.

Singh, G., Singh, A.K., Rajan, S. and Bhriguvanshi, S.R. (2002) Strategy for crop regulation in guava (*Psidium guajava* L.) through foliar urea sprays and its effect on different N-forms in leaves. *Journal of Applied Horticulture* 4(2), 93–98.

Singh, H., Mehrotra, N.K. and Shergill, T.S. (1994) Effect of urea spray on the crop regulation of guava cv. Allahabad Safeda. *Indian Journal of Horticulture* 51(4), 331–336.

Singh, N., Sharma, D., Kumari, S. and Kalsi, K. (2018) Techniques for crop regulation in guava – a review. *International Journal of Farm Sciences* 8(2), 131–135.

Singh, R.R. and Joon, M.S. (1984) Seasonal influence on physicochemical composition of guava fruit. *South Indian Horticulture* 32, 88–89.

Singh, U.P. (1986) Crop regulation studies in guava. PhD thesis, Rajendra Agricultural University, Pusa, India.

Singh, U.P., Sharma, R.K. and Hoda, M.N. (1993) Regulation of cropping in guava cv. Allahabad Safeda. *The Horticulture Journal* 6, 1–6.

Sharma, S., Sehrawat, S.K., Sharma, S. and Sharma, K.D. (2013) Impact of environment on time of anthesis, dehiscence and stigma receptivity of guava (*Psidium guajava* L.) under semi-arid region of India. *Annals of Biology* 29(3), 257–263.

Siqueira, K.M.d., Kiill, L.H.P., Martins, C.F. and Silva, L.T. (2012) Pollination ecology of *Psidium guajava* L. (Myrtaceae): richness, frequency and time of activities of floral visitors in an agricultural system. *Magistra* 24, 150–157.

Smith, P.F. (1970) A comparison of nitrogen sources and rates on old high yielding Valencia orange trees in Florida. *Journal of the American Society for Horticultural Science* 95, 15–17.

Srivastava, O.P. (1974) Studies on flowering habit, blooming period, anthesis, dehiscence and pollen grain of *Psidium guajava* L. varieties Apple Colour, Chittidar and Red Fleshed. *Progressive Horticulture* 6(1), 71–77.

Susanto, S., Melati, M. and Aziz, S.A. (2019) Pruning to improve flowering and fruiting of 'Crystal' guava. *Journal of Agricultural Science* 41(1), 48–54.

Tiwari, J.N. (1969) Studies of pollen germination in guava. *HortScience* 1(1), 54–60.

Tiwari, J.P. and Lal, S. (1984) *Research Report*. Indian Council of Agricultural Research Workshop on Fruit Research, Lucknow, India.

Tuler, A.C., Silva, T.d., Carrijo, T.T., Garbin, M.L., Mendonca, C.B.F. *et al.* (2017) Taxonomic significance of pollen morphology for species delimitation in *Psidium* (Myrtaceae). *Plant Systematics and Evolution* 303(3), 317–327.

Usman, M., Samad, W.A., Fatima, B. and Shah, M.H. (2013) Pollen parent enhances fruit size and quality in intervarietal crosses in guava (*Psidium guajava* L.). *International Journal of Agriculture, Biology and Engineering* 15(1), 125–129.

Vinod, M. and Sattagi, H.N. (2018a) Foraging activity of pollinators in guava under organic and conventional farming systems. *Journal of Entomology and Zoology Studies* 6(5), 865–872.

Vinod, M. and Sattagi, H.N. (2018b) Pollinator fauna and their relative abundance in guava under organic and conventional farming systems. *Journal of Experimental Zoology, India* 21(2), 1173–1179.

Wang, T.H. and Pech, M. (2011) *Taiwan Guava Production Manual*, 1st edn. Horticulture Crop Training and Demonstration Center, Technical Mission of Taiwan.

Wang, U.C. (1988) The use of chemicals for regulating production periods in processing guava. *Journal of the Chinese Society for Horticultural Science* 34(3), 211–217.

Wang, U.C. (1989) The studies on the methodology of regulation of harvesting periods in processing guava. *Journal of Taiwan Agricultural Research* 38(4), 438–445.

Yadava, L.U. (1996) Guava production in Georgia under cold-protection structure. In: Janick, J. (ed.) *Progress in New Crops*. ASHS Press, Arlington, Virginia, pp. 451–457.

10 Fruit Set, Development and Maturation

Rosemary J. du Preez*
Agricultural Research Council, Mbombela, South Africa

10.1 Introduction

Guava flowers are bisexual and found only in the leaf axils of newly emerging vegetative shoots of the current season's growth, irrespective of time of year. Consequently, fruit set can be very erratic depending on environmental conditions (Menzel, 1985).

Flowers are borne singly or in cymes of two to four in the axils of the current season's growth (Menzel and Paxton, 1986). Reproductive buds are formed and develop only in the axils of the leaves on the first to third nodes distal to the first or second nodes of the shoot, which are vegetative but remain dormant (Lötter, 1988). Often, only two of these reproductive nodes form flowers and it is not known what factors give rise to these limitations in reproductive bud formation despite conditions which appear favourable for bud formation (Lötter, 1988). Although lateral vegetative buds are influenced by apical dominance, the differentiation and development of the lateral reproductive buds are not limited by apical dominance (du Preez, 1997).

Flowering is variable and in unpruned trees, more than one flowering period can occur. In tropical regions, guava will flower and set fruit more than once a year whereas in regions that are more temperate, usually only one flowering and fruit set will occur. Studies in India have reported two flowering periods in a year, namely during late spring and summer (Sehgal and Singh, 1967; Adhikari and Kandel, 2015; Sahoo *et al.*, 2017), whereas Rathore and Singh (1974) reported three flowering seasons in early summer, late summer and autumn. In other countries including Australia (Menzel and Paxton, 1986), only a single flowering season is reported. It is probably due to climatic conditions that promote vegetative growth which results in the different flowering patterns.

Because flowers are always borne on newly emerging vegetative shoots, irrespective of time of year, manipulation of flowering and fruiting by stimulating vegetative growth, and thus flower development, is possible. Pruning, bending of shoots or application of growth regulators to promote vegetative growth are standard practices in most commercial guava orchards. Guava trees are pruned to:

- obtain the desired tree shape;
- obtain maximum production;
- induce rejuvenation;
- extend the harvesting period; and
- ensure acceptable fruit quality.

*E-mail: rosedup@arc.agric.za

DOI: 10.1079/9781789247022.0010

Flowering usually occurs after the plant completes a dormant period induced by cold temperatures, water stress, pruning or application of chemicals. Higher temperature, application of water after a stress period (Sweet, 1986), fertilization and pruning, or defoliation (Shigeura and Bullock, 1976) can all result in a vegetative flush that will flower. By pruning or applying chemicals that result in defoliation at different times of the year, the production season can be altered to the most suitable time to shorten the harvest period and to ensure good yields of quality fruit (Menzel, 1985; du Preez and Welgemoed, 1987; du Preez, 1997; Adhikari and Kandel, 2015; da Silva *et al.*, 2016).

After pruning, vegetative growth occurs for between 5 and 6 weeks after which flower buds become discernible. It is not possible to determine which shoots will bear flowers until the flower buds have emerged. Peak flowering usually occurs 10 to 14 weeks after pruning or defoliation (Menzel and Paxton, 1986; du Preez, 1997).

The time that guava flowers take to develop from emergence to anthesis varies but does not appear to be dependent on when pruning or defoliation takes place. In Australia, it is reported that flowers take 28 days to reach anthesis (Menzel and Paxton, 1986), which is similar to reports from Indonesia (Widyastuti *et al.*, 2019) where flowers took between 25 and 38 days to develop. Other reports indicated that flowers took between 27 and 40 days (Sahoo *et al.*, 2017), 30 to 38 days (Adhikari and Kandel, 2015), 38 to 42 days (Sehgal and Singh, 1967) and 28 to 48 days (Lötter, 1988; du Preez, 1997).

Guavas are primarily self-pollinated, although some cultivars seem to produce more fruits when cross-pollinated with another cultivar (Pommer and Murakami, 2009). The primary pollinator of guavas is the honeybee although wind pollination is also thought to play a role in fruit set.

10.2 Fruit

The guava fruit is a berry and is highly variable in shape (spherical to pyriform), size (25 to 100 mm diameter), skin colour (green to bright yellow), pulp colour (white, yellow, pink, salmon, red or orange), flavour (sweet or acid) and aroma (weak to pungent). The fruit has a thin skin (exocarp) and flesh (mesocarp) of varying thickness. The outer layer consists of finely granular pulp containing stone cells, inside is softer pulp in which the many small seeds are embedded. The round shape tends to predominate with the fruit slightly pointed at the stem end. The length varies from 50 to 70 mm and the breadth between 40 and 65 mm. The average fruit mass of the most common cultivars is usually between 80 and 160 g.

10.3 Fruit Set and Fruit Growth

Fruit set is the developmental stage which marks the transition of the flower (ovary) into a young fruit, which develops until maturity. Initial fruit set occurs after anthesis. Natural fruit set in guava is quite high (80–90%) of which only 34–36% of fruits reach maturity (Mitra and Bose, 1990). In seedless guavas, only 6% of the fruits attain maturity (Kundu and Mitra, 1994). Fruit set in triploid guava is satisfactory when cultivated together with diploid cultivars as a pollen source (Nakasone and Paull, 1998).

Fruit growth can be divided into three different types: fruit have a single, double or triploid sigmoidal growth curve. Growth of most pip fruits tends to show a smooth sigmoid form. Fruits such as date, banana, avocado, orange, mango and melon all have single sigmoidal fruit growth curves (Schroeder, 1953; Bollard, 1970). In contrast, stone fruit, grapes, olives and figs appear to have double sigmoidal growth curves (Crane and Baker, 1953; Coombe, 1960; Bollard, 1970). These observations have led to the concept of cyclic growth in fruit. In the three growth phases, which are distinguished and usually designated as periods 1, 2 and 3, the various components of the fruit do not develop simultaneously. In fruit with a double sigmoidal curve, phase 1 is the initial period of rapid growth and usually the pericarp and seeds increase in size and mass through cell division. In phase 2, the overall growth rate slows down markedly and there is an initial hardening of the endocarp mainly characterized by cell elongation. This observed slowing down of fruit growth is essentially a retardation of mesocarp growth. The embryo usually develops rapidly during this phase. Phase 3 is a period of final swelling and ripening of the fruit and usually occurs towards the end of this phase which is characterized by cell elongation and intercellular cavity development. All fruits that have double sigmoidal growth are similar, but some differences occur; for example, the proportion of growth achieved during phase 3 varies from over 80% to under 40% (Bollard, 1970).

The growth of guava fruit is represented graphically by a double sigmoidal curve (Menzel and Paxton, 1986; Fourie, 1988; du Preez, 1997; Cañizares *et al.*, 2003; da Silva *et al.*, 2016). The growth of the fruit can be divided into three phases. Phase 1 commences at anthesis and is the stage during which the seeds and the pericarp develop. Phase 1 was reported to last approximately 45 to 55 days irrespective of time of pruning or defoliation (du Preez, 1997). During phase 2 the overall growth slows down, and main growth is rapid development of the embryo and initial hardening of the endocarp (du Preez, 1997; Pereira and Couto, 1997). The observed slowing down of the fruit growth is essentially a retardation in mesocarp growth. The length of phase 2 is usually dependent on pruning time and is usually between 50 and 100 days in length, with pruning in early spring having a shorter period than trees pruned during late summer. Phase 3 is the final development stage of the fruit and is a period of 'final swell', ripening occurs towards the end of this period. It is characterized by cell elongation and intercellular cavity development. Phase 3 is the shortest period of guava fruit growth and is usually between 40 and 50 days. No significant differences occurred in this phase whether trees were unpruned, pruned, defoliated or treated with chemicals at different times of the year. Differences that do occur in the length of phase 3 can be attributed to different cultivars and varieties.

By representing growth during each phase as a percentage of the final fruit mass, the relative growth in each stage can be clearly seen (du Preez, 1997):

- phase 1, 14%;
- phase 2, 7%; and
- phase 3, 79%.

Fruit from trees pruned in late summer took longer to reach maturity than fruit from trees pruned or treated during late autumn and early spring. Trees pruned during late autumn showed the same growth pattern as fruit development on untreated trees (Fig. 10.1).

Yusof and Mohamed (1987) reported that guava had a single sigmoidal growth curve and described the curves of weight increase, reduction in firmness, total pectin content, and changes in glucose and ascorbic acid levels during the fruit development of a Vietnamese guava variety. Paull and Goo (1983) observed fruit development of 'Kahua Kula' and concluded that the increase in fresh weight showed a simple sigmoid curve. Srivastava and Narasimhan (1967) reported data typical of a double sigmoidal curve for seeded cultivars. However, seedless cultivars showed a pattern more typical of a single sigmoidal curve with the exception that the rapid growth continued until the fruit was harvested. This is unlike a true single sigmoid developmental pattern which has three growth phases, namely an initial slow increase in fruit mass, followed by a rapid exponential increase in fruit mass and a final stage when there is a decline in fruit growth.

Guavas that are reported as having a single sigmoidal curve are possibly double sigmoidal curves that do not have the initial rapid phase 1 growth possibly due to seedlessness. These fruits therefore only exhibit the second and third stages of fruit growth. The fruit growth of guavas all shows the same trend irrespective of time of flowering or treatments applied. Fruits from trees pruned during late summer take longer to reach maturity than those from trees pruned in early spring and unpruned trees; the difference being in phase 2, which is 5 to 7 weeks longer. This is possibly due to fruit growth occurring during the winter months. A total of 1450 to 1600 degree-days is necessary before the final

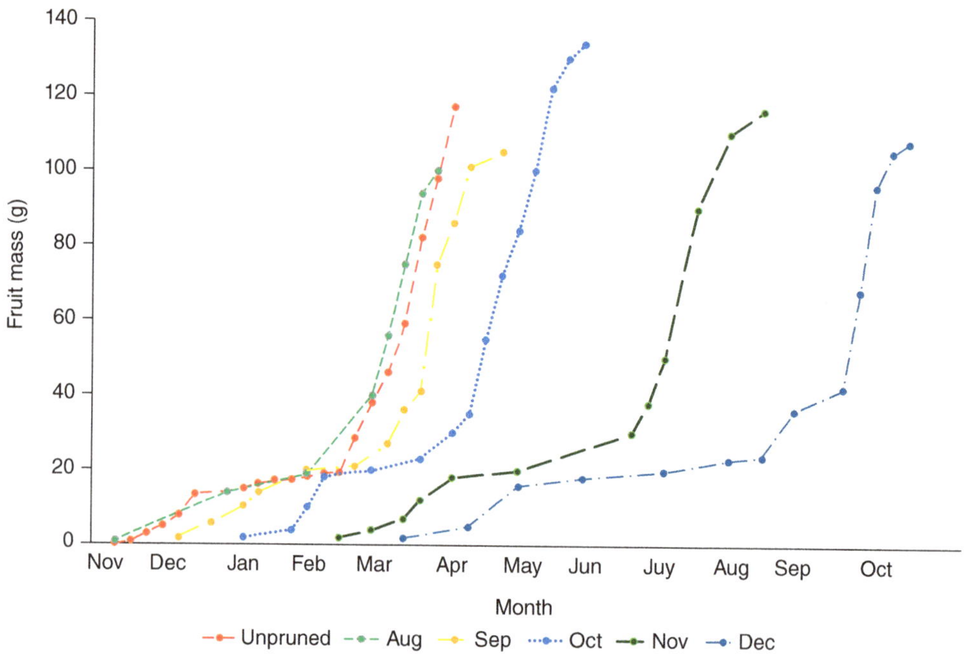

Fig. 10.1. Fruit growth curves of 'Fan Retief' guava subjected to different pruning times. From du Preez (1997).

rapid growth stage begins. The actual length can vary depending on cultivar.

10.4 Fruit Drop

Fruit drop is the premature abscission of fruits and is a natural thinning process that occurs in most fruit trees. Fruit drop in guava occurs at 3 to 9 weeks after flowering irrespective of the time of pruning or defoliation. Fruit drop in guava may occur due to various physiological and environmental factors. Nakasone and Paull (1998) stated that post-set drop may occur as a result of factors other than pollination, such as blossom end rot caused by calcium deficiency.

In a study carried out by du Preez (1997), the percentage fruit drop varied and was between 50 and 54% in South Africa (Fig. 10.2). Fruit drop between 44 and 82% was reported in Australia (Menzel and Paxton, 1986). In India, differences between different cultivars were noted. The fruit drop during the winter-season crop (39–60%) was also higher than fruit drop in the rainy-season crop (33–53%) (Sahoo *et al.*, 2017). Micronutrient foliar sprays were reported to reduce fruit drop. The control trees had 58% fruit drop whereas with different micronutrient spray treatments the fruit drop recorded was lower (34–47%) (Yadav *et al.*, 2014). High fruit retention (73%) was recorded using 500 ppm paclobutrazol spray treatments, resulting in higher yields in treated trees (Jain and Dashora, 2007). Fallen fruit usually has a mass of between 5 and 10 g.

10.5 Harvesting

Traditionally, fruits have been classified as climacteric or non-climacteric according to their respiration patterns and ethylene production rates during ripening (Biale, 1964). Climacteric fruit are those whose ripening process is accompanied by peaks in respiration and ethylene production rates, whereas non-climacteric fruit do not exhibit such increases (Giovannoni, 2001). In climacteric fruit, ethylene plays a key role in governing the physiological and biochemical changes that occur during ripening, whereas in non-climacteric fruit, ethylene is not necessary

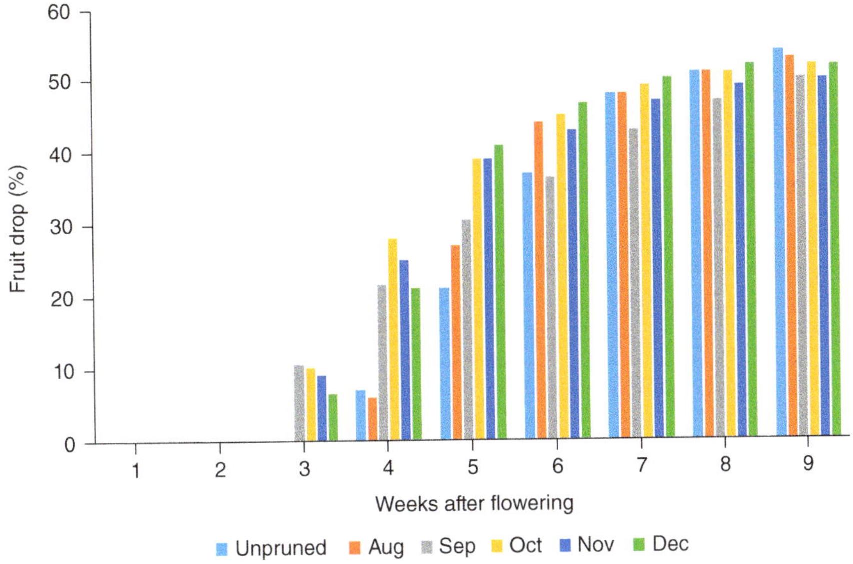

Fig. 10.2. Cumulative fruit drop for different pruning times of 'Fan Retief' guava. From du Preez (1997).

for the coordination and completion of the ripening process (Lelievre *et al.*, 1997; Barry and Giovannoni, 2007), defining the harvest time as well as postharvest handling and storage (Biale, 1964; Paul *et al.*, 2012). Ripe fruit are usually soft and delicate and cannot withstand the rigours of transport and repeated handling. Climacteric fruit are harvested before they are ripe although they are fully mature. Non-climacteric fruit do not ripen any further after harvesting, produce only very small quantities of ethylene and do not respond to ethylene treatment. Non-climacteric fruits include citrus, grapes, lychees and strawberries.

Guava is traditionally defined as a climacteric fruit (Biale, 1964); however, the available information on the respiration pattern of guava is contradictory. Guava presents some non-climacteric fruit characteristics and is therefore considered by some authors as a fruit belonging to that group (Biale and Barcus, 1970; Paul *et al.*, 2012), while other authors (Akamine and Goo, 1979; Brown and Wills, 1983; Mercado-Silva *et al.*, 1998; Azzolini *et al.*, 2005; Mendonça *et al.*, 2007) classify guava as a climacteric fruit for presenting an increase in the respiration rate and high postharvest ethylene production (climacteric peak), which results in its high perishability under ambient conditions. Reyes and Paull (1995) reported that guavas respond to the exogenous application of ethylene depending upon their maturity. Cultivar 'Beaumont' showed enhanced changes in skin colour and softening in response to exogenous application of ethylene only if the fruits were harvested at an immature-green stage but not when harvested at mature-green and quarter-yellow stages (Paul *et al.*, 2012). Climacteric or non-climacteric behaviour of guava fruit appears to be a varietal characteristic.

A study by Azzolini *et al.* (2005) showed that 'Pedro Sato' exhibits a gradual increase in the rates of respiration and ethylene production after harvest and it completes its ripening with changes in quality attributes. Ethylene was found to be necessary for skin colour changes and firmness loss during ripening. These characteristics classify guava as a climacteric fruit. However, maximum respiration activity as well as ethylene production were observed when the fruits were already ripe. In addition to this, the exogenous application of ethylene to these fruits at the mature-light green stage had no effect on the ripening process (Azzolini *et al.*, 2005). Studies have shown that guava cultivars have distinctive climacteric behaviours and cultivars that have a suppressed climacteric behaviour, and to a greater extent the non-climacteric varieties, have longer storage lives and are more sensitive to 1-methylcyclopropene (1-MCP) treatment to delay ripening than the traditional climacteric varieties (Porat *et al.*, 2009). It is thus evident that different guava varieties/cultivars have distinct climacteric behaviours which affect their natural postharvest longevity.

Pruning, defoliation or other chemical treatments usually shorten the peak harvest period of guava. Unpruned trees tend not to have a well-developed ripening peak. In subtropical areas, for unpruned trees, the harvest period is spread over 12 to 15 weeks (Shigeura, 1980; Menzel, 1985; du Preez, 1997). In tropical areas, the guava tree tends to flower more than once a year resulting in a number of harvest peaks with lower yields per harvest time (Adhikari and Kandel, 2015).

10.6 Yield

Annual yield of guava is recorded as being between 25 and 30 t ha^{-1}. However, guava yield is affected by time and severity of pruning. Unpruned trees give good yields, the disadvantage being that fruit harvesting is over a long period. Higher yields were obtained from trees pruned during spring. Adhikari and Kandel (2015) reported that the highest yields were obtained from trees pruned in mid-May (mid-summer) with a yield of between 85 and 122.5 kg per tree in the winter months. Variations in yield were dependent on the severity of pruning, with the more severe pruning resulting in lower yields. In South Africa, the highest yields were obtained from trees pruned in mid-September (early summer) with an average of 65.8 kg per tree. The later pruning in November and December (summer) had a significantly lower yield of 31 and 20 kg per tree, respectively (du Preez, 1997). The low yields were due to the fact that when the trees were pruned in November and December some fruit had already set. Pruning therefore resulted in reserves being lost when the trees were pruned. When the trees flowered for a second time after pruning, fewer reserves were available and thus less fruit was set. Manica *et al.* (1982) reported that pruning decreased yields, whereas Gopikrishna (1981) found that mild pruning increased yields. These seemingly conflicting results are probably due to the time as well as the manner in which pruning was applied.

10.7 Fruit Quality

Pruning, bending or chemical treatment of guava trees is an important management tool in guava production as it allows control over the flowering and harvesting times. The effect of this on yield and fruit quality is, however, critical to ensure acceptable fruit quality is obtained. There is a wide range of size and shape in different guava cultivars. However, the effect of various treatments to control harvest time also affects the size and quality of the fruit.

The fruit size tends to vary depending on pruning time or when chemical treatments are applied. In South Africa, the smallest fruits were obtained from trees pruned in mid-September (early summer) (du Preez and Welgemoed, 1990; du Preez, 1997) and trees pruned during mid-October had the largest fruits (Fourie, 1988; du Preez and Welgemoed, 1990; du Preez, 1997). The smaller fruit in September is probably because of the higher yield obtained from early summer pruning. Fruits from the December pruning were smaller than fruits from the trees pruned in October. The smaller fruit is probably due to climatic factors as the main fruit growth period is during the winter months and also, as the trees pruned during December had already flowered and set fruit when pruned, this resulted in loss of reserves which could have had an effect on fruit size. In Brazil, guava trees pruned during mid-September had the largest fruit but there was no significant difference in fruit size from trees pruned in August and late September. Fruits from trees pruned in October were, however, significantly smaller (da Silva *et al.*, 2016). Fruit size in India was influenced by the severity of pruning, with 300 mm terminal shoot pruning giving larger fruit in both the rainy (summer) and winter season. It is thus evident that fruit size is influenced by the severity as well as the time of pruning.

10.7.1 Total soluble solids

A gradual increase in total soluble solids (TSS) occurs in all genotypes throughout the development and ripening stages of guava fruits (Patel *et al.*, 2015). However, TSS varies between different cultivars which is probably due to the genetic characteristics of the cultivar. Pruning and application of chemicals have an effect on the final TSS content of guava fruit. In South Africa, trees pruned during October had the highest TSS values whereas those pruned during September had the lowest values. There was no significant difference between trees pruned during August, November and December compared with unpruned trees (du Preez and Welgemoed, 1988; du Preez, 1997). Bajpai *et al.* (1973) and da Silva *et al.* (2016) reported that fruit from unpruned trees had lower TSS values than fruit from pruned trees. Trials in India showed no difference in TSS levels from different pruning times; however, TSS levels increased with the severity of the pruning (Adhikari and Kandel, 2015). Spraying with urea, gibberellic acid and defoliation increased TSS significantly when compared with fruit from untreated trees in studies carried out in Egypt (Atawia *et al.*, 2017).

Although it is evident that cultural practices influence final TSS levels in guava, TSS is also influenced by external factors and the variation reported could be due to fruit maturation periods during high rainfall and cloudy weather as this has been shown to result in lower TSS content. The time of pruning and defoliation as well as the concentration of applied chemicals can therefore have a significant effect on the TSS levels in guava fruit. However, the application time of the treatments also influences the TSS levels and this should be taken into consideration before any treatments are applied.

10.7.2 Total acids

Guava cultivars show significant variation in acid content. Total acids (TA) increase during fruit development and the peak reached is dependent on cultivar. During the final stages of fruit development, the TA content drops (Mercardo-Silva *et al.*, 1998; Patel *et al.*, 2015). The initial increase in acidity could be attributed to increased synthesis of organic acids during fruit growth, whereas the decrease during the final fruit growth stage and maturation is considered to be due to the conversion of organic acids into sugars (Patel *et al.*, 2015). Late-summer pruning times in South Africa resulted in higher acidity levels in the fruit (Fourie, 1988; du Preez, 1997). However, Adhikari and Kandel (2015) reported that acidity was not influenced by time or severity of pruning in India. Urea sprays in trials in Egypt resulted in higher TA content whereas spraying with gibberellic acid resulted in lower TA content when compared with untreated trees (Atawia *et al.*, 2017). It is thus clear that the final TA content of fruit is influenced by cultivar as well as the treatments applied. The reasons for the differences are not clear however and more detailed physiological studies would be necessary to explain and facilitate accurate prediction of final quality of the fruit.

10.7.3 Ascorbic acid

Guavas are rich in ascorbic acid (vitamin C), containing two to five times the ascorbic acid content of oranges. Ascorbic acid content is low at initial development stages of the fruit, increases with the development of the fruit and reaches a maximum in fully mature fruit, then declines rapidly as fruit become overripe. The skin and outer flesh contain the most ascorbic acid (Le Riche, 1951). Patel *et al.* (2015) reported an increase in ascorbic acid until fruit are fully ripe in most cultivars. However, some cultivars showed an increase in ascorbic acid until day 120 of fruit development after which a slight decrease was observed.

There is a wide variation in ascorbic acid content between different cultivars. Lakshminarayana and Rivera (1979) report ascorbic acid level in Mexican cultivars of above 500 mg 100 g^{-1}. Cultivars in South Africa have a range of ascorbic acid level, between 150 and 260 mg 100 g^{-1} (du Preez and Welgemoed, 1990). Similar ascorbic acid content has been recorded for cultivars in India, with levels ranging between 130 and 240 mg 100 g^{-1} (Patel *et al.*, 2015).

Pruning did not have a significant influence on ascorbic acid content of fruit when compared with unpruned trees. Trees pruned in late summer tended to have lower ascorbic acid content whereas trees pruned in early spring and unpruned trees had similar ascorbic acid content. Spraying with paclobutrazol (PP_{333}) increased ascorbic acid content compared with untreated trees and trees that were defoliated (Atawia *et al.*, 2017). Seasonal effects appear to influence ascorbic acid level; this could be due to the degree to which the fruit are exposed to sunlight, as this has been shown to affect ascorbic acid concentration (Mapson, 1970). A fall in the ascorbic acid concentration can also be observed if parts of the plant are shaded (Mapson, 1970). Variations observed are thus probably due to a combination of climate and orchard conditions. Genotype appears to be the most important determining factor for ascorbic acid content in guava fruit.

10.8 Conclusion

Guava flowers are borne only on new emerging shoots, which facilitates manipulation of the flowering process and hence the harvest time. This enables controlled timing of production and ensures good yield and quality fruit as well as shortening the harvest period (flowering and manipulation of flowering time are discussed in detail in Chapter 9, this volume). Flowering generally occurs 8 to 12 weeks after pruning or chemical application and flowers generally take between 35 and 48 days to develop from bud emergence to anthesis, although shorter periods of flower development have been reported and are probably due to different cultivars and environmental conditions. Guava fruit have cyclic growth patterns in which there are two phases of active growth separated by a period during which little growth occurs. The absolute and relative durations of these growth periods differed for different pruning times and severity as well as with the application of different chemicals. This is indicative that growth phases and fruit development can be influenced by treatment applications. Harvest period is also shortened by the application of these cultural practices. Unpruned trees have a longer harvest period and in tropical conditions, more than one harvest occurs in a year. By application of different treatments, this can be controlled and harvest time regulated. Higher yields are generally achieved from trees that are pruned, defoliated or sprayed compared with untreated trees.

Quality of ripe fruit is also influenced by applied treatments. Generally, fruits from trees pruned at certain times or after the application of chemicals have a better quality with higher TSS values and acceptable TSS/TA ratio. Ascorbic acid was influenced only by the application of PP_{333}. The most important factor, however, appears to be the time that the treatment is applied to ensure good yields of quality fruit.

Understanding the phenological cycle of guava and the effect of management practices on yield and fruit quality allows management practices to be modified to develop strategies which should lead to productivity gains for different pruning or chemical application times. As the effects of various treatments and times of application on yield and fruit quality for different cultivars and climatic zones are known, it is possible to decide on the optimum time to apply treatments knowing what the quality of the fruit will be, the time that the fruit will be harvested, expected yield and the critical times to apply management practices such as irrigation, fertilization and pest control. This facilitates effective management of guava production.

References

Adhikari, S. and Kandel, T.P. (2015) Effect of time and level of pruning on vegetative growth, flowering, yield, and quality of guava. *International Journal of Fruit Science* 15(3), 290–301.

Akamine, E.K. and Goo, T. (1979) Respiration and ethylene production in fruits of species and cultivars of *Psidium* and species of *Eugenia*. *Journal of the American Society for Horticultural Science* 104, 632–635.

Atawia, A.A.R., El-Gendy, F.M.A., Bakry, Kh.A.I., Abd El-Ghany, N.A. and Singer, M.A.A. (2017) Physiological studies on flowering and fruiting of guava trees. *Middle East Journal of Agriculture Research* 6(1), 143–151.

Azzolini, M., Jacomino, A.P., Bron, H.U., Kluge, R.A. and Schiavinato, M.A. (2005) Ripening of 'Pedro Sato' guava: study on its climacteric or non-climacteric nature. *Brazilian Journal of Plant Physiology* 17, 299–306.

Bajpai, P.N., Shukla, H.S. and Chaturvedi, A.M. (1973) Effect of pruning on growth, yield and quality of guava (*Psidium guajava* L.). *Progressive Horticulture* 5(1), 73–79.

Barry, C.S. and Giovannoni, J.J. (2007) Ethylene and fruit ripening. *Journal of Plant Growth Regulation* 26, 143–159.

Biale, J.B. (1964) Growth, maturation and senescence in fruit. *Science* 146, 880–888.

Biale, J.B. and Barcus, D.E. (1970) Respiratory patterns in tropical fruit of the Amazon basin. *Tropical Science* 12, 93–104.

Bollard, E.G. (1970) The physiology and nutrition of developing fruits. In: Hulme, A.C. (ed.) *Food Science and Technology, A Series of Monographs*, Vol 1. Academic Press, London, pp. 387–425.

Brown, B.I. and Wills, R.B.H. (1983) Post-harvest changes in guava fruit of different maturity. *Scientia Horticulturae* 19, 237–243.

Cañizares, A., Laverde, D. and Puesme, R. (2003) Growth and development of guava (*Psidium guajava* L.) fruit in Santa Bárbara, Monagas State, Venezuela. *Revista Científica UDO Agrícola* 3(1), 34–38.

Coombe, B.G. (1960) Morphogenesis, growth and changes in sugars, auxins and gibberellins in the fruit of *Vitis vinifera*. *Australian Journal of Science* 22, 481. https://doi.org/10.1104/pp.35.2.241

Crane, J.C. and Baker, E.E. (1953) Growth comparisons of the fruits and fruitlets of figs and strawberries. *Proceedings of the American Society for Horticultural Science* 62, 257–266.

da Silva, M.J.R., Tecchio, M.A., Domiciano, S., Leonel, S. and Balestrero, R.I. (2016) Phenology, yield and fruit quality of 'Paluma' guava tree at different pruning times. *Ciência e Agrotecnologia* 40(3), 317–325.

du Preez, R.J. (1997) Phenological cycle of the 'Fan Retief' guava as influenced by time of pruning. MSc thesis, University of Pretoria, Pretoria, South Africa.

du Preez, R.J. and Welgemoed, C.P. (1987) Guavas: harvest time can be manipulated by pruning. *CSFRI Information Bulletin* 184, 6–7.

du Preez, R.J. and Welgemoed, C.P. (1988) Flowering and fruit development of the guava (*Psidium guajava* L.) subjected to different pruning treatments. *Subtropic* 9(4), 17–20.

du Preez, R.J. and Welgemoed, C.P. (1990) Variability in fruit characteristics of five guava selections. *Acta Horticulturae* 275, 351–359.

Fourie, A.J. (1988) Waterbehoeftes van die koejawelboom (*Psidium guajava* L.) soos onder werp aan verskillende snoeitye. MSc thesis, University of Pretoria, Pretoria, South Africa.

Giovannoni, J. (2001) Molecular biology of fruit maturation and ripening. *Annual Review of Plant Physiology and Plant Molecular Biology* 52, 725–749.

Gopikrishna, N.S. (1981) Studies on the effects of pruning on vegetative growth, flowering and fruiting in 'Sadar' guava (*Psidium guajava* L.). PhD thesis, University of Agricultural Science, Dharwar, India.

Jain, M.C. and Dashora, K., (2007) Growth, flowering, fruiting and yield of guava (*Psidium guajava* L.) cv. Sardar as influenced by various plant growth regulators. *International Journal of Agricultural Science* 3(1), 4–7.

Kundu, S. and Mitra S.K. (1994) Studies on flowering and fruiting of some guava cultivars in the laterite tract of West Bengal. *Haryana Journal of Horticultural Sciences* 23(3), 213–218.

Lakshminarayana, S. and Rivera, A.M. (1979) Promising Mexican guava selections rich in vitamin C. *Proceedings of the Florida State Horticultural Society* 92, 300–303.

Le Riche, F.J.H. (1951) *Chemical Changes During the Development of Some Guava Varieties*. Fruit Research Technical Series No. 21. Department of Agriculture, South Africa.

Lelievre, J.M., Latche, A., Jones, A., Bouzayen, M. and Pech, J.C. (1997) Ethylene and fruit ripening. *Physiologia Plantarum* 101, 727–739.

Lötter, J.d.V. (1988) Sekere aspekte van reproduktiewe groei van die koejawel (*Psidium guajava* L. cv. Fan Retief) onder toestande in die Wes-Kaap. Unpublished progress report, Department of Horticulture, University of Stellenbosch, South Africa.

Manica, I., Passos, L.P., Mundstock, E.C., Chaves, J.B. and Stringetha, P.C. (1982) Effect of four pruning dates on yield of two guava (*Psidium guajava* L.) cultivars in Minas Gerais. *Proceedings of the Tropical Region, American Society for Horticultural Science Annual Meeting* 25, 259–262.

Mapson, L.W. (1970) Vitamins in fruits. In: Hulme, A.C. (ed.) *The Biochemistry of Fruits and Their Products*, Vol. 1. Academic Press, London, pp. 369–384.

Mendonça, R.D., Ferreira, K.S., De Souza, L.M., Marinho, C.S. and Teixeira, S.L. (2007) Physical and chemical characteristics of 'Cortibel 1' and 'Cortibel 4' guavas stored in environmental conditions. *Bragantia* 66, 685–692.

Menzel, C.M. (1985) Guava: an exotic fruit with potential in Queensland. *Queensland Agricultural Journal* 111(2), 93–98.

Menzel, C.M. and Paxton, B.F. (1986) The pattern of growth, flowering and fruiting of guava varieties in subtropical Queensland. *Australian Journal of Experimental Agriculture* 26, 123–128.

Mercado-Silva, E., Bautista, P.B. and Garcia-Velasco, M.A. (1998) Fruit development, harvest index and ripening changes of guava produced in central Mexico. *Postharvest Biology and Technology* 13, 143–150.

Mitra, S.K. and Bose, T.K. (1990) Guava. In: Bose, T.K. and Mitra, S.K. (eds) *Fruits: Tropical and Subtropical*. Naya Prokash, Kolkata, India, pp. 280–303.

Nakasone, H.Y. and Paull, R.E. (eds) (1998) Guava. In: *Tropical Fruits*. CAB International, Wallingford, UK, pp. 149–172.

Paul, V., Pandey, R. and Srivastava, G.C. (2012) The fading distinctions between classical patterns of ripening in climacteric and non-climacteric fruit and the ubiquity of ethylene – an overview. *Food Science and Technology* 49(1), 1–21.

Paull, R. and Goo, T. (1983) Relationship of guava (*Psidium guajava*) fruit detachment force to the stage of fruit and chemical composition. *HortScience* 18(1), 65–67.

Patel, R.K., Maiti, C.S., Deka, B.C., Deshmukh, N.A., Verma, V.K. and Nath, A. (2015) Physical and chemical changes in guava (*Psidium guajava* L.) during various stages of fruit growth and development. *International Journal of Agriculture, Environment and Biotechnology* 8(1), 63–70.

Pereira, W.E. and Couto, F.A.D. (1997) Stem and fruit growth of six guava tree (*Psidium guajava* L.) cultivars under soil water stress conditions. *Acta Horticulturae* 452, 87–91.

Pommer, C.V. and Murakami, K.R.N. (2009) Breeding guava (*Psidium guajava* L.). In: Jain, S.M. and Priyadarshan, P.M. (eds) *Breeding Plantation Tree Crops: Tropical Species*. Springer, New York, pp. 83–120.

Porat, R., Weiss, B., Zipori, I. and Dag, A. (2009) Postharvest longevity and responsiveness of guava varieties with distinctive climacteric behaviours to 1-methylcyclopropene. *HortTechnology* 19(3), 580–585.

Rathore, D.S. and Singh, R.N. (1974) Flowering and fruiting in three cropping patterns of guava. *Indian Journal of Horticulture* 31(4), 331–336.

Reyes, M.U. and Paull, R.E. (1995). Effect of storage temperature and ethylene treatment on guava (*Psidium guajava* L.) fruit ripening. *Postharvest Biology and Technology* 6, 357–365. https://doi.org/10.1016/0925-5214(95)00007-S

Sahoo, J., Tarai, R.K., Sethy, B.K., Sahoo, A.K., Swain, S.C. and Dash, D. (2017) Flowering and fruiting behaviour of some guava genotypes under east and south east coastal plain zone of Odisha, India. *International Journal of Current Microbiology and Applied Sciences* 6(11), 3902–3911.

Schroeder, C.A. (1953) Growth and development of Fuerte avocado fruit. *Proceedings of the American Society for Horticultural Science* 61, 103–109.

Sehgal, O.P. and Singh, R. (1967) Studies on the blossom biology of guava (*Psidium guajava* L.). *Indian Journal of Horticulture* 22, 25–32.

Shigeura, G.T. (1980) The use of fertilizers in the production of guava. In: *Proceedings of the Hawaii Fertilizer Conference, Cooperative Extension Service, College of Tropical Agriculture and Human Resources*. Miscellaneous Publication No. 177. University of Hawaii, Honolulu, Hawaii, pp. 17–19.

Shigeura, G.T. and Bullock, R.M. (1976) Flower induction and fruit production of guava (*Psidium guajava* L.). *Acta Horticulturae* 57, 247–252.

Srivastava, H.C. and Narasimhan, P. (1967) Physiological studies during the growth and development of different varieties of guavas (*Psidium guajava* L.). *Journal of Horticultural Science* 42, 97–104.

Sweet, C. (1986) Guavas: new popularity reported. *California Grower* 10(10), 32–33.

Widyastuti, R.D., Susanto, S., Melati, M. and Kurniawati, A. (2019) Effect of pruning time on flower regulation of guava (*Psidium guajava*). *Journal of Physics: Conference Series* 1155, 012013.

Yadav, R.K., Ram, R.B., Kumar, V., Meena, M.L. and Singh, H.D. (2014) Impact of micronutrients on fruit set and fruit drop of winter season guava (*Psidium guajava* L.) cv. Allahabad Safeda. *Indian Journal of Science and Technology* 7(9), 1451–1453.

Yusof, S. and Mohamed, S. (1987) Physico-chemical changes in guava (*Psidium guajava* L.) during development and maturation. *Journal of the Science of Food and Agriculture* 38, 31–39.

11 Physiological Disorders

Nor Elliza Tajidin[1,2]*, Munirah Mohamad[2], Azimah Hamidon[3], Hamizah Hassan[2] and Siti H. Ahmad[2]

[1]Universiti Malaysia Sabah Sandakan Campus, Sandakan, Sabah, Malaysia; [2]Universiti Putra Malaysia, Serdang, Selangor, Malaysia; [3]Universiti Sultan Idris, Muallim, Perak, Malaysia

11.1 Introduction

Guavas are highly perishable fruits where the production and postharvest quality of guava are affected by micronutrient deficiencies that can lead to physiological disorders. Environmental variables such as temperature, light, aeration and nutritional imbalances can also result in a disturbance in the plant metabolic activities and cause physiological disorders (Singh *et al.*, 2019). It is crucial to understand several technical terms which identify the physiological and nutritional disorder symptoms, especially on guava fruit; for example, bronzing, chlorosis, lesion, scorching and others. Primary physiological disorders discussed in this chapter include chilling injury, external and internal browning, sunscald, bronzing, fruit drop and nutrient deficiency.

Guavas are sensitive to chilling injuries that could cause abnormal ripening, skin discoloration and increased decay (González-Aguilar *et al.*, 2004). They are sensitive towards rough handling during harvesting, which causes skin abrasion and browning on internally bruised areas (Kader, 1999). Mechanically injured areas of the skin and flesh are very susceptible to decay. Sunscald on guava occurs when exposed to direct sunlight. Bronzing in guava is a complex nutritional disorder (Sandhu and Gill, 2013). In fruit crops, especially guava, the deficiency of micronutrients causes many more disorders than that of macronutrients (Kumar and Kumar, 2016). Fruit drop is a severe concern in guava as it results in about 45–65% fruit loss, which is attributed to environmental factors (Sharma, 2005; Sandhu and Gill, 2013).

11.2 Bronzing

Bronzing of guava leaves is a nutritional disorder that affects production and may cause the death of the plant. This disorder is observed in guava orchards with low soil fertility and low pH (acidic soil) (Singh and Choudhary, 2012). Affected plants are characterized by initial development of purple to red specks scattered all over the leaf blades which later evolve into a distinct bronze coloration. In more advanced stages, the tree will show sickness by stunted growth and partially defoliated twigs and bronzed leaves (Gopi Kumar *et al.*, 1985). The affected parts will not bear flowers, and

*E-mail: elliza.tajidin@upm.edu.my

DOI: 10.1079/9781789247022.0011

if some flowers form fruits, they later become dry and cracked. Fruits from severely bronzed trees are characterized by brown-coloured patches on the fruit skin (Dahifale *et al.*, 2009).

Guava plants that are severely affected by this disorder are generally found in orchards with soil that is acidic and low in available potassium content. The exchangeable calcium and carbon/nitrogen ratio are also low compared with those in normal orchards. Nutrient deficiencies usually related to this physiological disorder are nitrogen, phosphorus, potassium, calcium, boron, magnesium, zinc and copper. Analysis of soil and leaves collected from guava orchards (healthy or affected by varying degrees of bronzing) revealed that poor nutrient status of the soil with regard to nitrogen, phosphorus and zinc was the main factor contributing to bronzing. Soil application of nitrogen, phosphorus, potassium and zinc at 200, 80, 150 and 50 kg per plant per year, respectively, or fortnightly foliar application of nitrogen, phosphorus, potassium and zinc each at 2% for 4 months, markedly enhanced the respective leaf concentrations of these nutrients and reduced bronzing. Nitrogen applied as ammonium sulfate was found more effective than urea in reducing bronzing (Gopi Kumar *et al.*, 1985). Bronzing disorder has also been noticed in guava orchards planted in deep and loamy soil (Kumar *et al.*, 2007). Foliar application of 0.5% diammonium phosphate and zinc sulfate in combination at weekly intervals for 2 months reduces bronzing in guava (Ramanjini Gowda *et al.*, 1992; Dahifale *et al.*, 2009).

11.3 Chilling Injury

Chilling injury can be categorized as a physiological disorder of fruit that results in reduced quality and loss of product function caused by storage at below tolerance but non-freezing temperature. Storage at low temperature is commonly used to delay ripening and most subtropical fruits are very sensitive to temperature. While the postharvest storage of produce at low temperature is beneficial, all aspects of the metabolism of the produce are not suppressed to the same extent. Continued activity of some metabolic systems at low temperature can lead to cellular dysfunction and collapse (Brown, 1981). Susceptibility to chilling injury and its manifestation vary widely among different fruits, different varieties of the same fruit and the same crop grown in different areas (Wills *et al.*, 1981). Chilling injury causes the release of metabolites such as amino acids, sugars and mineral salts from cells which, together with the degradation of the cell structure, provide an excellent substrate for the growth of pathogenic organisms – a serious postharvest problem even with similar fruit of tropical and subtropical origin (Brown, 1981).

The problem with guavas, and with most tropical fruits, is the high susceptibility to chilling (Wang, 1989). Chilling injury symptoms of guava include abnormal ripening, skin browning or discoloration, and increased incidence of decay upon returning the fruit to ripening temperatures. Disorder symptoms associated with fruits that were stored in a modified atmosphere are surface blackening and development of off-flavour probably because of anaerobiosis (Ali and Lazan, 1997). In general, a temperature of 8–10°C is regarded to be the critical limit for chilling injury for several cultivars (Singh and Mathur, 1954; Wills *et al.*, 1983; Augustin and Osman, 1988; Vazquez-Ochoa and Colinas-Leon, 1990; Alba-Jiménez *et al.*, 2018).

Guava being a highly perishable fruit, is susceptible to chilling injury when it is exposed to a temperature lower than 10°C (González-Aguilar *et al.*, 2004). Dalal *et al.* (1971), however, stated that guava can be stored at 8–10°C for 15 days and can only suffer chilling injury if the temperature is below 8°C. Narayana and Singh (1997) also stated that guava storage can be extended for up to 3 weeks at 5°C without any chilling injury symptoms, which could still develop at a temperature of 3°C. Guava cultivar 'Media China' showed chilling injury when stored below 8°C, and this malady was found to be associated with changes in cell membrane lipids. There was a decrease in microsomal membrane volume and microsomal protein and phospholipids content (5, 47 and 22%, respectively). A reduction in monogalactosyldiacylglycerol, digalactosyldiacylglycerol, phosphatidylinositol, phosphatidic acid and cardiolipin contents and a substitution of saturated fatty acids for unsaturated fatty acids in the microsomal membrane were observed (Alba-Jiménez *et al.*, 2018).

At low temperature, the tissues of guava fruit weaken because they are unable to carry on normal metabolic processes such as the production of ethylene, respiration and energy increase as well as the alteration of cellular structure. Under prolonged storage at low temperature, abnormal ripening, skin browning (Fig. 11.1) or discoloration, and decay commonly anthracnose caused by *Colletotrichum gloeosporioides* were observed (Lim and Khoo, 1990). Such alteration and dysfunction will lead to failure of mature-green or partially ripe guavas to ripen, browning of the flesh and the skin in severe cases, and increased decay incidence and severity upon transfer to higher temperatures. The various chilling injuries depend not only on the cultivar but also on the stage of maturity and the storage environment (Kader, 1999).

Different cultivars of guava show variation in the sensitivity or resistance to chilling temperature (Tiwari *et al.*, 2006). The chemical composition and metabolic status of the guava tissue at the time of chilling can affect the resistance of its tissues to chilling. Chilling-resistant tissues tend to have a higher volume of unsaturated fatty acids in the membrane lipids than chilling-sensitive tissues (Tabacchi *et al.*, 1979). Temperature is the main factor that can influence the degree of chilling injury. Other than temperature, light also can enhance chilling injury under certain circumstances (Van Hasselt, 1990). Relative humidity (RH) in the storage environment is another factor that affects the severity of the chilling injury. Generally, chilling injury symptoms are more severe under low RH than under high RH (Fig. 11.2).

The main aim of chilling injury research is to find effective techniques to alleviate injury induced by chilling. Reduction of chilling injury can be achieved either by increasing the tolerance of commodities to chilling stress or by retarding the development of chilling injury symptoms in the chilling-sensitive fruits. Wang (2010) found that low-temperature conditioning, controlled atmosphere, intermittent warming, growth regulator and chemical applications are effective in reducing chilling injury. The use of controlled atmosphere has been proven to delay ripening significantly and alleviate chilling injury in guava (Singh and

Fig. 11.1. Development of chilling injury symptoms in guava fruits stored at 5 ± 2°C during 14 days of storage without any proper packaging material. Photograph courtesy of Muhammad Zul Izlan Zulkifli.

Fig. 11.2. Chilling injury resulting from exposing guava to a temperature at 3°C (chilling temperature) for 3 days. Photograph courtesy of Munirah Mohamad.

Pal, 2008). Recently Alba-Jiménez *et al.* (2018) found that storage of guava at 10°C under 5 kPa carbon dioxide reduced chilling injury and prolonged storage life. Galactomannan (0.75%) and carnauba wax (0.9%) coating of fruits reduce chilling injury of 'Paluma' guava. The treatment also maintains firmness and colour under refrigeration (11°C) for 15 days of storage (Germano *et al.*, 2019). Treatment with methyl jasmonate (MJ) at 10^{-4} and 10^{-5} M before storage at 5°C for up to 15 days plus 2 days at 20°C reduce the chilling injury index and the ion leakage percentage of both red and white cultivars of guava (González-Aguilar *et al.*, 2004). Fumigation with MJ at 100 mg l^{-1} for 16 h markedly reduces peel browning area caused by chilling injury of 'Jen-Ju Bar' guava fruits after storage for 7 and 14 days at 0°C followed by 3 days at 20°C (SzuWei and ChunTa, 2013).

11.4 External (Skin) and Internal (Flesh) Browning

Postharvest handling of guava can result in physical damages such as skin abrasion and browning of bruised areas (Kader, 1999). Postharvest operations which include fruit handling, sorting, grading, packing and transportation can result in browning after-effects from mechanical damage (Montero *et al.*, 2009). The severity of such damage to guava is commonly related to several factors such as the height of fall, initial velocity, number of impacts, type of impact, size and surface and maturity of the fruit (Singh, 2011). Browning caused by bruising appears as a result of impacts and compressions of the fruit against other fruits, parts of the tree, containers, parts of any grading and treatment machinery and on any non-cushioned surfaces.

Browning is the most common effect of bruising caused by postharvest handling of guava. This damage results from cell breakage (Schoorl and Holt, 1983), which leads to stress and distortion of individual cells (Ruiz-Altisent and Moreda, 2011). Harker *et al.* (2010) reported that the breakage mechanism included three steps of cell fracture, rupture and cell-to-cell debonding. The cell membrane breakage leads to the release of cytoplasmic enzymes into the intercellular spaces which react with vacuolar contents (Mitsuhashi-Gonzalez *et al.*, 2010). This damage does not appear immediately after impact, however. The symptoms appear as browning over time and several serious damages usually accompany the changes. The browning does not just affect appearance but also significantly affects the respiration process and moisture loss (Kumar *et al.*, 2016).

Additionally, mechanical injuries that cause external guava browning may also penetrate the internal part of the fruit. The reaction of the browning catalysis enzyme polyphenol oxidase (PPO) happens in almost all fruit, including guava (Augustin *et al.*, 1985). The browning was generally thought to be a rapid degradation of phenolic compounds caused by PPO and peroxidase (Zhang *et al.*, 2005). Increased PPO activity during storage of guava has been observed (Vishwasrao and Ananthanarayan, 2016) that caused flesh browning. The fracture of cell membranes caused by injuries may result in mixing of phenolic compounds with PPO leaking out of vacuoles, leading to skin browning. The resulting browning affects the appearance of fruit, which is usually not acceptable in the market and contributes to postharvest losses (Fig. 11.3).

There are several factors that affect bruising such as produce maturity, ripening, time of harvest and also time lapse after harvest. Another factor that contributes to browning and the susceptibility of guava to bruising during postharvest handling is the type of cultivar (Hussein *et al.*, 2020). Other factors are alterations of physiological and biochemical properties in the guava, environmental effects such as humidity and temperature, and postharvest treatments. Hence, it is essential to practise good harvesting management, whether harvested manually or mechanically. Handling conditions (cool chain management) such as temperature and humidity also need to be taken care of to reduce damage.

Storage of fruits at low temperature may cause symptoms in the form of skin surface pitting and flesh browning (Wang *et al.*, 2009). Guava fruit is less susceptible to internal flesh browning following heat

Fig. 11.3. Abrasion resulting from skin scuffing against other fruits or surfaces of handling equipment or shipping boxes during postharvest handling. Photograph courtesy of Munirah Mohamad.

treatment provided with radio frequency energy that is used to control Mexican fruit fly larvae. There was also no commercial damage to the internal quality of guava fruit (Monzan *et al.*, 2005).

11.5 Fruit Drop

The basic condition of fruit set (synchronic bloom, transfer of pollen, fertilization, etc.) still does decide the fate of the flower (Cano-Medrano and Darnell, 1998) despite best weather conditions. Beyond a set quantity of fruits, the tree is unable to carry a larger load. A system of autoregulation works in the background and causes the drop of some fruits despite the accomplished fertilization and the equality of physiological precedents (Soltesz, 1997). The further development of fruits maintained on the tree depends mainly on the growing conditions (water, supply of nutrients, weather adversities, pruning, fruit thinning, biotic damages), which may cause fruit drop especially at the time of approaching maturity (Racskó *et al.*, 2007).

The natural fruit setting in guava is about 80–90% in most cultivars studied in India, of which only 30–35% of fruits reach maturity (Mitra and Bose, 1990). In triploid cultivar 'Seedless', however, only 40–45% of flowers set fruits, of which about 6% of fruits attain maturity; the rest drops down (Kundu and Mitra, 1994). Nakasone and Paull (1998) stated that post-setting drop may occur as a result of factors other than pollination, such as blossom end rot caused by calcium deficiency. Flower bud and fruit abscission in guava is a continuing problem from when the flower buds emerge to the time the fruits are about 2–3 cm in diameter (Shigeura and Bullock, 1983) that may continue however if the soil is deficient in moisture or nutrients and may also be due to biotic damages.

Very little information is available to understand the physiology of fruit drop in guava. The natural occurrence of cytokinins was examined in seeded and seedless fruit of guava. Five different cytokinins were isolated and three of them were tentatively identified as zeatin, zeatin riboside and zeatin nucleotide (Nagar and Raja Rao, 1981). Plant growth regulators are applied to control/minimize premature fruit drop and to increase fruit set (Curry and Greene, 1993; Sugiyama and Yamaki, 1995). Spraying of gibberellins (GA_3) at 10–50 mg l^{-1} before flowering or at the beginning of fruit set was reported to minimize fruit drop and increase fruit set in guava (Lal *et al.*, 2013; AbdurRab *et al.*, 2017; Purohit *et al.*, 2019). Reduction in fruit drop due to GA_3 treatment might be due to an increase in the initial growth of ovaries and reduce the peak of abscission (Agusti, 2000).

11.6 Nutrient Imbalances and Deficiency Disorders

Nutrient deficiency/imbalance affects growth, yield and fruit quality and, in certain cases, develops typical symptoms/disorders in fruit. The fruit size is reduced under nitrogen deficiency while the fruit quality is affected when potassium is deficient. Deficiency of zinc causes fruit cracking which is also observed under boron deficiency (Yahia *et al.*, 2019). The nutrient deficiency disorders and control measures have been dealt with in detail in Chapter 7 of this volume.

11.7 Sunscald

Sunscald, a disorder related to heat injury, occurs when the fruits are exposed to high light intensities of direct sunlight or high temperatures at the field or during storage or transport, either before or after harvest. The temperature of a fruit exposed to the sun is usually 5 to 10°C higher than that of the surrounding air (Woolf *et al.*, 2000). These high-temperature exposures increase the metabolic activity and cause a rapid loss of moisture which leads to shorter shelf-life of the fruit (Yahia *et al.*, 2019).

Sunscald in guava is a non-pathogenic disorder and symptoms normally appear on the sun-exposed side of affected fruit. The symptoms of sunscald appear as bleaching or yellowing patches with irregular, coarse and leathery appearance, a little sunken below the fruit surface and necrotic spots where flesh underneath the peel is brown due to thermal death of epidermal and sub-epidermal cells. As fruit suffer thermal death, this situation may subsequently lead to exposure to additional stress, including insect or pathogen attack (Lurie, 2009).

Sunscald is more prevalent in young orchards in which the canopy is thin and the fruits are more exposed. The key method for reducing sunscald is protecting the developed fruit from direct sun. Use of shade nets or spraying of white kaolin to cover the plant as well as fruits can decrease exposure to strong sunlight and protect them from overexposure (Gindaba and Wand, 2005). Pruning can also help to ensure that the leaves shade the fruit since sunscald is induced by ultraviolet-B radiation (Schrader *et al.*, 2003). Bagging the fruit with paper bags is a good control measure to protect the fruit from direct sunlight and insect infestation while on the tree (Lurie, 2009) (Fig. 11.4).

Fig. 11.4. Guava tree where the fruits are covered using recycled old newspaper to protect them from sunscald and pest and disease infestation. Photograph courtesy of Nor Elliza Tajidin.

11.8 Other Disorders

Physiological disorders related to herbicide or pesticide phytotoxicity have also been reported (Lim and Koo, 1990); for example, the use of copper fungicide may result in the formation of corky russets on the fruit epidermis. Injury due to improper spraying of paraquat results in irregular, reddish brown blotches on the fruit cuticle, which later develop into bigger patches. These affected fruits, being inferior in quality, will have a poor market value (Lim and Koo, 1990).

11.9 Conclusion

Guava suffers from a range of disorders that reduce returns to commercial growers. The disorders that may manifest on the tree are caused by nutrient deficiency/imbalance disorders that could lead to bronzing. Deficiencies of zinc and boron cause fruit cracking while on trees growing in acid soil where potassium uptake is low and some other nutrients are deficient, bronzing may appear. The fruit disorders include sunscald, flesh browning and chilling injury (under low-temperature storage). Fruit drop is another physiological disorder that causes huge loss to guava growers. These disorders are more common in certain cultivars and growing areas than others, while their incidence and severity also vary from season to season. The physiological mechanisms behind these disorders are not completely understood, although they often appear to be related to the weather or tree agronomy. It would be useful for researchers to generate technology to control these disorders through biotechnological interventions, breeding strategies and by understanding the physiological basis to reduce losses.

References

AbdurRab, S., Shah, S.H.A., Uttah, I., Bibi, F. and Zeb, I. (2017) Influence of de-blossoming and GA_3 application on fruit drop and growth of winter guava. *Sarhad Journal of Agriculture* 33(4), 526–531.

Agusti, M. (2000) Regulation of citrus cropping and improvement of fruit quality using exogenous plant growth regulators. In: *Proceedings of the 9th International Citrus Congress, Orlando, Florida, 3–7 December 2000*, pp. 351–356.

Alba-Jiménez, J.E., Benito-Bautista, P., Nava, G.M., Rivera-Pastrana, D.M., Vázquez-Barrios, M.E. and Mercado-Silva, E.M. (2018) Chilling injury is associated with changes in microsomal membrane lipids in guava fruit (*Psidium guajava* L.) and the use of controlled atmospheres reduce these effects. *Scientia Horticulturae* 240, 94–101.

Ali, Z.M. and Lazan, H. (1997) Guava. In: Mitra, S. (ed.) *Postharvest Physiology and Storage of Tropical and Subtropical Fruits*. CAB International, Wallingford, UK, pp. 145–165.

Augustin, M.A. and Osman, A. (1988) Postharvest storage of guava (*Psidium guajava* L. var. Taiwan). *Pertanika* 11, 45–50.

Augustin, M.A., Ghazali, H.M. and Hashim, H. (1985) Polyphenoloxidase from guava (*Psidium guajava* L.). *Journal of the Science of Food and Agriculture* 36(12), 1259–1265.

Brown, B.I. (1981) Temperature management and chilling injury of tropical and subtropical fruit. *Acta Horticulturae* 175, 339–342.

Cano-Medrano, R. and Darnell, R. (1998) Effect of GA_3 and pollination on fruit set and development in rabbiteye blueberry. *HortScience* 33(4), 632–635.

Curry, E.A. and Greene, D.W. (1993) CPPU influences fruit quality, fruit set, return bloom, and preharvest drop of apple. *HortScience* 28, 115–119.

Dahifale, H.N., Durgude, A.G. and Deshpande, A.N. (2009) Nutrient diagnosis for bronzing of leaves in guava orchards. *Journal of Maharashtra Agricultural Universities* 34(1), 6–10.

Dalal, V.B., Eipeson, W.E. and Singh, N.S. (1971) Wax emulsion for fresh fruits and vegetables to extend their storage life. *Indian Food Packer* 25(5), 9–15.

Germano, T.A., Aguiar, R.P., Bastos, M.S.R., Moreira, R.A., Ayala-Zavala, J.F. and Miranda, M.R.A.d. (2019) Galactomannan–carnauba wax coating improves the antioxidant status and reduces chilling injury of 'Paluma' guava. *Postharvest Biology and Technology* 149, 9–17.

Gindaba, J. and Wand, S.J.E. (2005) Comparative effects of evaporative cooling, kaolin particle film, and shade net on sunburn and fruit quality in apples. *HortScience* 40(3), 592–596.

González-Aguilar, G., Tiznado-Hernández, M., Zavaleta-Gatica, R. and Martínez-Téllez, M. (2004) Methyl jasmonate treatments reduce chilling injury and activate the defense response of guava fruits. *Biochemical and Biophysical Research Communications* 313(3), 694–701.

Gopi Kumar, K., Sulladmath, U.V. and Badiger, M.K. (1985) Nature, causes and remedies for bronzing in guava (*Psidium guajava* L.). *Singapore Journal of Primary Industries* 13(2), 112–127.

Harker, F.R., Redgwell, R.J., Hallett, I.C., Murray, S.H. and Carter, G. (2010) Texture of fresh fruit. In: Janick, J. (ed.) *Horticultural Reviews*. Wiley, Oxford, pp. 121–224.

Hussein, Z., Fawole, O.A. and Opara, U.L. (2020) Harvest and postharvest factors affecting bruise damage of fresh fruits. *Horticultural Plant Journal* 6(1), 1–13.

Kader, A.A. (1999) Guava: Recommendations for maintaining postharvest quality. Available at: http://postharvest.ucdavis.edu/Commodity_Resources/Fact_Sheets/Datastores/Fruit_English/?uid=26&ds=798 (accessed 15 March 2020).

Kumar, N., Suresh, J. and Anbu, S. (2007) Problems and prospects of guava cultivation in Tamil Nadu conditions. *Acta Horticulturae* 735, 329–334.

Kumar, R. and Kumar, V. (2016) Physiological disorders in perennial woody tropical and subtropical fruit crops – a review. *Indian Journal of Agricultural Sciences* 86(6), 703–717.

Kumar, V., Purbey, S.K. and Anal, A.K.D. (2016) Losses in litchi at various stages of supply chain and changes in fruit quality parameters. *Crop Protection* 79, 97–104.

Kundu, S. and Mitra, S.K. (1994) Studies on floral biology of different guava cultivars. *Crop Research* 8(1), 80–85.

Lal, N., Das, R.P. and Verma, L.R. (2013) Effect of plant growth regulators on flowering and fruit growth of guava (*Psidium guajava* L.) cv. Allahabad Safeda. *The Asian Journal of Horticulture* 8(1), 54–56.

Lim, T.K. and Khoo, K.C. (1990) *Guava in Malaysia: Production, Pests and Diseases*. Tropical Press, Kuala Lumpur.

Lurie, S. (2009) Stress physiology and latent damage. In: Florkowski, W.J., Shewfelt, R.L., Brueckner, B. and Prussia, S.E. (eds) *Postharvest Handling: A Systems Approach*, 2nd edn. Academic Press, San Diego, California, pp. 443–459.

Mitra, S.K. and Bose, T.K. (1990) Guava. In: Bose, T.K. and Mitra, S.K. (eds) *Fruits: Tropical and Subtropical*. Naya Prokash, Kolkata, India, pp. 280–303.

Mitsuhashi-Gonzalez, K., Pitts, M.J., Fellman, J.K., Curry, E.A. and Clary, C.D. (2010) Bruising profile of fresh apples associated with tissue type and structure. *Applied Engineering in Agriculture* 26(3), 509–517.

Montero, C.R.S., Schwarz, L.L., Santos, L.C.d., Andreazza, C.S., Kechinski, C.P. and Bender, R.J. (2009) Postharvest mechanical damage affects fruit quality of 'Montenegrina' and 'Rainha' tangerines. *Pesquisa Agropecuária Brasileira* 44(12), 1636–1640.

Monzan, M.E., Biasi, B., Wang, S., Tang, J., Hallman, G.J. and Mitcham, E.J. (2005) Persimmon and guava fruit response to radio frequency heating as an alternative quarantine treatment. *Acta Horticulturae* 687, 349–350.

Nagar, P.K. and Raja Rao, T. (1981) Studies on endogenous cytokinins in guava (*Psidium guajava* L.). *Annals of Botany* 48, 845–852.

Nakasone, H.Y. and Paull, R.E. (eds) (1998) Guava. In: *Tropical Fruits*. CAB International, Wallingford, UK, pp. 149–172.

Narayana, C.K. and Singh, B.P. (1997) Chilling injury and enzymatic changes during low temperature storage of guava. *Phala Samskarna* 95, 79–81.

Purohit, H.P., Bhutani, A.M., Chitroda, R.L. and Parmer, P. (2019) Response of pre-harvest spray of calcium nitrate and gibberellic acid on fruiting characters of guava cv.L-49. *Journal of Pharmacognosy and Phytochemistry* 8(4), 607–609.

Racskó, J., Leite, G.B., Petri, J.L., Zhongfu, S., Wang, Y. *et al.* (2007) Fruit drop: the role of inner agents and environmental factors in the drop of flowers and fruits. *International Journal of Horticultural Science* 13(3), 13–23.

Ramanjini Gowda, P.H., Sulladmath, U.V. and Nache Gowda, V. (1992) Causes and remedies of bronzing in guava (*Psidium guajava* L.). *Acta Horticulturae* 321, 898–902.

Ruiz-Altisent, M. and Moreda, G.P. (2011) Fruits, mechanical properties and bruise susceptibility. In: Gliński, J., Horabik, J. and Lipiec, J. (eds) *Encyclopedia of Agrophysics*. Encyclopedia of Earth Sciences Series. Springer, Dordrecht, The Netherlands, pp. 318–321.

Sandhu, S. and Gill, B.S. (2013) *Physiological Disorders of Fruit Crops*. New India Publishing Agency, New Delhi, p. 189.

Schoorl, D. and Holt, J.E. (1983) Mechanical damage in agricultural products: a basis for management. *Agricultural Systems* 11(3), 143–157.

Schrader, L., Zhang, J. and Sun, J. (2003) Environmental stresses that cause sunburn of apple. *Acta Horticulturae* 618, 397–405.

Sharma, R.R. (2005) *Physiological Disorders of Tropical & Subtropical Fruits – Causes and Control, Problems and Solution*. Indian Agricultural Research Institute, New Delhi, pp. 310–312.

Shigeura, G.T. and Bullock, R.M. (1983) *Guava (*Psidium guajava *L.) in Hawaii – History and Production*. Hawaii Institute of Tropical Agriculture and Human Resources, Research Station Series No. 035. University of Hawaii, Honolulu, Hawaii.

Singh, K.K. and Mathur, P.B. (1954) Cold storage of guavas. *Indian Journal of Horticulture* 11, 1–5.

Singh, N., Sharma, D.D., Singh, G., Thakur, K.K. and Kumari, S. (2019) Physiological disorders and their management in apple and pear fruits production. In: Singh, A. (ed.) *Advanced Botany*. AkiNik Publications, New Delhi, pp. 41–66.

Singh, S.P. (2011) Guava (*Psidium guajava* L.). In: Yahia, E.M. (ed.) *Postharvest Biology and Technology of Tropical and Subtropical Fruits – Cocona to Mango*. Woodhead Publishing, Cambridge, pp. 213–246.

Singh, S.P. and Choudhary, M.R. (2012) *Production Technology of Fruit Crops in Wasteland*. Scientific Publishers, Jodhpur, India.

Singh, S.P. and Pal, R.K. (2008) Controlled atmosphere storage of guava (*Psidium guajava* L.) fruit. *Postharvest Biology and Technology* 47(3), 296–306.

Soltesz, M. (ed.) (1997) Termeskotodes es-ritktias. In: *Integralt gyumolesztes*. Mezogazda Kiado, Budapest, pp. 309–331.

Sugiyama, N. and Yamaki, Y.T. (1995) Effect of CPPU on fruit set and fruit growth in Japanese persimmon. *Scientia Horticulturae* 80, 337–343.

SzuWei, Y. and ChunTa, W. (2013) Effects of methyl jasmonate fumigation on chilling injury of harvested 'Jen-Ju Bar' guava fruit. *Journal of the Taiwan Society for Horticultural Science* 59(1), 15–28.

Tabacchi, M.H., Hicks, J.R., Ludford, P.M. and Robinson, R.W. (1979) Chilling injury tolerance and fatty acid composition in tomatoes. *HortScience* 14, 424.

Tiwari, S., Tandon, D.K. and Esguerra, E.B. (2006) Chilling injury as an indicator of critical temperature for cold storage of guava (*Psidium guajava* L.) cv. Allahabad Safeda. *International Journal of Postharvest Technology and Innovation* 1(2), 170.

Van Hasselt, P.R. (1990) Light-induced damage during chilling. In: Wang, C.Y. (ed.) *Chilling Injury of Horticultural Crops*. CRC Press, Boca Raton, Florida, pp. 113–128.

Vazquez-Ochoa, R.I. and Colinas-Leon, M.T. (1990) Changes of guavas of three maturity stages in response to temperature and relative humidity. *HortScience* 25, 86–87.

Vishwasrao, C. and Ananthanarayan, L. (2016) Postharvest shelf-life extension of pink guava (*Psidium guajava* L.) using HPMC-based edible surface coatings. *Journal of Food Science and Technology* 53(4), 1966–1974.

Wang, C.Y. (1989) Chilling injury of fruits and vegetables. *Food Review International* 5, 209–236.

Wang, C.Y. (2010) Alleviation of chilling injury in tropical and subtropical fruits. *Acta Horticulturae* 864, 267–273.

Wang, Z., Duan, H. and Hu, C. (2009) Modelling the respiration rate of guava (*Psidium guajava* L.) using enzyme kinetics, chemical kinetics and artificial neural network. *European Food Research and Technology* 229(3), 495–503.

Wills, R.B.H., Lee, T.H., Graham, D., McGlasson, W.B. and Hall, E.G. (1981) *Postharvest: An Introduction to the Physiology and Handling of Fruits and Vegetables*. NSW University Press Ltd, Kensington, Australia, pp. 153–169.

Wills, R.B.H., Mulholland, E.E., Brown, B.I. and Scott, K.J. (1983) Storage of two new cultivars of guava for processing. *Tropical Agriculture* 60, 175–178.

Woolf, A.B., Wexler, A., Prusky, D., Kobiler, E. and Lurie, S. (2000) Direct sunlight influences postharvest temperature responses and ripening of five avocado cultivars. *Journal of the American Society for Horticultural Science* 125(3), 370–376.

Yahia, E.M., Carrillo-López, A. and Sañudo, A. (2019) Physiological disorders and their control. In: Yahia, E.M. (ed.) *Postharvest Technology of Perishable Horticultural Commodities*. Woodhead Publishing, Cambridge, pp. 499–527.

Zhang, Z., Pang, X., Xuewu, D., Ji, Z. and Jiang, Y. (2005) Role of peroxidase in anthocyanin degradation in litchi fruit pericarp. *Food Chemistry* 90(1), 47–52.

12 Photosynthesis and Productivity

Vinod K. Singh* and Manoj K. Soni
ICAR–Central Institute for Subtropical Horticulture, Lucknow, Uttar Pradesh, India

12.1 Introduction

Guava exceeds most of the tropical and subtropical fruit trees in terms of adaptability, productivity and tolerance to cold temperatures bordering freezing and light frosts. Guava is successfully cultivated in a wide range of environmental and edaphic conditions because of its tolerance to drought and salinity as compared with most of the warm-climate fruit plants (Samson, 1986).

Guava plants tolerate a wide range of frost-free climatic conditions, although mature trees survive light frosts (Rathore, 1976; Martin *et al.*, 1987; Jaiswal and Amin, 1992). Trees thrive well in humid and dry climates. The optimum conditions for guava cultivation and high yield of good-quality fruit include temperature between 20 and 30°C and well-distributed rainfall; however, fruit quality is poor in areas of high rainfall and high relative humidity (Samson, 1986). The best-quality guava fruits are harvested when they mature during the dry period (November to March). Thus, it is always beneficial to give trees a rest (an off season) by withholding irrigation water periodically. Guava bears more heavily in areas with a distinct winter season than in the deep tropics (Morton, 1987).

The optimum leaf number and area required for the development of individual fruit may be determined in guava, as standardized in other fruit crops (Cohen, 1975; Chittiraichelvan *et al.*, 1985). Such studies will yield valuable information with regard to the extent to which fruit thinning is required to obtain optimum fruit size. Since fruit development utilizes carbohydrates, either currently produced or stored as reserves, it would be useful to understand the leaf area–fruit growth relationship in guava where flowers are borne on the current year's shoots which may be with limited carbohydrates.

Weather aberration affects the production of guava significantly but lack of understanding about the role of environmental factors on quality of guava is underscored. The low productivity of guava is also due to no availability of genuine planting materials, low adoption of technologies and significant prevalence of old and unproductive orchards with declining yield pattern. Poor yield and low quality of fruits are largely due to poor photosynthetic efficiency and poor distribution of light, coupled with other compounding factors like plant architecture, canopy density, selection of varieties and adoption of plant protection technologies.

*E-mail: singhvk_cish@rediffmail.com

DOI: 10.1079/9781789247022.0012

Guava trees have a higher proportion of 'shade' to 'sun' leaves and their leaves are photosynthetically inactive under deeper shade and act as an unproductive sink. Light interception by several fruit tree canopies has been studied (Jackson, 1980; Porpiglia and Barden, 1981; Kappel and Flore, 1983; Marini and Marini, 1983). However, the level of irradiance required in guava canopies for quality yield has received little attention.

Increased interest in photosynthesis research in fruit crops has been noticed recently. This has resulted from an attempt to increase production efficiency through a better physiological understanding of growth and productivity. However, it is not clear that photosynthetic rate is directly limiting yield or that an increase in photosynthetic rate will result in increased yield of guava. This is because in fruit trees like guava, the carbon must not only be produced but also partitioned efficiently to fruit and flower buds for the next crop that may be in the same year, since guava is multiple bearing in habit and flowers come on the current year's shoots. Varietal characterization for possession of photosynthetically efficient higher ribulose bisphosphate activity is lacking in guava. In this chapter, effort is made to highlight the current understanding and the possible effects of environmental factors on the whole physiology of guava trees as expressed by growth, yield, fruit quality and photosynthetic features and to discuss possible areas of future research.

12.2 Physiology

Common guava (*Psidium guajava* L.) is a small tree with a slender trunk. In its leaves, veins are prominent below and lateral veins are not prominent, forming an intra-marginal vein from the edge of the leaves. Young twigs are four-angled, ridged and pubescent. The guava tree produces an abundance of suckers that should be removed and a framework of preferably four strong branches should be established. The crotch angle between the branches and the main stem should be wide enough to facilitate adequate penetration of light and provide physical strength to support fruit load at maturity.

Guava leaves exhibit the classic characteristics, having the C_3 biosynthetic pathway of photosynthesis. Maximum reported rates of leaf carbon dioxide (CO_2) assimilation are around 5.0–15 µmol m^{-2} s^{-1} (Singh and Singh, 2007; Singh *et al.*, 2017). Leaf carboxylation efficiency appears to saturate well beyond intercellular CO_2 concentration of 450 ppm and estimated intercellular CO_2 concentrations have been reported to be between 200 and 300 ppm. Individual leaf CO_2 assimilation appears to approach light saturation at 700–1000 µmol m^{-2} s^{-1}, depending on the leaf's position in the canopy. Leaf stomatal conductance has been reported to be closely coupled to leaf photosynthesis and its value was recorded in the range of 0.059 to 0.614 mol m^{-2} s^{-1} depending upon whether the leaf is from a flowering or non-flowering branch. In guava leaf photosynthesis appears to be fairly tolerant to water stress in the field and leaves can maintain photosynthetic capacity significantly. There is no quantitative involvement of photoperiod for flowering in guava as it is a day-neutral species. However, knowledge on responses of the crop to environmental factors such as light, temperature, water availability and CO_2 concentration is essential to determine the effect of suboptimal environmental conditions and to manage the crop for maximum productivity (Schaffer and Andersen, 1994).

The fruits emanate a strong, sweet, musky odour when ripe and they are generally round, ovoid or pear-shaped, with four or five protruding floral remnants (sepals) at the top. The fruit is a berry consisting of a fleshy pericarp and seed cavity with fleshy pulp and numerous small seeds. The fruit is commonly consumed fresh. The growth of guava fruit exhibits a double sigmoid curve with climacteric habit in ripening. Guava cultivars can be classified as red-fleshed (red pulp) or white-fleshed (white pulp). Red-fleshed guavas have more potential value as antioxidant sources than white-fleshed ones (Hassimotto *et al.*, 2005). Lycopene, an antioxidant, was found in red-fleshed guava cultivars 'Arka Kiran', 'Lalit', 'Arka Rashmi' and 'Punjab Pink' (Ravi *et al.*, 2018). Shape and flesh colour cannot be differentiated from the appearance of red and white guava fruits, because in some guava varieties like 'Lalima' the fruits are red in appearance but the flesh is white. Fruit colour in guava is mainly attributed to anthocyanin pigments which are genetically regulated. However, there is no clear study reported on the anthocyanin biosynthesis mechanism in guava, probably due to some genes involved in its biosynthesis having pleiotropy and being affected by many internal and external factors.

12.2.1 Leaf growth and photosynthesis

The leaf growth rate in guava (*P. guajava* L.) follows a single sigmoid curve which describes three distinct development stages: stage I, which includes the first 15 days; stage II during days 16–40; and stage III, from days 41 to 70, for the leaves reach physiological maturity (Nava *et al.*, 2014). The importance of understanding these stages in guava is that when the leaf is in stages I and II (young leaves still expanding) it has low photosynthetic capacity, but this increases progressively with age until the leaf has grown from 70 to 100% of its maximum size, which is also when it reaches its maximum net photosynthetic rate (Lakso and Flore, 2003). Thereafter leaf photosynthesis gradually decreases to become senescent when the leaf is unable to photosynthesize due to chlorophyll degradation and loss of functional chloroplast. Normally in fruit crops photosynthetic rate increases with leaf age, peaks just before full leaf expansion and then remains steady for a period of time before declining. However, the rate of decline is inhibited by the presence of fruit (Downton *et al.*, 1987a). Kennedy and Johnson (1981) reported that the largest changes in photosynthetic rate with increased carboxylation efficiency took place during the earliest stage of leaf development, coinciding with the period of greatest leaf expansion and chlorophyll synthesis.

Although it is widely accepted that leaf age strongly affects the photosynthetic capacity of individual leaves of most species (Flore and Lakso, 1989), there are very few data available specifically for guava. Tree canopies of guava grow quite rapidly, and the leaf CO_2 assimilation capacity is strongly influenced by internal canopy shading as reported in other fruit species (Marini and Marini, 1983). Therefore it is difficult to separate leaf age effects from changes to leaf light exposure in naturally growing field canopies. It is apparent that leaves in the most exposed part of the tree canopy tend to maintain relatively constant photosynthetic capacity in the absence of major stress during different phenological stages in different cropping seasons. Further, carbohydrate metabolism change is an important event in leaf development; while young (heterotrophic) leaves depend in part on the carbohydrate imported from other areas of the plant, mature (autotrophic) leaves produce excess photosynthates and act as the plant's principal source of translocated sugar (Turgeon, 1989). On the other hand, photosynthates produced by leaves are either used immediately by the nearest sink – the flowers and fruits (Finazzo *et al.*, 1994) – or in other situations are stored throughout the tree if the sink is limited. Hence, photosynthates production plays an important role in the plant productivity. However, it varies from location to location and from genotype to genotype. Although photosynthesis value is a concern in fruit species in general and guava in particular, the value of photosynthesis differs greatly (Singh and Rajan, 2009), with diverse measuring methods and equipment being used depending on the prevailing environmental conditions. The condition of the plant and the prior environmental history usually have a very important influence on the photosynthetic rate. The time of season or the relative stage of development of various plant organs affects the photosynthesis in a variety of ways. The vegetative vigour and cropping status are also important. Many of these factors are discussed later in this chapter.

12.2.2 Photosynthetic aptitude of cultivars

Guava prefers full sun and can survive in a dry summer, but regular irrigation is required for better performance. In adaptability, it exceeds the majority of tropical and subtropical fruit trees (Yadava, 1996). It is fairly cold hardy and can survive as low as 5°C for a short period of time at night. Guava does well in different types of soil and tolerates a pH range from 4.0 to 9.4. On the basis of growth and physiological characterizations such as photosynthetic rate, carboxylation efficiency, transpiration rate and water-use efficiency in guava cultivars 'Allahabad Safeda', 'L-49', 'Lalit' and 'Shweta' under hot arid conditions, cultivar 'Shweta' showed better adaptability followed by 'Lalit' (Singh *et al.*, 2017). Among the two cultivars tested for photosynthetic rate and associated gas exchange attributes at different light regimes, the 'Sardar' guava was found more efficient compared with 'Allahabad Safeda' (Singh and Singh, 2007). However, all the common varieties of guava are able to adjust their photosynthetic capacity across a wide range of light environments and this may be a consequence of a large physiological plasticity (Pattison *et al.*, 1998).

In order for trees to function efficiently, they must balance gaseous exchange between the inside and outside of the leaf to maximize CO_2 uptake for photosynthetic carbon assimilation and to minimize water loss through transpiration (Nilson and Assmann, 2007). In this context the co-relationship between stomatal density and gas exchange traits in guava was studied (Shiva *et al.*, 2017), which revealed that leaf stomatal density was positively correlated with stomatal conductance to net photosynthesis rate and that leaf net assimilation and number of stomata were negatively correlated with transpiration. Thus, these differences among genotypes can be taken into consideration as selection criteria for guava to be grown in regions with varying environmental conditions. However, cultivars that proceed with balanced reproductive, vegetative and rest phases are more likely to have sufficient carbon resources to meet periods of critical demand and therefore will sustain high yields. Menzel and Paxton (1986) studied the pattern of growth, flowering and fruiting of seven guava cultivars in the subtropics. They did not find any obvious growth pattern related to high production; however, some higher yielder varieties were identified for cultivation under suitable environmental conditions.

Guava plants experience a very wide range of temperature stress, especially under north Indian conditions where temperature often exceeds 40°C. Furthermore, higher temperature disrupts photosynthetic pigments and reduces the gas exchange, leading to a reduction in plant growth and productivity (Wahid *et al.*, 2007). High temperatures also influence photosynthetic capacity and stomatal conductance by decreasing the activation state of Rubisco (ribulose-1,5-bisphosphate carboxylase-oxygenase). Photosynthesis, being the energy-producing process, plays a vital role in energy signalling and cellular energy balance and could serve as a key environmental stress sensor in plant systems. Changes in several photosynthetic characteristics under high temperatures are excellent indicators of plant tolerance to heat stress (Wahid *et al.*, 2007). Changes in Fv/Fm ratio under heat stress conditions was also found a good indicator in screening heat-resistant cultivars of fruit crops (Kadir *et al.*, 2007). On this background, 22 cultivars of guava were evaluated for their photosynthetic rate, stomatal conductance, transpiration rate, internal CO_2, water-use efficiency and light-harvesting pigments (Shiva *et al.*, 2017). Wide variations in these parameters were obtained which can be utilized to identify the suitable growth conditions and plant adoption strategies of guava at different environmental conditions for maximizing gas exchange attributes and yield. Net CO_2 assimilation rate is reduced by partial decrease in both stomatal conductance and instantaneous carboxylation efficiency at temperatures above or below the optimum range (28–32°C) as observed in citrus species (Machado *et al.*, 2005). Hence, knowledge about temperature levels is useful in physiological research as well as horticultural crop production. However, optimum temperature levels for the photosynthesis in guava have not been exactly worked out.

Guava is sensitive for photosynthesis, being a C_3 plant in which CO_2 fixation takes place at one site and the first product is C_3, phosphoglyceric acid. The fluctuation in photosynthetic rate with higher values in the morning than the afternoon has been noted under full sunlight conditions (Downton *et al.*, 1987a; Yadav and Singh, 1995). Generally, assimilate supply is dependent on photosynthesis. The partitioning of carbohydrate determines the amount and patterns of plant growth and yield (Lakso and Flore, 2003). In the absence of optimum environmental conditions for a long period during the important phenological phase, the rate of net photosynthesis and stomatal conductance are affected badly and finally a decline in net CO_2 assimilation reduces quality and yield.

Fruit photosynthesis

It is suggested that photosynthesis in the flower, specifically in the green persistent sepals (calyx), might play an important role in initial fruit set and fruit growth. However, no information is available for photosynthesis in guava flowers. A virtually negligible rate of photosynthesis in fruit was reported (Chacko *et al.* 1982; Pavel and Dejong, 1993) and has been estimated to account for 5 to 9% of the total carbohydrates budget of developing fruits, depending on the position of fruit in the canopy (Pavel and Dejong, 1993). Photosynthesis rate and its contribution to carbohydrate requirements depend on the duration of exposure to sunlight and the growth stage of fruits. Photosynthesis of mature fruits contributed <5.0% of the total fruit carbohydrate requirements compared with mid-season fruits (>8.0%) depending on fruit exposure to light, as shaded fruits (less than 10% of sunlight) supplied significantly less of the weekly carbohydrate requirements through photosynthesis than exposed fruits. A decline in photosynthetic contribution in the late stage fruit was related to decreasing photosynthetic activities in association with declining chlorophyll contents.

12.2.3 Biochemical regulation of photosynthesis and fruit yield

Carbohydrate balance and seasonal trends

Crop productivity is dependent on photosynthetic efficiency and the efficiency with which photosynthates are partitioned. Tree growth and productivity are strongly dependent on the efficiency of carbon assimilation and on the efficiency and effectiveness of the distribution and use of carbohydrates for the maintenance of tree growth and the production of quality fruits. Research specific to the leaf photosynthetic processes and carbohydrate partitioning/distribution in guava is not as plentiful as for many other crops.

Carbohydrate translocated from leaves or reserve organs is most important for the growth and development of sink organs (mainly fruits). Increasing the leaf/fruit ratio generally increases fruit growth and carbohydrate content. Photosynthesis increases with fruit load and the leaves next to fruits are a strong source of carbohydrate (Ficher *et al.*, 2012). Its availability was observed to have a positive effect on reproductive growth of strawberry guava (*Psidium cattleyanum*) during the inductive period and growing fruits seemed to prevent vegetative development (Normand and Michels, 2007). It was also noted that carbohydrate has a crucial role in flowering and fruit growth in guava. However, in most fruit trees, the carbohydrate availability and carboxylation decrease continuously from flowering to harvest and then increase after harvest (Scholefield *et al.*, 1985; Singh, 2016). Generally, assimilate supply to fruits is dependent on photosynthesis and the partitioning of carbohydrate determines the amount and patterns of plant growth and yield (Lakso and Flore, 2003). Further, as carbohydrate is removed with the fruit during harvest and leaves are the organs of high carbon intake by the tree, after harvest, all practices that favour carbon uptake such as light and nutrition should be optimized (Fig. 12.1). Fixing the crop load is an important operation for a better balance between carbohydrate sources and sinks for promoting uniform fruiting during the next production cycle.

Guava is different from other subtropical trees in having many growing points that may be initiated within a short period of time and the early growth of shoots (meristem) does not depend on the current season's photosynthesis, which results in very rapid canopy development; this may be one of

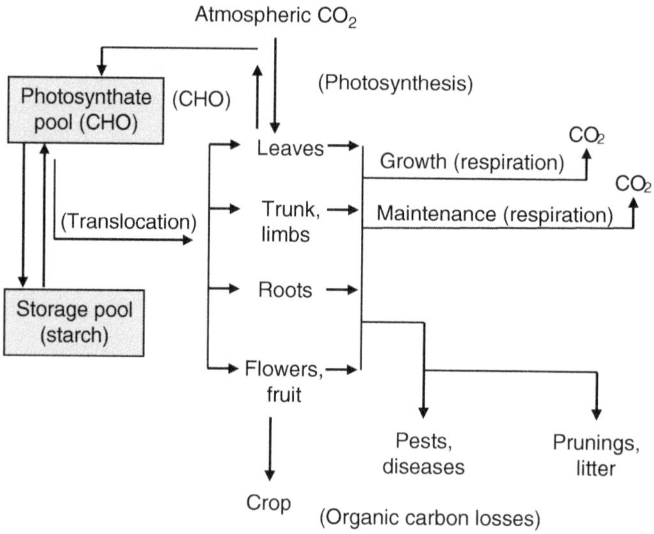

Fig. 12.1. Diagram of the carbon economy in fruit trees. CHO, carbohydrate. After DeJong and Ryugo (1998).

the reasons for guava being more responsive to pruning (Singh, G. *et al.*, 2007). Cropping pattern, type of shoots, light exposure as well as seasonal variation influence the pattern of photosynthetic rate in guava (Singh, 2016; Palit *et al.*, 2017). Photosynthetic rates and stomatal conductance were observed to be substantially lower in leaves proximal to inflorescences than in leaves on vegetative shoots (Singh, 2016). However, their values in leaves closer to fruitlets were intermediate between those in leaves of vegetative terminals and leaves close to growing fruit. Reduction in these parameters could also be due to differences in the leaf water status of shoots, which creates water stress for inducing flowering (Shigeura and Bullock, 1976). Such results were also reported in sweet cherry (Rom and Ferree, 1986). However, early work by Crew *et al.* (1975) in peach noted the higher rates of photosynthesis in leaves in close proximity to fruits compared with leaves measured further down the same branch and that leaf photosynthesis peaked during the period of peak fruit growth (i.e. maximum fruit sink strength). Our understanding of carbohydrate metabolism during guava fruit growth is not clear and there is confusion about the preferential primary source of carbohydrate transported to the fruit.

12.3 Hormonal Regulation

12.3.1 Plant growth

Phytohormones are produced naturally in higher plants, controlling growth and other physiological functions. Apart from this, they also regulate expression of the intrinsic genetic potential of plants at both transcriptional and translational levels. In guava, no systematic work on endogenous levels of different types of hormones at critical phenological periods was found. Researchers have concentrated only on the effect of different synthetic phytohormones on different attributes in guava. Plant hormones like gibberellic acid (especially GA_3) and naphthalene acetic acid (NAA) were reported to be very effective in increasing the vegetative growth, flowering and yield of guava (Gollagi *et al.*, 2019). Some of the hormones were also found to improve the fruit quality with reduced number of seeds and seed weight and to increase the pulp/seed ratio (Brahmachari *et al.*, 1995). Regulation of flowering and fruiting was also reported by use of a growth retardant in cultivar 'Sardar' guava (Brahmachari *et al.*, 1995).

12.3.2 Fruit characteristics

The common guava is a diploid tree fruit species with $2n$ equal to 22. However, natural and artificial triploids, including some aneuploid cultivars or selections, also are known to exist. Triploids generally produce seedless fruit (Jaiswal and Amin, 1992), but economically they are mostly shy bearers (Menzel, 1985) and thus their production needs to be improved. Nagar and Rao (1983) reported that the fruits of seeded 'Allahabad Safeda' grow much larger than those of the seedless 'Allahabad Seedless'. These scientists attributed such differences in seedlessness to the fluctuating endogenous auxin contents during guava fruit growth. Later on, it was concluded that auxin is most effective in inducing parthenocarpy in multi-seeded fruits, while gibberellins seem to be most effective when there are relatively few ovules (Leopold and Nooden, 1984). In guava, however, the gibberellins were reported to be effective to induce parthenocarpic fruit (Teaotia *et al.*, 1961). The absence of seeds is usually appreciated by consumers and producers because it increases fruit quality and shelf-life (Pandolfini, 2009).

12.4 Physiological Growth of Fruit

In a mild tropical or cool subtropical climate, guava can flower and fruit continuously throughout the year if water and temperature do not become limiting factors (Rathore, 1976). The growth curve for guava fruit is double sigmoid (Rathore, 1976; Nava *et al.*, 2014), presenting three distinct periods: (i) stage I, a period of rapid fruit growth starting a few days after anthesis and continuing through the next 45 to 60 days, depending on the fruiting season of the year in areas with multiple cropping seasons. At this stage, there is a higher division and cell elongation, processes that are responsible for the increased weight and volume of the fruit; (ii) stage II, a period of relatively slow fruit growth lasting only for 30 to 60 days. During this time, seeds attain full maturity and become very hard; and (iii) stage III, a period of exponential increase in fruit size lasting another 30 to 60 days following slow growth and ending at the time of fruit maturity. Fruit length and diameter increase markedly during this period, leading the fruit to 'eating ripeness'. Moreover, the duration of all three growth periods of guava fruit appeared to be inversely proportional to the prevailing temperature. Depending on guava cultivar or type and the growing conditions, it takes 100 to 150 days from bloom to fruit harvest (Rathore, 1976; Samson, 1986). During the fruit growth period, the major activities appear to be synthesis of cell-wall material and sugars, while pectin increases rapidly during the ripening period (Sastry, 1965). Thus, the sigmoid and double sigmoid growth in leaves and fruit, respectively, have three steps: for leaves, stage II was the fastest whereas for fruit, stages I and III were faster compared with stage II. The periods between one phase and another may vary with cultivar and location.

On the final phase of maturation, once the fruit reaches its maximum size and it is ready to be harvested, the fruit ripening takes place. During the ripening, physiological, biochemical and structural changes occur which involve the degradation of starch or other storage polysaccharides, the production of sugars, the synthesis of pigments and volatile compounds and the partial solubilization of cell walls (Jain *et al.*, 2003). Since guava is a climacteric fruit (Singh and Pal, 2008), the above changes take place very quickly and fully ripe guavas bruise easily and are highly perishable. Therefore it is very difficult to maintain the final quality of guava fruits during handling and postharvest techniques. Compositional variation in mineral contents in fruits of different guavas was noticed (Chauhan *et al.*, 1991). Tree age and canopy position of fruits also affect the quality of guava. Upper-canopy fruit mature early irrespective of the age of the tree. Wide variations in vitamin C content and mineral composition were reported with tree age (Asrey *et al.*, 2007), with fruits in young trees being rich in iron.

12.5 Pruning Physiology and Phenological Aberration

The productivity of guava is presently much below the productive potential due to traditional practices and the prevalence of dense, old and unproductive orchards with declining yield efficiency. High planting density has been identified to increase guava production in India in order to be competitive in the world market. Light disruption in the high-density canopy is a problem in enhancing the flowering and yield and the canopy should be exposed to receive the minimum level of light for efficient photosynthesis (Singh and Singh, 2007). The response of guava to pruning for canopy modification is well known. Modifications in pruning and training techniques influence plant spacing and production decisions. Numerous workers have shown that shading reduces specific leaf weight (Barden, 1974; Jackson and Palmer, 1977) along with changing the chloroplast structure (Skene, 1974), which ultimately alter the rate of net photosynthesis. However, Barden (1977) found that 'shade' leaves have similar levels of photosynthesis to 'sun' leaves and even in heavily shaded positions they make a positive contribution. The leaves of pruned trees, particularly the flowering branches, exhibited a significantly higher photosynthetic rate in 'Allahabad Safeda' and 'Sardar' guavas (Singh and Singh, 2007), which may be due to stimulation of stomatal opening as indicated by relatively higher stomatal conductance in pruned leaves. This supports the findings by Widyastuti *et al.* (2019) in 'Crystal' guava and earlier reports in other fruit crops (Chen and Lia, 1991). Higher photosynthesis with higher carbon/nitrogen ratio as noted in flowering branches of pruned trees of 'Crystal' guava may also be due to the involvement of a compensatory mechanism for induction of flowering that alters the balance between supply and sink demand for assimilate within the plant, which in turn influences the leaf photosynthesis. Interestingly, the trees with eight pairs of leaves produce larger size fruit than those with four pairs of leaves after pruning. Under natural conditions, the guava tree produces flowers and fruits three times a year in northern India (Rathore, 1976). The spring-season flowering begins during February–March for harvesting during July–September (rainy season). Likewise, monsoon-season flowering begins during June–July and fruiting occurs during November–January (winter season) while winter flowering occurs during October–November and fruiting during March–April. Guava trees bear flowers and fruit on the current season's recently matured shoots, either from lateral buds on older wood or shoot terminals (Crane and Balerdi, 2005; Pratibha and Goswami, 2013; Thakre *et al.*, 2013). Therefore, increasing the number of current-season new shoots by pruning has a significant impact on the production. However, it influences the normal phenological stages (Salazar *et al.*, 2006); in particular, aberrations in the normal BBCH (Biologische Bundesanstalt, Bundessortenamt und Chemische Industrie) codes (Singh *et al.*, 2015a). The aberrated scale is helpful for assessing the best period for field operations to get high yield with quality fruits in guava (Fig. 12.2).

12.6 Physiological Responses to Elevated Carbon Dioxide After Pruning

Global increase in CO_2 concentration is likely to continue and the crops of tomorrow are likely to grow under higher levels of atmospheric CO_2. Based on this, when guava cultivars were evaluated under elevated CO_2 levels, the pruned trees of 'Allahabad Safeda' were found more efficient for photosynthesis under varying CO_2 levels with higher yield compared with other cultivars (Singh and Singh, 2007). Increase in the rate of photosynthesis in pruned trees may be due to increased efficiency of Rubisco under elevated CO_2 or reduction in photorespiration (Acock and Allen, 1985). Similar increase in rate of photosynthesis due to CO_2 enrichment has been reported (Booker *et al.* 1997; Campbell *et al.*, 1998). Increase in higher-molecular-weight protein under elevated CO_2 has been reported by Evans (1989) and is correlated with increased light-harvesting

Fig. 12.2. Phenological stages in guava cultivar 'Lalit'. (A:00) Bud appears: bud is greenish brown, completely closed and firmly linked to the twig. (B1:01) Bud swelling: bud swells and becomes greenish. (B2:02) Bud growth begins: bud elongates, scales start opening. (C:11) First leaves sprouting: first leaf appears and is visible. (D2:15) More leaves unfold: leaves not attain full size. (D4:19) Leaves completely developed: leaf growth completed. (E1:51) Appearance of flower buds: flower buds become visible, internode lengthening stops. (E2:55) Flower buds visible: sepals still closed. (E3:57) Flower petals elongating: sepals slightly open, petals just visible. (E5:59) Sepals totally opened: sepals fully extended so that petals can open. (F:65) 50% open flowers: first petals falling. (G:67) Petal fall: flowers fading and most petals collapse. (H1:71) Fruit setting: fruit size up to 10 mm. (I1:78) Fruit growth: fruit increases up to 80% of final size. (J:81) Fruit colour changing: beginning of ripening, fruit reaching the final volume, colour changes from green to pale green. (K:89) Fruit ripening: fruit becomes completely yellow, releases a pleasant aroma and is ready for harvest.

capacity of thylakoid membranes. High rate of photosynthesis at higher level of CO_2 in pruned trees increased the level of organic carbon and nitrogen in the leaves, as a result production efficiency of the trees increased. Elevated CO_2 (390–780 ppm) was also found (Rezende *et al.*, 2015) to be favourable to physiological growth of *P. guajava* L. seedlings due to accumulation of starch as a result of increased photosynthetic rate. The effect of different concentrations of CO_2 on photosynthesis and carboxylation efficiency in leaves of flowering and non-flowering shoots was studied in mango (Singh and Ravishankar,

2011) and results suggested that the change in photosynthesis and associated attributes with flowering is reversible.

12.7 Influence of Light Intensity on Growth and Development of Guava

12.7.1 Light and tree physiology

Light is a driving factor for all plants on the planet. Dry mass production of a plant depends on the amount of light intercepted (Monteith, 1977), but in fruit trees the relationship between light intercepted and yield is fairly complex because it depends on the photosynthetic characteristics of the species and the amount of light intercepted and distributed within the canopy.

12.7.2 Photosynthetic characteristics

Sunlight is not only the energy source for photosynthesis, but also the most important factor affecting productivity in horticultural crops (Gregoriu *et al.*, 2007). Carbon exchange rate (CER) is strongly dependent on irradiance, absorption and utilization of photon energy (Jackson, 1980; Gregoriu *et al.*, 2007). Low irradiance and insufficient light penetration into the canopy influence CER directly by reducing photon energy utilization, thus decreasing productivity (Hampson *et al.*, 1996; Gregoriu *et al.*, 2007). There is no report of whole-tree CER for saturation of photosynthetically active radiation (PAR) in guava. CER is also dependent on leaf mass at different light levels. Leaf mass per area (LMA; the ratio of leaf dry weight to leaf area) is lower for shaded than for non-shaded leaves and was found to be a biological integrator of cumulative light exposure for a leaf (Marini and Barden, 1982; Kappel and Flore, 1983).

12.7.3 Light interception and distribution

Canopy management as a routine activity in horticultural crops is aimed at increasing light interception and productivity, stabilizing yield and improving fruit quality (Hampson *et al.*, 1996). Given that they need sunlight for flowering and fruit bud formation, fruit tree crops keep a balance between light interception and light distribution (Auchter *et al.*, 1976; Huett, 2004). The amount of light intercepted by a tree sets its maximum potential for yield; however, the actual yield produced by a tree will also reflect the influence of temperature, the tree's vegetative/reproductive equilibrium, its hormonal balance, and its nutritional and water status. From the standpoint of light alone, orchard productivity is influenced by the relationship between orchard design and available light and its distribution throughout the canopy.

The effect of light intensity and photomorphogenesis is certainly significant in the long-term growth and development of fruit trees. Photosynthetic radiation is also dependent upon the quantity of radiation. In photosynthetic responses to radiation, the term 'light' usually refers to photosynthetic photon flux (PPF) rather than total radiation (Janick, 1989). Since the guava is a C_3 crop, the short-term asymptotic photosynthetic light response is important in which the maximum rate, the saturation point in terms of PPF and the light compensation point are more worthy from a practical point of view.

Extensive research work on spacing trials of guava has been done by various workers (Rathore, 1976; Chundawat *et al.*, 1992; Kalra *et al.*, 1994; Lal *et al.*, 1996; Araujo *et al.*, 1999; Singh, G. *et al.*, 2007; Mitra *et al.*, 2018) but in contrast to temperate fruit crops (Heinicke, 1963; Looney, 1968; Tustin *et al.*, 1989; Wagenmaker and Callesen, 1989; Lal *et al.*, 1996), its effects on the radiation interception behaviour of trees have not been studied much (Singh and Dhaliwal, 2007; Singh and Singh, 2007). Preliminary observations in guava cultivars revealed that trees spaced at 6 m × 6 m intercepted significantly higher radiation than more closely spaced trees, with maximum radiation on the upper part irrespective of planting distance (Singh and Dhaliwal, 2007). Radiant energy distribution pattern of guava under different densities of planting was worked out (Brar *et al.*, 2009) and the spacing of 6 m × 4 m with 420 plants ha^{-1} was reported to be the best due to maximum absorption of solar radiation for higher yield of better-quality fruits. Higher availability of net photosynthates under maximum radiation may be the reason for trees to produce fruits with good quality. Higher translocation rates of photosynthates and fruit growth with exposure to full sun conditions, in contrast to deficient light conditions within the canopy, was worked out in other fruit crops long back (Jackson, 1980).

12.7.4 Light interception and quality yield of guava in canopy-modified trees

Different fruit crops (tropical, subtropical, temperate) require different periods of light or dark for flowering (Ghosh *et al.*, 2016). Although guava is a day-neutral plant and a certain photoperiod is required to develop the generative shoot, if that period falls short then the plants do not produce full flowering. Sunlight-use efficiency (i.e. converting light energy to dry matter) has long been the main research focus to obtain sustainable fruit production and quality in orchard systems. Hence, more technological innovations are required for adequate light management in fruit trees (Blanke, 2011; Palmer, 2011).

Canopy management in guava is very important, particularly in high-density planting systems. The primary aim of canopy management is to make the best use of the land, increase the productivity per unit area, increase the quality of the fruit and reduce the cost of production. The crux of canopy management lies in how best to manipulate the tree vigour and use the available sunlight to increase the productivity and minimize the adverse effects of weather parameters. In canopy management, major emphasis is usually required to reduce the excessive canopy shading in the fruiting region. Based on the correlation of tree morphological characters with fruit yield and quantity, dwarf and spreading trees with larger trunks are the ideotypes in guava.

Modification of the canopy through pruning helps improve the quantity of light (i.e. amount of PAR) intercepted and distributed in the orchard, finally enhancing the orchard efficiency (Jackson, 1980; Palmer, 1989; Corelli-Grappadelli and Lakso, 2007; Soni and Singh, 2020). In dense orchards of guava, a small fraction of the leaves (10–12%) are exposed to full sun and the rest receive small patches of light (12–15% of maximum sunlight) that pass through gaps in the leaf canopy. Numerous workers have shown that shading reduces specific leaf weight (Avery, 1975; Jackson and Palmer, 1977) along with changing the chloroplast structure (Skene, 1974), which ultimately alters the rate of net photosynthesis.

Improved light interception within tree canopies after pruning influences the vegetative growth, photosynthetic efficiency, flower initiation, fruit set, fruit colour and fruit quality (Robinson *et al.*, 1983; Syvertsen, 1984; Schaffer and Gaye, 1989). 'Allahabad Safeda' and 'Sardar' guavas exhibited saturation level of photosynthesis at 900–1000 μmol quanta m^{-2} s^{-1} (PAR) irradiance (Singh and Singh, 2007). Thus the density of orchards has to be so managed that leaves get light at this intensity to maximize the photosynthetic rate and yield. Increase in quality production with maximum 'A' grade fruit by improving the rate of photosynthesis and changing the light distribution pattern through canopy modification in mango has been done in detail recently (Soni and Singh, 2020). The variety which develops large crowns with dense foliage with high leaf area index (LAI) caused poor light penetration and low-quality production (Rajan *et al.*, 2001). Little information about the LAI is known in guava. However, LAI in conjunction with sunlight interception mainly in high-density planting systems is useful for analysing canopy productivity. The evergreen habit and wider adaptability of guava could be explained in terms of nutrient conservation, improved carbon balance in the leaf and as a general adaptation to environmental stress.

12.8 Light Quality Management in Orchards

Light quality (sunlight spectrum) management promises to provide a new technological alternative for sustainable quality production in horticultural crops. However, little information exists about physiological and technological aspects of light quality management in guava.

12.8.1 Light quality composition

In general, the light environment inside the tree canopy is made up by two components: (i) the unfiltered solar radiation (direct and diffuse) that has passed through gaps in the vegetation; and (ii) the filtered radiation that has been attenuated by the optical properties of reflectance and transmittance of the leaves, which has crucial importance in the light spectrum modification in fruit trees (Grant, 1997). The spectrum composition (percentage of total available light) of sunlight at different phenological stages has been worked out in temperate fruit crops such as peach and apple (Erez and Kadman-Zahavi, 1972; Palmer, 1977). In guava the influence of solar radiation and its effect on the physical properties of fruits at different levels of canopy have been reported previously (Singh and Dhaliwal, 2007).

12.8.2 Leaf morphology and function

Many authors have reported that 'sun' leaves present greater LMA, stoma density and palisade-cell thickness compared with 'shade' leaves. Greater thickness of the 'sun' leaves is accounted for by the greater length of cells and larger number of cell layers in the palisade (Skene, 1974). Although anatomical differences in 'sun' and 'shade' leaves can be attributed to light intensity changes, variation in chlorophyll content was detected under different light quality conditions in citrus (Tustin *et al.*, 1992).

12.8.3 Influence of light quality on growth and development

In fruit crops, long-term far-red light exposures initiate events that result in more carbohydrates being partitioned to stem and less to leaves and roots as compared with plants that receive red light, affecting the allocation to developing fruits (Glenn and Puterka, 2007). The essential role of light quantity in carbohydrate partitioning patterns has been studied in fruit trees, especially in apple and peach (Erez and Kadman-Zahavi, 1972; Corelli-Grappadelli *et al.*, 1994; Corelli-Grappadelli, 2003), but information on the effects of light quality conditions is not available.

Fruit colour development

Light also plays an important role in colour development in fruits, particularly in red-skinned fruits. The red skin of guava is due to anthocyanin pigments, mainly cyanidin-3-galactoside (Layne *et al.*, 2001); besides low temperature, low light condition is also required for its biosynthesis. It was reported that bagging (Singh, B.P. *et al.*, 2007) increases light sensitivity of coloured-skin guava fruits and stimulates more anthocyanin synthesis during maturation. Jackson (1967) and Jackson *et al.* (1971) also reported that the fruits of apple from the inner part of the canopy differed from those from the outer part in size, colour and postharvest quality behaviour, and similar differences were observed when trees were artificially shaded. The fruit yield is related to light interception, whereas fruit quality is a function of light distribution. Light intensity decreases within the tree canopy as the outer portion shades the inner canopy. Light exposure influences the fruit quality; that is, fruit size, firmness and soluble solids are increased under high light conditions and are decreased in deeply shaded areas of the tree canopy. At least 50–60% of full sunlight is required for fruits to develop and acquire maximum red colour. Inadequate light penetration and within-tree shade can limit yield by reducing fruit number, size and quality.

12.9 Manipulation of Light Quality in Orchards

The effect of light on fruit quality and how manipulation of light can be achieved through different practices in orchards towards quality fruit production under changing climatic conditions have been reviewed (Bastias and Corelli-Grappadelli, 2012; Purbey *et al.*, 2019). Pruning and training (Mitra *et al.*, 2018) as well as use of photo-selective shade nets (Shahak *et al.*, 2004) and reflective films in orchards (Layne *et al.*, 2001; Singh *et al.*, 2009) were suggested as management approaches to harness light for quality production of guava fruits. Thus, alteration of light quality makes significant differences in guava trees and could be a useful tool for sustainable manipulation of yield and quality of fruits. However, more knowledge is required in future about the interaction of tree and light quality under orchard systems.

12.10 Photosynthesis and Fruit Quality

Photosynthesis is an essential process for growth and survival of the plant and there may be specific stages in fruit in which photosynthate (carbon) availability to the fruit may be limiting. However, this does not mean that photosynthetic rate is directly limiting yield or that an increase in photosynthetic rate will result in increased yield (Flore and Lakso, 1989). There are several indications that photosynthetic rate may be self-limiting or controlled by sink demand, rather than driving sink activity. On the other hand, quality production of fruits depends on an adequate source–sink relationship. Generally increasing the leaf/fruit ratio increases the growth and quality of fruits. Photosynthesis increases with fruit load and the leaves next to fruits are a strong source of carbohydrate; however, all these phenomena depend on species, cultivar and agroclimatic region of crop cultivation. In guava, low quality of fruits was largely due to poor distribution of light in the dense canopy causing their poor photosynthetic efficiency (Singh and Singh, 2007). Thus, through pruning and training, the distribution of light and photosynthetic efficiency of the canopy can be improved, with increased yield and fruit quality. Manipulating other cultural practices in the orchard such as maintaining tree height, spacing, fruit thinning, bending of shoots, and application of nutrients and plant growth regulators (PGRs) may also influence photosynthesis and sink activity (fruit growth) of the plant.

12.10.1 Photosynthesis as influenced by sink activity (crop load)

In guava, fruit load has shown a significant effect on the growth and quality of fruit (Palit *et al.*, 2017). Excessive crop loading reduces fruit coloration through direct shading of neighbouring fruit, or through competition for assimilates needed for coloration especially in high-density planted guava. Increased sink strength (fruit load) has been associated with an increase in the rate of leaf photosynthesis in fruit crops (Monselise and Lenz, 1980; Choma *et al.*, 1982; Fujii and Kennedy, 1985; Marquard, 1987). It was also illustrated that fruited plants have greater dry matter production per unit leaf area than non-fruiting trees (Flore and Lakso, 1989). Interestingly, in guava during high fruit/leaf ratio, the leaves cannot induce enough pigment to synthesize enzymes and carbohydrates to colour or size the fruits to their full potential (Tahir and Hamid, 2002). Under some circumstances, the presence of fruit actually caused an increase in total dry matter production per plant, which presumably results from an increase in photosynthesis rate. In other studies, an increased fruit/leaf ratio did not result in an increase in total dry matter, but there was redistribution of assimilate into fruit at the expense of roots, shoots and leaves (Wardlaw, 1968, 1990). In some cases, overcropping can limit carbohydrate storage and vegetative growth and adversely affect the positive effects on photosynthesis due to the presence of fruit (Downton *et al.*, 1987b; Singh, 2016). Higher photosynthesis in fruited branches may also be due to the involvement of a compensatory mechanism for fruiting induction that alters the balance between supplies and sink demand for assimilate within plants. However, if assimilate production is controlled by sink consumption, then there are several factors that should be considered before making the comparison between fruiting and non-fruiting trees.

12.10.2 Photosynthesis as influenced by pruning

There are a number of methods which directly or indirectly influence photosynthesis and sink activity (fruit growth). Among these, the most important are tree height, spacing, fruit load/thinning, pruning, application of PGRs, and fertilization and irrigation schedule (Pawar and Rana, 2019).

In guava, pruning along with training alters the balance between vegetative and reproductive growth by the allocation of resources (partially carbohydrates) and hormonal movement (Myers, 2003). This crop has a higher proportion of 'shade' to 'sun' leaves and its leaves are found to be photosynthetically inactive under deeper shade and act as an unproductive sink. After pruning in guava, the ratio of diffuse light and incidence of light to the canopy increase, with increasing trend in photosynthetic rate of the whole tree and translocation of photosynthates to fruits leading to higher fruit weight (Anonymous, 2017). However, deep pruning in guava diminishes leaf area, tree photosynthesis and translocation of photosynthates to fruits and increases the root/shoot ratio (Casierra-Posada and Fischer, 2012), favouring vegetative growth. Increased flowering response due to pruning was also reported to be supported by the rate of stomatal conductance and a greater number of stomata than non-pruning (Singh and Singh, 2007). In 'Crystal' guava it was reported that pruning at different levels accelerates the appearance of flowers and increases the number of generative shoots, the number of flowers per tree and the amount of fruit harvested (Widyastuti *et al.*, 2019). Thus, in guava, medium/light pruning standardized as per location may provide greater fruit weight in contrast to heavy pruning (Lal *et al.*, 1996). It is thus clear that in the modified canopy of a high-density plantation, the vegetative growth, fruit yield and fruit quality of the orchard are functions of light interception and its distribution in the canopy, which regulates physiological balance (vegetative–reproductive) to ensure constant high-quality production per unit area (Saini *et al.*, 2018). Removal of upright water sprouts is also important in the pruning of guava for efficient translocation of photosynthates to the growing shoot apex for the emergence of healthy shoots and quality fruits. Pruning intensities, time and fruit load also influence the availability of photosynthates, nutritional competition among the developing fruits and fruit weight (Mahesh *et al.*, 2016; Guguloth and Rajkumar, 2018).

12.10.3 Photosynthates translocation as influenced by shoot bending

Shoot bending and fruit thinning were reported to be the best methods for quality production of guava (Sarkar *et al.*, 2005; Mamum *et al.*, 2012). In the case of bending of branch wood, tension of the branch is increased and phloem formation is decreased; as a result, photosynthates move slowly from the bent branch, maintaining increased carbon/nitrogen ratio and inducing more flowering and fruit set with higher yield of quality fruit in terms of total soluble solids, total sugars and ascorbic acid (Nandi *et al.*, 2017). Bending also forces dormant generative buds into flower through altering hormone levels, particularly the zeatin type of cytokinin in lateral buds (Ito *et al.*, 1999).

12.10.4 Photosynthetic process as influenced by plant growth regulators

PGRs play a key role in guava production processes such as flowering, fruit set, growth, size, appearance and fruit quality. The effects of various PGRs on different aspects of guava have been reviewed (Bhardwaj *et al.*, 2005; Gollagi *et al.*, 2019). In reviews it was clearly mentioned that PGRs also help in maintaining the desired tree growth for canopy management under high-density orcharding (Singh and Bal, 2006). Some (the auxin group) were mentioned to have high potential to increase the sink strength (Agnihotri *et al.*, 2012), leading to quality yield. Positive effect of growth retardant (paclobutrazol) in restriction of vegetative growth of guava plants (Brar, 2010) and of peaches in the meadow orchard system (Erez, 1985) was also reported. PGRs have also been shown to be involved in the regulation of photosynthesis and the movement of photosynthetic products from the source (leaf) to the sink (fruit). A number of studies have demonstrated that the increased growth and dry weight of plants by different PGRs were attributed to changes in photosynthetic rate, carbon fixation and enhanced accumulation of photosynthates (Treharne and Stoddart, 1968; Arteca, 1996; Dong and Arteca, 1996; Fletcher and Gilley, 2000). Stimulation of nutrient uptake was recorded by the triacontanol formulation (Misra and Srivastava, 1991). Similar response of growth retardant in mango was also found, which showed that increased photosynthetic rate with the chemical has a significant role in enhancing the fruit weight (Singh and Singh, 2003). Thus, PGRs play a pivotal role in physiological activities and contribute to improving the productivity and quality of crops.

12.10.5 Photosynthesis and fruit quality as influenced by mineral nutrients and irrigation

Perusal of the literature reveals that, among the various factors which affect the production and productivity of guava, nutrients and water assume much more significance. Inadequacy of one or other nutrient at a critical stage of fruit development adversely affects the productivity and quality of fruit. Response of guava is reported to nitrogen, phosphorus and potassium, and the application of calcium, magnesium, zinc and boron is also noticed to be essential in a particular situation at different phenological stages for quality yield (Mitra and Bose, 1985; Kotur *et al.*, 1997; Khan *et al.*, 2018). Variation in the seasonal nutrient status in guava cultivar 'Allahabad Safeda' was found (Shikhamany *et al.*, 1986). Leaf nutrient standards were recommended to be used for diagnosis of nutrient needs (Chadha *et al.*, 1973; Kumar and Pandey, 1979). Water is one of the most important inputs essential for crop production. Drip irrigation was proved to be the best system for irrigation in guava (Singh *et al.*, 2015b).

Several reviews have dealt with nutrient factors which affect photosynthesis directly or indirectly. Direct effects of nutrients on photosynthesis are normally easily reversible when the nutrient has a direct role in photosynthesis, for example manganese, as resupply of the deficient nutrient gives a rapid recovery of photosynthetic rate. Indirect effects include those which affect the synthesis of photosynthetic enzymes before the recovery of net photosynthesis (Barker, 1979). Studies have revealed that deficiency of manganese, potassium, molybdenum, nitrogen, magnesium and calcium all depress photosynthesis (Bottrill *et al.*, 1970). The net rate of CO_2 uptake was reduced when plants were grown at low concentration of nitrogen, phosphorus or potassium (Longstreth and Nobel, 1980). Application of moderate nitrogen was found to recover the reduced values of photosynthesis, transpiration rate, leaf chlorophyll content and shoot growth with saline water in two (white and red flesh) Sudanese guavas (Dinar *et al.*, 1999). Among the nutrients, potassium and magnesium are mineral nutrients that are required in large quantities by plants. Both elements critically contribute to the process of photosynthesis and the subsequent long-distance transport of photoassimilates. If they are not absorbed in sufficient quantities, a decline in the rate of photosynthetic carbon assimilation was observed which may finally affect the quality of fruits (Trankner *et al.*, 2018). In Brazilian guava cultivar 'Paluma' the most required nutrients are nitrogen, followed by potassium and calcium, with great participation in growth and physiology (Cavalcante *et al.*, 2018). The addition of calcium and potassium to soil increased chlorophyll a, b and total indices as well as variable (Fv) and maximum (Fm) fluorescence, quantum efficiency of photosystem II (Fv/Fm), stomatal conductance, transpiration rate and internal concentration of carbon in this crop. However, potassium was recorded as a key nutrient in the photosynthetic response of guava: when it was nourished with potassium, both water use and photosynthetic rate were enhanced due to regular stomatal opening, which provokes greater carbon assimilation (Pardo and Natale, 2008). More work is needed in this direction for a clear understanding.

12.11 Environmental Effects on Productivity

Weather aberration affects production and quality of guava and change in climate is the greatest concern of mankind in the 21st century. Climate change is projected to cause an increase in temperature, variations in rainfall and an increase in the frequency of extreme events such as heat waves, cold waves, frost days, droughts and floods. Various plant processes like vegetative growth, flowering, fruiting and fruit quality are highly vulnerable to environmental changes (Laxman and Srinivasa Rao, 2005; Hazarika, 2013; Ravishankar *et al.*, 2013; Rezende *et al.*, 2015; Mitra, 2018). Two major parameters of climate change that have far-reaching implications on plants are more erratic rainfall patterns and unpredictable spells of high temperature, which are consequently expected to reduce crop productivity. Delay in monsoon, dry spells of rain and untimely rain during the water stress period, supra-optimal temperature during flowering and fruit growth, and hailstorms are some of the extreme environmental conditions experienced by guava growers at some places in the guava-growing belt in India (authors' personal observation). Summer rains during the pre-monsoon period, particularly in areas where winter crop is taken as a major crop, affect the phenology and result in early end of the winter crop. Prevalence of high summer temperatures in the previous year suppresses blossom bud differentiation and promotes extension growth in major parts of the guava-growing region.

Several physiological disorders in fruit develop due to high temperature. Attractive, red-coloured guava varieties develop anthocyanin in the fruit peel at low temperature during cool nights at fruit maturation stage. An analysis of weather data showed that areas with a minimum temperature of 8–10°C or even cooler nights during winter are more suitable for quality production, while the rise in winter temperature during the night results in poor red colour development. Even varieties without red colour on the peel developed better fruit qualities during cooler nights.

High temperature and moisture stress also increase the cracking in guava fruit and increase in temperature at maturity will lead to early fruit maturity with low sugar content. Drought also reduced fruit set and increased fruit cracking in guava. Increases in atmospheric temperature and an extended rainfall period affected the production of winter-season crop of guava in northern India.

The moisture stress and high temperature during flowering for summer guava crops strongly influence ovule quality and consequently influence the fruit set and fruit growth causing low yield with small fruit. Flower drop is quite common in guava, if low temperature prevails during flowering (Dutta, 2013; Mitra, 2018). Studies have clearly shown that the populations and pollination activities of important pollinators are declining considerably with increasing temperature. The rise in temperature would also lead to higher respiration rate, altering the rate of photosynthesis and the partitioning of photosynthates to economic parts (sink). It could also alter the phenology of guava (Singh *et al.*, 2015a), shortening the crop duration and days to flowering and fruiting. There would also be a change in the availability of growing degree-days (GDD) leading to hastening of fruit maturity, ripening and senescence, and finally reducing the yield and quality of fruits. The temperature rise may not be evenly distributed between day and night and between different seasons (Rao *et al.*, 2010). For monitoring of environmental change, the BBCH scale for phenological studies in guava has been modified (Singh *et al.*, 2015a) which will help to assess the best timing of pruning for canopy management under specific climatic conditions for higher yield.

In guava the canopy architecture is a strong determinant of productivity and yield. But the links between the architectural properties of the plant and its mechanical properties, particularly its response to wind, are relatively unknown. As a result, biologically relevant data relating canopy architecture, light dynamics and short-scale photosynthetic responses in the canopy setting are scarce. The impact of wind on plants largely depends on speed, duration and the extent to which wind can penetrate canopy layers. Sufficient wind speeds can affect plant development, form and function, resulting in reduction in leaf size and plant size as well as damage to plant surfaces (Onoda and Anten, 2011). Strong winds, low soil moisture, high temperature and low humidity can accelerate flower drop of guava in India (Rathore and Singh, 1974). High winds also cause stem breakage and lodging, affect insect activity and population growth, and influence the development and dispersal of pests and diseases within cropping systems (Shaw, 2012). Wind speeds can alter transpiration rates, indirectly affecting photosynthesis via changes in stomatal conductance and leaf temperature (Smith and Ennos, 2003), but this would be dependent upon the local environmental conditions (Grace, 1988).

Air pollutants are one of the important environmental factors which affect the quality of guava directly or indirectly (through physiological changes). Guava is rich in antioxidant compounds and flavonoids, particularly guaijaverine, quercetin and two quercetin diglycosides. The tendency of a decrease in flavonoid composition and quantity in guava after exposure to air pollution containing high concentrations of nitrogen dioxide, sulfur dioxide and particulate matter was observed (Furlan *et al.*, 2010), which may lead to a reduction in ethno-medicinal and neutraceutical properties (Qian and Niharimbere, 2004) of guava.

Guava plants were also shown to be effective accumulators of sulfur and fluoride in biomonitoring studies, and sensitive to ozone (Rezende and Furlan, 2009). Ozone is currently assumed worldwide to be the most important air pollutant. The ozone-induced foliar injury in guava cultivar 'Paluma' has been considered an important bioindicator for areas suffering from high ozone concentrations (Furlan *et al.*, 2007). Interveinal red stippling appeared on guava plants even after short ozone exposure. Inhibition of photosynthesis and alteration of carbon allocation, growth and productivity reduction, changes in crop quality and increased sensitivity to abiotic and biotic stresses were also observed with ozone exposure (Zhang *et al.*, 2001). In metropolitan areas, the problem of air pollution is one of the most serious threats to human quality of life. In order to survive in highly adverse conditions, plants have evolved a genetic capacity to compensate for drastic environmental changes by producing antioxidants that neutralize the effect of oxidant pollutants. Recent findings showed that in guava, urban air pollution acts as an elicitor factor for the production of certain phenolic compounds (Sandre *et al.*, 2014) to alleviate the damage caused due to oxidative stress, but weather conditions were crucial for impacting pollution influence.

Fruit flies, tea mosquito bugs and other sucking pests in guava (Haseeb, 2007), along with diseases like blight, are becoming favourable due to warm and humid conditions (Misra, 2007). Anthracnose disease of guava in rainy-season crop is greatly influenced by the number and frequency of rainy days (Prakash and Pandey, 2007).

Development of new cultivars with higher yield potential and resistance to multiple stresses (drought, flood, salinity) should be the key to maintain yield. Improvement in germplasm of guava for heat stress tolerance with low seed content should be one of the targets of breeding programmes. Location-specific soil and water conservation models, development of biotic and abiotic stress-tolerant rootstocks, developing new cultivars tolerant to high temperature and producing good yield under stress conditions, etc., are some of the strategies to mitigate the impacts of climate change (Singh *et al.*, 2009; Hazarika, 2013; Mitra, 2018). Some biotic and abiotic stress-tolerant rootstocks of guava – namely wilt-resistant (*Psidium molle* × *P. guajava*), tolerance to drought and sodic soil (*Psidium cujavillis*), and dwarfing nematode-tolerant and wilt-tolerant Chinese guava (*Psidium friedrichsthalianum*) – have been developed (Singh, 2010). *P. cattleyanum* rootstock has been considered for grafting common guava to improve plant tolerance to wilt and low temperature (Teotia and Phogat, 1971; Singh *et al.*, 1976). However, more concerted efforts are needed to provide genuine planting materials for commercialization in farmers' fields in different agroecological zones for horizontal growth of the guava industry.

Carbon sequestration has a crucial role to mitigate the environmental impact of climate change (Rajatiya *et al.*, 2018) as fruit crops have a high potential for carbon sequestration. Despite the negative impacts of increasing temperature and atmospheric CO_2 level, this also benefits some crops. The photosynthesis rate of C_4 plants increases as compared with C_3 plants with increasing temperature and CO_2 (Hamilton *et al.*, 2008). Whereas it was observed that C_4 plant species are quickly saturated (Naresh *et al.*, 2008) as CO_2 concentration rises, in the C_3 species like guava, photosynthetic responses continue to rise with increasing CO_2 concentration (Singh and Singh, 2007; Rezende *et al.*, 2015) which may lead to increased growth (Rezende *et al.*, 2015) and higher production (Wheeler *et al.*, 1996; Daymond *et al.*, 1997; Wurr *et al.*, 1998).

12.12 Summary

The guava is cultivated successfully in a wide range of environmental and edaphic conditions; it grows rapidly and begins fruiting within two years, but commercial production starts from the third year. The guava is a C_3 photosynthesis-sensitive plant. There is no quantitative involvement of photoperiod for flowering in guava as it is day neutral. However, knowledge on the responses of this crop to environmental factors such as light, temperature, water availability and CO_2 concentration is essential to determine the effect of suboptimal environmental conditions and to manage the crop for maximum productivity. The leaf growth rate in guava is a single sigmoid curve which describes the three distinct development stages: stage I, which includes the first 15 days; stage II, for days 16–40 (young leaves still expanding); and stage III, from days 41 to 70 (leaves reach physiological maturity). Changes in photosynthetic rate with increased carboxylation efficiency take place during the earliest stage of leaf development, coinciding with the period of greatest leaf expansion and chlorophyll synthesis. On the other hand, the growth curve for guava fruit is double sigmoid and also presents three distinct periods of growth. Being climacteric, the physiological, biochemical and structural changes in guava fruit take place very quickly during ripening. Therefore, it is very difficult to maintain the final quality of fruits during handling postharvest. Guava plants experience a very wide range of temperature stress; higher temperature disrupts photosynthetic pigments and reduces gas exchange, leading to a reduction in plant growth and productivity. Increasing the leaf/fruit ratio generally increases fruit growth and carbohydrate content. Fixing the crop load is an important operation for a better balance between carbohydrate sources and sinks for promoting uniform fruiting. Guava is different from other subtropical trees in having many growing points that may be initiated within a short period of time and the early growth of shoots (meristem) does not depend on the current season's photosynthesis; as a result rapid canopy development can take place. Guava trees bear flowers and fruit on the current-season's recently matured shoots, either from lateral buds on older wood or shoot terminals. Therefore, increase in the number of current-season new shoots by pruning has a significant impact on the production. In this crop, no systematic work on variation in different types of endogenous levels of hormones at critical phenological stages was noticed. Sunlight is the most important factor affecting productivity. The poor yield and low-quality fruit of guava is largely due to poor photosynthetic efficiency and poor distribution of light, coupled with other compounding factors. Low irradiance and insufficient light penetration into the canopy influence the CER directly by reducing photon energy utilization, thus decreasing productivity. The radiant energy distribution pattern of guava under different densities of planting was worked out and use of photo-selective shade nets and reflective films in orchards were suggested as a management approach to harness light for quality production of fruits. Alteration of light quality makes a significant difference and could be a useful tool for sustainable manipulation of yield and quality of guava orchards. Various plant processes like vegetative growth, flowering, fruiting and fruit quality are highly vulnerable to environmental changes.

The current chapter presents findings about the whole physiology of guava crops. A clear understanding of environmental factors and their interaction with physiological processes is suggested to be extremely important for improving horticultural practices, optimizing photosynthetic carbon assimilation, and increasing fruit productivity and crop quality.

12.13 Future Lines of Work

Possible areas of future research are suggested as follows:

1. Widen the genetic base of guava for effective breeding and selection programmes to evolve photosynthetically efficient varieties with fewer and soft seeds. Wilt resistance and dwarfing rootstock is important particularly for high-density plantations.
2. Characterize inter-organ signals and their effects on photosynthesis to crucially explain how the root signals and affects a photosynthetic response in the source leaf, or how sink strength, either vegetative or reproductive, affects photosynthesis in the leaf.
3. Precisely establish associations and links between phytohormones and carbohydrates in evocation and floral initiation in current-season shoots.
4. Define the role and contribution of stored and current-season photoassimilates in fruit development, with due focus on source–sink relationships. Targeted studies on root dynamics are essential, particularly in high-density plantations where plant growth is restricted and regular pruning is done.
5. Establish if assimilation rate, total carbon fixed or photosynthesis stability can be used as a selection criterion for new variety development.
6. Develop ideal canopy configurations for harnessing natural resources, especially light, to optimize productivity and quality.
7. Map agroclimatic regions to profile genotype–environment interactions, develop projections for area expansion and determine the extent of plant adaptation to changes in environment.
8. Conduct sensitivity analysis to determine which components of photosynthetic carbon gain are most important in limiting photosynthesis under different environmental conditions.

References

Acock, B. and Allen, L.H. (1985) Crop response to elevated CO_2 concentration. In: Strain, B.R. and Cure, J.D. (eds) *Direct Effects of Increasing CO_2 on Vegetation*. US Department of Energy, Washington, DC, pp. 53–95.

Anonymous (2017) *Annual Report*. Precision Farming Development Centre, ICAR–Central Institute for Subtropical Horticulture, Lucknow, India, pp. 18–21.

Araujo, F.J., Urdaneta, T., Salazar, N. and Simancas, R. (1999) Effect of planting density on guava (*Psidium guajava* L.) yield in the Maracaibo plain, Venezuela. *Revista de la Facultad de Agronomia, Universidad del Zulia* 16, 13–16.

Arteca, R.N. (1996) Manipulation of growth and photosynthesis process by plant growth regulation. *Plant Growth Substances*. Springer, Boston, Massachusetts, pp. 240–272.

Ashutosh, A., Tiwari, R. and Singh, O.P. (2012) Effect of crop regulators on growth, yield and quality of guava. *Annals of Plant and Soil Research* 15, 54–57.

Asrey, R., Pal, R.K., Sagar, V.R. and Patel, V.P. (2007) Impact of tree age and canopy position on fruit quality of guava. *Acta Horticulturae* 735, 259–262.

Auchter, E.C., Schrader, A.C., Lagasse, F.S. and Aldrich, W.W. (1976) The effect of shade on growth, fruit bud formation and chemical composition of apple trees. *Proceedings of the American Society for Horticultural Science* 23, 368–382.

Avery, D.J. (1975) Maximum photosynthetic rate: a case study in apple. *New Phytology* 78, 55–63.

Barden, J.A. (1974) Net photosynthesis, dark respiration and specific leaf weight, and growth of young apple tree growth as influenced by light regime. *Journal of the American Society for Horticultural Science* 99, 547–551.

Barden, J.A. (1977) Apple tree growth, net photosynthesis, dark respiration and specific leaf weight as affected by continuous and intermittent shade. *Journal of the American Society for Horticultural Science* 102, 391–394.

Barker, A.V. (1979) Nutritional factors in photosynthesis of higher plants. *Journal of Plant Nutrition* 1, 309–342.

Bastias, M. and Corelli-Grappadelli, L. (2012) Light quality management in fruit orchards: physiological and technological aspects. *Chilean Journal of Agricultural Research* 72, 574–581.

Bhardwaj, R.L., Meena, R.R. and Mukherjee, S. (2005) Role of plant growth regulators in guava (*Psidium guajava* L.) – a review. *Agricultural Review* 26, 281–287.

Blanke, M. (2011) Managing open field production of perennial horticultural crops with technological innovations. *Acta Horticulturae* 916, 121–128.

Booker, F.L., Reid, C.D., Harti, S.B., Fiscus, E.C. and Miller, J.E. (1997) Photosynthesis and photorespiration in soybean exposed to elevated CO_2 and O_3. *Journal of Experimental Botany* 48, 1843–1852.

Bottrill, D.E., Possingham, J.V. and Kredemann, P.E. (1970) Effect of nutrient deficiencies on photosynthesis and respiration in spinach. *Plant and Soil* 32, 424–438.

Brahmachari, V.S., Mandal, A.K., Kumar, R. and Ray, R. (1995) Effect of growth substances on fruit set and physico-chemical characteristics of Sardar guava (*Psidium guajava* L.). *Recent Horticulture* 2, 127–131.

Brar, J.S. (2010) Influence of paclobutrazol and ethephon on vegetative growth of guava (*Psidium guajava* L.) plants at different spacing. *Notulae Scientia Biologicae* 2, 110–113.

Brar, J.S., Bal, J.S. and Singh, S.P. (2009) Radiant energy distribution in guava (*Psidium guajava* L.) plant at different spacing. *Journal of Agrometeorology* 11, 135–139.

Campbell, W.J., Allen, L.H. and Bowes, G. (1998) Effect of CO_2 concentration on rubisco activity, amount and photosynthesis in soybean leaves. *Plant, Cell & Environment* 14, 807–818.

Casierra-Posada, F. and Fischer, G. (2012) Poda de arboles frutales. In: Fischer, G. (ed.) *Manual para el Cultivo de Frutales en el Tropico*. Produmedios, Bogota, pp. 169–185.

Cavalcante, A.C.P., Cavalcante, L.F. and Cavalcante, A.G. (2018) Physiology of Paluma guava (*Psidium guajava*) plants fertilized with potassium and calcium. *Idesia (Chile)* 36, 163–172.

Chacko, E.K., Reddy, Y.T.N. and Ananthanarayanan, T.V. (1982) Studies on the relationship between leaf number and area and fruit development in mango (*Mangifera indica* L.). *Journal of Horticultural Science* 57, 483–492.

Chadha, K.L., Arora, J.S., Ravel, P. and Shikhamany, S.D. (1973) Variation in the mineral composition of the leaves of guava (*P. guajava* L.) as affected by leaf position, season and sample size. *Indian Journal of Agricultural Sciences* 43, 555–561.

Chauhan, K.S., Pudir, J.P.S. and Singh, S. (1991) Studies on the mineral composition of certain fruits. *Haryana Journal of Horticultural Sciences* 20, 210–213.

Chen, C.Y. and Lia, W.S.I. (1991) Effects of source–sink alteration on stomatal movement in different soybean cultivars. *Photosynthetica* 25, 437–439.

Chittiraichelvan, R., Shikhamany, S.D. and Chadha, K.L. (1985) Contribution of leaf area towards bunch development in Thompson seedless grape (*Vitis vinifera* L.). *Indian Journal of Horticulture* 42, 156–160.

Choma, M.E., Garner, J.L., Marini, R.P. and Barden, J.A. (1982) Effect of fruiting on net photosynthesis and dark respiration of 'Hecker' strawberries. *Horticultural Sciences* 17, 212–213.

Chundawat, B.S., Kikani, K.P., Verma, L.R. and Jadav, R.G. (1992) Studies on hedgerow plantation of guava cv. Allahabad Safeda. *Indian Journal of Horticulture* 49, 134–137.

Cohen, A. (1975) Physiological studies of the decline and aging of citrus tree. *Annual Report of Institute of Horticulture Science Activities (1971–1974)*, Bet Degan, Israel.

Corelli-Grappadelli, L. (2003) Light relations. In: Ferree, D.C. and Warrington, I.J. (eds) *Apple: Botany, Production and Uses*. CAB International, Wallingford, UK, pp. 195–216.

Corelli-Grappadelli, L. and Lakso, A.N. (2007) Is maximizing orchard light interception always the best choice? *Acta Horticulturae* 732, 507–518.

Corelli-Grappadelli, L., Lakso, A.N. and Flore, J.A. (1994) Early season pattern of carbohydrate partitioning in exposed and shaded apple branches. *Journal of the American Society for Horticultural Science* 119, 596–603.

Crane, J.H. and Balerdi, C.F. (2005) *Guava Growing in the Florida Home Landscape*. Horticultural Sciences Department Document No. HS4. Florida Cooperative Extension Service, Institute of Food and Agricultural Sciences, University of Florida, Gainesville, Florida.

Crews, C.E., William, S.L. and Vines, H.M. (1975) Characteristics of photosynthesis in peach leaves. *Planta* 126, 97–104.

Daymond, A.J., Wheeler, T.R., Hadley, P., Ellish, R.H. and Morison, J.I.L. (1997) Effect of temperature, CO_2 and their interaction on the growth, development and yield of two varieties of onion (*Allium cepa* L.). *Journal of Experimental Botany* 30, 108–118.

Dejong, T.M. and Ryugo, K. (1998) Carbohydrate assimilation, translocation, and utilization. In: Ramos, D.E. (ed.) *Walnut Production Manual*. UC ANR Publication No. 3373. University of California, Oakland, California, pp. 109–114.

Dinar, H.M., Ebert, G. and Ludders, P. (1999) Growth, chlorophyll content, photosynthesis and water relation in guava (*Psidium guajava* L.) under salinity and different nitrogen supply. *Gartenbauwissenschaft* 64, 54–59.

Dong, C.N. and Arteca, R.N. (1982) Changes in photosynthetic rates resulting from phytohormones treatments to the roots of tomato plants. *Photosynthesis Research* 3, 45–52.

Downton, W.J.S., Grant, W.J.R. and Loveys, B.R. (1987a) Diurnal changes in the photosynthesis of field grown grape vines. *New Phytology* 105, 71–80.

Downton, W.J.S., Grant, W.J.R. and Loveys, B.R. (1987b) Carbon dioxide enrichment increased yield of Valencia orange. *Australian Journal of Plant Physiology* 14, 493–501.

Dutta, S. (2013) Effect of climate change in Indian horticulture – a review. *International Journal of Science and Environmental Technology* 2, 661–671.

Erez, A. (1985) Growth control with paclobutrazol of peaches grown in a meadow orchard system. *Acta Horticulturae* 160, 26–32.

Erez, A. and Kadman-Zahavi, A. (1972) Growth of peach plants under different filtered sunlight conditions. *Physiologia Plantarum* 26, 210–214.

Evans, J.R. (1989) Photosynthesis and nitrogen relationships in leaves of C_3 plants. *Oecologia* 78, 9–19.

Faust, M. (1989) *Physiology of Temperate Zone Fruits*. Wiley, New York.

Ficher, G., Almanza-Marchan, P.J. and Ramirez, F. (2012) Source sink relationship in fruit species: a review. *Revista Colombiana de Ciencias Horticolas* 6, 238–253.

Finazzo, S.K., Davenport, T.L. and Sckffen, B. (1994) Partitioning photosynthates in avocado (*Persea americana* Mill.) during flowering and fruit set. *Tree Physiology* 14, 153–164.

Fletcher, R.A. and Gilley, A. (2000) Triazoles as a plant growth regulators and stress protectants. *Horticultural Reviews* 24, 55–137.

Flore, J.A. and Lakso, A.N. (1989) Environmental and physiological regulation of photosynthesis in fruit crops. *Horticultural Reviews* 11, 111–157.

Fujii, J.A. and Kennedy, R.A. (1985) Seasonal changes in the photosynthetic rate in apple trees. A comparison between fruiting and non fruiting trees. *Plant Physiology* 78, 519–524.

Furlan, C.M., Moraes, R.M., Bulbovas, P., Domongos, M. and Sanz, M.J. (2007) *Psidium guajava* 'Paluma' (the guava plant) as a new bio-indicator of ozone in the tropics. *Environmental Pollution* 147, 691–695.

Furlan, C.M., Santos, D.Y.A.C., Matta, I.B., Domingos, M. and Salatino, A. (2010) Guava flavonoids and the effect of industrial air pollutants. *Atmospheric Pollution Research* 1, 30–35.

Ghosh, A., Dey, K., Dash, S. and Dutta, P. (2016) Effect of light on flowering of fruit crops. *Advances in Life Sciences* 5, 2597–2603.

Glenn, D.M. and Puterka, G.J. (2007) The use of plastic films and sprayable reflective particles films to increase light penetration in apple canopies and improved apple colour and weight. *HortScience* 42, 91–96.

Gollagi, S.G., Ravi, G.K., Veena, G.L. and Murlidharan, B.M. (2019) Role of plant growth regulators in guava (*Psidium guajava* L.) cultivation – a review. *Journal of Pharmacognosy and Phytochemistry* 8, 805–808.

Grace, J. (1988) Plant response to wind. *Agriculture, Ecosystems & Environment* 22/23, 71–88.

Grant, R.H. (1997) Partitioning of biologically active radiation in plant canopies. *International Journal of Biometeorology* 40, 26–40.

Gregoriu, K., Pontikis, K. and Vemmos, S. (2007) Effects of reduced irradiance on leaf morphology, photosynthetic capacity and fruit yield in olive (*Olea europaea* L.). *Photosynthetica* 47, 172–181.

Guguloth, L. and Rajkumar, M. (2018) Effect of pruning intensities and fruit load on yield and quality of guava under high density planting system. *International Journal of Current Microbiology and Applied Science* 7, 1853–1860.

Hamilton, E.W., Heckathorn, S.A., Joshi, P., Wany, D. and Barua, D. (2008) Interactive effect of elevated CO_2 and growth temperature on the tolerance of photosynthesis to acute heat stress in C_3 and C_4 species. *Journal of Integrated Plant Biology* 50, 1375–1387.

Hampson, C.R., Azarenko, A.N. and Potter, J.R. (1996) Photosynthetic rate, flowering and yield component alteration in hazelnut in response to different light environments. *Journal of the American Society for Horticultural Science* 121, 1103–1111.

Haseeb, M. (2007) Current status of insect pest problem in guava. *Acta Horticulturae* 735, 453–468.

Hassimotto, N.M.A., Genovese, M.I. and Lajolo, F.M. (2005) Antioxidant activity of dietary fruits, vegetables, and commercial frozen fruit pulps. *Journal of Agricultural and Food Chemistry* 53, 2928–2935.

Hazarika, T.K. (2013) Climate change and Indian horticulture: opportunities, challenges and mitigation strategies. *International Journal of Environmental Engineering and Management* 46, 629–630.

Heinicke, D.R. (1963) The microclimate fruit trees, II. Foliage and light distribution patterns in apple trees. *Proceedings of the American Society for Horticultural Science* 83, 1–11.

Huett, D.O. (2004) Macadamia physiology review: a canopy light response study and literature review. *Australian Journal of Agricultural Research* 55, 609–624.

Ito, A., Yaegaki, H., Hayama, H., Kusaba, S., Yamaguchi, I. and Yoshioko, W. (1999) Bending shoots stimulates flowering and influences hormones levels in lateral buds of Japanese pear. *HortScience* 34, 1224–1228.

Jackson, J.E. (1967) Variability in fruit size and colour within individual trees. *Annual Report of East Malling Research Station for 1966*, pp. 110–115.

Jackson, J.E. (1980) Light interception and utilization by orchards. *Horticultural Reviews* 2, 208–267.

Jackson, J.E. and Palmer, J.W. (1977) Effects of shade on growth and cropping of apple trees. I. Experimental details and effects on vegetative growth. *Journal of Horticultural Science* 52, 245–252.

Jackson, J.E., Sharples, R.O. and Palmer, J.W. (1971) The influence of shade and within-tree position on apple fruit size, colour and storage quality. *Journal of Horticultural Science* 46, 277–287.

Jain, N., Dhawan, K., Malhatotra, S. and Singh, R. (2003) Biochemistry of fruit ripening of guava (*Psidium guajava* L.): compositional and enzymatic changes. *Plant Foods for Human Nutrition* 58, 309–315.

Jaiswal, V.S. and Amin, M.N. (1992) Guava and jackfruit. In: Hammerschlag, F.A. and Litz, R.E. (eds) *Biotechnology of Perennial Fruit Crops (Biotechnology in Agriculture)*. CAB International, Wallingford, UK, pp. 421–431.

Janick, J. (1989) Environmental and physiological regulation of photosynthesis in fruit crops. *Horticultural Review* 11, 110–157.

Kadir, S., Weihe, M.V. and Al-Khatib, K. (2007) Photochemical efficiency and recovery of photosystem II in grapes after exposure to sudden and gradual heat stress. *Journal of the American Society for Horticultural Science* 132, 764–769.

Kalra, S.K., Sindhu, P.S., Dhaliwal, G.S. and Singh, R. (1994) Effect of different spacing on yield of guava cv. Allahabad Safeda. *Indian Journal of Horticulture* 51, 272–274.

Kappel, F. and Flore, J.A. (1983) Effect of shade on photosynthesis, specific leaf weight, leaf chlorophyll content of leaves and morphology of young peach trees. *Journal of the American Society for Horticultural Science* 108, 541–544.

Kennedy, R.A. and Johnson, D. (1981) Changes in photosynthetic characteristics during leaf development in apple. *Photosynthesis Research* 2, 213–233.

Khan, S., Kumar, A., Dahiya, D.S., Baloda, S. and Malik, A. (2018) Significance of nutrient application on growth, yield and quality of guava: a review. *International Journal of Chemical Studies* 6, 2936–2942.

Kotur, S.C., Kumar, R. and Singh, H.P. (1997) Influence of nitrogen, phosphorus and potassium on composition of leaf and its relationship with fruit yield in Allahabad Safeda guava (*Psidium guajava* L.) on an Alfisol. *Indian Journal of Agricultural Sciences* 67, 568–570.

Kumar, P. and Pandey, R.M. (1979) Sampling for mineral content in leaves of guava cultivar Lucknow-49. *Scientia Horticulturae* 11, 163–174.

Lakso, A.N. and Flore, J.A. (2003) Carbohydrate partitioning and plant growth. In: Baugher, T.A. and Singha, S. (eds) *Concise Encyclopedia of Temperate Tree Fruit*. Food Products Press, New York, pp. 21–30.

Lal, S., Tiwari, J.P. and Mishra, K.K. (1996) Effect of plant spacing and pruning intensity on flowering and fruiting of guava. *Annals of Agricultural Research* 17, 83–89.

Laxman, R.H. and Srinivasa Rao, N.S. (2005) Influence of temperature on phenology, yield and quality characteristics of grapes cv. Sharad Seedless. Presented at *National Seminar on Impact, Adaptation and Vulnerability of Horticultural Crops to Climate Change, Indian Institute of Horticultural Research, Bangalore, India, 9 December 2005*.

Layne, D., Jiang, Z. and Rushin, J.W. (2001) Tree fruit reflective film improves red skin coloration and advances maturity in peach. *HortTechnology* 11, 234–242.

Leopold, A.C. and Nodden, A.C. (1984) Hormonal regulatory system in plants. In: Scott, T.K. (ed.) *Hormonal Regulation of Development II. The Functions of Hormones from the Level of the Cell to the Whole Plant.* Encyclopedia of Plant Physiology (New Series), Vol. 10. Springer, Berlin/Heidelberg, pp. 4–22.

Longstreth, D.J. and Nobel, P.S. (1980) Nutrient influence on leaf photosynthesis. *Plant Physiology* 65, 541–543.

Looney, N.E. (1968) Light regime within standard size apple tree as determined spectrophotometrically. *Proceedings of the American Society for Horticultural Science* 93, 1–6.

Machado, E.C., Schmidt P.T., Medina C.L. and Ribeiro R.V. (2005) Photosynthetic responses of three citrus species to environmental factors. *Pesquisa Agropecuária Brasileira* 40, 1161–1170.

Mahesh, R.K., Jholgiker, P., Mahatma, N.P., Ravi, P., Shinanand, P. and Kallapna, S.N. (2016) Effect of time and level of pruning on growth and yield of guava cv. Sardar under high density planting. *Research on Environmental Life Sciences* 9, 849–853.

Mamun, A.A., Rehman, M.H. and Rahim, M.A. (2012) Effect of shoot bending and fruit thinning on productivity of guava. *Journal of Environmental Science and Natural Resources* 5, 167–172.

Marini, R.P. and Barden, J.A. (1982) Light penetration on overcast and clear days, and specific leaf weight in apple trees as affected by summer or dormant pruning. *Journal of the American Society for Horticultural Science* 107, 39–40.

Marini, R.P. and Marini, M.C. (1983) Seasonal changes in specific leaf weight, net photosynthesis, and chlorophyll content of peach leaves as affected by light penetration and canopy position. *Journal of the American Society for Horticultural Science* 108, 600–605.

Marquard, R.D. (1987) Influence of leaf to fruit ratio on nut quality, shoot carbohydrates and photosynthesis of pecan. *HortScience* 22, 256–257.

Martin, F.W., Campbell, C.W. and Ruberte, R.M. (1987) *Perennial Edible Fruits of the Tropics: An Inventory*. Agriculture Handbook No. 642. US Department of Agriculture, Agricultural Research Service, Washington, DC.

Menzel, C.M. (1985) Guava: an exotic fruit with potential in Queensland. *Queensland Agriculture Journal* 111(2), 93–98.

Menzel, C.M. and Paxton, B.F. (1986) The pattern of growth, flowering and fruiting of guava varieties in subtropical Queensland. *Australian Journal of Experimental Agriculture* 26, 123–128.

Misra, A.K. (2007) Present status of important diseases of guava in India with special reference to wilt. *Acta Horticulturae* 735, 507–523.

Misra, A. and Srivastava, N.K. (1991) Effect of the triacontanol formulation 'Miraculan' on photosynthesis, growth, nutrient uptake and essential oil yield of lemongrass (*Cymbopogon flexuosus*) Steud. Watts. *Plant Growth Regulation* 10, 57–63.

Mitra, S.K. (2018) Climate change: impact and mitigation strategies for tropical and subtropical fruits. *Acta Horticulturae* 1216, 1–12.

Mitra, S.K. and Bose, T.K. (1985) Effect of varying levels of nitrogen, phosphorus and potassium in yield and quality of guava (*Psidium guajava* L.) var. L-49. *South Indian Horticulture* 33, 286–292.

Mitra, S.K., Gosh, B. and Pathak, P.K. (2018) High density orcharding and canopy management in guava. *Acta Horticulturae* 1205, 123–129.

Monselise, S.P. and Lenz, F. (1980) Effect of fruit load on photosynthesis of budded apple trees. *Gartenbauwissenschaft* 45, 220–224.

Monteith, J.L. (1977) Climate and the efficiency of crop production in Britain. *Philosophical Transactions of the Royal Society, London, Series B* 281, 277–294.

Morton, J.F. (1987) *Fruits of Warm Climates*. Julia F. Morton Publishers, Miami, Florida.

Myers, S.C. (2003) Training and pruning principles. In: Baugher, T.A. and Singha, S. (eds) *Concise Encyclopedia of Temperate Tree Fruit*. Food Products Press, New York, pp. 339–345.

Nagar, P.K. and Rao, T.R. (1983) Endogenous auxins in seeded and seedless fruits of guava. *Scientia Horticulturae* 18, 323–331.

Nandi, P., Roy, D., Ghosh, B. and Kundu, S. (2017) Effect of bending of shoots on flowering, yield and quality of guava cv. Khaja. *Journal of Applied and Natural Science* 9, 1365–1368.

Naresh, K., Kasturi Bai, K.V., Rajgopal, V. and Agrawal, P.K. (2008) Simulating coconut growth, development and yield with the Infocrop–coconut model. *Tree Physiology* 28, 1049–1058.

Nava, A.D., Jaimes, M.N., Victor, A.G.H. and Castro, E.H. (2014) Growth kinetics of vegetative and reproductive organs of guava (*Psidium guajava* L.) in Iquala Guerrer, Mexico. *Agricultural Sciences* 5, 1468–1475.

Nilson, S.E. and Assmann, S.M. (2007) The control of transpiration. Insights from *Arabidopsis*. *Plant Physiology* 143, 19–27.

Normand, F. and Michels, T. (2007) Vegetative and reproductive development of strawberry guava in relation to carbohydrate status of the tree. *Acta Horticulturae* 735, 223–227.

Onoda, Y. and Anten, N.P.R. (2011) Challenges to understand plant responses to wind. *Plant Signaling & Behavior* 6, 1057–1059.

Palit, P., Kumar, A.K., Bhagwan, A. and Sreedhar, M. (2017) Studies on crop load, fruit thinning and their effects on growth attributes of guava (*Psidium guajava* L.) cv. Allahabad Safeda under meadow planting system. *Agriculture Update* 12, 804–811.

Palmer, J.W. (1977) Light transmittance by apple leaves and canopies. *Journal of Applied Ecology* 14, 505–513.

Palmer, J.W. (1989) Canopy manipulation for optimum utilization of light. In: Wright, C.J. (ed.) *Manipulation of Fruiting*. Butterworths, London, pp. 245–262.

Palmer, J.W. (2011) Changing concept of efficiency in orchards system. *Acta Horticulturae* 903, 41–50.

Pandolfini, T. (2009) Seedless fruit production by hormonal regulation of fruit set. *Nutrients* 1, 168–177.

Pardo, R.M. and Natale, W. (2008) Effect of liming on the mineral nutrition and yield of growing guava trees in a Typic Hapludox soil. *Communications in Soil Science and Plant Analysis* 39, 2191–2204.

Pattison, R.R., Goldstein, G. and Ares, A. (1998) Growth, biomass allocation and photosynthesis of invasive and native Hawaiian rainforest species. *Oecologia* 117, 449–459.

Pavel, E.W. and Dejong, T.M. (1993) Estimating the photosynthetic contribution of developing peach fruits to their growth and maintenance of carbohydrates requirements. *Physiologia Plantarum* 8, 331–338.

Pawar, R. and Rana, V.S. (2019) Manipulation of source–sink relationship in pertinence of better fruit quality and yield of fruit crops: a review. *Agricultural Reviews* 40, 200–207.

Porpiglia, P.J. and Barden, J.A. (1981) Effects of pruning of penetration of photosynthetically active radiation and leaf physiology in apple trees. *Journal of the American Society for Horticultural Science* 106, 752–754.

Prakash, O. and Pandey, B.K. (2007) Current status of guava diseases in India and their integrated management. *Acta Horticulturae* 735, 495–505.

Pratibha, Lal, S. and Goswami, A.K. (2013) Effect of pruning and planting systems on growth, flowering, fruiting and yield of guava cv. Sardar. *Indian Journal of Horticulture* 70, 496–500.

Purbey, S.K., Singh, S.K. and Pongener, A. (2019) Management of light for quality production of litchi. *International Journal of Bio-resource and Stress Management* 10, 529–538.

Qian, H. and Niharimbere, V. (2004) Antioxidant power of phytochemical from *Psidium guajava* leaf. *Journal of Zhejiang University of Science* 5, 676–683.

Rajan, S., Kumar, R. and Negi, S.S. (2001) Variation in canopy characteristics of mango (*Mangifera indica* L.) cultivars from diverse eco-geographical regions. *Journal of Applied Horticulture* 3, 95–97.

Rajatiya, J., Varu, D.K., Gohil, P., Solanki, M., Mishra, P. and Solanki, R. (2018) Climate change: impact, mitigation and adaptation in fruit crops. *International Journal of Pure & Applied Bioscience* 6, 1161–1169.

Rao, A.V.M.S., Santhibhusan Chowdary, P.S., Manikandan, N., Rao, G.G.S.N., Rao, V.U.M. and Ramkrishna, Y.S. (2010) Temperature trends in different regions of India. *Journal of Agrometeorology* 12, 187–190.

Rathore, D.S. (1976) Effect of season on growth and chemical composition of guava (*Psidium guajava* L) fruits. *Journal of Horticultural Science* 51, 41–47.

Rathore, D.S. and Singh, R.N. (1974) Flowering and fruiting in the three cropping pattern of guava. *Indian Journal of Horticulture* 31, 331–336.

Ravi, G.K., Jholgiker, P., Thippanna, K.S., Kumar, B.N. and Pattepur, S. (2018) Evaluation of red fleshed guava (*Psidium guajava* L.) varieties for their processing potential. *International Journal of Current Microbiology and Applied Science* 7, 7475–7483.

Ravishankar, H., Tarun Adak, Singh, V.K., Pandey, B.K., Singh, A.K. and Salvi, B.R. (2013) Empirical appraisal of some weather parameters dynamics for their possible implications on mango production in some important mango growing regions with special reference to Lucknow region of Uttar Pradesh. In: *Proceedings of the National Seminar on Climate Change and Indian Horticulture: Exploring Adaptation and Mitigation Strategies for Expeditious Resilience, BAU, Sabour, Bihar, India, 25–27 May 2013*, pp. 159–183.

Rezende, F.M. and Furlan, C.M. (2009) Anthocyanins and tannins in ozone fumigated guava trees. *Chemosphere* 76, 1445–1450.

Rezende, F.M., De Souza, A.P., Buckeridge, M. and Furlan, C.M. (2015) Is guava phenolic metabolism influenced by elevated atmospheric CO_2? *Environmental Pollution* 196, 483–488.

Robinson, T.L., Seeley, E.J. and Barritt, B.H. (1983) Effect of light environment and spur age on 'Delicious' apple fruit size and quality. *Journal of the American Society for Horticultural Science* 108, 855–861.

Rom, C.R. and Ferree, D.C. (1986) Influence of fruit on spur leaf photosynthesis and transpiration of 'Golden Delicious' apple. *HortScience* 21, 1028–1029.

Saini, H., Baloda, S.V. and Saini, P. (2018) Effect of pruning on productivity of guava under high density plantation – a review. *Current Journal of Applied Science and Technology* 27, 1–12.

Salazar, D.M., Melgarejo, P., Martinez, R., Martinez, J.J., Hernandez, F. and Burguera, M. (2006) Phenological growth stages of guava tree (*Psidium guajava* L.). *Scientia Horticulturae* 108, 157–161.

Samson, J.A. (1986) *Tropical Fruits*, 2nd edn. Tropical Agriculture Series. Longman Scientific & Technical, New York.

Sandre, A.A., Pina, J.M., Moraes, R.M. and Furlan, C.M. (2014) Anthocyanins and tannins: is the urban air pollution an elicitor factor? *Brazilian Journal of Botany* 37, 9–18.

Sarkar, A., Ghosh, B., Kumar, S. and Sukul, P. (2005) Effect of shoot pruning and bending on yield and fruit quality of guava cv. L-49. *Environmental Ecology* 23, 621–623.

Sastry, M.V. (1965) Biochemical studies in the physiology of guava. II. Major chemical changes. *Indian Food Packer* 19, 5–10.

Schaffer, B. and Gaye, G.O. (1989) Gas exchange chlorophyll and nitrogen content of mango leaves as influenced by developmental light environment. *HortScience* 24, 507–509.

Schaffer, B. and Andersen, P.C. (1994) *Handbook of Environmental Physiology of Fruit Crops,* Vol. II. *Sub-tropical and Tropical Crops*. CRC Press, Boca Raton, Florida.

Scholefield, P.B., Sedgley, M. and Alexander, D. (1985) Carbohydrate cycling in relation to shoot growth, floral initiation and development and yield in the avocado. *Scientia Horticulturae* 25, 99–110.

Shahak, Y., Gussakovsky, E., Cohen, Y., Lurie, S., Stern, R. *et al.* (2004) ColorNets: a new approach for light manipulation in fruit trees. *Acta Horticulturae* 636, 609–616.

Shaw, R.H. (2012) Wind management within canopies. In: Hatfield, J. (ed.) *Biometeorology in Integrated Pests Management.* Elsevier, London/New York, pp. 17–41.

Shigeura, G.T. and Bullock, R.M. (1976) Flower induction and fruit production of guava (*Psidium guajava* L.). *Acta Horticulturae* 57, 247–252.

Shikhamany, S.D., Iyer, C.P.A., Hariprakash, R.M. and Subramanian, T.R. (1986) Variation in the seasonal nutrient status in guava cv. Allahabad Safeda. *Indian Journal of Horticulture* 43, 73–78.

Shiva, B., Nagaraja, A., Srivastava, M. and Goswami, A.K. (2017) Genotypic variability in photosynthetic performance and gas exchange indexes of guava cultivars. *International Journal of Agriculture Sciences* 9, 4689–4692.

Singh, A. and Dhaliwal, G.S. (2007) Solar radiation interception and its effect on physical characteristics of fruits of guava cv. Sardar. *Acta Horticulturae* 735, 297–302.

Singh, B.P., Singh, R.A., Singh, G. and Killadi, B. (2007) Response of bagging on maturity, ripening and storage behavior of winter guava. *Acta Horticulturae* 735, 597–601.

Singh, G., Singh, A.K. and Mishra, D. (2007) High density planting in guava. *Acta Horticulturae* 735, 235–238.

Singh, H.J. and Bal, J.S. (2006) Effect of pruning and growth regulators on physico-chemical characters of guava during rainy season planted at different spacing. *International Journal of Agricultural Science* 2, 533–537.

Singh, H.P. (2010) Impact of climate change on horticultural crops. In: Singh, H.P., Singh, J.P. and Lal, S.S. (eds) *Challenges of Climate Change in Horticulture*. Westville Publishing House, New Delhi, pp. 1–8.

Singh, S.K., Malhotra, S.K., Bhargava, R., Singh, R.S. and Shukla, A.K. (2017) Morphological and physiological characterization of guava (*Psidium guajava* L.) under hot-arid zone of Rajasthan. *Indian Journal of Agricultural Sciences* 87, 491–495.

Singh, S.P. and Pal, R.K. (2008) Controlled atmospheric storage of guava (*Psidium guajava* L.) fruit. *Postharvest Biology and Technology* 47, 296–306.

Singh, U.R., Pandey, I.C., Upadhyay, N.P. and Tripathi, B.M. (1976) Effect of different rootstocks on growth, yield and quality of guava. *The Punjab Horticultural Journal* 16, 121–124.

Singh, V.K. (2016) Physiology of flowering in mango (*Mangifera indica* L.). In: Peter, K.V. (ed.) *Innovations in Horticultural Sciences*, Vol. III. New India Publishing Agency, New Delhi, pp. 329–347.

Singh, V.K. and Rajan, S. (2009) Changes in photosynthetic rate, specific leaf weight and sugar contents in mango (*Mangifera indica* L.). *The Open Horticulture Journal* 2, 34–37.

Singh, V.K. and Ravishankar, H. (2011) Photosynthetic aptitude of leaves on flowering and non flowering terminals in mango (*Mangifera indica* L.) cv. Dashehari. *Progressive Agriculture* 11, 174–179.

Singh, V.K. and Singh, A. (2003) Effect of paclobutrazol on regularity of bearing in mango (*Mangifera indica* L.). *Physiology and Molecular Biology of Plants* 9, 239–248.

Singh, V.K. and Singh, G. (2007) Photosynthetic efficiency, canopy micro climate and yield of rejuvenated guava trees. *Acta Horticulturae* 735, 249–257.

Singh, V.K., Singh, G. and Bhriguvanshi, S.R. (2009) Effect of polyethylene mulch on soil nutrient level and root, leaf and fruiting characteristics of mango (*Mangifera indica* L). *Indian Journal of Agricultural Sciences* 79, 11–17.

Singh, V.K., Ravishankar, H., Singh, A. and Soni, M.K. (2015a) Pruning in guava (*Psidium guajava* L.) and appraisal of consequent flowering phenology using modified BBCH scale. *Indian Journal of Agricultural Sciences* 85, 1472–1476.

Singh, V.K., Soni, M.K. and Singh, A. (2015b) Effect of drip irrigation and polyethylene mulching on fruit yield and quality of guava cv. Allahabad Safeda under meadow orcharding. *Indian Journal of Horticulture* 72, 479–484.

Skene, D.S. (1974) Chloroplast structure in mature apple leaves grown under different levels of illumination and their responses to changed illumination. *Proceedings of the Royal Society B: Biological Sciences* 186, 75–78.

Smith, V.C. and Ennos, A.R. (2003) The effects of air flow and stem flexure on the mechanical and hydraulic properties of the stem of sunflowers *Helianthus annuus* L. *Journal of Experimental Botany* 54, 845–849.

Soni, M.K. and Singh, V.K. (2020) Response of drip irrigation on different tree architecture of mango cv. Dashehari for quality production. *Journal of Eco-friendly Agriculture* 15, 14–17.

Syvertsen, J.P. (1984) Light acclimatization in citrus leaves. II. CO_2 assimilation and light, water, and nitrogen use efficiency. *Journal of the American Society for Horticultural Science* 109, 812–817.

Tahir, F.M. and Hamid, K. (2002) Studies of physico-chemical changes due to fruit thinning in guava (*Psidium guajava* L.). *Journal of Biological Sciences* 2, 744–745.

Teotia, S.S. and Phogat, K.P.S. (1971) Effect of root stock on growth, yield and quality of guava (*Psidium guajava*). *Progressive Horticulture* 2, 37–45.

Teaotia, S.S., Pandey, I.C. and Mathur, R.S. (1961) Gibberellin induced parthenocarpy in guava (*Psidium guajava* L.). *Current Science* 30, 312.

Thakre, M., Goswami, A.K., Pratibha and Lal, S. (2013) Effect of various methods of crop regulation in guava under double-hedge row system of planting. *Indian Journal of Horticulture* 70, 211–216.

Trankner, M., Tavakol, E. and Jakil, B. (2018) Functioning of potassium and magnesium in photosynthesis, photosynthate translocation and photoprotection. *Physiologia Plantarum* 163, 414–443.

Treharne, K.J. and Stoddart, J.L. (1968) Effect of gibberellins on photosynthesis in red clover (*Trifolium pratense* L.). *Nature* 220, 457–458.

Turgeon, R. (1989) The sink–source transition in leaves. *Annual Review of Plant Physiology and Plant Molecular Biology* 40, 119–138.

Tustin, S., Hurst, P., Warrington, I. and Stanley, J. (1989) Light distribution and fruit quality through multi-layered trellis apple canopies. *Acta Horticulturae* 243, 209–212.

Tustin, S.L., Grapadelli, C. and Ravaglia, G. (1992) Effect of previous season and current light environments on early season spur development and assimilate translocation in golden delicious apple. *Journal of Horticultural Science* 67, 351–360.

Wagenmaker, P.S. and Callesen, O. (1989) Influence of light interception on apple yield and fruit quality related to arrangement and tree height. *Acta Horticulturae* 243, 149–158.

Wahid, A., Gelani, S., Ashraf, M. and Foolad, M.R. (2007) Heat tolerance in plants: an overview. *Environmental Experimental Botany* 61, 199–223.

Wardlaw, I.F. (1968) The control and pattern of movement of carbohydrates in plants. *Botanical Review* 34, 79–105.

Wardlaw, I.F. (1990) The control of carbon partitioning in plants. *New Phytology* 116, 341–381.

Wheeler, T.R., Ellis, R.H., Hadley, P., Morison, J.I.L., Batts, G.R. and Daymond, A.J. (1996) Assessing the effect of climate change on field crop production aspects. *Applied Biology* 45, 49–54.

Widyastuti, R.D., Susanto, S., Melati, M. and Kurniawati, A. (2019) Effect of pruning time on flower regulation of guava (*Psidium guajava* L.). *Journal of Physics, Conference Series* 1155, 012013.

Wurr, D.C.E., Hand, D.W., Edmondson, R.N., Fellows, J.R., Hannah, M.A. and Cribb, D.M. (1998) Climate change: a response surface study of the effect of CO_2 and temperature on growth of the beetroot, carrot and onion. *The Journal of Agricultural Science* 131, 125–133.

Yadav, R.B.R. and Singh, V.K. (1995) Selection of leaf and time for measurement of photosynthesis in mango trees (*Mangifera indica* L.). *Indian Journal of Plant Physiology* 38, 186–187.

Yadava, U.L. (1996) Guava (*Psidium guajava* L.): an exotic tree fruit with potential in the south eastern United States. *HortScience* 31, 789–794.

Zhang, J., Ferdinand, J.A., Vanderheyden, D.J., Skelly, J.M. and Innes, J.L. (2001) Variation of gas exchange within native plant species of Switzerland and relationship with ozone injury: an open-top experiment. *Environmental Pollution* 113, 177–185.

13 Pests

Rodrigo Lasa[1]*, Andrea Birke[1], Larissa Guillén[1], Martín Aluja[1] and Daniel Carrillo[2]
[1]Instituto de Ecología A.C., Xalapa, Veracruz, Mexico; [2]University of Florida IFAS, Homestead, Florida, USA

13.1 Introduction

Guava is widely distributed in all tropical and subtropical regions of the world, where it is cultivated in commercial orchards, gardens, backyards and abandoned lots in urban areas, along roadsides or growing wild, as birds and mammals disperse its seeds. Many insect and mite species feed on guava tree parts throughout the world. However, despite the diversity of arthropods associated with guava, there are not more than a dozen species considered major pests on each continent. The remaining species are considered minor pests and casual feeders that can occasionally impact production at the local or regional level. Although the number of major guava pests is small, global warming and the globalization of international trade could change this situation rapidly (Ni *et al.*, 2012; Lehmann *et al.*, 2020).

This chapter focuses on major pests of guava in the different production areas of the world. The most important insects and mites have been grouped into six categories that have been divided according to different species and their importance across the different guava-producing regions. We have also emphasized the basic features of each group related to their feeding damage, biology, behaviour, ecology and biorational pest management strategies. Minor pests are also listed.

13.2 Fruit Flies (*Diptera: Tephritidae*)

Fruit flies are undoubtedly the most important pest of guavas worldwide in terms of their economic impact, as measured by direct and indirect losses caused to infested fruit and severe quarantine restrictions established by international export protocols.

13.2.1 Distribution

The spread and invasion of pestiferous fruit flies, greatly enhanced by globalization, is one of the most challenging threats faced by most fruit commodities, including guava, forcing countries to establish strategies to prevent the introduction and establishment of exotic flies through international trade. Monitoring, control strategies and trade agreements that impose rigorous sanitation treatments prior to guava export represent a major economic loss to countries that harbour these pests.

*E-mail: rodrigo.lasa@inecol.mx

DOI: 10.1079/9781789247022.0013

As guava originated in the Americas, it is attacked by several native fruit flies, mainly of the genus *Anastrepha*. However, tephritid species within the genera *Bactrocera* and *Ceratitis* also infest guava in different tropical and subtropical regions of the world. Given that a detailed description of all fruit fly species that infest guava worldwide would exceed the scope of this chapter, only general aspects of their distribution, damage, behaviour, ecology, biology and management are considered here. More detailed information on each fruit fly species can be found on CABI datasheets (CABI, 2020) or by consulting the extended bibliography on the biology, ecology and management of these pests summarized in several books (White and Elson-Harris, 1992; Aluja and Norrbom, 2000; Aluja *et al.*, 2009; Shelly *et al.*, 2014; Ekesi *et al.*, 2016; Clarke, 2019). As mentioned before, the most important fruit flies infesting guava are distributed in three genera as follows.

Fig. 13.1. Females of (A) *Anastrepha striata* and (B) *Anastrepha fraterculus* (*Diptera: Tephritidae*) on a guava fruit. Photograph of *A. fraterculus* courtesy of A. Luiz Marsaro.

Anastrepha

The most important *Anastrepha* pests of guava include the guava fruit fly, *Anastrepha striata* (Schiner) (Fig. 13.1A), the Caribbean fruit fly, *Anastrepha suspensa* (Loew), the South American fruit fly, *Anastrepha fraterculus* (Wiedemann) (Fig. 13.1B) and *Anastrepha sororcula* Zucchi (see Section 13.7, 'Other Minor Pests', for additional species occasionally attacking guava). Different *Anastrepha* species predominate in different regions from North to South America with no current reports of their establishment in any other guava-growing areas of the world. *Anastrepha suspensa* is the most important guava fruit fly in Florida and the Caribbean, *A. striata* predominates in Mexico, Central America and northern South America, whereas *A. fraterculus* (actually a complex of cryptic species) occurs from northern Mexico to southern areas of South America with the exception of Chile (CABI, 2020). Damage by *A. sororcula* is mainly limited to Brazil.

Bactrocera

The most important *Bactrocera* pests of guava include the oriental fruit fly, *Bactrocera dorsalis* (Hendel), the peach fruit fly, *Bactrocera zonata* (Saunders), the guava fruit fly, *Bactrocera correcta* (Bezzi) and *Bactrocera carambolae* (Drew and Hancock). All these flies are of Asian origin, but some species are also well established in Africa. In the Americas, no *Bactrocera* species are present except for *B. carambolae* that was accidentally introduced into Suriname and thereafter expanded to some regions of northern South America (Marchioro, 2016).

Among the *Bactrocera* species, *B. dorsalis* is considered the most destructive, invasive and widespread. It is a major pest of guava in several regions of China, India, Pakistan and Thailand and was accidentally introduced into Hawaii, the Pacific Islands and recently expanded to most sub-Saharan African regions (CABI, 2020). *Bactrocera zonata* is less widespread than *B. dorsalis*, but its strong preference to attack guava makes this pest more important than *B. dorsalis* in some regions of India and Pakistan. In Egypt, *B. zonata* has become more destructive in guava than the highly polyphagous native fruit fly, *Ceratitis capitata* (Wiedemann) (Amro and Abdel-Galil, 2008). Finally, *B. correcta*, first identified in India, is also considered one of the most important guava pests in South-East Asia, Pakistan and China, where it is currently expanding to the north (Liu *et al.*, 2019).

Ceratitis

The most important *Ceratitis* pest species of guava include the Mediterranean fruit fly, *C. capitata*, the Natal fruit fly, *Ceratitis rosa* (Karsch) and the mango fruit fly, *Ceratitis cosyra* (Walker). *Ceratitis anonae* (Graham) has also been reported infesting guavas in central West Africa. *Ceratitis capitata* is considered an important pest of guava crops in western North Africa (site of origin). Although *C. capitata* has become an exotic guava pest in the Americas, Asia, the Indian Ocean and the Pacific Islands, it has been displaced many times from the 'guava niche' by other native tephritid species. *Ceratitis rosa*, considered dominant in temperate areas, became the predominant fly infesting guava orchards in Swaziland, islands of the Indian Ocean and has been reported infesting guava in some regions of Kenya and South Africa (Ekesi *et al.*, 2016).

13.2.2 Biology, ecology and behaviour

Tephritid fruit flies cause direct damage by puncturing the guava fruit to lay eggs under the skin of fruit after being attracted by guava odours and colour (Aluja and Norrbom, 2000). The biology of these flies is similar among all the species that attack guava. Depending on the species, females usually lay one or a cluster of eggs in mature green or ripening guava fruits, but occasionally in green, unripe fruits. Damage can vary depending on fly species and regions, but up to 90% of fruits may be infested (Jalaluddin 1996; Ekesi *et al.*, 2016; Gundappa *et al.*, 2018).

After egg hatching, small holes are visible as larvae drill galleries to feed, and fruits become soft, ripen prematurely and decay easily due to rapid maturation. Larvae feed on the fruit pulp and at the third instar (10 to 15 days after hatching) leave the fruit to pupate in the soil, subsequently emerging as adults. Females reach sexual maturity at 2 weeks or earlier and can live for several months under suitable conditions. Generally, these tropical flies have a lek-mating system (i.e. groups of males meet to engage in courtship and attract virgin females to mate). Courtship involves the production of a sexual pheromone and wing waving behaviour (Aluja *et al.*, 1993). The life cycle lasts from 3 to 5 weeks depending on the fruit species and temperature. Tropical and subtropical fruit flies are multivoltine and can complete six to eight overlapping generations per year. Their population dynamics are strongly influenced by ecological factors such as host fruit availability (guava or alternative hosts) and climatic factors such as temperature, relative humidity and precipitation (Aluja and Rull, 2009). Although guava grows in similar environmental and ecological conditions around the world, each region has special particularities that interact with the biology of each fly species and demands different management scenarios (Aluja and Norrbom, 2000).

13.2.3 Management

The control of fruit flies can be achieved using several techniques through area-wide integrated pest management schemes, which involve multiple orchards and surrounding areas (Aluja and Rull, 2009).

Cultural practices

The most generalized cultural practice used for tephritid fly control is to remove infested decaying fruits to reduce pest populations in the larval and pupal stages. Fallen fruit should be removed at frequent intervals and buried with lime, or exposed to the sun in black bags, to kill larvae before they leave fruit to pupate in the soil. Inspection and removal of fruit from other susceptible host species around the orchard are especially important to reduce initial infestations, particularly if alternative host maturation precedes that of guava. Harvesting fruit before they become very ripe can also reduce damage and is a cheaper strategy than covering developing fruit with paper bags to preclude oviposition (Morera *et al.*, 2010).

Tolerant cultivars

This is the most desirable strategy to reduce direct crop losses caused by fruit flies. Resistance can be associated with female oviposition preference or with the fruit chemical properties that inhibit or reduce the capacity of larvae to develop (Aluja and Mangan, 2008). Several studies have explored the susceptibility of various guava cultivars to different species of *Bactrocera* (Rana *et al.*, 1990; Jalaluddin and Sadakathulla, 1999; Reddy and Vasugi, 2002) and *Anastrepha* (Raga *et al.*, 2006; Galli *et al.*, 2019). Smooth-skinned cultivars seem to enhance or facilitate fruit fly infestation, whereas cultivars with low pH and high concentrations of phenolic compounds may increase tolerance to attack (Jalaluddin and Sadakathulla, 1999; Reddy and Vasugi, 2002; Raga *et al.*, 2006).

Trapping

Monitoring adults in orchards, using specific traps and baits optimized for each fruit fly species, can facilitate selection of the most appropriate pest control strategy. Different trap models and attractants for different fly species have been reviewed by the Food and Agriculture Organization of the United Nations (IPPC, 2012) and the International Atomic Energy Agency (IAEA, 2013). The highly specific male pheromonal precursor methyl eugenol is recommended as a lure for *Bactrocera* species infesting guava, whereas the male pheromonal precursor Trimedlure is recommended for *C. capitata* and *C. rosa*. Protein attractants, or synthetic lures that release ammonia, are less specific but could be used for attraction of females of many species, particularly *Anastrepha* spp., a genus for which specific male attractants have not yet been developed.

13.2.4 Control strategies

Direct control strategies

Direct control strategies are mainly focused on the control of adult flies, due to the difficulty of reaching the larvae and pupae that are physically protected within fruits or the soil, respectively. Although broad-spectrum synthetic insecticide sprays are still widely used, other biorational control techniques have been developed to target fruit flies more precisely, reduce risks to agricultural workers, and decrease the presence of insecticide residues in the environment and agricultural produce. Bait sprays, bait stations and mass trapping are techniques used at the orchard level (Shelly *et al.*, 2014). These techniques are mainly targeted at females to reduce oviposition and direct damage and should be optimized for each fruit fly species and production system. Other tactics used in area-wide integrated pest management programmes include the sterile insect technique (SIT), which consists of inducing sterility in a wild pest population through the mass release of sterile males (Enkerlin, 2005). The SIT approach is often combined with male annihilation to reduce male populations using male attractants (Shelly *et al.*, 2014) and biological control through classical and augmentative releases of fruit fly parasitoids (Table 13.1). The use of entomopathogens such as nematodes and fungi to control guava fruit flies and other guava pests was reviewed in detail by Maniania *et al.* (2016).

Postharvest treatments

Gamma irradiation is a highly effective postharvest phytosanitary treatment to disinfest guavas prior to their export. A generic dose of 150 Gy has been established for tephritid flies by the International Plant Protection Convention (IPPC, 2009). Higher doses of radiation (400 Gy) have been recommended to eliminate whiteflies, lepidopterous larvae, scales and mealybugs (Hallman, 2011). Even doses of 600 Gy do not induce changes in the organoleptic

Table 13.1. Parasitoid species that attack fruit flies considered major pests of guava with potential use for biological control (BC) in integrated pest management (IPM) programmes. From Purcell *et al.*, 1998; Malavasi and Zucchi, 2000; Vargas *et al.*, 2007.

Fly species	Parasitoid species	Approach/Place	Notes
Anastrepha fraterculus	*Aganaspis pelleranoi*	Natural BC/Argentina, Mexico	Parasitoid attacking fly larvae in fallen fruit
	Doryctobracon areolatus	Natural BC/Brazil	Larval parasitoid
	Diachasmimorpha longicaudata	Classical BC/Argentina, Mexico	Larval parasitoid
	Asobara anastrephae, Doryctobracon brasiliensis, Lopheucoila anastrephae, Odontosema anastrephae, Opius bellus, Utetes anastrephae	Natural BC/Parasitoids attacking *A. fraterculus* fly larvae in guava fruit in low percentages in different American countries. More research is needed before suggesting for IPM programmes	
Anastrepha striata	*D. longicaudata*	Classical BC/Mexico	Larval parasitoid
	D. areolatus	Natural BC/Brazil, Mexico	Larval parasitoid
	A. pelleranoi	Natural BC/Argentina, Mexico	Parasitoid attacking fly larvae in fallen fruit
	Doryctobracon crawfordi, O. anastrephae, U. anastrephae	Natural BC/Parasitoids attacking *A. striata* fly larvae in low percentages in different American countries. More research is needed before suggesting for IPM programmes	
Anastrepha suspensa	*D. longicaudata*	Classical BC/Florida	Larval parasitoid
	D. areolatus	Classical BC/Florida	Larval parasitoid
	Psyttalia concolor	Classical BC/Florida	Larval parasitoid
	U. anastrephae	Natural BC/Larval parasitoid recovered from different fruit infested by *A. suspensa*, but not from guava	
	Aceratoneuromyia indica, Aganaspis daci, P. concolor	Classical BC in Florida/Parasitoids attacking *A. suspensa* fly larvae in low percentages. More research is needed before suggesting for IPM programmes	
Bactrocera correcta	*D. longicaudata*	Natural BC/Thailand	Larval parasitoid
	Fopius arisanus	Natural BC/Thailand and other Asian countries	Egg parasitoid
	Fopius vandenboschi	Natural BC/Thailand	Larval parasitoid
	Fopius persulcatus, Psyttalia makii, Psyttalia incisi, Utetes bianchii	Natural BC in Thailand/Parasitoids recovered from guava and other fruit infested by *B. correcta* and other *Bactrocera* spp. More research is needed about their abundance and performance as a natural agent of BC	
	Psyttalia fletcheri	Natural BC in Thailand/Recovered from different fruit infested by *B. correcta* and other *Bactrocera* spp. but not from guava. More research is needed about their abundance and performance as a natural agent of BC	
Bactrocera zonata	*F. arisanus*	Natural BC in Thailand Classical BC in different countries of America and Réunion Island	Egg parasitoid
	D. longicaudata, F. vandenboschi, P. makii	Natural BC in Thailand/Recovered from different fruit infested by *B. zonata* and other *Bactrocera* spp. More research is needed about their abundance and performance as a natural agent of BC	
Bactrocera dorsalis	*D. longicaudata*	Natural BC in several Asian countries Classical BC in several American and African countries	Larval parasitoid

Continued

Table 13.1. Continued.

Fly species	Parasitoid species	Approach/Place	Notes
	F. arisanus	Natural BC in several Asian countries Classical BC in Hawaii and several American and African countries	Egg parasitoid
	F. vandenboschi	Natural BC/Thailand	Larval parasitoid
	P. makii	Natural BC/Thailand	Larval parasitoid
	Dirhinus giffardii	Classical BC	
	Fopius ceratitivorus	Natural BC in South Africa	Egg–larval parasitoid
	Psyttalia humilis	Natural BC in South Africa	Larval parasitoid
	F. persulcatus, P. fletcheri, P. incisi, U. bianchii	Natural BC in Thailand/Recovered from different fruit infested by *B. dorsalis* and other *Bactrocera* spp. More research is needed about their abundance and performance as a natural agent of BC	
	Tetrastichus giffardianus	Classical BC in Hawaii	Larval parasitoid
Bactrocera carambolae	*D. longicaudata*	Natural BC/Thailand	Larval parasitoid
	F. arisanus	Natural BC/Thailand	Egg parasitoid
	F. vandenboschi	Natural BC/Thailand	
	P. incisi, P. makii	Natural BC in Thailand/Recovered from different fruit infested by *B. carambolae* and other *Bactrocera* spp. More research is needed about their abundance and performance as a natural agent of BC	
Ceratitis capitata	*Aceratoneuromyia indica*	Classical BC in several American countries	Larval parasitoid
	D. longicaudata	Classical and augmentative BC in Hawaii and Mexico Classical BC in several American countries and Spain	Larval parasitoid. Good performance, but is affected by presence of *F. arisanus* in guava orchards in Hawaii
	D. giffardii	Classical BC/Hawaii, Bolivia	Pupal parasitoid
	Diachasmimorpha tryoni	Classical and augmentative BC in Hawaii and Mexico Classical BC in Argentina, Mexico, Guatemala, etc.	Larval parasitoid with good performance in presence of other parasitoids such as *F. arisanus*
	F. arisanus	Classical BC in Hawaii and several American and African countries	Egg parasitoid. Good performance in guava orchards
	F. ceratitivorus	Natural BC in South Africa	Egg–larval parasitoid
	P. humilis	Natural BC in South Africa	Larval parasitoid
	A. indica, T. giffardianus	Classical BC in Hawaii and several American countries/Larval parasitoids recovered in low numbers	
	Bracon celer, Psyttalia perproxima, T. giffardianus, Tetrastichus giffardii	Natural BC in South Africa/Larval parasitoids recovered from different coffee berries	
Ceratitis cosyra	*F. ceratitivorus*	Natural BC in South Africa	Egg–larval parasitoid
	P. humilis	Natural BC in South Africa	Larval parasitoid associated to *C. cosyra* in different fruit to guava
	Fopius caudatus	Natural BC in Senegal	Parasitoid recovered from different fruit to guava

characteristics or nutritional content of some guava cultivars (Zhao *et al.*, 2016).

13.3 *Coleoptera*

Pests of economic importance in the order *Coleoptera* belong to the families *Curculionidae*, *Cerambycidae* and *Scarabaeidae*. Species attacking guava within these families have a limited geographical distribution that varies according to soil, climate and vegetation characteristics.

13.3.1 *Conotrachelus* spp. (*Coleoptera: Curculionidae*)

Distribution

The two main pest weevils are *Conotrachelus dimidiatus* Champion (Fig. 13.2), distributed from southern Florida to Honduras, and *Conotrachelus psidii* Marshall distributed through the guava-growing regions of Colombia, Venezuela, Peru, Bolivia, Paraguay and Brazil. In central Mexico, another species, *Conotrachelus copalensis*, also infests pink guava (Salas-Araiza and Romero-Nápoles, 2012).

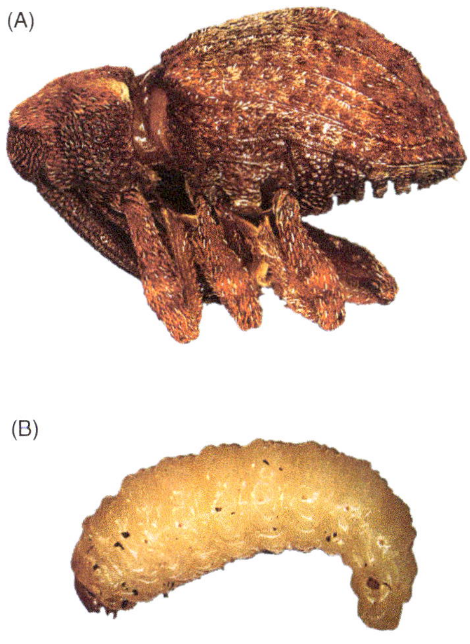

Fig. 13.2. *Conotrachelus dimidiatus* (*Coleoptera: Curculionidae*) (A) adult and (B) larva.

Biology, ecology and behaviour

Oviposition by *Conotrachelus* spp. females in fruitlets, or small unripe guavas, results in dark cork-like invaginations which are externally visible. Internally, pulp becomes dark, dries and after several weeks, fruit deform and abscise. Fruit infestation results in yield losses that can reach 50–100% in a single season (Salas-Araiza and Romero-Nápoles, 2012). *Conotrachelus* species are univoltine, their incidence fluctuates from one year to the next depending on climatic conditions and can be extremely high during one season but almost absent during the following season (González-Gaona, 2004; Pinchao and Muñoz, 2019). *Conotrachelus* species have also been recorded from pineapple guava (*Acca sellowiana* Berg Burret) and other species within *Psidium*.

Newly emerged adults have low mobility and always feed on flower buds and small fruits; activity increases once adults mature sexually at ~2 weeks post-emergence. Females mate several times during their life. *Conotrachelus dimidiatus* females only lay one egg per oviposition hole, whereas *C. psidii* lay two or three eggs (Aragón-García *et al.*, 2015; Machado da Rosa *et al.*, 2015). After oviposition, the puncture hole is sealed with a mixture of chewed plant material and saliva (Aragón-García *et al.*, 2015). Egg eclosion lasts about 1 week and larvae pass through four instars (20–50 days) before leaving the fruit and burrowing into the soil to pupate. Fourth instar larvae can remain in the soil for over 4 months without pupating (Bailez *et al.*, 2003; Aragón-García *et al.*, 2015).

Management

Weevil populations are assessed by determining the prevalence of fruit with cork-like invaginations in each tree or by counting adults after shaking stems. Monitoring traps with specific attractants such as guava semiochemicals (e.g. α-pinene, limonene, β-caryophyllene, acetol, propionic acid) (Tafoya *et al.*, 2011) and the aggregation pheromone papayanal (Romero-Frías *et al.*, 2016) are still under development.

Chemical control is mainly based on the application of broad-spectrum organophosphate insecticides, although the use of such substances is increasingly restricted in several countries. More environmentally friendly chemicals that proved to be effective against *Conotrachelus* spp. are novaluron, indoxacarb and phosmet (Rodriguez-Saona *et al.*, 2013). Biological control strategies rely mainly on the use of entomopathogenic nematodes or vegetable oils mixed with entomopathogenic fungi applied as soil drenches beneath guava trees (Dolinski, 2016; Maniania *et al.*, 2016).

13.3.2 *Aristobia* spp. (*Coleoptera: Cerambycidae*)

Distribution

Guava stem borers, *Aristobia testudo* Voet and *Aristobia reticulator* (Fabricius), are important coleopteran pests of guava in the north-eastern Hill region of India and China (Thakur *et al.*, 2012; Agarwala and Bhattacharjee, 2015).

Biology, ecology and behaviour

Larval stem-boring *Aristobia* spp., also known as lychee stem borers, mainly affect lychee, guava and pigeon pea (Firake *et al.*, 2013). Beetle damage can be easily identified by the presence of small holes regularly located at 20–30 cm intervals surrounded by straw-coloured faecal matter. Signs of adult activity are detected when stems and branches have 1–1.5 cm diameter exit holes (Gundappa *et al.*, 2018). Infested stems and branches wilt and die. Although this pest prefers older guava trees, reaching up to 80–100% infestation levels, young trees can also be infested at levels of 30–70%.

Females usually cut a slit into the bark, readily identifiable by a U-shaped mark, and lay eggs under it (Kumawat *et al.*, 2017). The egg incubation period ranges between 2 and 3 weeks. Newly hatched larvae normally feed subcortically and, after a few days, bore into the sapwood. Larval to pupal development usually lasts about 9 months, with an additional month for adult emergence.

Management

Regular inspection of trees during adult activity allows growers to kill beetles manually, by destroying eggs and larvae using a wire, or by removing and burning infested branches and stems. Severe infestations can be controlled by inserting plugs of cotton soaked in organophosphate insecticides, kerosene or petrol into the holes and sealing them with clay (Firake *et al.*, 2013).

13.4 Moths and caterpillars

Although several lepidopteran species are associated with guava, only two groups, *Indarbela* spp. and *Strepsicrates* spp., are considered major pests and are widely distributed. However, given their more localized presence, other lepidopteran guava pests have been included in Section 13.7 on minor pests, some of which may be considered to be major pests in some regions, such as *Deudorix isocrates* (Fabricius), *Rapala varuna* Horsfield, *Conogethes punctiferalis* Guenee, *Microcolona technographa* Meyric, *Othreis fullonia* (Clerck) in India, Pakistan and Bangladesh, and *Simplicivalva ampliophilobia* Davis in Colombia.

13.4.1 *Indarbela* spp. (*Lepidoptera: Cossidae*)

Distribution

Two bark-eating caterpillars, *Indarbela quadrinotata* (Walker) and *Indarbela tetraonis* (Moore), have been associated with damage to guava orchards in South and South-East Asia, mainly in India and adjoining countries (CABI, 2020).

Biology, ecology and behaviour

Indarbela spp. are polyphagous pests damaging several fruit crops, such as citrus, mango and lychee, among others. Females lay clutches of 15–20 eggs on the bark and larvae bore a tunnel, about 15–25 cm deep, into the trunk or branches. Larvae feed on bark during the night and seek refuge in the tunnel during the day. Because of this activity, large dark brown chewed wood particles with faecal matter are noticeable in infested branches. The larvae develop over a period of 9–11 months and pupate inside the tunnel, completing one generation per year. Stem damage reduces guava growth and fruiting due the interruption of sap translocation. Infestation levels, with death of severely infested branches, can reach 82% of trees in unmanaged orchards in India. Infestation at young stage also leads to death of the tree (Gundappa *et al.*, 2018).

Management

Regular visual monitoring of guava branches, especially near the forks, is essential in order to detect the initial stages of infestation and to implement control activities in a timely manner. Important differences in the number of holes per tree produced by the bark-eating caterpillars have been reported for different guava cultivars (Rao and Prasad, 2004); thus avoiding planting susceptible cultivars could reduce pest damage. Although larvae can be manually killed with a wire inserted into the hole, other control strategies have proved to be effective, such as the injection of organophosphate insecticides into bored tunnels before sealing the holes or plugging them with insecticide-soaked cotton plugs. Application of entomopathogenic fungi *Beauveria bassiana* and *Aspergillus candidus* has also proven effective for the biological control of *Indarbela* spp. (Gundappa *et al.*, 2008).

13.4.2 *Strepsicrates* spp. (*Lepidoptera: Tortricidae*)

Distribution

The guava leaf roller, *Strepsicrates smithiana* (Walsingham), has been reported in guava orchards in Florida, the Caribbean islands, Central America and South America, including Colombia and Ecuador. Other species, such as *Strepsicrates ejectana* (Walker), *Strepsicrates rothia* (Walker), *Strepsicrates tetropsis* (Busck) and *Strepsicrates semicanella*, have been reported in guava in the South Pacific, Malaysia, Trinidad and Tobago and Japan, respectively (Gould and Raga, 2002; Wakamura *et al.*, 2005).

Biology, ecology and behaviour

Females of the guava leaf roller mainly lay eggs on tender shoots, so the main damage occurs during the vegetative period, 15–45 days after pruning, although damage can be observed periodically through the year and in other phenological stages (Peña *et al.*, 1999). Larvae feed on leaves and shoots, causing yellowing and subsequent death of apical shoots, with severe damage inflicted by the final larval instars. Final instars draw young leaves together using silk and pupate within the resulting rolled leaves. The life cycle varies between 36 and 52 days, depending on weather conditions, with several generations per year. High populations of the insect can decrease production by up to 50%, especially in pear guava cultivars in Colombia (Canacuán and Carabalí, 2015).

Management

The biological insecticide, *Bacillus thuringiensis*, is extremely effective for the control of leaf rollers if applied when larvae are young, because larval susceptibility decreases as larvae develop and because it is more difficult to control them when they are inside rolled-up leaves. The naturally derived insecticide spinosad is effective against leaf rollers (Smirle *et al.*, 2003). Sprays should be applied shortly after larval hatching to reduce damage, so flushes of new foliage need to be monitored in search of egg masses, feeding injury or the presence of small larvae.

13.5 Mealybugs and Scales

13.5.1 Distribution

Mealybugs and scales are the largest and most diverse group of species associated with guava, albeit not necessarily the most damaging. This group is composed of plant-sucking insects that share similarities in their biology and ecology and are widely distributed along all tropical and subtropical regions of the world. Most species are considered secondary pests and do not only infest guava, but also attack other fruit crops and many ornamental plants. Unusual weather conditions like drought or misuse of pesticide may cause population outbreaks triggering major problems. Due to the complexity of distribution and species diversity, general aspects of their biology, ecology and management of main species within each group are summarized below. Other important hemipteran species infesting guava, including those mentioned here, are considered in Section 13.7 on minor pests, among them, the tea mosquito bug, *Helopeltis antonii* Signoret (*Hemiptera: Miridae*). The feeding damage caused by this pest results in necrotic brown spots that cause important reductions in the commercial value of fruit in central and southern India (Jayanthi and Verghese, 2007).

Fig. 13.3. (A) A colony and (B) an individual adult of the striped mealybug, *Ferrisia virgata* (*Hemiptera: Pseudococcidae*), a minor pest that infests guava shoots, leaves and fruit.

13.5.2 Biology, ecology and behaviour

Mealybugs

The common mealybugs that infest guava belong to two families, *Pseudococcidae* and *Margarodidae*. Each growing region in the world has a different mealybug complex composed of endemic species and species with a broad distribution in tropical and subtropical regions around the world. Common species reported across the globe include *Dysmicoccus brevipes* (Cockerell), *Ferrisia virgata* (Cockerell) (Fig. 13.3), *Planococcus citri* (Risso), *Pseudococcus nipae* (Cockerell) and *Pseudococcus longispinus* (Targioni) (Gould and Raga, 2002). Mealybugs are soft-bodied insects that damage twigs, stems and leaves, but can also infest the fruit. A few species like *D. brevipes* also infest roots. A severe infestation can result in misshapen fruit. Mealybugs often produce abundant honeydew, which serves as a growth substrate for black sooty mould. The sooty mould can diminish the plant's photosynthetic capabilities and reduce the fruit market value, although it grows superficially and can be removed by scrubbing. Females produce white wax that looks like cotton or snow and usually covers the colonies and protects their egg masses. The first instars or crawlers are the dispersal stage. Crawlers disperse passively by drifting on wind currents or actively by walking to a new host. The last nymphal and adult stages have a more sedentary habit and form large colonies around the original colonizer.

Scales

Scales that infest guava belong to three families, *Coccidae*, *Diaspididae* and *Eriococcidae*. They share similarities with mealybugs in their dispersal mechanisms and host infestation habits. Most scales usually infest twigs, stems, leaves and fruit of guava and other fruit and ornamental plants. Coccids or soft scales also produce abundant honeydew and are often associated with black sooty mould. Common soft scales include *Parasaissetia nigra* (Neitner), *Coccus acuminatus* Signoret, *Coccus viridis* Green, *Hemiberlesia lataniae* (Signoret) (Fig. 13.4) and *Pulvinaria psidii* (Maskell) (Gould and Raga, 2002). Diaspidids or armoured scales usually do not produce honeydew and are protected by a rigid cover. *Aspidiotus destructor* Signoret is an example of a widespread armoured scale often reported infesting guava. In Venezuela and Colombia, the guava cottony scale, *Capulinia linarosae* Kondo and Gullan (*Eriococcidae*), has become one of the most destructive guava pests since 1993 (Chirinos, 2000). The genus *Capulinia* appears to be oligophagous, feeding primarily on plants in the family *Myrtaceae* (Kondo *et al.*, 2016).

13.5.3 Management

Although most mealybugs and scales are considered secondary pests, they can become serious pests in areas that lack effective natural enemies or where natural control is disrupted by pesticides or the presence of ants (Kwee and Chong, 1990; Gould and Raga, 2002). Parasitoids and predators usually provide natural control of scales and mealybugs. Examples of predators of scales and mealybugs are lacewing larvae, syrphid fly larvae, predatory midges and coccinellid beetles (Mani, 2016; Shylesha and Mani, 2016). A great diversity of parasitic wasps (i.e. hymenopteran families *Encyrtidae*, *Aphelinidae*, *Eulophidae*, *Pteromalidae*, *Eupelmidae* and *Signiphoridae*, etc.) also parasitize scales and mealybugs and are often more specific than predators (Viggiani, 1997; Mani, 2016; Shylesha and Mani, 2016). Natural control can be disrupted by ants that feed on the honeydew produced by mealybugs and scales. In return for the food, ants protect these pests from predators and parasitoids. Bagging guavas to protect them from fruit flies also disrupts natural control by predators and parasites and forms a favourable microclimate for mealybugs inside the bag.

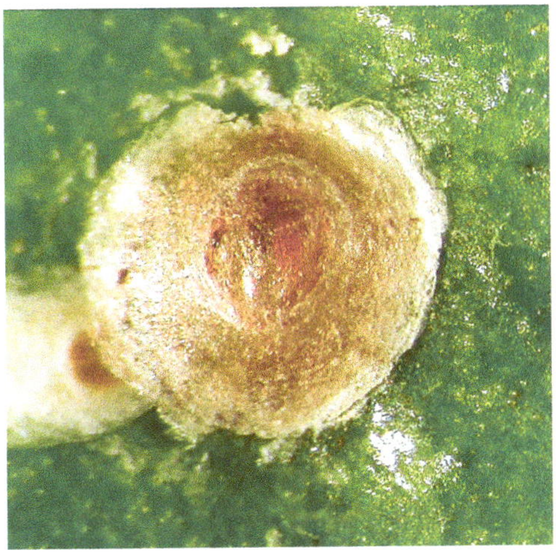

Fig. 13.4. The latania scale, *Hemiberlesia latanie* (*Hemiptera: Coccidae*), infests guava leaves, stems and fruit.

Oil or soap sprays are effective in controlling mealybugs and scales and minimizing damage to natural enemies. Chemical control with contact insecticides requires excellent coverage and should be applied to the crawler stage. Crawlers lack the outer waxy layer that protects later stages and they are more mobile on the plant. Eggs usually escape insecticide treatments because they are protected by the females and/or waxy coverings. Therefore, a follow-up treatment is required to control the newly hatched crawlers. Insect growth regulators and systemic insecticides are commonly used and can provide effective management of mealybugs and scales. Not all systemic insecticides, however, are effective against all scales. Imidacloprid, a neonicotinoid systemic insecticide, controls mealybugs and soft scales, but does not control armoured scales (Grafton-Cardwell *et al.*, 2008). Another important consideration is that the adults remain firmly attached to the plant even after their death. This may give a false impression of control failures (Kwee and Chong, 1990) that can lead to unnecessary additional treatments that harm natural enemies.

13.6 Thrips

13.6.1 Red-banded thrip, *Selenothrips rubrocinctus*

Distribution

First identified in cacao in the West Indies, the red-banded thrip, *Selenothrips rubrocinctus* (Giard), is now widely distributed across all the main guava production regions of the world.

Biology, ecology and behaviour

This insect is well identified by a characteristic red pigmentation in the first three and last abdominal segments. It is polyphagous and affects several other tropical fruit trees such as cashew, mango and avocado. Adults feed in colonies with nymphs and pupae near leaf veins on the underside of leaves, mainly in the outer middle foliage of the tree. Adults usually lay single eggs into the lower epidermis of young leaves, or in a protected area of epidermis on the fruit. The oviposition hole is covered with a black drop of faecal matter to protect the egg (Cooper, 1977). Nymphs feed by sucking cell contents on the underside of young foliage and severe infestations cause a characteristic leaf silver colour that can cause abundant loss of leaves. They also reduce the value of commercial fruit by russeting skin due the piercing of the fruit epidermis during feeding. Honeydew excretory products also favour the development of black sooty mould (Brown and Chin, 2013). The life cycle lasts between 2 and 4 weeks, depending on the temperature, with up to eight overlapping generations per year.

Management

Adults can be monitored using a 10× magnifier, with early morning or late afternoon the best time to examine leaves for monitoring counts. A grading system involving counting samples in 20% of the guava trees of the orchard has been recommended (Brown and Chin, 2013). Chemical controls are often not necessary for thrips, as natural predators such as spiders, mites, lacewings, predatory thrips and predatory bugs can effectively prevent thrips outbreaks (Brown and Chin, 2013). However, populations should be monitored for possible outbreaks and if chemical control is necessary, initial spot treatments to infested areas with biorational insecticides such as azadiractin, spinosad, spinetoram, and insecticidal soaps and oils are recommended instead of broad-spectrum insecticides such as pyrethroids or organophosphates (Guastella *et al.*, 2018).

13.7 Other Minor Pests

There are many minor pests of guava that occasionally cause local or regional outbreaks and that can negatively impact guava production in different regions of the world (Table 13.2). For example, the beetle *Costalimaita ferruginea* (Fabricius) (*Coleoptera: Chrysomelidae*), is a pest of eucalyptus that occasionally causes damage to guavas in Brazil (Fig. 13.5). The spiralling whitefly, *Aleurodicus dispersus* Russel, is another example of a broadly distributed polyphagous pest that impacts guava production in some regions of India and for which different management options should be considered (Gundappa *et al.*, 2013). However, no specific strategies have been developed to manage most of these pests in guava crops. As all these species are polyphagous and commonly attack other tropical fruit crops, knowledge of their biology, ecology and management should be gathered from the main host plants and adapted for use in guava.

13.8 Conclusions

Although the level of damage caused by most guava pests is not exceptionally high in most guava-producing countries, the climatic conditions and crop diversity of tropical and subtropical ecosystems in which guava is distributed enhance the risk of

Table 13.2. Minor pests of guava. From Mallikarjunappa *et al.* (1990); Gould and Raga (2002); Molina *et al.* (2002); Rowland (2003); Sarwar (2006); Hill (2008); Kaul *et al.* (2009); Kumar *et al.* (2009); Ghoshal and Barman (2012); Abou-Awad *et al.* (2016); Mani (2016); Carrillo *et al.* (2017); Devi and Jha (2017); Hawkeswood and Sommung (2017); Lasa *et al.* (2017); Gundappa *et al.* (2018); Pulido *et al.* (2019).

Order/Scientific name	Common name	Family	Damage	Main impact region
INSECTS				
Diptera				
Anastrepha obliqua Macquart	West Indian fruit fly	*Tephritidae*	Larvae in fruit	Brazil
Anastrepha zenildae Zucchi		*Tephritidae*	Larvae in fruit	Brazil
Bactrocera papayae Drew and Hancock		*Tephritidae*	Larvae in fruit	South-East Asia
Bactrocera psidii (Froggatt)	South Sea guava fruit fly	*Tephritidae*	Larvae in fruit	Oceania
Bactrocera tau Walker		*Tephritidae*	Larvae in fruit	Central and South-East Asia
Ceratitis anonae Graham		*Tephritidae*	Larvae in fruit	Western and Central Africa
Drosophila suzukii (Matsumura)	Spotted wing drosophila	*Drosophilidae*	Larvae in fruit	Mexico, Hawaii, Argentina
Neosilba spp.		*Lonchaeidae*	Larvae in fruit	Brazil
Lepidoptera				
Amorbia emigratella Busck	Mexican leaf roller	*Tortricidae*	Leaf defoliation	North and Central America
Argyresthia eugeniella Busck	Guava moth	*Yponomeutidae*	Fruit borer	Florida, Pakistan
Attacus atlas (L.)	Atlas moth	*Saturniidae*	Leaf defoliation	South-East Asia
Conogethes punctiferalis Guenee	Peach yellow moth	*Crambidae*	Fruit borer	India, Pakistan
Deudorix isocrates (Fabricius); *Virachola isocrates* (Fabricius)	Common guava blue, pomegranate butterfly	*Lycaenidae*	Fruit borer	India, Bangladesh
Othreis fullonia (Clerck); *Eudocima phalonia* (L.)	Mango fruit-piercing moth	*Noctuidae*	Fruit piercing, leaf defoliation	Sri Lanka, Philippines, India, Malaysia. Present in Africa, the Indian Islands, Asia, Australasia, the Pacific Islands
Microcolona technographa Meyric	Tender shoot borer	*Agonoxidae*	Shoot borer	India
Rapala varuna Horsfield	Indigo flash	*Lycaenidae*	Fruit borer	South India
Simplicivalva ampliophilobia Davis	Guava borer worm	*Cossidae*	Stem borer	Colombia
Thaumatotibia leucotreta (Meyrick); *Cryptophlebia leucotreta* (Meyrick)	False codling moth	*Tortricidae*	Leaf defoliation	Africa
Coleoptera				
Anomala bengalensis (Blanchard)		*Scarabaeidae*	Root, leaf and flower damage	India
Anthonomus irroratus Dietz		*Curculionidae*	Fruit borer	Florida
Apogonia ferruginea (Fabricius)		*Scarabaeidae*	Leaf and flower damage	India

Continued

Table 13.2. Continued.

Order/Scientific name	Common name	Family	Damage	Main impact region
Apogonia rauca (Fabricius)		*Scarabaeidae*	Leaf and flower damage	India
Batocera rufomaculata (De Geer)	Mango stem borer	*Cerambycidae*	Fruit borer	India, China
Costalimaita ferruginea Fabricius	Eucalyptus yellow beetle	*Chrysomelidae*	Leaf damage	Brazil
Costalimaita lurida Lefébre	Eucalyptus beetle	*Chrysomelidae*	Leaf damage	Brazil
Cyclocephala spp.		*Scarabaeidae*	Root and fruit damage	Mexico, Caribbean and Panama
Diaprepes abbreviatus L.	Citrus weevil	*Curculionidae*	Leaf defoliation and root damage	Caribbean and Florida
Maladera insanabilis (Brenske)		*Scarabaeidae*	Leaf and flower damage	Middle East and North Africa
Trirachys holosericeus (Fabricius); *Aeolesthes holosericea* (Fabricius)	Apple stem borer	*Cerambycidae*	Stem borer	India and South-East Asia
Hemiptera				
Aleurocanthus woglumi Ashby	Citrus blackfly	*Aleyrodidae*	Leaf and shoot damage	Worldwide tropics and subtropics
Aleurodicus dispersus Russel	Spiralling whitefly	*Aleyrodidae*	Leaf and shoot damage	India, Philippines
Aleurothrixus floccosus (Maskell)	Woolly whitefly	*Aleyrodidae*	Leaf and shoot damage	Worldwide tropics and subtropics
Aleurotuberculatus psidii Singh		*Aleyrodidae*	Leaf and shoot damage	Asia, Malaysia
Aonidiella aurantii (Makell)	California red scale	*Diaspididae*	Leaf, shoot and fruit damage	Worldwide tropics and subtropics
Aphis gossypii (Glover)	Cotton aphid	*Aphididae*	Leaf and shoot damage	Worldwide tropics and subtropics
Aspidiotus destructor Signoret	Coconut scale	*Diaspididae*	Leaf and shoot damage	Hawaii
Capulinia linarosae Kondo and Gullan	Guava cottony scale	*Eriococcidae*	Leaf and shoot damage	Venezuela, Colombia
Ceroplastes cirripediformis Comstock		*Coccidae*	Leaf and shoot damage	Egypt
Ceroplastes destructor Newstead		*Coccidae*	Leaf and shoot damage	Africa
Ceroplastes floridensis Comstock	Soft scale	*Coccidae*	Leaf and shoot damage	Worldwide tropics and subtropics
Ceroplastes rusci L.	Fig wax scale	*Coccidae*	Leaf and shoot damage	Africa, Asia, South America
Chinavia hilaris (Say); *Acrosternum hilare*	Green stink bug	*Pentatomidae*	Fruit damage	Pakistan
Coccus hesperidum L.	Brown soft scale	*Coccidae*	Leaf and shoot damage	Worldwide tropics and subtropics
Coccus viridis (Green)	Coffee green scale	*Coccidae*	Leaf and shoot damage	Worldwide tropics and subtropics
Dysmicoccus brevipes (Cockerell)	Pineapple mealybug	*Pseudococcidae*	Leaf, shoot and fruit damage	Worldwide tropics and subtropics

Continued

Table 13.2. Continued.

Order/Scientific name	Common name	Family	Damage	Main impact region
Exallomochlus hispidus (Morrison)	Cocoa mealybug	*Pseudococcidae*	Leaf, shoot and fruit damage	Indonesia and Malaysia
Ferrisia virgata (Cockerell)	Striped mealybug	*Pseudococcidae*	Leaf, shoot and fruit damage	Worldwide tropics and subtropics
Greenidea psidii Van der Goot; *Greenidea formosana* (Maki)		*Aphididae*	Leaf, shoot and fruit damage	Asia, Hawaii
Greenidea ficicola Takahashi		*Aphididae*	Leaf, shoot and fruit damage	Asia
Helopeltis antonii Signoret	Tea mosquito bug	*Miridae*	Leaf, shoot and fruit damage	India
Hemiberlesia lataniae (Signoret)	Latania scale; palm scale	*Diaspididae*	Leaf, shoot and fruit damage	Worldwide tropics and subtropics
Icerya aegyptiaca (Douglas)	Egyptian fluted scale	*Margarodidae*	Leaf and shoot damage	India, South-East Asia, Africa, Australia
Icerya seychellarum (Westwood)	Giant mealybug	*Margarodidae*	Leaf and shoot damage	Asia
Maconellicoccus hirsutus (Green)	Pink hibiscus mealybug	*Pseudococcidae*	Leaf and shoot damage	Worldwide tropics and subtropics
Myzus persicae Sulzer	Green peach aphid	*Aphididae*	Leaf and shoot damage	Worldwide tropics and subtropics
Nipaecoccus nipae (Maskell)	Spiked mealybug	*Pseudococcidae*	Leaf and shoot damage	Central America, Asia, South Africa
Nipaecoccus viridis (Newstead)	Spherical mealybug	*Pseudococcidae*	Leaf, shoot and fruit damage	India
Lepidosaphes laterochitinosa Green	Armoured scale	*Diaspididae*	Leaf and shoot damage	China, Southern Pacific, Florida
Leptoglossus spp.	Leaf-footed bug	*Coreidae*	Fruit damage	Florida, Central America, South America
Paracoccus marginatus Williams and Granara de Willink	Papaya mealybug	*Pseudococcidae*	Leaf and shoot damage	Worldwide tropics and subtropics
Parasaissetia nigra (Neitner)	Nigra scale	*Coccidae*	Leaf and shoot damage	Southern Pacific, Mexico
Phenacoccus psidiarum Cockerell		*Pseudococcidae*	Leaf, shoot and fruit damage	Mexico, Madagascar
Philephedra tuberculosa Nakahara and Gill	Soft scale	*Coccidae*	Leaf and shoot damage	Florida, Hawaii
Planococcus citri (Risso)	Citrus mealybug	*Pseudococcidae*	Leaf and shoot damage	Worldwide tropics and subtropics
Planococcus lilacinus (Cockerell)	Cacao mealybug	*Pseudococcidae*	Leaf and shoot damage	India, Malaysia, Philippines
Planococcus pacificus Cox		*Pseudococcidae*	Leaf and fruit damage	India, Malaysia
Planococcus minor (Maskell); *Planococcus psidii* Cox	Guava mealybug	*Pseudococcidae*	Leaf and shoot damage	India, Madagascar, Mexico, Pacific Islands
Pseudococcus spp.		*Pseudococcidae*	Leaf and shoot damage	South-East Asia
Pulvinaria floccifera (Westwood)	Cottony camellia scale	*Coccidae*	Leaf and shoot damage	Egypt

Continued

Table 13.2. Continued.

Order/Scientific name	Common name	Family	Damage	Main impact region
Pulvinaria psidii Maskell; *Chloropulvinaria psidii* Borchsenius	Green shield scale; guava mealy scale	*Coccidae*	Leaf and shoot damage	Worldwide tropics and subtropics
Rastrococcus iceryoides (Green)	Mango mealybug	*Pseudococcidae*	Leaf and shoot damage	India
Rastrococcus invadens (Green)	Fruit tree mealybug	*Pseudococcidae*	Leaf and shoot damage	South-East Asia, West Africa
Saissetia oleae (Olivier)	Black scale; olive scale	*Coccidae*	Leaf, shoot and fruit damage	India, China
Trialeurodes floridensis Quaintance	Avocado whitefly	*Aleyrodidae*	Leaf, shoot and fruit damage	Mexico, Florida
Trialeurodes vaporariorum Westwood	Greenhouse whitefly	*Aleyrodidae*	Leaf, shoot and fruit damage	Worldwide tropics and subtropics
Triozoida limbata (Enderlein)	Guava psyllid	*Psyllidae*	Leaf and shoot damage	Brazil
Thysanoptera				
Liothrips anonae Moulton		*Phlaeothripidae*	Leaf damage, stained fruit	Brazil
Liothrips similis Bagnall		*Phlaeothripidae*	Leaf damage, stained fruit	Venezuela
Rhipiphorothrips cruentatus Hood	Mango thrip	*Thripidae*	Leaf damage	Asia
Mites				
Abacarus uruetae Keifer		*Eriophyidae*	Leaf damage	Venezuela, Colombia
Aculops guajavae Abou-Awad, Al-Azzazy & Afia	Guava mite	*Eriophyidae*	Leaf damage	Egypt
Aculus spp.		*Eriophyidae*	Bud damage	Venezuela
Brevipalpus phoenicis (Geijskes)	False spider mite	*Tenuipalpidae*	Leaf damage	Globally
Brevipalpus californicus (Banks)	Citrus flat mite	*Tenuipalpidae*	Fruit, leaves, stems, bud damage	Globally
Eotetranychus hicoriae (McGregor)		*Tetranychidae*	Leaf damage	South India
Oligonychus psidium Estebanes & Backer		*Tetranychidae*	Leaf damage, stained fruit	Mexico, Venezuela, Colombia, Brazil
Rhynacus haramotonis Keifer		*Eriophyidae*	Leaf damage	Venezuela
Tegolophus guavae (Boczeck)		*Eriophyidae*	Rust on terminal leaves	Egypt, Florida, Caribbean Islands, Colombia
Tenuipalpus pernicis Donnadieu	False spider mite	*Tenuipalpidae*	Leaf damage	India
Tuckerella ornata (Tucker)		*Tuckerellidae*	Leaf, bud and flower damage	Venezuela

establishment of exotic invasive pests. Indeed, the risk of introduction of exotic pests has increased significantly over recent decades due to globalization and growing international trade. This applies to many pests, particularly fruit flies and other polyphagous insects such as scales, mealy bugs, whiteflies and mites that have experienced

Fig. 13.5. *Costalimaita ferruginea* (*Coleoptera: Chrysomelidae*) damage on (A) leaves, (B) fruit and (C) flowers. Photograph courtesy of A. Luiz Marsaro.

relatively large range expansions in recent years, including towards guava-producing regions with less stringent phytosanitary control measures that are commonly applied in highly developed regions. Additionally, global climate warming is impacting pest population dynamics, geographical range expansion, life history and trophic interactions that can influence pest status and the severity of damage across a range of crops. Given that the incidence of pest species and damage vary according to biogeographic conditions, each country would be advised to promote the development of tailor-made pest management packages adapted to local scenarios. Monitoring techniques and biorational control strategies can be improved or optimized in many, if not all, cases. Research should focus on developing diversified guava horticultural production systems that foster a reduction in pest impact and a greater economic return than traditional monoculture systems. As mentioned in this chapter, the main focus of pest control measures should be targeted towards fruit flies, some coleopteran and lepidopteran species that bore fruit and stems, and some minor pests that include scales, mealybugs, thrips, whiteflies and mites that increase guava production costs and reduce fruit quality and yields.

Acknowledgements

We acknowledge the valuable help and contribution of Dr Jorge Peña and the figures courtesy of Dr Alberto Luiz Marsaro Júnior.

References

Abou-Awad, B.A., Alazzazy, M.M. and Afia, S.I. (2016) Biology of *Aculops guajavae*, a new species (Acari: Eriophyidae) infesting guava trees. *International Journal of ChemTech Research* 9, 108–113.

Agarwala, B.K. and Bhattacharjee, P.P. (2015) Redescription of *Aristobia reticulator* (F., 1781) (Coleoptera: Cerambycidae: Lamiinae), with a taxonomic note and record of a new food plant for adults in northeastern India. *The Coleopterist's Bulletin* 69, 205–212.

Aluja, M. and Mangan, R.L. (2008) Fruit fly (Diptera: Tephritidae) host status determination: critical conceptual, methodological, and regulatory considerations. *Annual Review of Entomology* 53, 473–502.

Aluja, M. and Norrbom, A.L. (2000) *Fruit Flies (Tephritidae), Phylogeny and Evolution of Behaviour*. CRC Press, Boca Raton, Florida.

Aluja, M. and Rull, J. (2009) Managing pestiferous fruit flies (Diptera: Tephritidae) through environmental manipulation. In: Aluja, M., Leskey, T. and Vincent, C. (eds) *Biorational Tree Fruit Pest Management*. CAB International, Wallingford, UK, pp. 171–213.

Aluja, M., Jácome, I., Birke, A., Lozada, N. and Quintero, G. (1993) Basic patterns of behavior in wild *Anastrepha striata* (Diptera: Tephritidae) flies under field-cage conditions. *Annals of the Entomological Society of America* 86, 776–793.

Aluja, M., Leskey, T.C. and Vincent, C. (2009) *Biorational Tree Fruit Pest Management*. CABI, Wallingford, UK.

Amro, M.A. and Abdel-Galil, F.A. (2008) Infestation predisposition and relative susceptibility of certain edible fruit crops to the native and invading fruit flies (Diptera: Tephritidae) in the new valley oases, Egypt. *Assiut University Bulletin of Environmental Research* 11, 89–97.

Aragón-García, A., Pérez-Torres, B.C., Vera-Cano, D.A., Trejo-Vazquez, R. and Mota-Nava, H.B. (2015) Hábitos reproductivos del picudo de la guayaba *Conotrachelus dimidiatus* (Coleoptera: Curcuclionidae) en Calvillo, Aguascalientes. *Entomología Mexicana* 2, 613–618.

Bailez, O.E., Viana-Bailez, O.E., de Lima, J.O.E. and Moreira, D.D.O. (2003) Life-history of the guava weevil, *Conotrachelus psidii* Marshall (Coleoptera: Curculionidae), under laboratory conditions. *Neotropical Entomology* 32, 203–207.

Brown, H. and Chin, D. (2013) Red-banded thrips on fruit trees (*Selenothrips rubrocinctus*). Agnote No. I34. Northern Territory Government, Australia. Available at: https://dpir.nt.gov.au/__data/assets/pdf_file/0019/233614/719.pdf (accessed 15 May 2020).

CABI (2020) Invasive Species Compendium: detailed coverage of invasive species threatening livelihoods and the environment worldwide. CAB International, Wallingford, UK. Available at: https://www.cabi.org/ISC (accessed 15 May 2020).

Canacuán, D.E. and Carabalí, A. (2015) *Strepsicrates smithiana* (Walsingham, 1891), enrollador de hojas de *Psidium guajava*. Identificación, daño y ciclo biológico. *Corpoica Ciencia y Tecnología Agropecuaria* 16, 279–292.

Carrillo, D., Peña, J. and Duncan R. (2017) *Guava Pests and Beneficial Insects*. Publication No. ENY-412. University of Florida, IFAS Extension, Gainesville, Florida.

Chirinos, D.T. (2000) *Biología de la mota blanca del guayabo, Capulinia sp., cercana a jaboticabae von Ihering (Hemiptera: Eriococcidae) y su potencial de desarrollo de poblaciones sobre varias especies de Psidium.* Universidad Central de Venezuela, Caracas.

Clarke, A. (2019) *Biology and Management of Bactrocera and Related Fruit Flies*. CABI, Wallingford, UK.

Cooper, B. (1977) Identifying red-banded thrips, *Selenothrips rubrocinctus* (Giard), resistance in guava (*Psidium guajava* L.). MSc in Horticulture thesis, University of Hawaii, Honolulu, Hawaii.

Devi, A.R. and Jha, S. (2017) Incidence pattern of guava shoot borer, *Microcolona technographa* Meyrick (Lepidoptera: Agonoxidae) in relation to weather conditions in West Bengal, India. *Pest Management in Horticultural Ecosystems* 23, 1–6.

Dolinski, C. (2016) Entomopathogenic nematodes against the main guava insect pests. *BioControl* 61, 325–335.

Ekesi, S., Mohamed, S.A. and De Meyer, M. (2016) *Fruit Fly Research and Development in Africa. Towards a Sustainable Management Strategy to Improve Horticulture*. Springer, Cham, Switzerland.

Enkerlin, E.R. (2005) Impact of fruit fly control programmes using the sterile insect technique. In: Dyck, V.A., Hendrichs, J. and Robinson, A.S. (eds) *Sterile Insect Technique: Principles and Practice in Area-Wide Integrated Pest Management*. Springer, Dordrecht, The Netherlands, pp. 651–676.

Firake, D.M., Behere, G.T., Deshmukh, N.A., Firake, P.D. and Thakur, A.N.S. (2013) Recent scenario of insect-pests of guava in North East India and their eco-friendly management. *Indian Journal of Hill Farming* 26, 55–57.

Galli, J.A., Michelotto, M.D., Bernardes-Soares, M.B., Mello-Martins, A.L. and Fischer, I.H. (2019) Population fluctuation of fruit fly *Anastrepha* spp. (Diptera: Tephritidae) in guava (*Psidium guajava* L.) accesses produced in organic system. *Acta Agronómica* 68, 9–15.

Ghoshal, S. and Barman, S. (2012) Population dynamics and feeding potentiality of *Tenuipalpus pernicis* (Chaudhri, Akbar and Rasool) on guava (*Psidium guajava*). *International Journal of Life Sciences, Biotechnology and Pharma Research* 1, 220–226.

González-Gaona, E. (2004) Control de plagas insectiles. In: González-Gaona E., Padilla-Ramírez, S., Reyes-Muro, L., Perales M.A. and Esquivel-Villagrana, F. (eds) *Guayaba: su Cultivo en México*. INIFAP, Zacatecas, México, pp. 86–111.

Gould, W.P. and Raga, A. (2002) Pests of guava. In: Peña, J.E., Sharp, J.L. and Wysoki, M. (eds) *Tropical Fruit Pests and Pollinators*. CAB International, Wallingford, UK, pp. 295–313.

Grafton-Cardwell, E.E., Lee, J.E., Robillard, S.M. and Gorden, J.M. (2008) Role of imidacloprid in integrated pest management of California citrus. *Journal of Economic Entomology* 101, 451–460.

Guastella, D., Cocuzza, G.E.M. and Ripisarda, C. (2018) Integrated pest management in tea, cocoa and coffee. In: Cocuzza, G.E.M. and Ripisarda, C. (eds) *Integrated Pest Management in Tropical Regions*. CAB International, Wallingford, UK, pp. 285–312.

Gundappa, B., Jayanthi, P.D.K. and Verghese, A. (2013) Management of spiralling whitefly, *Aleurodicus dispersus* (Russel) in guava, *Psidium guajava* L. *Pest Management in Horticultural Ecosystems* 19, 102–105.

Gundappa, B., Rajkumar, M.B., Singh, S. and Rajan, S. (2018) Pests of guava. In: Omkar (ed.) *Pests and Their Management*. Springer, Singapore, pp. 491–516.

Hallman, G.J. (2011) Phytosanitary applications of irradiation. *Comprehensive Reviews in Food Science and Food Safety* 10, 143–151.

Hawkeswood, T.J. and Sommung, B. (2017) A record of the nocturnal/diurnal scarab beetle, *Xylotrupes mniszechi* Thomson, 1859 (Coleoptera: Scarabaeidae: Dynastinae) from the farming district of Ubon Ratchathani, Ubon Ratchathani Province, Thailand, with notes on adult feeding on guava fruits. *Calodema* 581, 1–5.

Hill, D.S. (2008) *Pests of Crops in Warmer Climates and Their Control*. Springer, Dordrecht, The Netherlands.

IAEA (2013) FAO/IAEA trapping manual for area-wide fruit fly programmes. International Atomic Energy Agency, Vienna. Available at: http://www-naweb.iaea.org/nafa/ipc/public/FruitFlyTrapping.pdf (accessed 18 January 2021).

IPPC (2009) International Plant Protection Convention. ISPM 28: Phytosanitary treatments for regulated pests. PT7: Irradiation treatment for fruit flies of the family Tephritidae. Food and Agriculture Organization of the United Nations, Rome. Available at: https://www.ippc.int/static/media/files/publication/en/2016/06/PT_07_2009_En_2016-04-22_PostCPM11_InkAm.pdf (accessed 15 May 2020).

IPPC (2012) International Plant Protection Convention. ISPM 35: Systems approach for pest risk management of fruit flies (Tephritidae). Food and Agriculture Organization of the United Nations, Rome. Available at: https://www.ippc.int/static/media/files/publication/en/2018/10/ISPM_35_2012_En_FF_Post-CPM-13_InkAm_2018-10-01.pdf (accessed 15 May 2020).

Jalaluddin S.M. (1996) Bioecology and management of guava fruit fly *Bactrocera correcta* (Bezzi). Thesis, Tamil Nadu Agricultural University, Coimbatore, India.

Jalaluddin, S.M. and Sadakathulla, S. (1999) Development and survival of *Bactrocera correcta* (Bezzi) (Diptera: Tephritidae) on selected guava cultivars. *Pest Management in Horticultural Ecosystems* 5, 24–27.

Jayanthi, P.D.K. and Verghese, A. (2007) Management of tea mosquito bug, *Helopeltis antonii* Sign. using neem seed kernel spray in guava crop. *Journal of Entomological Research* 31, 15–18.

Kaul, V., Shankar, U. and Khushu, M. (2009). Bio-intensive integrated pest management in fruit crop ecosystem. In: Peshin, R. and Dhawan, A.K. (eds) *Integrated Pest Management: Innovation–Development Process*. Springer, Dordrecht, The Netherlands, pp. 631–666.

Kondo, T., Gullan, P.J. and Cook L.G. (2016) A review of the genus *Capulinia* Signoret (Hemiptera: Coccoidea: Eriococcidae) with description of two new species. *Zootaxa* 111, 471–491.

Kumar, S., Sankar, M., Sethuraman, V. and Musthak, A. (2009) Population dynamics of white grubs (Coleoptera: Scarabaeidae) in the rose environment of Northern Bangalore, India. *Indian Journal of Science and Technology* 2, 46–52.

Kumawat, M.M., Singh, K.M. and Wangchu, L. (2017) First report of an invasive longhorn beetle, *Aristobia reticulator* (Voet) (Coleoptera: Cerambycidae) in litchi, *Litchi chinensis* Sonn. (Sapindaceae), in India. *The Coleopterist's Bulletin* 71, 131–136.

Kwee, L.T. and Chong, K.K. (1990) *Guava in Malaysia: Production, Pests and Diseases*. Tropical Press Sdn Bhd, Kuala Lumpur.

Lasa, R., Tadeo, E., Dinorín, L.A., Lima, I. and Williams, T. (2017) Fruit firmness, superficial damage and location modulate infestation by *Drosophila suzukii* and *Zaprionus indianus* (Diptera: Drosophilidae): the case of guava *Psidium guajava* L., in Veracruz, Mexico. *Entomologia Experimentalis et Applicata* 162, 4–12.

Lehmann, P., Ammunét, T., Barton, M., Battisti, A., Eigenbrode, S.D. *et al.* (2020) Complex responses of global insect pests to climate warming. *Frontiers in Ecology and the Environment* 18, 141–150.

Liu, X., Zhang, L., Haack, R.A., Liu, J. and Ye, H. (2019) A noteworthy step on a vast continent: new expansion records of the guava fruit fly, *Bactrocera correcta* (Bezzi, 1916) (Diptera: Tephritidae), in mainland China. *BioInvasions Records* 8, 530–539.

Malavasi, A. and Zucchi, R.A. (2000) *Moscas das Frutas de Importância Econômica no Brasil. Conocimiento Básico e Aplicado*. Holos Editora Ltda, Ribeirão Preto, Brazil.

Mallikarjunappa, S., Nageshchandra, B.K. and Kumar, P. (1990) A note on the nutritional preference of mite, on guava. *Current Research, University of Agricultural Sciences (Bangalore)* 19, 45–47.

Machado da Rosa, J., Carissimi Boff, M.I., Zanelato-Nunes, M., Agostinetto, L. and Boff, P. (2015) Damage caused by *Conotrachelus psidii* (Coleoptera: Curculionidae) to the fruits of feijoa (*Acca sellowiana*). *Revista Colombiana de Entomología* 41, 12–17.

Mani, M. (2016) Fruit crops: guava. In: Mani, M. and Shivaraju C. (eds) *Mealybugs and Their Management in Agricultural and Horticultural Crops*. Springer, New Delhi, pp. 377–384.

Maniania, N.K., Ekesi, S. and Dolinski, C. (2016) Entomopathogens routinely used in pest control strategies: orchards in tropical climate. In: Lacey, M. (ed.) *Microbial Control of Insect and Mite Pests*. Academic Press, London, pp. 269–282.

Marchioro, C.A. (2016) Global potential distribution of *Bactrocera carambolae* and the risks for fruit production in Brazil. *PLoS One* 11, e0166142.

Molina, J.C., Pereira, P.G. and Gonzalez, M.Q. (2002) Insectos y ácaros del guayabo (*Psidium guajava* L.) en plantaciones comerciales del estado Zulia, Venezuela. *Revista de la Facultad de Agronomía* 19, 140–148.

Morera, R., Blanco, H. and Luis, C.R. (2010) Evaluation of different bagging materials for the control of the fruit fly *Anastrepha* sp. (Diptera, Tephritidae) and fruit pathogens in Taiwanese guava fruits (*Psidium guajava* L.). *Acta Horticulturae* 849, 283–292.

Ni, W.L., Li, Z.H., Chen, H.J., Wan, F.H., Qu, W.W. *et al.* (2012) Including climate change in pest risk assessment: the peach fruit fly, *Bactrocera zonata* (Diptera: Tephritidae). *Bulletin of Entomological Research* 102, 173–183.

Peña, J.E., Hennessey, M., Duncan, R. and Vasquez, T. (1999) Arthropod seasonality and control of fruit flies on guava in south Florida. *Proceedings of the Florida State Horticultural Society* 112, 206–209.

Pinchao, E.C. and Muñoz, A.C. (2019) Mapping the spatial distribution of *Conotrachelus psidii* (Coleoptera: Curculionidae): factors associated with the aggregation of damage. *Neotropical Entomology* 48, 678–691.

Pulido, V.C., Insuasty, O.I., Sarmiento, Z.X. and Ramirez, J. (2019) Guava borer worm (Lepidoptera: Cossidae), a limiting pest in guava: biology, lifecycle and management alternatives. *Heliyon* 5, e01252.

Purcell, M.F., Herr, J.C., Messing, R.H. and Wong, T.T.Y. (1998) Interactions between augmentatively released *Diachasmimorpha longicaudata* (Hymenoptera: Braconidae) and a complex of opine parasitoids in a commercial guava orchard. *Biocontrol Science and Technology* 8, 139–151.

Raga, A., de Souza Filho, M.F., Prestes, D.A.O., de Azevedo-Filho, J.A. and Sato, M.E. (2006) Susceptibility of guava genotypes to natural infestation by *Anastrepha* spp. (Diptera: Tephritidae) in the municipality of Monte Alegre do Sul, State of São Paulo, Brazil. *Neotropical Entomology* 35, 121–125.

Rana, J.S., Parkash, O. and Verma, S.K. (1990) A note on relative susceptibility of some guava cultivars to the fruitfly *Dacus zonatus* (Sounders). *Haryana Journal of Horticultural Sciences* 19, 131–133.

Rao, S.R.K. and Prasad, D.M. (2004) Relative preference of guava by bark eating caterpillar, *Indarbela quadrinotata* Walk. *Journal of Applied Zoological Researches* 15(1), 66.

Reddy, P.V.R. and Vasugi, C. (2002) Evaluation of guava accessions for resistance to the fruit fly *Bactrocera dorsalis* (Hendel) in relation to certain morphological characters. *Pest Management in Horticultural Ecosystems* 8, 27–32.

Rodriguez-Saona, C.R., Wise, J.C., Polk, D., Leskey, T.C. and Vandervoort, C. (2013) Lethality of reduced-risk insecticides against plum curculio (Coleoptera: Curculionidae) in blueberries, with emphasis on their curative activity. *Pest Management Science* 69, 1334–1345.

Romero-Frías, A., Murata, Y., Simoes-Bento, J.M. and Osorio, C. (2016) (1*R*,2*S*,6*R*)-Papayanal: a new male-specific volatile compound released by the guava weevil *Conotrachelus psidii* (Coleoptera: Curculionidae). *Bioscience, Biotechnology, and Biochemistry* 80, 848–855.

Rowland, J.M. (2003) Male horn dimorphism, phylogeny and systematics of rhinoceros beetles of the genus *Xylotrupes* (Scarabaeidae, Coleoptera). *Australian Journal of Zoology* 51, 213–258.

Salas-Araiza, M.D. and Romero-Nápoles, J. (2012) Species of *Conotrachelus* (Coleoptera: Curculionidae: Molytinae) associated to the guava and new species description. *Revista Colombiana de Entomología* 38, 124–127.

Sarwar, M. (2006) Occurrence of insect pests on guava (*Psidium guajava*) tree. *Pakistan Journal of Zoology* 38, 197–200.

Shelly, T., Epsky, N., Jang, E.B., Reyes-Flores, J. and Vargas, R. (2014) *Trapping and The Detection, Control and Regulation of Tephritid Fruit Flies: Lures, Area-Wide Programs and Trade Implications*. Springer, Dordrecht, The Netherlands.

Shylesha, A.N. and Mani, M. (2016) Natural enemies of mealybugs. In: Mani, M. and Shivaraju C. (eds) *Mealybugs and Their Management in Agricultural and Horticultural Crops*. Springer, New Delhi, pp. 149–171.

Smirle, M.J., Lowery, D.T. and Zurowski, C.L. (2003) Susceptibility of leaf rollers (Lepidoptera: Tortricidae) from organic and conventional orchards to azinphosmethyl, spinosad, and *Bacillus thuringiensis*. *Journal of Economic Entomology* 96, 879–884.

Tafoya, F., Velasco-Olvera, J.G., Perales-Segovia, C., González-Gaona, E. and Escoto-Rocha, J. (2011) Evaluación de compuestos volátiles para estimar poblaciones del picudo de la guayaba *Conotrachelus dimidiatus*. *Acta Universitaria* 21, 65–69.

Thakur, N.A., Firake, D.M., Behere, G.T., Firake, P.D. and Saikia, K. (2012) Biodiversity of agriculturally important insects in north eastern Himalaya: an overview. *Indian Journal of Hill Farming* 25, 37–40.

Vargas, R.I., Leblanc, L., Putoa, R. and Eitam, A. (2007) Impact of introduction of *Bactrocera dorsalis* (Diptera: Tephritidae) and classical biological control releases of *Fopius arisanus* (Hymenoptera: Braconidae) on economically important fruit flies in French Polynesia. *Journal of Economic Entomology* 100, 670–679.

Viggiani, G. (1997) Eulophidae, Pteromalidae, Eupelmidae and Signiphoridae. In: Ben-Dov, Y. and Hodgson, C.J. (eds) *Soft Scale Insects – Their Biology, Natural Enemies and Control*. Elsevier, Amsterdam, pp. 147–158.

Wakamura, S., Arakaki, N. and Kinjo, K. (2005) Sex pheromone components of an olethreutid moth, *Strepsicrates semicanella* (Walker) (Lepidoptera: Tortricidae), a pest of guava and eucalyptus in Okinawa. *Applied Entomology and Zoology* 40, 637–642.

White, I.M. and Elson-Harris, M.M. (1992) *Fruit Flies of Economic Significance: Their Identification and Bionomics*. CAB International, Wallingford, UK.

Zhao, J., Ma, J., Wu, M., Jiao, X., Wang, Z. *et al.* (2016) Gamma radiation as a phytosanitary treatment against larvae and pupae of *Bactrocera dorsalis* (Diptera: Tephritidae) in guava fruits. *Food Control* 72, 360–366.

14 Nematodes

Regina M.D.G. Carneiro[1]*, Marcilene F.A. Santos[1] and José Mauro C. Castro[2]
[1]*Embrapa Recursos Genéticos e Biotecnologia, Brasília, Distrito Federal, Brazil;*
[2]*Embrapa Semiárido, Petrolina, Pernambuco, Brazil*

14.1 Introduction

Several plant-parasitic nematodes have been detected in association with guava trees in rhizosphere soil samples: *Meloidogyne* spp., *Hoplolaimus* spp., *Pratylenchus* spp., *Helicotylenchus* spp., *Tylenchus* spp., *Tylenchorhynchus* spp., *Hemicriconemoides* spp., *Aphelenchoides* spp., *Longidorus* spp., *Xiphinema* spp., *Rotylenchulus* spp. and others. These nematodes appear in studies on the diversity of the nematode community and are not correlated with damage to the guava crop (Ansari and Khan, 2012; Khan *et al.*, 2012). In addition, *Helicotylenchus dihystera*, *Pratylenchus brachyurus*, *Rotylenchulus reniformis*, *Mesocriconema* sp., *Hemicriconemoides strictathecatus*, *Pratylenchus* sp., *Tylenchorhynchus contractus*, *Xiphinema brevicolle* and *Monotrichodorus monohystera* have also been found associated with guava roots with no reference to damage caused (Castellano *et al.*, 2012). Indian authors have suspected that a complex (nematode × fungi) particularly *H. dihystera* and *Fusarium oxysporum* f.sp. *psidii* could be involved with guava wilt (Khan *et al.*, 2001). Surveys conducted in healthy and diseased orchards in five Indian states, as well as experiments under controlled conditions, obtained convincing evidence that *H. dihystera* is involved in guava wilt. In South Africa, guava wilt is caused by *Nalanthamala psidii*, without nematode detection (Schoeman *et al.*, 2017) and in India, guava wilt has been attributed to many fungi, with *F. oxysporum* f.sp. *psidii* being the most isolated from diseased trees, but no definitive proof of the causal agents of this disease and nematode relationship exists (see Chapter 15, this volume). The root-knot nematode (RKN) (*Meloidogyne* spp.) is a recognized limiting factor in commercial guava production in Central and South American and other countries (Carneiro *et al.*, 2001; Xu *et al.*, 2004; Iwahori and Ban, 2009; Humphreys *et al.*, 2012; Kumar and Rawat, 2018). In Cuba, Puerto Rico, Mexico, Venezuela and Brazil, guava production has declined during the past quarter of a century due to increasing pressure from *Meloidogyne* spp. (El-Borai and Duncan, 2005). Many detections made in the past, using the perineal region as an identification tool, have incorrectly shown the presence of *Meloidogyne incognita*, *Meloidogyne javanica* and *Meloidogyne arenaria* instead of *Meloidogyne enterolobii* (Moura and Moura, 1989; Villota and Agudelo, 1997; Crozzoli and Casassa, 1998;

*E-mail: regina.carneiro@embrapa.br

DOI: 10.1079/9781789247022.0014

Avelar-Mejía *et al.*, 2001; Gallegos-Morales *et al.*, 2009). In fact, for this nematode species, the perineal region presents very different profiles (Carneiro *et al.*, 2001; Brito *et al.*, 2004), leading to confusion with other species. More recently, surveys carried out in Costa Rica, Vietnam, India and South Africa using enzymatic and molecular markers have shown that the only species that occurs in guava is *M. enterolobii* (Carneiro *et al.*, 2001; Molinari *et al.*, 2005; Iwahori and Ban, 2009; Humphreys *et al.*, 2012; Castro, 2019; Rashidifard *et al.*, 2019), which is now called the guava RKN. In a greenhouse study of host suitability of guava cultivar Paluma, *M. incognita*, *M. javanica* and *M. arenaria* were not confirmed on the host. These *Meloidogyne* species did not parasitize guava. Only *M. enterolobii* reproduced in cultivars of this plant (Rossi *et al.*, 2002; Carneiro *et al.*, 2012). These results confirmed the misindentification of *M. enterolobii* (= *Meloidogyne mayaguensis*) using only morphological approaches.

The following review of the nematodes from cultivated guava is limited to the major problem caused by *M. enterolobii*, its idenfication and its management strategies.

14.2 Synonymization of *Meloidogyne enterolobii* Yang & Eisenback, 1983 with *Meloidogyne mayaguensis* Rammah & Hirschmann, 1988

This nematode was originally decribed from a population that caused severe damage on the pacara earpod tree (*Enterolobium contortisiliquum*) on Hainan Island in China. Based on female perineal patterns, it was preliminarily identified as *M. incognita*; however, from a morphological approach, the population was very different from *M. incognita* and a new species of RKN was described. The RKN *M. enterolobii* was described from a population sampled in China and isolated from a tree species (*E. contortisiliquum*). This nematode was also reported from other regions in China, but mainly isolated from guava (*Psidium guajava*) (Xu *et al.*, 2004).

A few years later, a new species of RKN was described from specimens recovered from galled roots of aubergine in Puerto Rico and named *M. mayaguensis*. In their original description, the authors indicated that this species resembles *M. enterolobii* but differs from it in some morphological features (Rammah and Hirschmann, 1988). In addition, the esterase phenotype of *M. mayaguensis* (as the population from Puerto Rico was called) is identical to *M. enterolobii* (Esbenshade and Triantaphyllou, 1985). A possible confusion and misidentification of *M. mayaguensis* as *M. incognita* has also been reported (Fargette *et al.*, 1994; Carneiro *et al.*, 2001; Brito *et al.*, 2004). However, in contrast to *M. enterolobii*, the importance of *M. mayaguensis* as a crop pest is well documented (Castagnone-Sereno, 2012). More recently, the taxonomic relationship of these two species was further questioned based on molecular data. Xu *et al.* (2004) demonstrated sequence identity of a mitochondrial DNA (mtDNA) region between these two nominal species and suggested that *M. mayaguensis* should be considered a junior synonym of *M. enterolobii* (Castagnone-Sereno, 2012). Hunt and Handoo (2009) considered *M. enterolobii* the valid name and *M. mayaguensis* a junior synonym. The official synonymization was finally established by Karssen *et al.* (2012) by comparing the holo and paratypes of these two species, using morphological and morphometrical approaches. In this chapter, most of the references cited take this synonymization into account.

14.3 Methods of Identifying *Meloidogyne* Species

The identification of RKNs at the species level using only morphological approaches presents many difficulties: conserved morphology, variable morphometry, host effects, intraspecific variation, mixture of species, parthenogenetic reproduction, existence of cryptic species, and the increasing number of described species whose diagnosis and descriptions are often doubtful (Hunt and Handoo, 2009). There is also a significant problem with the concept of species for predominantly parthenogenetic organisms (Trudgill, 1991; Hunt and Handoo, 2009).

Among the methods used in the diagnosis of *Meloidogyne* spp., we can highlight the perineal configuration of females. For many years, this character has been used in the characterization of *Meloidogyne* species. These patterns have caused many identification errors over many years, as is the case of *M. enterolobii* and *Meloidogyne paranaensis* being confused with *M. incognita* and other species (Carneiro *et al.*, 2016). Afterwards, the characteristics of the males (often rare) and second-stage juveniles (J2) were added to the diagnosis, such as the shape of the labial region, including annealing, and the shape of male and female stylets under scanning electron microscopy. However, with the increasing number of described species it has been difficult to identify several species using only morphological and morphometric criteria. The isoenzyme electrophoresis technique consists of assessing the relative mobility (Rm) of isoenzyme polymorphic bands. The mobility of enzymes in polyacrylamide gel under electric current varies according to their electrical charge and molecular weight, leading to the visualization of bands in different positions on the gel that are specific to most *Meloidogyne* species (Carneiro *et al.*, 2016). The main advantages of this technique include the recognition of several *Meloidogyne* spp., even in a mixed species population, and the characterization of atypical or non-identified species with efficiency and reliability (Carneiro *et al.*, 2000, 2016; Blok and Powers, 2009). Among the studied isoenzymes, esterases are the most used in the identification of RKNs, with more than 40 described phenotypes for different species (Blok and Powers, 2009; Carneiro *et al.*, 2016).

The use of molecular markers based on DNA has already allowed the correct identification of some species of *Meloidogyne* and has proved that these techniques are reliable (Blok and Powers, 2009). Although they are still restricted to a few species, techniques involving molecular tools are excellent diagnostic methods for *Meloidogyne* spp. with the advantage of being independent of the phenotypic variation of esterases, which sometimes involves more complex interpretation (Carneiro *et al.*, 2016). Molecular markers allow simple, accurate and rapid identification (Blok and Powers, 2009), although they do not allow the detection of new or cryptic species which are relatively frequent in the genus *Meloidogyne*.

Considering the difficulty of identifying *M. enterolobii* only by the perineal pattern (Carneiro *et al.*, 2001; Brito *et al.*, 2004) or host races (= *M. incognita* race 2 or 4), it is certain that *M. enterolobii* from guava was misidentified in many countries (Mexico, Venezuela, Brazil and Colombia). In these countries, the identification of *Meloidogyne* in guava has been based on morphological data and differential hosts (Moura and Moura, 1989; Villota and Agudelo, 1997; Crozzoli and Casassa, 1998; Avelar-Mejía *et al.*, 2001; Gallegos-Morales *et al.*, 2009). Studying a population from Venezuela using esterase phenotype, Molinari *et al.* (2005) identified *M. enterolobii* in this country.

The precise identification of *M. enterolobii* in all states of Brazil was done through the En2 esterase profile with two main bands (Rm = 0.7 and 0.9) and two weaker bands (Rm = 0.75 and 0.95) (Carneiro *et al.*, 2001, 2016). Several populations studied by Tigano *et al.* (2010) showed low intraspecific variability and a precise species-specific primer (MK7F) was developed for this species that was validated in the 15 studied populations of *M. enterolobii*, from different countries, previously identified with esterase phenotype.

As mentioned above, morphological identification of *M. enterolobii* is not a simple task, even for a well-qualified taxonomist (Carneiro *et al.*, 2000). By contrast, esterase patterns have been shown to be a valuable tool for specific identification of this species among other common species (Carneiro *et al.*, 2000, 2016). However, the limitation of this technique is that the J2 individuals cannot be reliably diagnosed, hindering its use in routine examinations of soil samples (Castagnone-Sereno, 2012). A number of molecular tools have been developed and demonstrated as efficient in differentiating *M. enterolobii* from other *Meloidogyne* species based on the presence/absence and/or size of the applications in polymerase chain reactions. The molecular target chosen in the various protocols available mainly includes mtDNA (Blok *et al.*, 2002; Brito *et al.*, 2004; Xu *et al.*, 2004), ribosomal DNA (Adam *et al.*, 2007), satellite DNA (Randig *et al.*, 2009) and sequence characterized amplified region (SCAR) markers (Tigano *et al.*, 2010).

14.4 Life Cycle and Host–Parasite Relationships

RKNs usually cause the formation of knots or galls on the roots of susceptible guava plants. The RKNs are sedentary endoparasites, whose female produces an average of 400 to 500 eggs in a gelatinous matrix called egg-mass, usually externally to the root but sometimes internally in the cortical parenchyma. After embryonic development, the first-stage juvenile (J1) undergoes the first ecdysis while still in the egg, giving rise to the J2 which hatches from the egg due to mechanical force exerted by its stylet, and also by the action of chitinases produced in the oesophageal glands and released through the stylet (Abad *et al.*, 2009). When migrating to the soil, the J2, which is the infectious phase, begins the search for roots to feed on, guided by the root exudates of the plants. With the help of degrading enzymes of the plant cell wall, J2 penetrates the tip of the root, migrating intercellularly until reaching the region of the vascular parenchyma where it establishes its feeding site, that is, the formation of multinucleated giant cells (Taylor and Sasser, 1983), preceded by the injection of secretions of substances from the oesophageal glands. The nematode ingests the cytoplasmic content of giant cells and it acts as a metabolic drain that diverts nutrients from the plant for its own benefit. The injection of secretions leads to hypertrophy and hyperplasia of cells, usually accompanied by the widening of the roots with the formation of galls (Moens *et al.*, 2009). During this process, the juvenile's width increases, and it undergoes new ecdyses with the formation of third- and fourth-stage juveniles (J3 and J4) and finally adults (female or male). When there are good conditions for the development of the nematode, in most cases the development of females occurs. During this post-embryonic development, the reproductive system develops and gonads grow (Eisenback and Triantaphyllou, 1991). However, in adverse conditions, such as a high population of nematodes in the root or resistance of the host plant, juveniles that would become females instead become males, as their sexual primordial development results in testicles instead of ovaries. This is the case in resistance presented by *Psidium cattleyanum* to *M. enterolobii*, with the formation of several males at 31 days after inoculation (DAI) (Freitas *et al.*, 2014). This phenomenon is known as sexual reversal and is one of the survival mechanisms of these nematodes, since fewer eggs will be produced and parasitism on the infected plant will be reduced, guaranteeing the survival of the nematode (Freitas *et al.*, 2014). There is no mating in parthenogenetic species like *M. enterolobii* and males remain in the soil or roots until death (Eisenback and Triantaphyllou, 1991). The duration of the life cycle is strongly affected by temperature, humidity and host plant. The females of *Meloidogyne* spp. produce eggs for 3 weeks; after that time, production ceases and they can live a little longer. Males live for weeks, and J2 can live for a few days to months (Taylor and Sasser, 1983). The life cycle of *M. enterolobii* was studied by Freitas *et al.* (2014) on susceptible *P. guajava* cultivar Paluma (commercial cultivar) at 25–30°C using histopathological observations. In the susceptible host, examination of acid fuchsin-stained guava roots and toluidine blue-stained sections showed that J2 were able to penetrate the root tips, migrate along with the sieve elements and develop normally after having initiated the differentiation of feeding sites (Freitas *et al.*, 2014). At 3 DAI, numerous J2 were observed localized in the subapical meristem of the roots, and J2 were present in the root cortex migrating towards the vascular cylinder; at this stage, cell wall damage was frequently seen. At 6–10 DAI, numerous juveniles were observed within the vascular cylinder, and feeding juveniles were visible inside the vascular cylinder close to giant cells. At 17–23 DAI, numerous J3/J4 had established feeding sites with 6–14 well-formed giant cells containing numerous nuclei, dense cytoplasm and small vacuoles (Freitas *et al.*, 2014). At this time, enlargement of the vascular cylinder region took place, and large galls were apparent. Giant cells displayed thickened cell walls. At 27 DAI, adult females with egg-masses were observed associated with severe injuries to the surrounding cells and disruption of the roots at the root surface (Freitas *et al.*, 2014). The optimal temperature for development of *M. enterolobii* was 28°C and was found to correspond to the geographical distribution of this nematode in tropical and subtropical regions. Studies on the development and life cycle of the nematode in India revealed that mature females were observed in guava roots at 24–28 days and started laying eggs at 28–30°C (Ashokkumar *et al.*, 2019).

14.4.1 Symptoms, damage and dissemination

M. enterolobii infections are associated with general chlorosis, discoloration of stems and branches, leaf fall, nutrient deficiency symptoms, and reduced flowering and fruiting (Fig. 14.1B and C). Severely infected trees decline rapidly, culminating in the death of the plant (Fig. 14.1C and D). Roots of infected trees show multiple galls and root necrosis, causing a drastic decrease in fine roots (Fig. 14.2A and B). The RKN infects the whole root system, from small rhizoids to more lignified pivotant roots (Fig. 14.2B). The fruits of infected plants are small, showing early maturation (Fig. 14.2C). The decline and death of guava trees parasitized by *M. enterolobii* in Brazil is a complex disease involving another

Fig. 14.1. (A) 'Paluma' guava tree grafted on a 'BRS Guaraçá' hybrid (30 months old) in an area highly infested by *Meloidogyne enterolobii* (Irrigated District Nilo Coelho, Petrolina, Pernambuco, Brazil). (B, C) Evolution of symptoms on 'Paluma' guava trees in the same area infected with *M. enterolobii*. (D) Wood from dead guava trees used as firewood. Photographs courtesy of José Egídio Flori.

Fig. 14.2. (A, B) Roots of 'Paluma' guava infected with *Meloidogyne enterolobii*. (C) Fruit of a healthy plant (larger) and fruit of an infected plant (smaller) showing early maturation. Photographs courtesy of Regina M.D.G. Carneiro.

microorganism, *Fusarium solani* (Gomes *et al.*, 2011, 2013, 2014). In other countries, other fungi were also reported to be related to the disease syndrome, including *F. oxysporum* (Rashidifard *et al.*, 2019), *Verticillium dahliae*, *Pythium aphanidermatum* and *Trichoderma* sp. (Avelar-Mejía *et al.*, 2001), but for these fungi, no studies have been performed that prove the pathogenicity with Koch's postulates.

In Brazil, *M. enterolobii* was initially considered the only causal agent responsible for the decline and death of infected guava trees. Later, the involvement of *F. solani* in the expression of symptoms was proved (Gomes *et al.*, 2011, 2013, 2014). The parasitism caused by the nematode facilitates the entrance of the fungus into the plant root system, causing extensive root deterioration. The pathogenesis was apparently aggravated by stress factors such as excess or lack of water or drastic pruning, suggesting that there is a component of physiological imbalance involved in guava decline associated with *M. enterolobii/F. solani*. Since then, the decline of guava was treated as a complex disease, resulting from the joint action of these two pathogens. It was also noted that the fungus is dependent on the nematode for penetration into root tissues (Gomes *et al.*, 2014). The synergistic interaction between *M. enterolobii* and *Fusarium* spp. has also been reported affecting guava in Venezuela and Mexico (Suarez *et al.*, 1999). For this reason, many research actions continue in the search for rootstocks resistant to the nematode.

The dissemination of *M. enterolobii* into several Brazilian states (Gomes *et al.*, 2017) took place by means of the commercialization of seedlings infected by the nematode (Souza *et al.*, 2018). However, the parasitism of native plants that act as an inoculum reservoir cannot be disregarded (Lima *et al.*, 2005) and in Paraná state, after preparing areas for planting seedlings, the parasitism of small guava plants by nematodes increased under irrigation (Carneiro *et al.*, 2006a).

In the San Francisco river region (in the state of Pernambuco, Brazil), the area cultivated with guava has decreased from 6000 to 1669 ha in 7 years (2000–2006), a reduction of more than 70% in guava production (Plantec/Codevasf), which means a reduction of 10% per year was estimated (Carneiro *et al.*, 2007). Currently, the problem continues, with areas being removed, the wood becoming firewood (Fig. 14.1D) and areas being replanted. Pereira *et al.* (2009) registered direct damage of 112.7 million reals (~US$61 million) in guava production in the Brazilian states of Rio de Janeiro, Ceará, Rio Grande do Norte, Pernambuco and Bahia.

In addition to this monetary damage, 3703 jobs were lost due to the decline and death of the orchards in these regions. As these regions are also important seedling producers, this nematode has spread to other Brazilian states (Carneiro *et al.*, 2007; Gomes *et al.*, 2017). In India, infestations of *M. enterolobii* in new/old guava orchards and nurseries were detected in Madhya Pradesh, Rajasthan and Uttar Pradesh. In these regions the role of nurseries in *M. enterolobii* spread in association with *F. oxysporum* was reported (Khan *et al.*, 2019a,b).

However, this species was detected in the state of Rio de Janeiro in native areas of Atlantic forest (Lima *et al.*, 2005), suggesting that the nematode was not introduced in Brazil from other countries, as mentioned when it was reported for the first time in the country (Carneiro *et al.*, 2001). In Paraná state it was also detected in guava without any introduction from other states (Carneiro *et al.*, 2006a).

This syndrome is making cultivation of guava in the infested areas unfeasible, causing serious economic problems to growers and the economy of the region (Carneiro *et al.*, 2007; Castro, 2019).

14.4.2 Host status of cover crops, maize and fruit plants for *Meloidogyne enterolobii*

Since the detection of *M. enterolobii* in guava in Brazil (Carneiro *et al.*, 2001), it has been observed that this species of nematode is very polyphagous and parasitizes a large number of plants with economic importance, including fruits, grains, vegetables, and a large number of plants that have resistance genes to other RKNs, such as tomato, pepper and potato (Carneiro *et al.*, 2006b). A compilation of these plants was initially made by Silva and Oliveira (2010) and complemented by Castro (2019).

In this section, we emphasize plants that are considered poor hosts or non-hosts (resistant) to *M. enterolobii* and that have the potential to be used in crop rotation in areas infested by *M. enterolobii*, in order to decrease the population of this nematode in highly infested areas. Carneiro *et al.* (2012) tested the reactions of 38 cover crop plants to *M. enterolobii* under greenhouse conditions. Tomato 'Rutgers' was used as control of the inoculum viability. Among 38 cover crops assessed, 26 were bad hosts (resistant): groundnut (*Arachis hypogaea*) 'Cavalo Vermelho'; white oat (*Avena sativa*) 'IAPAR 126'; black oat (*Avena strigosa*) 'IAPAR 61'; rapeseed (*Brassica napus*) 'CAN 420' and 'CAN 401'; foxtail millet (*Setaria italica*); rattlepod (*Crotalaria anagyroides, C. apiclolice, C. grantiana, C. juncea, C. ochroleuca*); hyacinth bean (*Dolichos lablab*); millet (*Eleusine coracana*); grey velvet bean (*Mucuna cinerea*); green velvet bean (*Mucuna* sp.); black velvet bean (*Mucuna aterrima*); forage radish (*Raphanus sativus* var. *oleiferus*) 'AL 1006', 'Seletina Nova', 'IPR116' and 'Jesuíta'; white tephrosia (*Tephrosia candida*); timbó (*Ateleia glazioveana*); castor bean (*Ricinus communis*) 'IAC80'; triticale (*Triticum aestivum* × *Secale cereale*); cowpea; and Australian cowpea (*Vigna unguiculata*). Eight species/genotypes of cover crops were non-host (immune): groundnut 'IAC Oirã', 'IAC Poitã' and 'IAC Tatuí'; Italian ryegrass (*Lolium multiflorum*); rye (*Secale cereale*) 'IPR 89'; butterfly pea (*Clitoria ternatea*); mung bean (*Vigna radiata*); and cooper (*Glycine wightii*).

Some crop systems alternating non-host or bad host plants may be effective in controlling *M. enterolobii*, taking into account different climatic regions where this nematode causes damage.

Six genotypes of maize ('NB-7361', 'SHS-5080', 'GNX-1020', 'GNX-3010', 'BRS-1031' and 'BM-1115') were bad hosts to *M. enterolobii* with reproduction factors (RFs) < 1.0 and can be used in crop rotation systems to reduce or eliminate nematode populations (Dias *et al.*, 2010). In India, onion and garlic caused significant reduction in *M. enterolobii* populations owing to higher phenol and resultant repellent activity and can be used as cover crops in guava orchards in this country (Vanitha *et al.*, 2018).

Fifteen tropical fruit trees behaved as non-hosts or poor hosts to *M. enterolobii*, including: assaí palm (*Euterpe oleracea*); atemoya (*Annona cherimola* × *Annona squamosa*); avocado

(*Persea americana*); two cultivars of cashew nut (*Anacardium occidentale*), eight citrus rootstocks (*Citrus* spp. and *Poncirus trifoliata*); three cultivars of coconut (*Cocos nucifera*); eight cultivars of grape (*Vitis* spp.); jabuticaba (*Myrciaria jaboticaba*); three cultivars of mango (*Mangifera indica*); mulberry (*Rubus* sp.); 12 cultivars of passion fruit (*Passiflora* spp.); three cultivars of sapodilla (*Manilkara zapota*); two cultivars of soursop (*Annona muricata*); nine cultivars of star fruit (*Averrhoa carambola*); and seven cultivars of strawberry (*Fragaria* × *ananassa*). Only five fruit trees are good hosts: fig (*Ficus carica*); 10 cultivars of banana (*Musa* spp.); six cultivars of Barbados cherry (*Malpighia* spp.); 'Chardonnay' and 'Solferino' grapes; and six cultivars of melon (*Cucumis melo*). Unfortunately, these non-host fruit trees were never tested under field conditions to confirm the results that were obtained in two greenhouse tests (Freitas *et al.*, 2017). These non-host fruit species and cultivars can replace eradicated guava trees in fields severely infested by *M. enterolobii* and become an economic option for growers in areas where the nematode is considered a serious problem (Freitas *et al.*, 2017).

14.5 Resistance in *Psidium* spp. to Root-knot Nematodes

The susceptibility of 43 and 52 wild *P. guajava* accessions to *M. enterolobii* was observed in our studies assessed in Brasília and Petrolina, respectively (Castro *et al.*, 2012; Freitas *et al.*, 2014). The same was detected by other authors to *M. enterolobii* (Maranhão *et al.*, 2001; Carneiro *et al.*, 2007; Almeida *et al.*, 2009; Scherer, 2009). In marked contrast with this work, Casassa *et al.* (1997) evaluated seven accessions of *P. guajava* in relation to *M. enterolobii* (incorrectly called *M. incognita* race 1). The resistance was observed only in *P. guajava* accession 'S3'. Milan (2007, 2010) also reported resistance on guava accessions. In the first work, *P. guajava* 'B-12' was classified as resistant to *M. enterolobii* (incorrectly called *M. incognita*) (RF = 0.88) and having the potential to control the nematode, so that it could be used as a rootstock for commercial guava clones. A total of more than 2000 F_1 seedlings were produced in the hybridization programme and these F_1 seedlings are in the nursery stages for further evaluation. These results also showed that *Psidium longipes* and *Psidium arayan* presented resistance to the guava RKN, but their potential as rootstocks for guava cultivation is still questionable. In the second work, this author observed that three accessions of *P. guajava* ('K-10', 'A-06', 'J-16') were resistant to *M. enterolobii* (incorrectly called *M. incognita*), with gall index (GI) < 2. Unfortunately, the RFs were not measured. These resistant accessions are considered graft-compatible with local commercial clones and thus could be used as rootstocks. No further information about commercial resistant materials was available.

Susceptibility of CISH wilt-resistant rootstock (*Psidium molle* × *P. guajava*) against *M. enterolobii* was detected in field conditions. A consistent association of *F. oxysporum* along with the nematode has been recorded in infested guava orchards (Khan and Singh, 2019).

Psidium friedrichsthalianum (Costa Rican wild guava) was considered resistant to *M. enterolobii* (= *M. mayaguensis* and incorrectly called *M. incognita* race 1 and 2, *M. arenaria* and *M. javanica*) in different countries: Venezuela (Casassa *et al.*, 1997, 1998; Molero *et al.*, 2010), Colombia (Villota and Agudelo, 1997) and Brazil (Carneiro *et al.*, 2007; Freitas *et al.*, 2014).

The performance of guava (*P. guajava*) grafted on to tolerant *P. guajava* rootstock and on to resistant *P. friedrichsthalianum* was evaluated. In the nursery, 100% successful grafts were obtained with guava and Brazilian wild guava (*Psidium guineense*) rootstocks, while success with *P. friedrichsthalianum* was 90% ('Arrayán', 'Criollo'). Initial vegetative growth was significantly higher in the guava rootstock treatment. In the field, this treatment also showed more vigorous growth. 'Arrayán' rootstock clearly showed a strong dwarfing effect, while the 'Brazilian' and 'Criollo' wild guava rootstocks exhibited an intermediate effect on vegetative vigour. Number and weight of fruits were higher when guava was used as a rootstock, and no differences were observed with wild guava rootstocks (Bogantes-Arias and Newcomer, 2010). In order to evaluate the susceptibility to nematodes of the rootstocks of adult plants, an additional control treatment was included in the field, which consisted of the application of nematicides to the guava rootstock treatment. The results indicated that the RKN was able to colonize the roots of the guava rootstocks with and without nematicides, as well as colonizing the roots of Brazilian wild guava. No nematodes were detected in the roots of the 'Criollo' and 'Arrayán' rootstocks, confirming the resistance of *P. friedrichsthalianum* (Bogantes-Arias and Newcomer, 2010).

In our studies, only Costa Rican wild guava *P. friedrichsthalianum*, *P. cattleyanum* (yellow wild guava), *Acca sellowiana* and *Psidium rufum* presented resistance or immunity to *M. enterolobii*. *Psidium acutangulum* and *P. guineense* (Brazilian guava) were susceptible (Freitas *et al.*, 2014). In contrast with our results, Lee *et al.* (1998) classified the only *P. guineense* accession tested as immune. Almeida *et al.* (2009) classified *P. friedrichsthalianum* (probably an incorrect identification) as susceptible to this pathogen (RF = 13.03). Additionally, these authors verified that three non-identified *Psidium* spp. accessions and *Eugenia stipitata* were resistant to *M. enterolobii*, with RF < 1. Ten plants of *P. friedrichsthalianum* and nine of *P. cattleyanum* were compatible with Paluma guava when grafted by budding technique in greenhouse conditions. These grafted plants were maintained in field conditions. One year later, five Costa Rican plants were still alive and producing buds, flowers and fruits, while the plants grafted on *P. cattleyanum* presented an incompatibility reaction (enlargement in grafting region); consequently, all nine plants died during the first year (Freitas *et al.*, 2014). Similar results were obtained for *P. cattleyanum* by Robaina *et al.* (2015).

'Plants using the *P. friedrichsthalianum* rootstock, planted in an infested *M. enterolobii* area in Petrolina, are still alive, with no nematode symptoms after more than 6 years of planting. However, the tree sizes are smaller and their productivity is lower' (J.E. Flori, Petrolina, Brazil, 2020, personal communication).

Although the sources of resistance to *M. enterolobii* in *Psidium* spp. have been successfully identified (Cuadra and Quincosa, 1982; Carneiro *et al.*, 2007; Castro *et al.*, 2012; Freitas *et al.*, 2014), incompatibility or limited compatibility was observed in different studied rootstocks. These studies to identify resistance sources did not result in commercial applications of the results. The development and assessment of a new strategy for an interspecific hybrid among plants of a resistant wild *Psidium* species and susceptible *P. guajava* has been proposed by Costa *et al.* (2012). These authors developed a hybrid (*P. guajava* 'GUA 161-PE' × *P. guineense* 'ARA 138-RR') resistant to the nematode and with high compatibility with guava Paluma and Pedro Sato (Fig. 14.1A). They suggested this hybrid rootstock as an alternative to control the guava RKN in field conditions (Souza *et al.*, 2014). In addition to high resistance to the nematode, the resulting hybrid presented growth similar to that of guava and good vigour when used as a rootstock for the cultivars Paluma and Pedro Sato.

Costa *et al.* (2017) have suggested, considering the segregation and heritability in the broad sense, that the resistant *P. guajava* × *P. guineense* hybrid presented the dominant resistance model, controlled by two genes with epistatic effects. The presence of only one dominant allele is the condition for resistance of the hybrid to *M. enterolobii*. New *P. guajava* × other wild *Psidium* must be developed and assessed in order to enlarge the sources of resistance to the pathogen, which would make it possible to gain effective control of this pathogen in commercial areas of guava orchards (Costa *et al.*, 2017).

14.6 New Prospects Using Genetic Resistance in Brazil

In April 2019, on the occasion of the 46th anniversary of the foundation of the Brazilian Agricultural Research Corporation (Embrapa), the Corporation launched an *M. enterolobii* resistant rootstock. The 'BRS Guaraçá' rootstock cultivar, obtained by 10 years of research carried out at the Embrapa Semi-arid Research Station, is the result of a single cross between guava accession 'GUA 161-PE' (*P. guajava*) and the accession 'ARA 138-RR' (*P. guineense*), carried out in 2010, in the Bebedouro Experimental Field, Petrolina, Pernambuco state. The success of this hybridization was confirmed by DNA markers, as well as by morphological characteristics such as leaf vein. 'BRS Guaraçá' evaluations were performed for resistance to *M. enterolobii* with artificial inoculation under greenhouse and field conditions at Embrapa Semi-arid Research Station and Bebedouro Experimental Field, and in growers' areas in Petrolina. The hybrid resulting from this cross has 50% of the guava genome, minimizing incompatibility with the guava cultivars, including the small size of *P. guineense*. 'BRS Guaraçá' presents a plant of great vigor, which makes it ideal for use as a rootstock' (C.A.F. Santos, Petrolina, Brazil, 2020, personal communication).

The 'Paluma' and 'Pedro Sato' guava cultivars, which are the ones most planted in Brazil, were grafted on hybrids to assess the compatibility. The full-cleft whip grafting method was used with a wedge-shaped cut that was inserted in the rootstock cleft (Costa *et al.*, 2012). In the evaluations of the 'BRS Guaraçá' graft with the main commercial varieties of guava, no exudations, no cracks in the stem of the grafted plants nor any differences in diameter at the grafting site were observed, indicating their compatibility. A high compatibility was observed, which showed fruit production of around 40 t ha^{-1} in harvests carried out 30 months after transplanting in producer areas. The height of the 'Paluma' and 'Pedro Sato' plants grafted on the 'BRS Guaraçá' was greater than 2.0 m at 30 months after transplanting (Costa *et al.*, 2017).

For the production of seedlings of the hybrid cultivar BRS Guaraçá, the protocol adopted for the production of guava seedlings is recommended by Embrapa, with setting rates greater than 75%. 'BRS Guaraçá' was registered in the National Register of Cultivars (RNC) of the Ministry of Agriculture, Livestock and Supply (MAPA) on 1 September 2016, under registration number 35849, with the Brazilian Agricultural Research Corporation (Embrapa) as applicant.

In November 2018, through a public sale, the Embrapa Semi-arid Research Station made available batches of seedlings from plant matrices for implantation of clonal gardens in nurseries, which will start the production of guava seedlings grafted on hybrid 'BRS Guaraçá'.

The biggest limit of this control strategy is the time required to obtain the rootstock matrices and prepare the grafted seedlings for planting in the infested areas. It is the only effective control method to maintain guava plants in areas infested with *M. enterolobii*.

14.7 Management Strategies

Considering that *M. enterolobii* is the major species of nematode (RKN) that parasitizes guava and causes severe damage on this crop (Carneiro *et al.*, 2001, 2012; Castro *et al.*, 2012), there are several control strategies that can be used in an integrated way.

1. In new areas free of the parasite, its introduction must be prevented by using certified seedlings that are not infected by the nematode. Nematological analyses of seedlings or soil samples should be performed to detect *M. enterolobii*, which is easily identified with esterases (roots) or molecular techniques (nematodes from soil).

2. In areas infested by the nematode, the destruction of attacked plants can represent an important strategy, with the elimination of nematode foci preventing them from spreading to healthy guava trees. After uprooting infected plants, all contaminated roots must be carefully piled up and burned inside the cultivation area.

3. In areas where the nematode foci are at the top of the terrain, the lower part must also be sampled, because the nematode tends to percolate with water. In this case, the diversion of excess water from irrigation or rain processes must be the first measure to be adopted.

4. There are no nematicides and biological control agents registered in Brazil for the management of *M. enterolobii* in guava crop (Carneiro *et al.*, 2020).

5. It is possible to clean the infested areas, keeping them free of weeds and using antagonistic plants (cover crops) provided in Carneiro *et al.* (2012). This rotation with cover crop plants, presented Section 14.4.2 above, must be done for at least 1 year. Field trials should be carried out to corroborate the greenhouse tests and recommend this strategy.

6. If the farmer wants to plant another fruit tree in an infested guava area, a list of fruit trees that are non-hosts to *M. enterolobii* was compiled by Freitas *et al.* (2017) and is provided in Section 14.4.2 above.

7. The most important method of control in areas infested by *M. enterolobii* is the use of resistant rootstock, namely 'BRS Guraraçá', which was made available by Embrapa Semi-arid to Brazilian nursery growers.

14.8 Conclusions

The worst nematode problem affecting guava is that caused by the RKN, *M. enterolobii*, and this nematode is a recognized limiting factor in commercial guava production in Central and South American and other countries. Considering the difficulty of identifying *M. enterolobii* (= *M. mayaguensis*) by the perineal pattern only, this species has been misidentified in different regions around the world and it was identified frequently as *M. incognita* or *Meloidogyne* spp. The species identification is possible using esterase phenotype and molecular markers. Considering the low variability among *M. enterolobii* isolates, genetic resistance could be considered the most effective method of control. The synergistic interaction between this nematode and *Fusarium* spp. (guava decline) has been reported as affecting guava in Brazil, Venezuela, Colombia and India. Screening of several *Psidium* species for possible resistant rootstocks to *M. enterolobii* has been performed. The rootstocks *P. friedrichsthalianum*, *P. cattleyanum* and some *P. guineense* were considered resistant, but incompatible with commercial cultivars of *P. guajava* in field conditions. *P. molle*, some *P. guineense* and several of *P. guajava* were considered susceptible. More recently, one hybrid (*P. guajava* × *P. guineense*) produced resistant rootstocks, 'BRS Guaraçá', developed by Embrapa Semi-arid Research Station, in Brazil. These rootstocks presented high resistance to *M. enterolobii* and high compatibility with commercial cultivars of guava in controlled and field conditions. This hybrid resistant rootstock cultivar was made available by Embrapa to Brazilian farmers to be cultivated in areas infested by the nematode. Although this RKN displays a very wide host range, studies showed that crop rotation is possible for cleaning areas infested with the nematode, using 35 available antagonistic plants. Some cultivars of maize are also very promising for use in reducing populations of *M. enterolobii* in infested fields. Fifteen fruit trees are nonhosts (resistant to *M. enterolobii*) and can be planted in infested areas. Considering all the available control methods, the use of the resistant rootstock is the most promising strategy because it makes it possible to plant guava trees grafted on 'BRS Guaraçá' rootstock in areas infested by *M. enterolobii*.

References

Abad, P., Castagnone-Sereno, P., Rosso, M.N., Engler, J.A. and Favery, B. (2009) Invasion, feeding and development. In: Perry, R.N., Moens, M. and Starr, J.L. (eds) *Root-knot Nematodes*. CAB International, Wallingford, UK, pp. 163–181.

Adam, M.A.M., Phillips, M.S. and Blok, V.C. (2007) Molecular diagnostic key for identification of single juveniles of seven common and economically important species of root-knot nematode (*Meloidogyne* spp.). *Plant Pathology* 56, 190–197.

Almeida, E.J., Santos, J.M. and Martins, A.B.G. (2009) Resistance of guava and araca to *Meloidogyne mayaguensis*. *Pesquisa Agropecuária Brasileira* 44, 421–423.

Ansari, R.A. and Khan, T.A. (2012) Diversity and community structure of phytonematodes associated with guava in and around Aligarh, Uttar Pradesh, India. *Trends in Biosciences* 5(3), 202–204.

Ashokkumar, N., Poornima, K. and Kalaiarasan, P. (2019) Embryogenesis, penetration and post penetration development of *Meloidogyne enterolobii* in guava (*Psidium guajava*). *Annals of Plant Protection Sciences* 27(1), 140–145.

Avelar-Mejía, J.J., Teliz-Ortiz, D. and Zavaleta-Mejía, E. (2001) Patógenos associados con el 'declinamiento del guayabo'. *Revista Mexicana de Fitopatologia* 19, 223–229.

Blok, V. and Powers, T.O. (2009) Biochemical and molecular identification. In: Perry, R.N., Moens, M. and Starr, J.L. (eds) *Root-knot Nematodes*. CAB International, Wallingford, UK, pp. 98–117.

Blok, V.C., Wishart, J., Fargette, M., Berthier, K. and Phillips, M.S. (2002) Mitochondrial DNA differences distinguishing *Meloidogyne mayaguensis* from the major species of tropical root-knot nematodes. *Nematology* 4, 773–781.

Bogantes-Arias, A. and Newcomer, E. (2010) Evaluation of four rootstocks for graft in guava (*Psidium guajava* L.). *Agronomia Mesoamericana* 21, 103–111.

Brito, J.A., Powers, T.O., Mullin, P.G., Inserra, R.N. and Dickson, D.W. (2004) Morphological and molecular characterization of *Meloidogyne mayaguensis* from Florida. *Journal of Nematology* 36, 232–240.

Carneiro, R.M.D.G., Almeida, M.R.A. and Quénéhervé, P. (2000) Enzyme phenotype of *Meloidogyne* spp. populations. *Nematology* 2, 645–654.

Carneiro, R.M.D.G., Moreira, W., Almeida, M.R.A. and Gomes, A.C.M.M. (2001) Primeiro registro de *Meloidogyne mayaguensis* em goiabeira no Brasil. *Nematologia Brasileira* 25, 223-228.

Carneiro, R.G., Mônaco, A.P.A., Moritz, M.P., Nakamura, K.C. and Scherer, A. (2006a) Identificação de *Meloidogyne mayaguensis* em goiabeira e em plantas invasoras, em solo argiloso no Estado do Paraná. *Nematologia Brasileira* 30, 293–298.

Carneiro, R.M.D.G., Almeida, M.R.A., Braga, R.S., Almeida, C.A. and Gioria, R. (2006b) Primeiro registro de *Meloidogyne mayaguensis* parasitando plantas de tomate e pimentão resistentes à meloidoginose no estado de São Paulo. *Nematologia Brasileira* 30, 81–86.

Carneiro, R.M.D.G., Cirotto, P.A., Silva, D.B. and Carneiro, R.G. (2007) Resistance to *Meloidogyne mayaguensis* in *Psidium* spp. accessions and their grafting compatibility with *P. guajava* cv. 'Paluma'. *Fitopatologia Brasileira* 32, 281–284.

Carneiro, R.M.D.G., Freitas, V.M., Mattos, J.C., Castro, J.M.C., Gomes, C.B. and Carneiro, R.G. (2012) Major guava nematodes and control prospects using resistance on *Psidium* spp. and non-host crops. *Acta Horticulturae* 959, 41–49.

Carneiro, R.M.D.G., Monteiro, J.M.S. and Gomes, G. (2016) Gênero *Meloidogyne*: diagnose através de eletroforese de isoenzimas e marcadores SCAR. In: Oliveira, C.M.G., Santos, M.A. and Castro, L.H.S. (eds) *Diagnose de Fitonematoides*. Millennium, Campinas, Brazil, pp. 47–70.

Carneiro, R.M.D.G., Monteiro, T.S.A., Eckstein, B. and Freitas, L.G. (2020) Controle de nematoides fitoparasitas. In: Fontes, E.M.G. and Valdares-Inglis, M.C. (eds) *Controle Biológico de pragas da Agricultura*. Embrapa, Brasília, pp. 371–414.

Casassa, A.M., Matheus, J., Crozzoli, R., Bravo, V. and Gonzalez, C. (1997) Response of some selections of guava to *Meloidogyne incognita*, in Mara County, Zulia State, Venezuela. *Fitopatologia Venezolana* 10, 5-8.

Casassa, A.M., Crozzoli, R., Matheus, J., Bravo, V. and Marin, M. (1998) Effect of the root-knot nematode, *Meloidogyne incognita*, on the growth of guava (*Psidium* spp.) in nurseries. *Nematologia Mediterranea* 26, 237–242.

Castagnone-Sereno, P. (2012) *Meloidogyne enterolobii* (= *M. mayaguensis*): profile of an emerging, highly pathogenic, root-knot nematode species. *Nematology* 14, 133–138.

Castro, J.M.C. (2019) *Meloidogyne enterolobii* e sua evolução nos cultivos brasileiros. *Informe Agropecuário* 40, 41-48.

Castro, J.M.C., Santos, C.A.F., Flori, J.E., Siqueira, S.V.C., Novaes, P.A.R. and Lima, R.G. (2012) Reaction of *Psidium* accessions to the *Meloidogyne enterolobii* root-knot nematode. *Acta Horticulturae* 959, 51–57.

Castellano, G., Casassa-Padron, A.M., Ramírez-Méndez, R., Pérez-Pérez, E., Burgos, M.E. and Crozzoli, R. (2012) Plant parasitic nematodes associated with strategic fruit crops in Barait Municipality of the Zulia state, Venezuela. *Fitopatologia Venezolana* 25, 2–6.

Costa, S.R., Santos, C.A.F. and Castro, J.M.C. (2012) Tolerance of *Psidium guajava* × *P. guineense* hybrids to *Meloidogyne enterolobii*. *Série Documentos* 247, 39.

Costa, S.R., Santos, C.A.F. and Castro, J.M.C. (2017) Reaction of *Psidium* accessions to the *Meloidogyne enterolobii* root-knot nematode. *European Journal of Plant Pathology* 148, 405–411.

Crozzoli, R. and Casassa, A.M. (1998) Species and strains of *Meloidogyne* in guava at Venezuela. *Revista de la Facultad de Agronomia* 15, 107–108.

Cuadra, R. and Quincosa, A. (1982) Comportamento de differentes espécies de *Psidium* como patrones para guayabos resistentes a *Meloidogyne*. *Ciencias de la Agricultura* 13, 19–26.

Dias, W.P., Freitas, V.M., Ribeiro, N.R., Moita, A.W. and Carneiro, R.M.D.G. (2010) Reação de genótipos de milho a *Meloidogyne mayaguensis* e *M. ethiopica*. *Nematologia Brasileira* 34, 99–105.

Eisenback, J.D. and Triantaphyllou, H.H. (1991) Root-knot nematode: *Meloidogyne* spp. and races. In: Nickle, W.R. (ed.) *Manual of Agricultural Nematology*. Marcel Dekker, New York, pp. 191–274.

El-Borai, F.E. and Duncan, L.W. (2005) Nematode parasites of subtropical and tropical fruits tree crops. In: Luc, M., Sikora, R.A. and Bridge, J. (eds) *Plant Parasitic Nematodes in Subtropical and Tropical Agriculture*. CAB International, Wallingford, UK, pp. 467–492.

Esbenshade, P.R. and Triantaphyllou, A.C. (1985) Use of enzyme phenotypes for the identification of *Meloidogyne* species. *Journal of Nematology* 17, 6–20.

Fargette, M.K.G., Davies, M.P. and Trudgill, D.L. (1994) Characterization of resistance breaking *Meloidogyne incognita*-like populations using lectins, monoclonal antibodies and spores of *Pasteuria penetrans*. *Fundamental and Applied Nematology* 17, 537–542.

Freitas, V.M., Correa, V.R., Motta, F.C., Sousa, M.G., Gomes, A.C.M.M. *et al.* (2014) Resistant accessions of wild *Psidium* spp. to *Meloidogyne enterolobii* and histological characterization of resistance. *Plant Pathology* 63, 738–746.

Freitas, V.M., Silva, J.G.P., Gomes, C.B., Castro, J.M.C., Correa, V.R. and Carneiro, R.M.D.G. (2017) Host status of selected cultivated fruit crops to *Meloidogyne enterolobii*. *European Journal of Plant Pathology* 148, 307–319.

Gallegos-Morales, G., Cepeda-Siller, M., Hernandez-Castillo, F.D., Acosta-Zamarripa, A.M., Velasquez-Valle, R. *et al.* (2009) Beneficent microorganisms associated to *Meloidogyne incognita* (Kofoid & White) Chitwood in guava (*Psidium guajava* L.) of Calvillo, Aguascalientes, Mexico. *Revista Mexicana de Fitopatologia* 27, 106–112.

Gomes, V.M., Souza, R.M., Mussi-Dias, V., Silveira, S.F. and Dolinski, C. (2011) Guava decline: a complex disease involving *Meloidogyne mayaguensis* and *Fusarium solani*. *Journal of Phytopathology* 159, 45–50.

Gomes, V.M., Souza, R.M., Silveira, S.F.d. and Almeida, A.M. (2013) Guava decline: effect of root exudates from *Meloidogyne enterolobii*-parasitized plants on *Fusarium solani in vitro* and on growth and development of guava seedlings under controlled conditions. *European Journal of Plant Pathology* 137, 393–401.

Gomes, V.M., Souza, R.M., Almeida, A.M. and Dolinski, C. (2014) Relationships between *M. enterolobii* and *F. solani*: spatial and temporal dynamics in the occurrence of guava decline. *Nematoda* 1, e01014.

Gomes, V.M., Ribeiro, R.M., Viana, A.P., Souza, R.M., Santos, E.A. *et al.* (2017) Inheritance of resistance to *Meloidogyne enterolobii* and individual selection in segregating populations of *Psidium* spp. *European Journal of Plant Pathology* 148, 699–708.

Humphreys, D.A., Williamson, V.M., Salazar, L., Flores-Chaves, L. and Gomez-Alpizar, L. (2012) Presence of *Meloidogyne enterolobii* Yang & Eisenback (= *M. mayaguensis*) in guava and acerola from Costa Rica. *Nematology* 14, 199–207.

Hunt, D.J. and Handoo, Z. (2009) Taxonomy, identification and principal species. In: Perry, R.N., Moens, N. and Starr, J.L. (eds) *Root-knot Nematodes*. CAB International, Wallingford, UK, pp. 55–97.

Iwahori, H. and Ban, V. (2009) First report of root-knot nematode *Meloidogyne enterolobii* on guava in Vietnam. *Plant Disease* 93, 675.3.

Karssen, G., Liao, J.L., Kan, Z., Van Heese, E. and Den Nijs, L. (2012) On the species status of the root-knot nematode *Meloidogyne mayaguensis* Rammah & Hirschmann, 1988. *ZooKeys* 181, 67–77.

Khan, A., Shaukat, S.S. and Samad, M.A. (2012) Species diversity and multivariate analysis of nematode communities associated with guava (*Psidium guajava* L.) in Karachi District, Sindh. *Indian Journal of Nematology* 42(1), 75–79.

Khan, R.M. and Singh, A. (2019) Response of guava, vegetable and wilt resistant rootstock against *Meloidogyne enterolobii*. *Annals of Plant Protection Sciences* 27(3), 380–382.

Khan, R.M., Reddy, P.P. and Kumar, S. (2001) Role of nematode and fungi in guava wilt. *Pest Management in Horticultural Ecosystems* 7(2), 152–161.

Khan, R.M., Ahmad, I., Kumar, K.H., Singh, A. and Shukla, P.K. (2019a) Infestation of *Meloidogyne enterolobii* in newly established/old guava orchards and nurseries in Madhya Pradesh, Rajasthan and Uttar Pradesh. *Annals of Plant Protection Sciences* 27(1), 170–171.

Khan, R.M., Ahmad, I., Kumar, K.H. and Singh, A. (2019b) Identification of *Meloidogyne enterolobii* infesting guava using mitochondrial DNA based analysis and host status. *Annals of Plant Protection Sciences* 27(2), 280–284.

Kumar, S. and Rawat, S. (2018) First report on the root-knot nematode *Meloidogyne enterolobii* (Yang and Eisenback, 1988) infecting guava (*Psidium guajava*) in Udham Singh Nagar of Uttarakhand, India. *International Journal of Current Microbiology and Applied Science* 7, 1720–1724.

Lee, M.D., Chen, C.H., Tsay, T.T. and Lin, Y.Y. (1998) Survey and control of guava nematode diseases. *Plant Protection Bulletin (Taipei)* 40, 265–276.

Lima, I.M., Souza, R.M., Silva, C.P. and Carneiro, RM.D.G. (2005) *Meloidogyne* spp. from preserved areas of Atlantic forest in the State of Rio de Janeiro, Brazil. *Nematologia Brasileira* 29, 31–38.

Maranhão, S.R.V.L., Moura, R.M. and Pedrosa, E.M.R. (2001) Reação de indivíduos segregantes de araçazeiro a *Meloidogyne incognita* raça 1, *M. javanica* e *M. mayaguensis*. *Nematologia Brasileira* 27, 173–178.

Milan, A.R. (2007) Breeding of *Psidium* species for root knot nematode resistance in Malaysia. *Acta Horticulturae* 735, 61–69.

Milan, A.R. (2010) Collection and evaluation of guava (*Psidium guajava* L.) for nematode resistance in Malaysia. *Acta Horticulturae* 849, 357–362.

Moens, M., Perry, R.N. and Starr, J. (2009) *Meloidogyne* species – a diverse group of novel and important plant parasites. In: Perry, R.N., Moens, M. and Starr, J.L. (eds) *Root-knot Nematodes*. CAB International, Wallingford, UK, pp. 1–17.

Molero, T., Molina, J. and Casassa-Padron, A.M. (2010) Advances in the genetic study of resistance of cultivars of *Psidium* spp. to *Meloidogyne* spp. in a tropical dry forest. *Acta Horticulturae* 849, 309–318.

Molinari, S., Lamberti, F., Crozzoli, R., Sharma, S.B. and Sanchez-Portales, L. (2005) Isozyme patterns of exotic *Meloidogyne* spp. populations. *Nematologia Mediterranea* 33, 61–65.

Moura, R.M. and Moura, A.M. (1989) Root-knot on guava: a severe disease in Pernambuco State, Brazil. *Nematologia Brasileira* 13, 13–19.

Pereira, F.O.M., Souza, R.M., Souza, P.M., Dolinski, C. and Santos, G.K. (2009) Estimativa do impacto econômico e social direto de *Meloidogyne mayaguensis* na cultura da goiaba no Brasil. *Nematologia Brasileira* 33, 176–181.

Rammah, A. and Hirschmann, H. (1988) *Meloidogyne mayaguensis* n. sp. (Meloidogynidae), a root-knot nematode from Puerto Rico. *Journal of Nematology* 20, 58–69.

Randig, O., Deau, F., Santos, M.F.A., Tigano, M.S., Carneiro, R.M.D.G. and Castagnone-Sereno, P. (2009) A novel species-specific satellite DNA family in the invasive root-knot nematode *Meloidogyne mayaguensis* and its potential use for diagnostics. *European Journal of Plant Pathology* 125, 485–495.

Rashidifard, M., Marais, M., Daneel, M.S., Mienie, C.M.S. and Fourie, H. (2019) Molecular characterisation of *Meloidogyne enterolobii* and other *Meloidogyne* spp. from South Africa. *Tropical Plant Pathology* 44, 213–224.

Robaina, R.R., Campos, G.S., Marinho, C.S., Souza, R.M. and Bremenkamp, C.A. (2015) Grafting guava on cattley guava resistant to *Meloidogyne enterolobii*. *Ciência Rural* 45(9), 1579–1584.

Rossi, C.E., Ferraz, L.C.C.B. and Montaldi, P.T. (2002) Resistance of fruit tree rootstocks to *Meloidogyne incognita* race 2 and *M. javanica*. *Arquivos do Instituto Biológico* 69, 43–49.

Scherer, A. (2009) Ocorrência e hospedabilidade de *Meloidogyne mayaguensis* em goiabeiras e em plantas de cobertura de solo no Paraná. PhD thesis, Universidade Estadual de Londrina, Brazil.

Schoeman, M.H., Labuschagne, N. and Calitz, F.J. (2017) Efficacy of fungicides, plant resistance activators and biological control agents against guava wilt disease caused by *Nalanthamala psidii*. *South African Journal of Plant and Soil* 34(2), 119–124.

Silva, R.V. and Oliveira, R.D. (2010) Ocorrência de *Meloidogyne enterolobii* (sin. *M. mayaguensis*) em goiabeiras no Estado de Minas Gerais, Brasil. *Nematologia Brasileira* 34, 172–177.

Souza, R.B.C., Santos, C.A.F., Flori, J.E., Castro, J.M.C., Costa, S.R. *et al.* (2014) Avaliação aos 6 meses de transplantio em áreas de produtores de híbrido interespecífico de *Psidium* resistente ao *Meloidogyne enterolobii*. *Jornada de Iniciação Científica da Embrapa Semiárido* 9.

Souza, R.R.C., Santos, C.A.F. and Costa, S.R (2018) Field resistance to *Meloidogyne enterolobii* in a *Psidium guajava* × *P. guineense* hybrid and its compatibility as guava rootstock. *Fruits* 73, 118–124.

Suarez, H.Z., Rosales, L.C. and Rondon, A. (1999) Synergistic effect of the fungi *Macrophomina* and *Fusarium* with root-knot nematode *Meloidogyne* spp. on decline of guava. *Nematologia Mediterranea* 27, 79–82.

Taylor, D.T. and Sasser, J.N. (1983) *Biologia, Identificación y Control de los Nematodos de Nódulo de la Raíz (Especies de Meloidogyne)*. Cooperative Publication. North Carolina State University, Raleigh, North Carolina.

Tigano, M., Siqueira, K., Castagnone-Sereno, P., Mulet, K., Queiroz, P. *et al.* (2010) Genetic diversity of the root-knot nematode *Meloidogyne enterolobii* and development for SCAR marker for this guava-damaging species. *Plant Pathology* 59, 1054–1061.

Trudgill, D.L. (1991) Resistance to and tolerance of plant parasitic nematodes in plants. *Annual Review of Phytopathology* 29, 167–192.

Vanitha, P., Poornima, K. and Somasundram, E. (2018) Cover cropping in guava orchards to reduce population of root-knot nematode, *Meloidogyne enterolobii*. *Pest Management in Horticultural Ecosystems* 24(1), 53–57.

Villota, F.J. and Agudelo, F.V. (1997) Evaluation of guava material (*Psidium guajava* L.) for the damage behaviour of *Meloidogyne incognita*. *Fitopatologia Colombiana* 21, 31–38.

Xu, J., Liu, P., Meng, Q. and Long, H. (2004) Characterization of *Meloidogyne* species from China using isozyme, phenotypes and amplified mitochondrial DNA restriction fragment length polymorphism. *European Journal of Plant Pathology* 110, 309–315.

15 Diseases

Ashok K. Misra*
Central Institute for Subtropical Horticulture, Lucknow, Uttar Pradesh, India

15.1 Introduction

There are many major diseases of guava. About 177 pathogens are reported on various parts of the guava plant or are associated with guava fruit, of which 167 are fungal, three are bacterial, three are algal, three are nematodes and one is an epiphyte. About 91 pathogens are reported on fruits, 42 on foliage, 18 on twigs, 18 on roots and 17 fungi are isolated via surface washing of fruits. These cause various diseases, namely pre- and postharvest rots of fruits, anthracnose, canker, wilt, dieback, twig drying, leaf spot, leaf blight, red rust, sooty mould, rust, seedling blight, damping off, etc. (Misra, 1995, 2001, 2003; Misra and Prakash, 1990). Wilt is one of the most destructive diseases of guava in many guava-growing countries and loss due to this disease is substantial. Fruit rots in the field (preharvest rots) or during transit and storage (postharvest rots) are the most serious diseases of guava, which cause maximum loss. Due to the perishable nature of the fruit and very short self-life, guava suffers badly from different rot pathogens. Diseases due to deficiencies of some nutrients like zinc, magnesium and others are also reported besides physiological disorders such as fruit firm rot and internal breakdown. Thus, although the guava is a hardy plant, it suffers from a number of important diseases, namely wilt and fruit rots, resulting in the high loss of guava production.

15.2 Wilt

Since wilt is a disease that is soilborne in nature, there are limitations for its management. The true 'Allahabad Safeda' guava in India seems to have become extinct from the famous Allahabad region due to wilt disease and only its divergent types are now available. The average productivity of guava can at least be doubled if wilt disease is managed effectively.

In South Africa, the cultivar 'Fan Retief' was mostly grown until 1981. Guava wilt disease (GWD) caused by *Nalanthamala psidii* became a serious problem during 1981 and within 10 years it spread throughout the guava-growing areas of Limpopo Province. Quarantine measures implemented in 1985 prevented the spread of the disease to Western Cape Province (Manicom, 1980; Grech, 1985, 1990).

15.2.1 Geographical distribution

Lim and Manicom (2003) stated that GWD was first reported in Taiwan by Kurosawa (1926).

*E-mail: misra_a_k@yahoo.co.in

DOI: 10.1079/9781789247022.0015

In India, the disease was first reported from a famous guava-growing area, Allahabad in Uttar Pradesh, during 1935. After the first report of guava wilt in 1935 from Babakkarpur, Allahabad, Das Gupta and Rai (1947) recorded the disease in Lucknow in severe form. Several more reports were made from Uttar Pradesh (Dey, 1948; Anonymous, 1949a, 1950; Prasad *et al.*, 1952; Mathur, 1956; Edward and Srivastava, 1957; Singh and Lal, 1953; Pandey and Dwivedi, 1985; Misra and Prakash, 1986; Misra, 1987). After the first few reports from Uttar Pradesh, the disease was reported in West Bengal (Chattopadhyay and Sengupta, 1955; Chattopadhyay and Bhattacharjya, 1968a,b). In India, the disease is also reported from Haryana (Suhag, 1976; Mehta, 1987), Punjab (Chandra Mohan *et al.*, 1986), Rajasthan (Katyal, 1972; Bhargava *et al.*, 2003), Delhi state (Anonymous, 1953), Jharkhand (Srivastava *et al.*, 2001), Andhra Pradesh (Jhooty *et al.*, 1984), Madhya Pradesh (in the Hatod area near Indore) and Tamil Nadu (Thanjavur district) (personal observation) and Orissa (Das and Bose, 1993).

Wilt in guava plants (with different pathogens and names) has also been reported in other countries, namely Florida, USA (Webber, 1928; Alfieri *et al.*, 1984), Taiwan (Kurosawa, 1926; Hsieh *et al.*, 1976; Leu and Kao, 1979; Leu *et al.*, 1979), Cuba (Rodriguez and Landa, 1977), the Philippines (Quimio and Quimio, 1975; Quimio *et al.*, 1984), South Africa (Grech, 1985; Schoeman *et al.*, 1997; Vos *et al.*, 2000), Brazil (Tokeshi *et al.*, 1980; Rodrigues *et al.*, 1987; Junqueira *et al.*, 2001), Mexico and Honduras (Grech, 1994), Pakistan (Ansar *et al.*, 1994), Bangladesh (Hamiduzzaman *et al.*, 1997; Hussain *et al.*, 2014), Canberra, Australia (Nakasone, 1998; Lim and Manicom, 2003), Thailand (Athipunyakom and Manoch, 1998; Athipunyakom and Luangsaard, 2008) and Malaysia (Schroers *et al.*, 2003, 2005; Ploetz, 2007).

Wilt of guava with almost similar symptoms to those in Taiwan was later reported from South Africa in 1980 and from the Johor Province of Malaysia in 1995 (Schoeman, 1997). The diseases in Taiwan, South Africa and Malaysia were probably caused by the same, or a closely related, fungus.

In South Africa, the disease first appeared in south-eastern Mpumalanga Province. Within 10 years, it spread throughout the guava-growing areas including Northern Province and drastically reduced the guava area. The areas have clonally propagated the pink cultivar, 'Fan Retief'. Due to quarantine measures taken during 1985, spread of the disease to the Western Cape Province was prevented. Disease is more of a soilborne nature in South Africa (Grech, 1985). In Taiwan, the disease is present in all guava-producing areas and the guava plants survive only 10–12 years. In Malaysia, the disease affected 42% of a commercial planting of the clonally propagated pink cultivar 'Beaumont' (Schoeman, 1997).

15.2.2 Losses

In India, as early as 1953, Singh and Lal (1953) estimated 5–15% loss amounting to almost 1 million rupees due to guava wilt every year in 12 districts of Uttar Pradesh. In West Bengal, it was estimated that the yield in guava is reduced by 80% due to wilt (Chattopadhyay and Sengupta, 1955). Jhooty *et al.* (1984) reported that in Andhra Pradesh land value was reduced to half by the presence of the disease. In Punjab and Haryana, about 150 and 300 acres of guava orchards, respectively, were uprooted during 1978–1981 due to heavy incidence of wilt (Jhooty *et al.*, 1984). In Uttar Pradesh, the once famous guava areas of Unnao, Gangaghat and Bithoor are now no longer dominated by guava; it has been replaced by annual crops. Misra and Shukla (2002) estimated that loss due to wilt in guava varied from 5 to 60% around Lucknow. Gupta *et al.* (2010b) stated that in guava, wilt is the only disease which is threatening guava cultivation in India. Zakir Hussain *et al.* (2014) reported 2907 plants in eight districts of Bangladesh were infected with guava wilt (*Fusarium oxysporum* f.sp. *psidii*) and an average incidence of 22.94% irrespective of guava cultivar grown.

15.2.3 Symptoms

Any guava plant dying cannot be termed as wilt. It is necessary to discriminate the two; that is, wilt and death of the plant. Death of the plant may be due to heavy infestation of insect pests like stem borers, bark-eating caterpillars, termites or white grubs, etc.; it can also be due to root-rotting fungi like *Pythium*, *Phytophthora*, *Rhizoctonia* or *Macrophomina*, etc.; or it can be due to cultural reasons/operations/water deficiency, etc.; or physiological or soil reasons. In recent reports a claim has been made that the disease is caused by nematodes. But this also cannot be grouped under wilt. It is also to be understood that guava wilt is a disease of adult plants and not a disease of guava seedlings or young guava plants.

True wilt is when the xylem vessels are affected, the movement of water and nutrients is restricted and finally the plant dies. Xylem vessels are blocked by the physical presence of pathogen mycelium/spores, gums or formation of tylose, etc. or due to the effect of toxins produced by the pathogen. Hence, it is essential to understand the basics before studying the wilt in guava.

Plants affected by wilt show varied symptoms. However, it can be grouped into two major symptoms: slow and quick wilt. Quick wilt can be termed as true wilt. In this, plants take from 2 weeks to 2 months for complete wilting after the appearance of first visible symptoms (Fig. 15.1A). In slow wilt, plants take more time for complete wilting (Misra and Pandey, 2000b). Sometimes partial wilting also takes place, which is a typical symptom of wilt in guava (Fig. 15.1B). Initially, one side or some portions show symptoms and later the other half is affected.

Wilt (quick wilt) affected plants show sudden drooping of leaves with yellow coloration. At the later stage shedding of leaves takes place. Twigs become bare and fail to bring forth new leaves or flowers and eventually dry up. Fruits of all the affected branches remain underdeveloped, become hard, black and stony, and remain on branches for some time. Later the entire plant becomes defoliated and eventually dies.

In north India, yellowing of the leaves with intervenial chlorosis is generally seen during the month of August in wilt-affected plants. At this stage, the leaves drop even with slight shaking of the plants. During September, general drooping of the leaves takes place. During October complete wilting of plants takes place with almost dried leaves and small dried black fruits hanging

Fig. 15.1. (A) Wilt-affected plant. (B) Partial wilt-affected plant.

(winter crop) on the branch. It is also seen that some plants show wilting of variable degree during different months but later escape/resist wilting. These plants start recovering from December onwards. It is generally observed that around 17% of plants, which initially show some symptoms of wilting, ultimately escape/resist wilting (Misra and Pandey, 2000b).

The finer roots show black streaks, which become prominent on removing the bark (Das Gupta and Rai, 1947). The roots also show rotting at the basal region and the bark is easily detachable from the cortex. The cortical regions of the stem and root show distinct discoloration and damage. Light brown discoloration is noticed in vascular tissues (Chattopadhyay and Bhattacharjya, 1968a,b). Wilted plants later show bark splitting.

15.2.4 Causal organism

Various pathogens are reported by different workers as the causal organism of guava wilt, but although some may be a true causal organism, others are only associated with the affected plants. The pathogens reported by researchers from India are *F. oxysporum* f.sp. *psidii*, *Fusarium solani*, *Macrophomina phaseoli*, *Rhizoctonia bataticola*, *Cephalosporium* sp., *Gliocladium vermoesenii*, *Gliocladium roseum*, *Verticillium albo-atrum*, *Acremonium* sp., etc. (Vestal, 1941; Das Gupta and Rai, 1947; Dey, 1948; Prasad *et al.*, 1952; Chattopadhyay and Sengupta, 1955; Edward and Srivastava, 1957; Chattopadhyay and Bhattacharjya, 1968a,b; Sohi, 1983a,b; Chandra Mohan, 1985; Pandey and Dwivedi, 1985; Opina, 1995; Misra and Pandey, 2000a).

The reports from other countries of the world are different. *Clitocybe tabescens*, *Myxosporium psidii*, *Gliocladium psidii*, *G. vermoesenii*, *Acremonium diospyri*, *Septofusidium* sp., *Penicillium vermoensenii*, *Colletotrichum gloeosporioides*, *F. oxysporum* f.sp. *psidii*, *Meloidogyne* sp., *Helicotylenchus* sp., *Pratylenchus* sp., *Pseudomonas* sp. and *Erwinia psidii* are reported by various workers (Kurosawa, 1926; Webber, 1928; Hsieh *et al.*, 1976; Rodriguez and Landa, 1977; Leu *et al.*, 1979; Tokeshi *et al.*, 1980; Grech, 1985; Rodrigues *et al.*, 1987; Ansar *et al.*, 1994; Hamiduzzaman *et al.*, 1997; Vos *et al.*, 2000; Junqueira *et al.*, 2001; Lim and Manicom, 2003; Schroers *et al.*, 2005).

N. psidii has been reported as the causal organism for wilt disease of guava from South Africa, Taiwan, Thailand and Bangladesh (Schroers *et al.*, 2005; Athipunyakom and Luangsaard, 2008; Alam *et al.*, 2019). *N. psidii* was successfully isolated from root and stem tissues of guava plants grown in orchards at Ishurdi (24°4′2.27352″N, 89°6′59.07312″E) in the north-eastern region of Bangladesh in 2017. Koch's postulates were proved using three isolates (NPB-001, NPB-002 and NPB-004). To confirm the identity of the causal fungus, the internal transcribed spacer (ITS) region of rDNA of the pathogen was amplified with primers ITS1/ITS4 and sequenced. The three isolates sequenced showed 98 to 100% similarity with the sequence of *N. psidii* (AY864836) in the National Center for Biotechnology Information's (NCBI) database using BLAST.

Taxonomic revisions of *M. psidii*, *Gliocladium psidii*, *G. vermoesenii* and *A. diospyri* were made (which are closely related pathogens) and their placement in the genus *Nalanthamala* was done (Schroers *et al.*, 2005).

It is possible that one or more pathogens may be responsible for the disease. Some may help as a cofactor or predisposing factor for the wilt syndrome. Gomes *et al.* (2011) advocated that the guava decline/wilt can be characterized as a complex problem caused by the synergistic effect of these organisms, in which parasitism by the nematode also predisposes the plants to root decay and then entry of the pathogen.

In India, the months of July and August were identified as a suitable period for inoculation, when humidity is high due to the onset of monsoon (Misra and Pandey, 2000a). Furthermore, at ICAR–Central Institute for Subtropical Horticulture (CISH), Lucknow, it was revealed that *G. roseum* was one of the pathogens of guava wilt, which reproduces symptoms of wilt on artificial inoculation besides other pathogens (Misra and Pandey, 2000a) (Fig. 15.2).

In a massive isolation programme at ICAR–CISH, isolations from different places yielded 43 fungal isolates. Some were identified as a causal organism, while others were only associated with the disease (Misra and Pandey, 2000a). Later, *F. oxysporum* f.sp. *psidii*, *F. solani*, *Fusarium chlamydosporum* and *G. roseum* successfully fulfilled Koch's postulates.

In an extensive survey and isolations from a severely wilt-affected area, it was further found that *F. oxysporum* f.sp. *psidii* and *F. solani* are the most frequently isolated pathogens (Misra, 2008). These two pathogens can therefore be considered as the two most important pathogens of wilt disease of guava in India and can be considered as the real pathogen of guava wilt causing typical wilt. The other pathogens are also responsible, but with lesser frequency. It was also found that out of the two pathogens, the frequency of *F. solani* was higher compared with *F. oxysporum* f.sp. *psidii* (Misra, 2008). In the collection of 176 isolates it was found that the isolates of *F. oxysporum* f.sp. *psidii* and *F. solani* collected from different places showed great variability. These cultures were deposited at ICAR–National Bureau of Agriculturally Important Microorganisms (NBAIM), Mau (Misra, 2008).

It was found that *F. oxysporum* f.sp. *psidii* and *F. solani* are highly variable pathogens. Data on the cultural and physiological (temperature and pH) characteristics revealed that maximum mycelial growth was obtained in potato–dextrose–agar (PDA) as semi-solid medium and malt extract broth as liquid medium. Maximum sporulation was recorded in oatmeal agar and mycological broth. The optimum temperature and pH for growth of both *Fusarium* spp. isolates were 28°C and 5.5, respectively. The isolates differed in their colony growth, mycelial mass, macroconidia and microconidia production (Gupta *et al.*, 2010a). On the basis of morphological variability, the isolates of *F. oxysporum* f.sp. *psidii* and *F. solani* were grouped on the basis of colony colour, texture, colour of metabolite released, frequency of their occurrence on culture, sporulation and macroconidia production, and type/percentage of wilting they

Fig. 15.2. Guava plants (5–6 years old), inoculated in the field by the stem hole inoculation technique with *Gliocladium roseum*, reproduced wilt symptoms.

caused. High degree of variation is probably the reason for the variable incidence and severity of the disease in different places.

Relative pathogenic ability was assessed in 50 *Fusarium* isolates: *F. oxysporum* f.sp. *psidii* (n = 14), *F. solani* (n = 32), *F. chlamydosporum* (n = 2) and *Fusarium moniliforme* (n = 2). Isolate F9 (*F. solani*) was found as the most virulent (Misra and Gupta, 2010).

Partial sequencing of *F. oxysporum* f.sp. *psidii* isolate VKGF01 (HM102500), *F. solani* isolates VKGFS1 and VKGFS2 (HMI02502 and HM102503) and *F. chlamydosporum* isolate VKGFC01 (HM102506) was also done. The submitted DNA sequence of *F. chlamydosporum* was compared for genetic position in the *Fusarium* spp. evolutionary phylogenetic tree (Gupta and Misra, 2012).

Guava (*Psidium guajava*) wilt, caused by *N. psidii*, has been a destructive disease in Taiwan, Thailand, Malaysia and South Africa (Hong *et al.*, 2015). A study was conducted to elucidate the importance of root infection by *N. psidii* in guava in Taiwan. The overall isolation frequency of *N. psidii* was significantly higher from roots than from trimmed and pruned twigs. When guava trees were inoculated by applying *N. psidii*-infested soil to injured roots, wilt symptoms developed within 6–13 months. *N. psidii* infection of guava seedlings via root contact with wilted guava trunks was recorded for the first time in Taiwan.

A total of 36 representative isolates of *F. oxysporum* f.sp. *psidii* (six from one location) and 36 representative isolates of *F. solani* (six from one location) collected from six different agroclimatic zones/guava-growing areas of India were subjected to molecular characterization. For *F. solani*, the data from random amplification of polymorphic DNA (RAPD) were used to construct a dendrogram using UPGMA (unweighted pair group method with arithmetic mean) software. This showed that isolates from Agra and Farukhabad regions were very similar and form one group. Punjab and Ranchi region isolates formed a second group. *F. solani* isolates of Allahabad region were distinctly related to isolates of other regions. Similarly, for *F. oxysporum*, RAPD data were used to construct a dendrogram using the UPGMA algorithm. It showed that isolates from Agra and Farukhabad regions were similar and form a group. Isolates from Punjab and Ranchi regions formed a second group.

In Brazil, *Meloidogyne mayaguensis* has become a major threat to guava cultivation (Gomes *et al.*, 2011). Approximately one-third of the cultivated area is infested with it, resulting in decimation of orchards. Gomes *et al.* (2011) stated that guava decline is a complex disease caused by the synergistic effect of *M. mayaguensis* and *F. solani* in which parasitism by the nematode predisposes the plant to root decay caused by the fungus. Gomes *et al.* (2015) further concluded that parasitism by *Meloidogyne* is a local effect on the pathogenicity of *F. solani*. In India Poornima *et al.* (2016) reported that guava orchards of Tamil Nadu are facing sudden decline due to incidence of root-knot nematode (*Meloidogyne enterolobii*). The association of the nematode *Heliocotylenchus dihystera* was found to be higher in the field of affected orchards and is considered to enhance the incidence of wilt (Khan and Misra, 2003). Prevalences of *Meloidogyne incognita*, *Meloidogyne javanica*, *Rotylenchulus reniformis*, *Helicotylenchus* spp., *Hoplolaimus* spp. and *Pratylenchus* spp. were recorded in Haryana, India (Madhu *et al.*, 2019). In addition, fungi associated with guava decline symptoms were *F. oxysporum*, *M. phaseoli* and *Rhizoctonia solani*. It was evident that *M. incognita* and *F. oxysporum* and their interaction were the predominant reason for the cause of guava decline in India.

Shukla *et al.* (2019) opined that early-age wilting of guava trees has mostly been observed due to *F. oxysporum* f.sp. *psidii*, *F. solani* and *M. enterolobii* interaction and this interaction needs to be worked out. Ganeshan *et al.* (2019) stated that guava decline complex is caused by *F. oxysporum* f.sp. *psidii* and is facilitated by *M. enterolobii*.

Guava orchards of Ratlam district, Madhya Pradesh, India are facing symptoms of sudden decline and loss in productivity due to heavy infestation of the highly pathogenic root-knot nematode, *M. enterolobii* (Singh, 2020). It is also causing havoc through predisposing the host to secondary attack by the wilt fungus *F. oxysporum* f.sp. *psidii*, causing a disease complex with synergistic effects on the common host guava. When uprooted, wilted plants show numerous galls in the vascular region of the host. Plants severely complexly infested with *M. enterolobii* and *F. oxysporum* f.sp. *psidii* show small leaves, leaf browning, leaf drop and growth inhibition, whereas roots are distorted with small and large multiple galls, leading to sudden death of the tree (the problem of nematodes in guava and their association with wilt disease have also been dealt in Chapter 14, this volume).

15.2.5 Epidemiology

In India there is a clear understanding on the period of higher disease incidence during the year. Misra and Pandey (1999c, 2000b) found that wilting generally starts at the onset of rains during August and September, with maximum wilting occurring during the month of October. Some plants, which show slight yellowing, start recovering from December onwards. On analysing weather data, it was found that higher rainfall during July–September, with maximum temperature ranging from 31.3 to 33.5°C, minimum temperature ranging from 23 to 25°C and humidity around 76%, favour the wilt incidence. In general, 2 months are required for the complete wilting of plants (from appearance of first visible symptom to complete wilting). However, the minimum period is only 16 days. Zakir Hussain *et al.* (2014) also found that after initiation of wilt symptoms in Bangladesh, it took 15 to 170 days for completion of wilt. In guava 25.51% of plants wilted within the first 30 days and 78.57% cumulative wilting was recorded after 60 days. Higher wilt incidence was recorded in August, September and October, whereas a remarkable decrease of the disease was recorded during winter, namely November and December. There are variable reports about the severity of disease at different soil pH levels and variable fertilizer levels. However, in general it was found that the disease is common in different types of soil and at different pH or different levels of fertilizers. Hence, it cannot be correlated with soil types or pH or different levels of fertilizers, etc.

15.2.6 Histopathology

Pathogens enter the host tissue through the breaks and openings in the epidermis (Fig. 15.3A and B). Necrosis in the internal tissue and vascular bundle restricts the movement of water and nutrients in the plant and thus results in wilting of guava plants.

Physical presence of the pathogen and tylose formation were also recorded in the

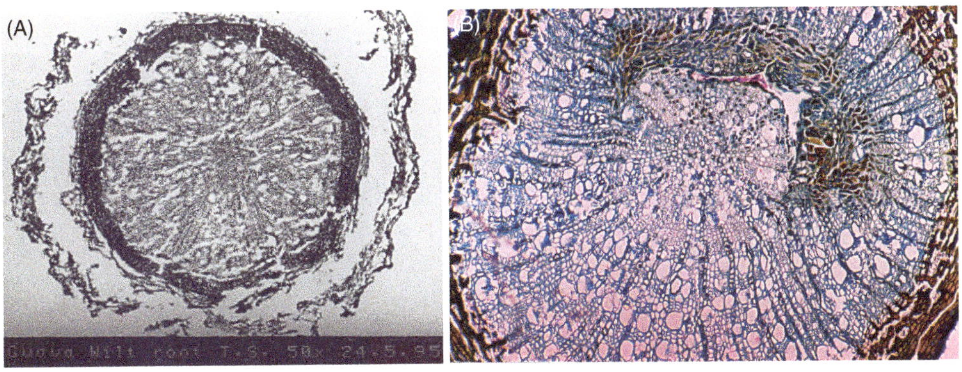

Fig. 15.3. (A, B) Transverse sections of diseased root showing disintegration of epidermis and cortex and necrosis of vascular tissue.

root tissues, which restrict the movement of water and nutrients in the plant (Fig. 15.4A and B).

15.2.7 Disease management

Disease management through chemicals

Different chemical management practices were suggested by different previous workers. These were Chaubatia paste, water-soluble 8-quinolinol sulfate, Benlate or Bavistin, Metasystox and zinc sulfate, thiophanate-methyl, captafol and thiabendazole (Anonymous, 1949a; Jain, 1956; Suhag, 1976; Leu *et al.*, 1979; Bhargava *et al.*, 2003). In South Africa, tebuconazole, propiconazole, prochoraz, triforine and carbendazim + flusilazole were found effective in *in vitro* evaluation (Joubert and Frean, 1993); however, none of these fungicides proved effective in the field. Misra and Pandey (1999b) reported that although different fungicides, namely Bavistin, Topsin M, Indofil M-45, thiram and Blitox, check the various wilt pathogens effectively in the laboratory, when applied in the field (soil) the wilt pathogens increase their aggressiveness and a spore mass is produced once the effect of these fungicides diminishes (Misra and Pandey, 1999a); hence they cannot be recommended. These chemicals are costly and repeated application is also not economical. Moreover, considering the soil mass, increasing aggressiveness of the pathogen, profuse spore production and economics, chemical management seems not very effective.

Besides fungicides some soil amendment chemicals/cakes/fertilizers were also evaluated for control of wilt. Mathur *et al.* (1964) found wilt control by soil treatment with 1.82 kg of lime or gypsum per tree although the control mechanism was not well understood. Oil cakes like neem cake, mahua cake and kusum cake supplemented with urea at 10 kg and 1 kg, respectively, also check the disease (Das Gupta and Ghoshal, 1977). At ICAR–CISH, Lucknow, it was found that wilt was checked by application of 6 kg neem cake + 2 kg gypsum per plant (Misra and Pandey, 1994–1995). These are useful means and can be integrated into integrated disease management practices as one of the components.

Disease management through cultural practices

Mathur (1956) advocated that wilt could be controlled by proper sanitation in the orchard. Wilted trees should be uprooted, burnt and a trench should be dug around the tree trunk. Edward (1960) suggested that while planting, roots of plants should not be severely damaged. Maintenance of proper tree vigour by timely and adequately manuring, interculture and irrigation enables them to withstand infection. The planting pits may be treated with formalin and kept covered for about 3 days and then planting should be done after 2 weeks. Khan and Misra (2003), Prasad *et al.* (2003), Misra

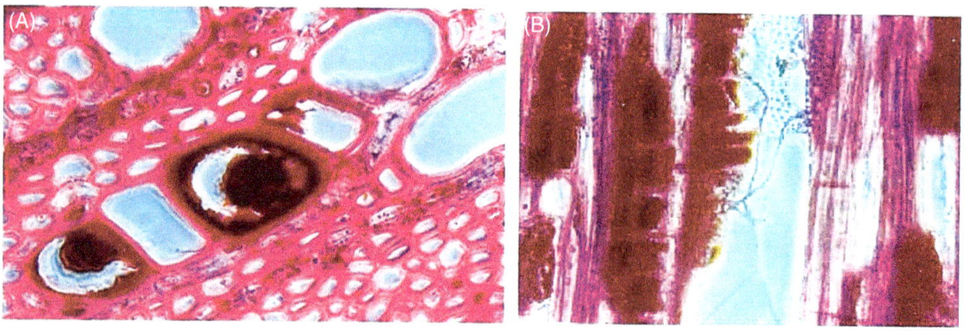

Fig. 15.4. (A) Transverse and (B) longitudinal section of stem showing tyloses and fungal mycelium causing blockage of vessels.

(2004) and Misra *et al.* (2004b) suggested intercropping with turmeric or marigold to restrict the wilting of guava. Misra (2017) also observed that the orchards which were tilled frequently had more incidence of wilt compared with less tilled orchards. Tillage during monsoon enhanced wilt incidence (personal observation). As the fungus is soilborne in nature, flood irrigation spreads the disease propagules. Hence, separate basin irrigation or drip irrigation should be encouraged for the management of disease.

In another pathosystem, Hong *et al.* (2015) compared the isolation frequency of *N. psidii* from the trimming wounds on twigs and the root system. They found that the isolation frequency of *N. psidii* was significantly higher in roots than in the trimming sites on twigs, hence recommended removing the infected roots to avoid further spread of *N. psidii* via root grafting or wounded roots.

Disease management through varietal resistance and rootstock

Edward (1961) reported that cultivars 'Chittidar', 'Hafsi', 'Safeda Riverside', 'Rolf' and 'Stone Acid' were susceptible and guava species *Psidium cattleyanum* var. *lucidium* and also *Syzigium cumini* (Jamun) were resistant to wilt. Edward and Gaurishankar (1964) further reported that *Syzigium cumini*, *Lagerstroemia indica*, *P. cattleianum* (*Psidium molle*), *Psidium quianense*, Chinese guava (*Psidium friedrichsthalianum*) and Philippine guava are resistant to wilt. The strawberry guava (*P. cattleyanum*) has been reported as a relatively hardy species from Réunion (Normand, 1994). At ICAR–CISH, Lucknow, Misra (1998–1999) studied the relative field tolerance of 20 guava cultivars and categorized them into different groups on the basis of their natural susceptibility. The cultivars 'Allahabad Safeda', 'Florida Seedling', 'Guinees', 'Hafsi', 'Karela', 'Mirzapuri Seedling', 'Nasik', 'Pear Shaped', 'Sindh', 'Superior' and 'White Fleshed' proved highly susceptible; 'Behat Coconut' and 'Portugal' as susceptible; and 'Apple Colour', 'Chittidar', 'Seedless', 'Spear Acid', 'Superior Sour Lucidium', 'Red Fleshed' and 'Smooth Green' as tolerant. The cultivar 'Chittidar' is graded as a reasonably good cultivar, which has a reasonable level of resistance as well as good quality and taste; thus can be used directly as resistant material (Fig. 15.5). Misra (2006) developed the stem hole inoculation technique for artificially reproducing the wilt in grown plants and tested some materials. Kamle *et al.* (2012) revealed that *Fusarium* sp. culture filtrate can be potentially employed as a potent selection agent for carrying out *in vitro* selection against wilt disease of guava.

Stock–scion compatibility was also evaluated. When evaluating cultivar 'Safeda' as scion and the resistant material reported by Edward and Gaurishankar (1964) – namely *S. cumini*, *L. indica*, *P. cattleianum* (*P. molle*), *P. quianense*, Chinese guava (*P. friedrichsthalianum*) and Philippine guava – as rootstock, *L. indica* proved incompatible, *S. cumini* as partially compatible and other guava species as compatible. They suggested use of resistant rootstock as a possible means for management of guava wilt. Later, at ICAR–CISH, Lucknow, Misra (2004) again tried *S. cumini* and *L. indica* as rootstock and found late incompatibility in them with

Fig. 15.5. Guava cultivar 'Chittidar', a cultivar tolerant to wilt.

scion *P. guajava*. Further working on the rootstock, Misra *et al.* (2003b) identified an F_1 population of *P. molle* × *P. guajava* free from wilt (Fig. 15.6) when grown in a wilt-sick plot and artificially inoculated repeatedly with wilt pathogens (*Fusarium* spp., *Acremonium* sp. and *G. roseum*). This is a rootstock compatible with any cultivar of guava. This material is now used at ICAR–CISH, Lucknow as resistant rootstock for multiplying plants of the choicest guava varieties.

Two resistant rootstocks, 'TS-G1' and 'TS-G2', were identified by the Agricultural Research Council's Institute for Tropical and Subtropical Crops (ARC–ITSC). These were developed by screening 30,000 guava seedlings *in vitro*. The released rootstock 'TS-G2' saved the guava industry in Mpumalanga and Limpopo provinces (South Africa) from extinction. Again in 2009, a second outbreak affected the resistant 'TS-G2' cultivar, placing the guava industry under threat once more (Schoeman *et al.*, 2012). In a study by Schoeman and Labuschagne (2014), selection 'MS44' showed some tolerance against the G2 isolate of the pathogen obtained from diseased 'TS-G2' trees, while selection 'MS70' showed some tolerance against the G1 isolate obtained from diseased 'TS-G1' trees. These selections were also resistant to the original 'Fan Retief' guava isolate of the pathogen. These selections are useful in wilt management programmes in South Africa.

Fig. 15.6. Rootstock of *Psidium molle* × *Psidium guajava* (F_1) free from wilt.

Disease management through bioagents

As wilt pathogens are primarily soilborne, any management material that multiplies itself in the soil is more useful for the management of the disease. Chemicals (fungicides) do not fulfil this requirement. Chemicals are also hazardous for the soil and the environment; moreover, when the effect of chemicals diminishes, the resistance developed enables the pathogen to become more virulent and aggressive (Misra and Pandey, 1999b). Hence, considering the above facts, it was considered more desirable to use bioagents for the management of the wilt disease of guava (Misra, 2019b).

At ICAR–CISH, Lucknow, three bioagents were found effective for the management of the wilt disease: *Aspergillus niger* strain AN17, *Trichoderma harzianum* and *Penicillium citrinum* (Misra *et al.*, 2000, 2004a; Prakash *et al.*, 2002).

A. niger is the fastest growing and most effective in controlling wilt disease in the field (Misra and Prasad, 2003; Misra, 2004). Besides this quality, it is also a growth enhancer and plants treated with *A. niger* develop faster with greater height, thickness and number of leaves (Misra *et al.*, 2000). These can be grown easily on any substrate like maize/bajra seeds, etc. and can also be multiplied on cheap substrates like *Saccharum* spp. (grass) and dry and green leaves of *P. guajava* (Shukla *et al.*, 2003). At village level these bioagents can be multiplied even in earthen pots (Misra and Prasad, 2004).

Techniques of mass multiplication of bioagents (Fig. 15.7A and B) and application in the field are standardized (Misra *et al.*, 2003a,c; Misra, 2004; Misra and Singh, 2005). Bioagents can be multiplied in farmyard manure (FYM) and applied to plants.

When planting new guava plants, each pit should be filled with 5 kg of this FYM enriched with *A. niger* and then the new guava plant planted in this pit. For adult plants, 10 kg of this enriched FYM should be added near the root zone, properly mixed into the soil by turning and then mulched with organic mulch (Misra, 2018) (Fig. 15.8).

Promising control for the bio-management of guava wilt in South Africa was achieved with the combination of rhizobacterial strains *Bacillus cereus* S7 and *Paenibacillus alvei* T29. This treatment also seems to have a plant growth-enhancing effect apart from disease suppression (Schoeman *et al.*, 2017).

In Egypt, Abdel-Monaim *et al.* (2014) reported that the application of plant growth-promoting rhizobacteria (PGPR) individually and/or mixed, when used as a soil drench treatment, reduced root rot/wilt incidence and severity. The mixed PGPR gave better results compared with individual PGPR.

Fig. 15.7. (A, B) Mass multiplication of bioagents in farmyard manure.

Fig. 15.8. Organic mulching after bioagent application.

Integrated eco-friendly approach

Considering the complexity of the problem, an integrated eco-friendly approach for the management of guava wilt is suggested (Misra *et al.*, 2003c, 2004b; Misra, 2005):

- Use resistant rootstock (*P. molle* × *P. guajava*).
- Apply bioagent (*A. niger* or *Trichoderma* spp. or *P. citrinum*) at the time of planting and regularly once every year in the form of enriched FYM before the monsoon season.
- Intercrop with marigold or turmeric.
- Apply neem cake and gypsum.
- Minimum tillage. Avoid tillage at least during the monsoon.
- Use a drip irrigation system in the orchard.
- Maintain the plant population.
- Maintain sanitation in the orchard.

It is very necessary to implement the complete management package so that effective management is achieved (Misra, 2019a). Following only one or two components does not give the desired result. If wilt disease is managed, the production of guava can easily be doubled.

15.3 Anthracnose

Anthracnose disease is a serious problem in India (Narsimhan, 1938; Mehta, 1951; Venkatakrishniah, 1952; Tandon and Agarwala, 1954; Misra and Prakash, 1986; Rawal, 1993). It causes dieback, twig blight, wither tip and fruit spots. The dieback phase is caused by *Gloeosporium psidii* resulting in the death of plants and was observed at Allahabad (Tandon and Agarwala, 1954).

The disease is reported from Puerto Rico (Lui, 1972), the Philippines (Quimio and Quimio, 1975), Taiwan (Yang and Chuang, 1994), Egypt (Wahid, 2001), Bangladesh (Rahman *et al.*, 2003; Islam *et al.*, 2015), Nigeria (Amusa *et al.*, 2005), Australia (Cooke and Drenth, 2009), Hawaii, USA (Keith and Zee, 2010) and Pakistan (Haider *et al.*, 2016). In Nigeria, 80% of guava plants were found infected with anthracnose disease and over 40% of the fruit produced on those trees were severely infected.

15.3.1 Pathogen (*Gloeosporium psidii* Delacroix = *Glomerella psidii* (Del.) Sheld./*Colletotrichum psidii* Curzi; *Colletotrichum gloeosporioides* teleomorph *Glomerella cingulata*)

Brown to dark brown acervuli form on the affected parts of the plants. Setae and conidia are formed in the acervuli. Mycelium is intercellular, branched and light brown in colour. Conidiophores are hyaline and small, setae are long, tapering at the end, dark brown to black in colour. Conidia are formed at the tip of the conidiophores and are sickle-shaped, unicellular, hyaline, measuring 11.24 µm × 4.5–5 µm. They germinate by germ tube. In moist weather, acervuli appear as black dots on twigs or fruits, which later produce a pinkish spore mass. Spores are disseminated by wind or rain and initiate fresh infection.

15.3.2 Dieback phase

Dieback phase of anthracnose was reported from Minto Park, Allahabad, India in 1952 and the intensity of the disease varies from half-dead to complete death of plants (Tandon and Agarwala, 1954).

Symptoms

The plant dies backwards from the top. Young shoots, leaves and fruits are readily attacked. The greenish colour of the growing tips changes to dark brown and later to black necrotic areas, which extend backwards causing dieback. The disease is more noticeable after a period of incubation in the infected buds and twigs. The brown spots, formed previously, change into silvery grey and ultimately develop at the junction of the diseased and healthy parts. The fungus develops from the infected twigs and then the petiole and young leaves are attacked. These may droop or fall, leaving the dried twigs without leaves. In moist conditions, acervuli of the fungus may be seen as black dots scattered throughout the dead parts of the twigs (Tandon and Agarwala, 1954).

Disease management

Although complete control is not possible, the application of 3:3:50 Bordeaux mixture and 0.22 or 0.33% Perenox gives encouraging results in reducing the development of dieback and mummies (Tandon and Agarwala, 1954). *In vitro* biological control of *C. gloeosporioides* by *Aspergillus flavus* gave good results, followed by *A. niger* and *T. harzianum*. Among fungicides Derosal gives effective control, followed by Bayleton, Daconil, Ridomil Gold, mancozeb and Alliet (Haider *et al.*, 2016).

15.3.3 Fruit and leaf infection phase

Fruit and leaf infection was reported from Saharanpur, India and is generally seen in the rainy-season crop.

Pathogen

Anthracnose of guava fruit is caused by *Gloeosporium psidii* = *Colletotrichum* sp., *Colletotrichum acutatum*. *Colletotrichum* sp. isolated from different fruits are pathogenic either to their original host only or to several hosts (Yang and Chuang, 1994). *C. acutatum*, a new fruit rotting pathogen of guava, was reported from Assam, where 90% of the stored guava fruit rotted in 5–10 days due to this fungus (Das and Bose, 1993).

Symptoms

Pinhead spots are first seen on the unripe fruits which gradually enlarge, measuring 5–6 mm in diameter. They are dark brown to black in colour, sunken, circular and have minute black stromata in the centre of the lesion which produce creamy spore masses in moist weather. Several spots coalesce to form bigger lesions. The infected areas on the unripe fruits become corky and hard, and often develop cracks in the case of severe infection. *Gloeosporium psidii* remains in dormant condition in the young, infected fruits and subsequently it resumes activity and causes rot when the fruit starts ripening (Tandon and Agarwala, 1954). On ripe fruits many small, shallow, water-soaked lesions are produced on the surface and with age they enlarge and become depressed. Subsequently they coalesce to form large spots, irregular in shape and size, and under humid conditions they develop salmon-coloured spore masses at the centre (Srivastava and Tandon, 1969a). On ripe fruits, the infection causes softening of tissues and lesions attain a diameter of 10 to 20 mm. Spot margins are dark brown with pink spores developing in the centre of spots (Fig. 15.9). Anthracnose caused by *Gloeosporium psidii* is also common at Lucknow, but in winter crops symptoms do not develop as well as in rainy-season crops (Misra and Prakash, 1986).

Unopened buds and flowers are also attacked, which causes their shedding. Spread of infection is very rapid on fully matured green fruits, whereas young fruits do not normally get infected (Midha and Chohan, 1968). On leaves, the fungus causes necrotic lesions at the tip or on the margin. These lesions are usually ashy grey and bear fruiting bodies. The tender twigs are also infected, which wither and die from the tip downwards giving a wither tip appearance (Tandon and Singh, 1969).

Disease management

Effective control of anthracnose can be achieved by sprays of Bordeaux mixture

Fig. 15.9. Anthracnose on guava fruit.

(3:3:50) at 7-day intervals. Copper oxychloride and cuprous oxide also significantly control the disease (Tandon and Singh, 1969), but Bordeaux mixture and other copper fungicides cause russeting of fruits especially in 'Allahabad Safeda' guava and reduce their market value (Sohi and Sridhar, 1969). Monthly sprays of Difolatan (0.3%) and Dithane Z-78 (0.2%) are effective in controlling the disease (Anonymous, 1974). Thiabendazole and aureofungin are also effective (Sharma *et al.*, 1983). Growth and acervulus formation are inhibited by thiophanate-methyl, benomyl and thiabendazole at 5–50 ppm.

Storage at 10°C prevents decay but when guava fruits are brought back to room temperature the decay is enhanced. Higher temperature favours decay (Bhargava *et al.*, 1965), so fruits should be stored at cooler temperature.

Guava cultivars 'Apple Guava' (deep red-fleshed), 'Apple Shaped Seedling', 'Behat Coconut', 'Red Chittidar', 'Muzaffarnagar', 'Bulandshahar' and 'Lucknow-49' as well as species *Psidium chinensis* Lodd., *P. cattleyanum* var. *lucidium*, *P. quianense* and *P. molle* are susceptible to anthracnose, while 'Apple Guava' (light red-fleshed) has moderate resistance (Tandon and Singh, 1969; Singh and Bhargava, 1977a,b). *P. chinensis* resists leaf infection whereas *P. molle* and 'Beaumont' are highly susceptible, and 'Allahabad Safeda' develops heavy infections on fruits (Anonymous, 1974). Sharma (1981) reported that anthracnose development is delayed by up to 4 days in cultivars 'Apple Colour', 'Lucknow-49', 'Dharwar', 'Chakaiya' and a hybrid between 'Banarasi Surkha' and 'Apple Colour'. Resistant hybrids were obtained from 'Allahabad Safeda' and 'Banarasi Surkha' guavas (Naresh *et al.*, 1987). A few hybrids from crosses with parent 'Apple Colour' were found resistant. Rahman *et al.* (2003) reported that the pear-shaped fruits had less susceptibility than elliptical round fruits. Local varieties were less susceptible than commercial ones.

15.4 Canker

Fruit canker caused by *Pestalotia psidii* was recorded from Bombay, India in 1911 (Chibber, 1911) and later from Mysore (Narsimhan, 1938; Venkatakrishniah, 1952), Thane, Dharwar, Poona (Patel *et al.*, 1950), Haryana (Kaushik *et al.*, 1972), Paonta Valley, Himachal Pradesh (Verma and Sharma, 1976), Lucknow (Misra and Prakash, 1986) and Rahuri, Maharashtra (Antu, 2013).

Canker due to *Pestalotiopsis psidii* was also reported from Australia, Burma, Ecuador, Malaysia, Mozambique, Nigeria, Puerto Rico, Venezuela, Zambia (Lim and Manicom, 2003) and Hawaii, USA (Keith *et al.*, 2006; Keith and Zee, 2010). In Hawaii, Keith *et al.* (2006) reported that scabby fruit canker, caused by *Pestalotiopsis psidii*, is one of the most common fruit diseases in guava-growing areas and affects all developmental stages of guava fruit. Scabby canker can drastically reduce fruit yield and quality of guava fruits during the preharvest stage and can also lead to fruit losses during postharvest storage. Results from Hawaii showed that there are four *Pestalotiopsis* species infecting guava. These are *Pestalotiopsis clavispora*, *Pestalotiopsis microspora*, *Pestalotiopsis* sp. GJ-1 and *Pestalotiopsis disseminata* (Keith *et al.*, 2006).

Solarte *et al.* (2018) studied guava canker/scab lesions on guava leaves and fruit in different regions of Colombia and identified them as *Pestalotiopsis* and *Neopestalotiopsis* spp. They further clarified that *Neopestalotiopsis* are with versicolorous conidia, while *Pestalotiopsis* have concolorous conidia.

15.4.1 Pathogen (*Pestalotia psidii* Pat., *Pestalotiopsis psidii* (Pat.) Mordue, *Pestalotiopsis clavispora* (Atk.) Steyaert, *Pestalotiopsis microspora* (Speg.) Batista & Peres, *Pestalotiopsis* sp. GJ-1 and *Pestalotiopsis disseminata* (von Thümen) Steyaert, *Neopestalotiopsis* spp.)

Fruit canker is caused by *Pestalotia psidii* (Chibber, 1911). Narsimhan (1940) and Venkatakrishniah (1952) found *C. psidii*, *Glomerella psidii* and *Pestalotia psidii* associated with canker. Venkatakrishniah (1952) advocated that *C. psidii* is a general parasite and *Pestalotia psidii* is specialized to guava. Both species are present on young green and mature fruits/leaves but *Pestalotia psidii* is considered the real cause of canker (Patel *et al.*, 1950). The pathogen develops dark black and circular pycnidia on culture media and fruits, and these contain conidiophores and conidia. The conidia are typically five-celled, oblong, clavate or elliptic-fusoid, erect, hardly constricted at septa, measuring 13–31 µm × 5–10 µm; the three median cells are guttulate, highly brownish; the central cell being the thickest and greatly bulged and other cells are comparatively hyaline; the apical conical or cylindrical cell grows out into three hyaline, slender, elongated appendages; the basal cell is obtuse, erect with a small pedicel. The mycelium of young culture is subaerial, serrate, thin, septate, cottony white to pinkish, irregularly branched and measuring up to 3 µm in diameter. In old cultures, the hyphae are more or less thickened (Patel *et al.*, 1950).

The maximum disease occurs at 25–30°C and at high relative humidity (RH) (Kaushik *et al.*, 1972). The fungus grows in a wide range of hydrogen ion concentrations, but optimum pH is 3.9–4.9 with maximum growth at pH 4.9. The spores of the fungus germinate at 10°C and increase with rise in temperature up to 32°C. Germination of spores of *Pestalotia psidii* is maximum at 30°C, spores do not germinate above 40°C (Ramaswamy *et al.*, 1984). High RH (98%) is required for germination.

Lim and Manicom (2003) and Antu (2013) reported that the disease is caused by *Pestalotiopsis psidii*, a weak parasite which occurs on fruits as an endophyte. It is an opportunist endophyte, which invades fruits through insect injuries (Verma and Sharma, 1976; Lim and Khoo, 1990). Characteristic conidia are produced in acervuli with three apical appendages on leaves and fruits.

Results from Hawaii showed that there are four *Pestalotiopsis* species infecting guava. These are *P. clavispora*, *P. microspora*, *P.* sp. GJ-1 and *P. disseminata* (Keith *et al.*, 2006). The genus *Pestalotiopsis* Steyaert is a heterogeneous group of coelomycetous fungi. *Pestalotiopsis* is characterized by spores having mostly four euseptate and pigmented median cells with two to four apical appendages arising as tubular extensions from the apical cell and a centric basal appendage. However, *Pestalotiopsis* is a complex genus and can be difficult to classify to the species level because characters such as growth rate, conidial morphology and fruiting structure characteristics tend to vary within species. Species are differentiated primarily on conidial characteristics such as size, septation, pigmentation, and presence or absence of appendages. From the studies by Keith *et al.* (2006), different species vary on the basis of minor morphological variation. Conidia vary from 21.7 ± 0.5 (standard error) to 27.7 ± 0.4 µm in mean length and from 5.5 ± 0.1 to 7.5 ± 0.2 µm in mean width. Basal appendages are hyaline, straight or slightly curved, and vary from 2.8 ± 0.1 to 5.3 ± 0.3 µm in mean length. Number of apical appendages ranges from two to four, with three being the most common. The appendages show the most size variation, with mean lengths that range from 11.8 ± 0.6 to 26.5 ± 1.1 µm. On the basis of the morphological and cultural characteristics using Guba's monograph, Keith *et al.* (2006) identified species as *P. microspora*, *P. clavispora*, *Pestalotiopsis psidii* and *P. disseminata*.

Solarte *et al.* (2018) studied 81 isolates obtained from guava canker/scab lesions on guava leaves and fruit in different regions of Colombia and identified them as *Pestalotiopsis* and *Neopestalotiopsis* spp. They analysed the morphology, pathogenicity and genetic diversity of the isolates based on the sequences of the ITS, β-tubulin and elongation factor genes. Isolates were morphologically, pathogenically and genetically diverse. Selected monosporic isolates included in the multiple-gene analysis were identified as belonging to two genera: *Neopestalotiopsis* (65 isolates with versicolorous conidia) and *Pestalotiopsis* (four isolates with concolorous conidia).

15.4.2 Symptoms

The disease generally occurs on green fruits and rarely on leaves. The first evidence of infection on fruit is the appearance of minute, brown- or rust-coloured, unbroken circular necrotic areas which, in advanced stage of infection, tear open the epidermis in a circinate manner. The margin of the lesion is elevated, and a depressed area is noticeable inside (Fig. 15.10). The crater-like appearance is more noticeable on fruits than on leaves. The canker is confined to a very shallow area and does not penetrate deep into the flesh of the fruit. In older cankers, white mycelium consisting of numerous spores is noticeable. Canker on the green fruits of different varieties exhibits considerable differences in appearance (Patel *et al.*, 1950). In severe cases, raised, cankerous spots develop in great numbers and the fruits break open to expose seeds. The infected fruits remain underdeveloped, becoming hard, malformed and mummified, and drop in great numbers. Sometimes small, rusty brown angular spots appear on the leaves (Venkatakrishniah, 1952). In winter, the cankerous spots are common but in the rainy season minute red specks are formed (Verma and Sharma, 1976).

Lim and Manicom (2003) described the disease from South Africa effected by *Pestalotiopsis psidii*, where distinctive dark brown to black, raised spots develop on fruits. The necrotic epidermis tears open to form a small, raised crater with an elevated margin and sunken centrum, the corky scab symptom. As the fruit expands, cracks emanate from the lesions. Keith *et al.* (2006) and Keith and Zee (2010) described it from Hawaii. The main diagnostic symptoms were grey/light brown lesions surrounded by dark brown borders on leaves and brown, raised, corky, necrotic lesions on the exocarp of fruit. Typical symptoms on fruit start as tiny, water-soaked spots. Later these spots darken in colour and become necrotic, and then the tiny spots expand to discrete, circular, dark brown to black spots. It is also observed that multiple lesions coalesce to form a bigger scabbed appearance. As the fruit develop, the small, corky lesions often tear open, giving rise to raised, corky scabs. Symptoms on leaves begin as small dark brown spots, which expand to become grey/light brown circles surrounded by a dark brown border. Lesions also develop on stems of fruits. Solarte *et al.* (2018), from Colombia, described characteristic symptoms of guava scab as corky, ovoid or round lesions on fruit surfaces. These lesions may thicken, affecting the flesh below and reducing fruit quality and commercial value.

Fig. 15.10. Fruit canker of guava.

15.4.3 Disease management

The spread of disease (in the early stage of infection) is controlled by three or four sprays of 1% Bordeaux mixture or lime sulfur (1:25) at 15-day intervals (Venkatakrishniah, 1952). Applications of insecticides to reduce feeding by insects are helpful. Pruning trees and wider spacing can suppress mite damage. Also, natural enemies, such as *Oecophylla smaragdina*, can reduce populations of these insects (Lim and Khoo, 1990).

In cultivar ‘Lucknow-49’, development of canker pustules is large, more elevated and numerous. On cultivar ‘Dhokla’, it is not well developed. On cultivar ‘Sindh’, the development of pustules is insignificant and inconspicuous, while cultivar ‘Nasik’ is almost immune (Patel *et al.*, 1950). ‘Safeda’ and ‘Apple Colour’ are highly resistant cultivars to canker.

15.5 Algal Leaf and Fruit Spot

The alga causes spots on leaves and fruits. Ruehle (1941) reported *Cephaleuros virescens* on the leaves, fruits, twigs and bark of guava from Florida, USA. Thirumalachar (1945) collected *Cephaleuros parasiticus* from Mysore, India, which is responsible for blemishes of guava fruits. Yadav (1953) reported *Cephaleuros* sp. on guava from Patna, India. Misra and Prakash (1986) found *C. virescens* affecting guava leaves in areas of Lucknow and Sitapur, India and the incidence was found as high as 30%. Nelson (2008) reported *Cephaleuros* leaf and fruit spots of guava by two species, *C. virescens* and *C. parasiticus*, which can cause substantial damage to guava cultivars in Hawaii. The disease occurs commonly in relatively wet conditions, such as found in the eastern half of the island of Hawaii or in many of Hawaii's coastal forests.

15.5.1 Pathogen (*Cephaleuros virescens* Kunze ex E.M. Fries (= *Cephaleuros mycoidea* Karst.), *Cephaleuros parasiticus Karsten*)

Ruehle (1941) reported algal leaf and fruit spot caused by *C. virescens* (*C. mycoidea*), while Thirumalachar (1945) reported that it is caused by *C. parasiticus*. Yadava (1953) and Marlatt and Campbell (1980a,b) reported it to be caused by *Cephaleuros* spp. Nelson (2008) reported both species, *C. virescens* and *C. parasiticus*, causing damage to guava varieties in Hawaii. These are aerophilic, filamentous green algae which require a film of water to complete their life cycle. The genus *Cephaleuros* is a member of the *Trentepohliales* and a unique order, *Chlorophyta*. They have photosynthetic ability and are known as green algae. The pathogen extends itself between the cuticle and epidermis and penetrates the epidermal cells. The affected cells eventually die. It is also called red rust because the upper surface of the thallus produces erect, yellow to red filaments and fruiting bodies. These are flat, short, closely crowded branched filaments, beneath which are irregularly branched rhizoids. Most obviously fruiting bodies consist of upright multicellular filaments bearing one to eight sharply bent pedicels. Each pedicel bears a pear-shaped or nearly spherical sporangium, which eventually emits from eight to 32 motile biflagellate spores. On guava, contrary to other hosts, no thallus is apparent on the upper leaf surface, the lesion extends through the entire lamina soon after it appears and the sporulation occurs on the lesion surface on the underside of the leaf (Marlatt and Campbell, 1980a). However, Lin (2005) reported that most of the signs of algal spot appear on the upper leaf surface and sometimes on the lower surface in Taiwan. The alga sporulates readily during the period of greatest rainfall.

15.5.2 Symptoms

Cephaleuros infects immature guava leaves during the early spring flush. Minute, shallow brown lesions appear on leaves; as the disease progresses, the lesions enlarge to 2–3 mm in diameter. On leaves the spots may vary from specks to big patches (Fig. 15.11). They may be crowded or scattered. Leaf tips, margins or areas near the midvein are most often infected.

On immature fruits the lesions are nearly black. As fruits enlarge, lesions become sunken. Cracks frequently develop on older blemishes as a result of enlargement of fruits. Penetration of fruit is confined to several layers of cells beneath the epidermis. Fruit lesions are usually smaller than leaf spots. They are darkish green to brown or black in colour (Ruehle, 1941; Marlatt and Campbell, 1980a,b).

Disease begins to appear from April and is more serious during May to August. The pathogen sporulates readily during the period of highest rainfall (July–September) and the disease incidence is greatest during September.

15.5.3 Disease management

The control of alga can be achieved by sprays of copper oxychloride (0.3%) three or four

Fig. 15.11. Red rust of guava leaf.

times at intervals of 15 days. Spraying copper oxychloride in the rainy season is more effective (Ruehle, 1941). An ascomycetous parasite closely resembling *Strigula astridiza* on *C. parasiticus* can be used for controlling the disease biologically (Thirumalachar, 1945).

Large fruit of the Peruvian cultivar 'Florida' is highly susceptible (Ruehle, 1941). The cultivars 'Patillo' and 'Blitch' are low disease cultivars; 'Ruby' × 'Supreme 6-29' is a moderate disease cultivar; and 'Webber' × 'Supreme' and 'Ruby' × 'Supreme 10-30' are high disease cultivars (Marlatt and Campbell, 1980a,b).

Cultural practices help in reducing the disease (Nelson, 2008). These may be:

- Remove the affected leaves, collect and discard fallen leaves.
- Prune the overcrowded branches and allow sunlight to penetrate. This will reduce RH and maintain dry conditions in the orchard.
- Use balanced fertilizer and keep plants healthy.
- Good drainage of the orchard.
- Control weeds to reduce RH in the plant canopy.
- Manage proper plant spacing to improve aeration and light exposure in the orchard.

15.6 Stem Canker/Bark Canker

Stem canker of guava was reported from Patharchatta, Nainital, India (Rana, 1981), also North Gujarat and Bombay Presidency, India (Uppal, 1936). Wang and Hsieh (2005) from Taiwan reported stem canker of guava by *Botryosphaeria rhodina*.

15.6.1 Pathogen (*Diplodia natalensis* Pole-Evans = *Lasiodiplodia theobromae* (Pat.) Griffon & Maubl.; perfect stage *Botryosphaeria rhodina* (Berk. & Curtis) Arx; *Physalospora psidii* Stevens & Pierce)

Disease can be caused by *Diplodia natalensis*. Pycnidia are grey to black, erumpent, globoid to irregular, ostiolate and measure 160–600 μm in diameter. The pycnidiospores are at first hyaline, later becoming brown, uniseptate and striate. They are oblong to elliptical and measure 19.2–32.0 μm × 11.8–16.0 μm (average: 27.1 μm × 14.8 μm) (Rana, 1981). The bark canker disease of guava has been reported from other parts of India also but the fungus causing the disease is reported as *Physalospora psidii*. Infection originates in the bark and spreads rapidly along the stem from one branch to another, resulting in desiccation, cracking, decortication, and death of the affected part and finally of the whole tree. Numerous perithecia of the causal organism are scattered over the dead bark.

The fungal pathogen *Botryosphaeria rhodina* was identified as the causal agent of a guava stem canker in Taiwan (Wang and Hsieh, 2005). Sexual stages of the pathogen isolated from infected tissues were identified as *B. rhodina*. Spermogonia (160–220 μm × 110–140 μm) were suboblate to conical in shape, with hyaline, rod-like spermatia (2.5–4 μm × 1–1.5 μm). Isolates produced asexual spores of *Lasiodiplodia theobromae* on water–agar medium with a piece of guava leaf under near-ultraviolet light.

15.6.2 Symptoms

Initial symptoms of the disease are longitudinal cracks in the bark on stems or branches, which are visible during the post-monsoon period in October–November. Usually the disease originates from injured bark. On scraping the bark, brown to black streaks or bands are present in the subcortical region. The affected bark turns dark brown to greyish and develops large vertical cracks. The disease spreads up and down from one branch to another and ultimately passes on to the main trunk and upper roots. Leaves on the engirdled portions lose their colour slowly and become purplish bronze. Completely engirdled trees decline and die gradually over the course of 2–3 years (Rana, 1981).

Wang and Hsieh (2005) from Taiwan reported stem canker of guava by *B. rhodina*. Symptoms of this stem canker show bark cracks on the trunks and branches of guava trees, retarding tree growth and leading to leaf wilting and progressive death. A brown discoloration is observed in the infected areas of cambium and xylem, extending to the pith tissues.

15.6.3 Disease management

In severe infection the disease can be prevented by removal and destruction of the infected stem. In mild infection pruning of infected stems and branches is done and the cut ends are painted with Bordeaux paste (1 part copper sulfate and 2 parts each of (lime and linseed oil) or Chaubatia paste (copper carbonate, 800 g; red lead, 800 g; linseed oil, 1 litre). Spraying the tree with copper oxychloride 0.2% after pruning reduces canker incidence.

15.7 Phoma Leaf Blight

The disease was first reported by Sridhar and Ullasa (1978) although it was observed during 1975 from Hessarghatta, Bangalore, India.

15.7.1 Pathogen (*Phoma jolyana* Priozy and Morg.; now *Didymella musae* (P. Joly) Q. Chen & L. Cai)

Leaf blight in guava is caused by *Phoma jolyana* (Sridhar and Ullasa, 1978), now renamed as *Didymella musae*.

15.7.2 Symptoms

First symptom of the blight is the appearance of small circular spots with a dark brown centre surrounded by a reddish margin. In advanced stages, these spots gradually enlarge and coalesce, resulting in large necrotic patches. The fruiting bodies (pycnidia) of the fungus appear in large numbers in the form of small, light brown to black pinheads on the necrotic spots. Both young and old leaves are susceptible to the infection and severely affected plants are completely defoliated (Sridhar and Ullasa, 1978).

15.7.3 Disease management

Two sprays of copper oxychloride or copper hydroxide at 0.3% at an interval of 15 days reduces the disease.

15.8 Cercospora Leaf Spot

Leaf spot of guava caused by *Cercospora sawadae* was recorded from Mauritius (Anonymous, 1965), Nainital, India (Bose and Muller, 1967) and Annamalai University, Tamil Nadu, India (Raghunathan and Prasad, 1969). The disease is severe during December to February.

15.8.1 Pathogen (*Cercospora sawadae* Yamamoto; now *Pseudocercospora sawadae* (W. Yamam.) Goh & W.H. Hsieh)

The disease is caused by *Cercospora sawadae*. The pathogen was renamed as *Pseudocercospora sawadae*.

Conidia are subhyaline to pale olivaceous, slightly curved, obclavate to obclavate-cylindric, one- to five-septate, base obconic, tip obtuse, measuring 3–4.2 µm × 18–87.4 µm. Mycelial patches are hypophyllus, fruiting effuse, irregular, stromae lacking; conidiophores are non-fasciculate, arise singly from hyphal threads, straight or slightly curved, geniculate, two- to five-septate, pale to olivaceous brown, tip conic, measuring 3–3.7 µm × 46–77 µm (Raghunathan and Prasad, 1969).

15.8.2 Symptoms

The disease appears as water-soaked, brown irregular patches on the lower surface and yellowish on the upper surface of the leaf (Fig. 15.12). Old leaves are mostly and severely affected, later curling and subsequently dropping off (Raghunathan and Prasad, 1969).

Fig. 15.12. Cercospora leaf spot of guava.

15.8.3 Disease management

The cultivars 'Red Fleshed', 'White Large', 'Lucknow-49' and 'Chittidar' are susceptible, while the cultivars 'Safeda' and 'Seedless' are resistant (Raghunathan and Prasad, 1969).

15.9 Pestalotia Leaf Spot

A new pathogenic species of *Pestalotia*, *Pestalotia jodhpurensis* sp. nov., that causes leaf spot of guava was reported by Bilgrami and Purohit (1971).

15.9.1 Pathogen (*Pestalotia jodhpurensis* Bilgrami and Purohit)

This leaf spot is caused by *P. jodhpurensis*. Fruiting pustule of typical pycnidium is amphigenous, globoid to ovoid, dark brown, occasionally with a prominent neck, ostiolate, 167–389 µm in diameter; conidia are five-celled, narrowly elliptical, slightly constricted at septa, measuring 17.55–24.30 µm × 3.51–5.40 µm; three intermediate coloured cells are light olive, concolorous, 10.8–13.5 µm; upper two coloured cells are usually smaller than the lowermost coloured cell; apical cell hyaline, short, conic, bearing two or three (usually three) setulae, 8.1–13.5 µm long; basal cell hyaline, long, conic to cylindric, pedicels oblique, 2.7–7.29 µm long (Bilgrami and Purohit, 1971).

15.9.2 Symptoms

The spots originate from the leaf tip or margin and gradually advance towards the base, assuming a dark brown/grey colour. Fruiting pustules develop profusely on the upper surface of the leaves. The pathogen is found consistently associated with such spots.

15.9.3 Disease management

Two sprays with fungicide Kocide WP (copper hydroxide) at 0.3% at an interval of 15 days (Moustafa *et al.*, 2015).

15.10 Curvularia Leaf Spot

Leaf spot of guava caused by *Curvularia siddiquii* was first reported from Naini, Allahabad, India during 1964 (Srivastava, 1963).

15.10.1 Pathogen (*Curvularia siddiquii* Ahmed et Quraishi)

The hyphae are white or olive green, septate, branched, 3.0–4.0 µm wide; conidiophores light brown, variable in size, unbranched, septate, 3.5–4.5 µm wide; conidia large, brown, curved, four-celled, two inner cells bigger than the distal cells, measuring 27.5–39.0 µm × 13.5–21.0 µm (average: 33.0 µm × 16.5 µm).

15.10.2 Symptoms

Dark brown spots on the leaves of guava develop during the months of September and October. The infection is restricted to the tips and margins only in the initial stages, but subsequently the spots cover the whole leaf lamina. Defoliation is observed in the case of severe infections.

15.10.3 Disease management

Two sprays of copper oxychloride or copper hydroxide at 0.3% at an interval of 15 days reduce the disease.

15.11 Pestalotiopsis Leaf Spot

15.11.1 Pathogen (*Pestalotiopsis psidii*) (Pat.) Mordue; *Pestalotiopsis versicolor* (Speg.) Steyaert

Pestalotiopsis psidii causes the Pestalotiopsis leaf spot on guava. *Pestalotiopsis versicolor* (Speg.) Steyaert, causing leaf spot of *Anogeissus latifolia*, was found pathogenic when inoculated on guava (Agarwal and Ganguli, 1959).

15.11.2 Symptoms

Violet spots with a necrotic centre are formed near the midrib and pycnidia are formed on the lower surface (Fig. 15.13) (Misra, 1987).

15.11.3 Disease management

Two sprays of copper oxychloride or copper hydroxide at 0.3% at an interval of 15 days reduce the disease. Singh *et al.* (2017) reported that garlic extract at 2–4% and the bioagents *T. harzianum*, *Trichoderma hamatum*, *P. citrinum* and *Penicillium glabrum* are effective in *in vitro* tests.

15.12 Pseudocercospora Leaf Spot

Pseudocercospora leaf spot is prevalent in the warm, humid and rainy guava-producing areas of South Florida, USA.

Fig. 15.13. Pestalotiopsis leaf spot of guava.

15.12.1 Pathogen (*Pseudocercospora psidii*)

The mycelium of *Pseudocercospora psidii* is internal, olivaceous, consisting of septate, branched, smooth hyphae, 3–4 μm wide. Conidiophores are aggregated in dense fascicles arising from the upper cells of a light brown stroma up to 50 μm wide; conidiophores light brown, smooth, one- to three-septate, subcylindrical, straight to variously curved and unbranched. Conidia solitary, olivaceous, smooth, non-guttulate subcylindrical to narrowly obclavate, apex subobtuse, base narrowly obconically truncate, straight to curved, one- to five-septate. Sporulation is greatest during warm, wet weather and most abundant from May to September. Spores are disseminated via wind, splashing rain, insects and irrigation.

15.12.2 Symptoms

Symptoms may occur on leaves, stems and fruit. Small lesions (2–8 mm) appear as irregular to subcircular, dark smoky brown on the upper leaf surface, with a darker brown, diffuse border. Under high humidity, sporulation of the causal fungus may be seen in lesion centres as greenish grey, felty tufts of mycelium. Individual lesions may coalesce to form large areas of necrotic tissue.

15.12.3 Disease management

Due to the favourable environment for disease development in South Florida, strategic chemical control is deemed necessary for successful guava production. Copper-based fungicides are labelled for use on guava in Florida.

15.13 Rust of Guava

The taxonomic position of *Puccinia psidii* has recently been revised (Beenken, 2017). Phylogenetic analyses as well as several former studies show that *Puccinia psidii* does not belong to the genus *Puccinia* and appears outside the family *Pucciniales*. It is, however, closely related to the genera *Dasyspora*, *Puccorchidium*, *Sphenorchidium* and *Sphaerophragmium*. Consequently, the new genus *Austropuccinia* has been erected and placed in the newly circumscribed family *Sphaerophragmiaceae*. As the information available in the literature was until recently associated with the pathogen name *Puccinia psidii* G. Winter (now *Austropuccinia psidii*), the former name is retained for convenience in this chapter.

Rust of guava caused by *Puccinia psidii* is a native of South and Central America, where it was first discovered on guava and originally described from infected *Psidium pomiferum* leaves (Winter, 1884); hence, its vernacular name is guava rust. Later, it was reported in 1949 in the distribution map released by the Commonwealth Mycological Institute (CMI) (Anonymous, 1949b). The description of *Puccinia psidii* is provided in the CMI description set 6 (Anonymous, 1965). *Puccinia psidii* is a very unusual rust with an extremely wide host range within the *Myrtaceae*. It is able to infect more than 15 genera and 30 species of *Myrtaceae* and poses a potential threat to *Myrtaceae*-based industries worldwide in *Puccinia psidii* disease-prone regions. Fourteen species of eucalyptus were recorded as hosts of *Puccinia psidii*. The pathogen is regarded as a major threat to eucalyptus plantations and other *Myrtaceae* worldwide (Glen *et al.*, 2007).

Rust is a serious problem to guava production in Brazil. It damages guava fruit in home gardens or commercial plantations in disease-prone regions of Brazil and is a minor pathogen of a range of *Myrtaceae*, including several *Psidium* species, in natural vegetation. It is a most common and serious disease (Ribeiro and Pommer, 2004). *Puccinia psidii* caused epidemics in Brazil and affected 'Paluma' cultivar of guava, the most important commercial guava variety in Brazil.

Since its first report from Brazil (Winter, 1884), the disease has extended its geographic range to other South and Central American states, including the continental USA (Florida) (Marlatt and Kimbrough, 1979). *Puccinia psidii* has reached the Pacific

Hawaiian island of Oahu, where it was reported on *P. guajava* (Uchida *et al.*, 2006) and other hosts including *Metrosideros polymorpha* (a new host species), *Syzygium jambos* and *Eugenia* spp., and subsequently it is now found on all other Hawaiian Islands except Niihau (Killgore and Heu, 2005).

On the Australian continent, *Puccinia psidii* infects various *Myrtaceae* species and is highly destructive to eucalyptus forest, having become very important (Coutinho *et al.*, 1998; Carnegie *et al.*, 2010). Kawanishi *et al.* (2009) reported rust disease on ohia and *Puccinia psidii* as the causal fungus in Japan.

Fig. 15.14. Rust of guava.

15.13.1 Geographical distribution

Puccinia psidii is known from South America, mainly east of the Andes, including northernmost Argentina, Uruguay, Paraguay, Brazil, Venezuela, Ecuador and Colombia; Central America and the Caribbean, including Cuba, Dominican Republic, Jamaica, Puerto Rico, Trinidad and Tobago; and Florida in the USA (Commonwealth Mycological Institute, 1987; Coutinho *et al.*, 1998; Booth *et al.*, 2000; Alfenas *et al.*, 2003).

15.13.2 Losses

Ferrari *et al.* (1997) reported that guava rust, caused by *Puccinia psidii*, attacks all the young tissues of the plant and can cause losses of up to 80–100% in Louveira, São Paulo state, Brazil. Martins *et al.* (2014) opined that rust incidence during flowering and fruiting directly influences guava yield and is the main cause of yield loss in Rio de Janeiro state, Brazil. Maximum yield loss was equivalent to 91% and the mean yield loss was 88%. Equivalent reduction in the number of fruits was also recorded.

15.13.3 Symptoms

Puccinia psidii infects young, actively growing leaves (Fig. 15.14) and shoots, as well as buds, inflorescences, flowers, sepals and young fruits of guava. It is most severe on young, unripe *Psidium* fruits and floral buds, rather than leaves and shoots. Lesions are brown to grey with masses of bright yellow or orange yellow urediniospores. Occasionally, lesions have sori containing dark brown teliospores or a mixture of the two spore types. Severe rust disease in young trees may kill shoot tips, causing loss of leaders and a bushy habit. However, depending on the level of resistance, punctiform pustules may be formed over the brown, necrotic lesions.

In the nursery, the shoots of guava seedlings are generally severely affected, while in orchards the damage is primarily on reproductive organs. Large rust pustules completely jeopardize the quality of mature guava fruits. Besides, the rust pustules predispose guava fruits to decay by secondary infections (rot pathogens) and opportunistic insects in maturing fruits.

15.13.4 Pathogen (*Puccinia psidii* G. Winter; now *Austropuccinia psidii* (G. Winter) Beenken)

Uredinia amphigenous, although mostly hypophyllous, caulicolous, on flowers and fruits; when on leaves in groups of brownish or blackish spots up to 5 mm diameter, subepidermal, becoming erumpent, pale yellow when young, later pale-yellow orange, pulverulent, 0.1–0.5 mm diameter; urediniospores globose, ellipsoid to obovoid, 19–27 μm × 15–26 μm, wall hyaline to yellowish, 1.5–2.5 μm thick, echinulate, germ pores obscure (Hernández, 2006).

Telia similar to uredinia or teliospores in uredinia; teliospores ellipsoid to oblong, rounded at apex, narrow below, slightly constricted at septum, 30–48 μm × 17–22 μm, wall 1.5–2.5 μm thick at sides, 2–4 μm thick at apex, pale yellowish, smooth; pedicel colourless, deciduous (Hernández, 2006).

Puccinia psidii is capable of infecting many species in the *Myrtaceae*. This fungus was originally described from infected *P. pomiferum* leaves (Winter, 1884) and initially only *Psidium* spp. were listed as hosts (Sydow and Sydow, 1904–1924). Arthur (1915, 1922) and Stevenson (1975) listed only *P. guajava* and *S. jambos* as hosts. Ferreira (1989) suggested that the host range has expanded within the *Myrtaceae* family. He speculated that the wild ancestor of *Puccinia psidii* was able to infect a number of species, such as *S. jambos*, *Eucalyptus* spp., *Myrcia jaboticaba*, *Callistemon speciosus*, and wild and commercial cultivars of *P. guajava*. De Castro (1983) and De Castro *et al.* (1985) reported that there are three physiological groups of *Puccinia psidii*. Group 1 infects *Eucalyptus* spp. and *S. jambos*, Group 2 infects *Eucalyptus* spp. and *P. guajava* and Group 3 is able to infect *P. guajava* only. *Puccinia psidii* has some physiological specialization to *P. guajava* since guava isolates are frequently non-virulent to eucalyptus and other hosts (Coelho *et al.*, 2001; Aparecido *et al.*, 2003a); however, this relationship is not absolute to designate formae speciales because certain guava isolates can induce mild symptoms in some eucalyptus genotypes and other hosts (Aparecido *et al.*, 2003a; Furtado *et al.*, 2005). *Puccinia psidii* and its anamorph *Uredo psidii* J.A. Simpson, K. Thomas & C.A. Grgurinovic have many synonyms. A review of species of *Uredinales* pathogenic on species of *Myrtaceae* (Simpson *et al.*, 2006) described eight rust species, including *U. psidii*, *Uredo rangelii* and *Uredo seclusa*, which are all anamorphs of *Puccinia psidii sensu lato* (Glen *et al.*, 2007).

Zhong *et al.* (2008) developed and characterized 15 polymorphic microsatellite markers present in the genome of the guava rust fungus, *Puccinia psidii*. The primers for these microsatellite markers were designed by sequencing clones from a genomic DNA library enriched for a simple sequence repeat (SSR) motif of (AG). All these 15 primer pairs successfully amplified DNA fragments from a sample of 22 *Puccinia psidii* isolates, revealing a total of 71 alleles. The observed heterozygosity at the 15 loci ranged from 0.05 to 1.00. The SSR markers developed would be useful for a population genetics study of the rust fungus.

15.13.5 Life cycle

Puccinia psidii is considered to be an autoecious species with an incomplete life cycle. With the exception of spermogonia, all stages are produced on the same myrtaceous host. Aecia and aeciospores are morphologically identical to uredinia and urediniospores (Figueiredo, 2001). Under natural conditions, *Puccinia psidii* produces abundant urediniospores. Teliospores and basidiospores are comparatively less. Frequency on all hosts is higher in warmer months (Ferreira, 1983). Aeciospores have not been observed or recognized in nature (Figueiredo, 2001). Production of teliospores is stimulated by temperature. Basidiospores are produced free of urediniospores. After about 18 days, aecia and aeciospores are produced, which are morphologically indistinguishable from uredinia and urediniospores. Spermogonia, however, have not been observed. Urediniospore germination and infection are affected by temperature, leaf wetness, light intensity and photoperiod (Ruiz *et al.*, 1989a,b). Several studies have agreed that high humidity or leaf wetness and low light for a minimum of 6 h following inoculation are necessary for successful germination and infection (Piza and Ribeiro, 1988; Ruiz *et al.*, 1989a,c). Several studies have determined different optimum temperatures for urediniospore germination.

Teliospores germinate and basidiospores are produced at temperatures ranging from 12 to 24°C with maximum basidiospore production at 21°C (Aparecido *et al.*, 2003b). Ninety per cent of the urediniospores germinate within 6 h of inoculation and 90% of these form appressoria within 18 h, whereas a low percentage enter through stomata without appressorium formation. Infection pegs from the appressoria penetrate between the anticlinal walls of the epidermal cells and colonize the mesophyll (Glen *et al.*, 2007).

15.13.6 Epidemiology

Humid regions with mild temperatures and long periods of leaf wetness during flowering and initial fruiting stages favour rust epidemics in guava (Rocabado, 2003).

15.13.7 Disease management

Various workers have tried various fungicides for the management of fungal rust of guava, which are summarized below:

- Spray of 1% Bordeaux mixture after winter rain has been recommended. The next spray is given when trees begin their active growth and subsequent sprays are at monthly intervals. B 03818 (a combination of nickel chloride and zineb) at 0.03% has also proven effective for the control of rust and is equivalent to 1% Bordeaux mixture (Andrade, 1959).
- Chlorothalonil, mancozeb, copper oxychloride, oxycarboxin and triforine give 100% control of *Puccinia psidii*. Triadimenol, triforine and oxycarboxin exhibit therapeutic effects (Ruiz *et al.*, 1991).
- Ferrari *et al.* (1997) from Brazil reported that products based on chlorothalonil are most efficient in the control of guava rust, presenting control above 85%. For management of guava wilt, frequent fungicide spraying should be made (Ferrari *et al.*, 1997; Goes *et al.*, 2004).
- Triadimenol and azoxystrobin were most effective in reducing the incidence of rust on fruit in the Northern Region of Rio de Janeiro state, Brazil. Treatment with triadimenol gives the best result (Martins *et al.*, 2011).
- *Fusarium decemcellulare* (*Nectria rigidiuscula*) was found as a hyperparasite on the urediniospores of *Puccinia psidii* in Recife, Pernambuco, Brazil and can be used as biocontrol (Amorim *et al.*, 1993).
- Sweet guava cultivars, both white- and red-pulped, are more susceptible to rust than the sour cultivars. Vigorous growth cultivars have high infection (Andrade, 1951).

15.14 Damping Off of Seedlings

Damping off of seedlings is a serious disease and often responsible for enormous losses in nurseries (Tandon, 1961; Gupta, 1978; Lim and Khoo, 1990).

15.14.1 Pathogen (*Rhizoctonia solani* Kühn; *Rhizoctonia* sp.)

Damping off is caused by *Rhizoctonia solani* (Gupta, 1978). Tandon (1961) reported seedling blight of guava caused by *Rhizoctonia* sp., but he did not specify the species.

15.14.2 Symptoms

Both pre-emergence and post-emergence phases of the disease are observed. In the pre-emergence phase, the infected seeds and seedlings show water-soaked discoloration, the seed becomes soft and ultimately rots. The affected young seedlings are killed before they reach the soil surface. In the post-emergence phase, the hypocotyl at ground level or the upper leaves are discoloured to a yellowish to brown colour, which spreads

Fig. 15.15. Damping off of guava seedlings.

downwards, and later turn soft and finally rot and constrict. The affected seedlings ultimately topple down and die (Fig. 15.15). Strands of mycelium may appear on the surface of the plants under humid conditions (Tandon, 1961; Gupta, 1978).

15.14.3 Disease management

For management of the disease, the following recommendations may be made:

- Diseased guava seedlings and weeds should be removed and burned immediately.
- Excessive use of water and close planting should be avoided as the organism is moisture loving. Seedbeds should be prepared with proper drainage arrangement.
- As the fungus survives on several hosts, planting of other susceptible hosts should be avoided.
- Seed treatment with captan or thiram (2 min dipping of guava seeds).
- Drenching of soil with copper oxychloride helps in reducing disease intensity in the nursery.
- Good control of the disease is achieved by Bavistin (carbendazim) and Brassicol (quintozene) at 3 and 5 g kg^{-1} seed, respectively (Gupta, 1979).

15.15 Clitocybe Root Rot

15.15.1 Pathogen (*Clitocybe tabescens* (Scop.) Bres.)

Root rot caused by *Clitocybe tabescens* has been observed most frequently in guava trees, killing them in old citrus groves, in various localities of Florida, USA (Rhoads, 1927; Webber, 1928). The disease is reported on *P. guajava*, *P. molle* and cattley guava (Rhoads, 1927; Webber, 1928; Rhoads, 1942).

15.16 Phytophthora Fruit Rot

Fruit rot of guava caused by *Phytophthora parasitica* was first reported by Mitra (1929) in Bihar, India. He described that the disease appears in the wet season (July–September). The disease was also reported to cause considerable losses in Mysore state, India during the rainy season and it is prevalent in other states as well (Rao, 1966; Sridhar *et al.*, 1975; Gupta *et al.*, 1977; Rawal, 1993). The disease was also reported from Cuba (Ariosa, 1982).

Cultivars 'Allahabad Safeda', 'Banaras', 'Banglore Local', 'Red Fleshed', 'Pink Fleshed' and 'Seedless' were found susceptible to the disease (Sohi and Sridhar, 1971; Singh *et al.*, 1976).

Fruit rot of guava caused by *Phytophthora citricola* was first reported by Ko *et al.* (1982) in an orchard of Waiakea-Uka on the island of Hawaii. The distribution of disease appears to be limited (Ko *et al.*, 1982). The fungus is self-inducing (homothallic), producing oospores in single culture.

15.16.1 Pathogen (*Phytophthora parasitica* Dastur/*Phytophthora nicotianae* var. *parasitica* (Dastur) G.M. Waterh.; *Phytophthora citricola* Sawada)

Fruit rot of guava is caused by *P. parasitica* (Mitra, 1929; Sohi and Sridhar, 1971) and *P. nicotianae* var. *parasitica* (Singh *et al.*, 1976) as well as *P. citricola* (Ko *et al.*, 1982).

15.16.2 Symptoms

Those fruits which have fallen on the ground or hang near ground level were reported to be mostly affected (Fig. 15.16), as are those which have been placed in storage. The disease starts at the stylar end. The whitish cottony growth develops very quickly as the fruit ripens and is able to cover almost the entire surface within a period of about 3–4 days during humid weather. The fruits near soil level covered with dense foliage under high RH are most severely affected. The skin of the fruit below the whitish cottony growth becomes a little soft, turns light brown to dark brown and emits a characteristic unpleasant smell. The diseased fruits generally retain their normal shape unless they are invaded by saprophytes, which cause rotting. These fruits either remain intact or drop off. When the disease appears on young and half-grown fruits, they shrink, turn dirty brown to dark brown, remain hard in texture, either remain intact as mummified fruit or drop off (Singh *et al.*, 1976).

In fruit rot of guava caused by *P. citricola*, fruit hanging close to the soil surface rot during rainy season. The infected area appears greyish brown and water soaked with a greyish black centre. Mature fruits appear to be less susceptible than mature-green or green fruits.

Fig. 15.16. Phytophthora fruit rot of guava.

15.16.3 Disease management

Diathane Z-78 (0.2%) or aureofungin (10 ppm) are reported effective in controlling the disease, while copper oxychloride is found toxic to the fruits (Sohi and Sridhar, 1971).

Cultivars 'Lucknow-49', 'Banarsi Surkha', 'Allahabad Safeda' and 'Mishri' were reported as highly susceptible, 'Tehsildar' as moderately susceptible, and 'Chittidar' and 'Apple Colour' were quite resistant (Gupta *et al.*, 1977).

15.17 Diplodia Dry Rot

The dry rot of guava fruits was observed during 1969 in India. In some of the infected trees more than 40% of the fruits were infected (Rajgopalan and Wilson, 1972a).

15.17.1 Pathogen (*Diplodia natalensis* Pole-Evans)

The dry rot of guava fruit is caused by *Diplodia natalensis*. The pycnidia of the fungus produced on guava fruits are erumpent, more or less globose, dark coloured and measure 175–475 μm × 90–185 μm. Pycnidiospores are initially hyaline, oblong and unicellular. On maturity they become oblong to elliptical, two-celled and dark brown, having longitudinal striation on the wall. They measure 21.5 to 32.2 μm long and 12.2 to 16.5 μm wide (Rajgopalan and Wilson, 1972a).

15.17.2 Symptoms

Initially light brown spots develop, mostly at the stalk end or at the calyx end of the fruit. In a few cases, infection spreads quickly and within 3–4 days the entire fruit is affected. Completely infected young and mature fruits become dark brown to almost black in colour and ultimately dry up. A number of dry fruits can be seen on infected trees. Numerous pycnidia of the pathogen appear as pinhead-like structures on the rind of the dried fruits (Rajgopalan and Wilson, 1972a).

15.17.3 Disease management

Ziride 3000 ppm controls Diplodia dry rot of guava fruits in orchards and does not cause any phytotoxic effect on flowers and fruits of guava when it is applied at 15-day intervals (Rajgopalan and Wilson, 1972b).

15.18 Phomopsis Fruit Rot

In India, the disease was reported from Saugar (Rao *et al.*, 1976). About 60% of the fruits of a plant were found affected.

15.18.1 Pathogen (*Phomopsis destructum; Phomopsis psidii*)

The disease is caused by *Phomopsis destructum*. The colony is white, forming zones in the medium. Mycelium is thin walled, hyaline, branched, septate; pycnidia dark coloured, leathery to carbonaceous, ovoid, thick walled, formed generally in 20-day-old cultures, 300–600 µm in diameter. Sporogenous cells are hyaline, simple, rarely branched, phialidic enteroblastic, arising directly from the innermost layer of cells lining the pycnidial cavity. Fungus produces two types of spores: stylospores, long, slender, often curved and bent at one side like a walking stick, 11–30 µm × 1.5–2.0 µm; and pycnidiospores, fusiform, small, hyaline, non-guttulate, 2.8–8.0 µm × 2.0–2.8 µm (Rao *et al.*, 1976). Another pathogen, *Phomopsis psidii*, is also reported as a cause of the disease (Khare *et al.*, 1994).

15.18.2 Symptoms

The infected fruits show disease symptoms near the stalk. Under favourable environmental conditions, infection centres are numerous. Lesions are dark brown, at first small and then increasing in size to 2 cm diameter. The tissues soften and the entire fruit rots within 8–12 days (Fig. 15.17). The rotten fruits fall from the parent plants, causing heavy loss in the yield of the crop.

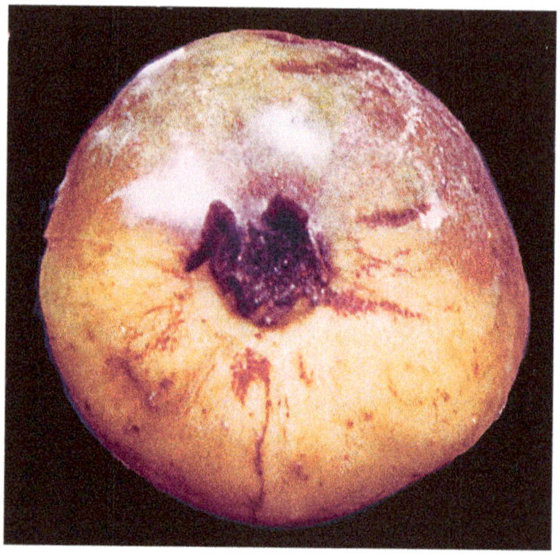

Fig. 15.17. Phomopsis fruit rot of guava.

15.18.3 Disease management

In *in vitro* tests, griseofulvin was found very effective in completely inhibiting the mycelial growth at 22 ppm, while aureofungin and nystatin inhibited it at 100 ppm. Tetracycline and chloramphenicol are also effective (Rao and Agarwal, 1976a). Among fungicides, Rao and Agarwal (1976b) found Blitane, Blitox and Cuman very effective against *Phomopsis* at a lower dose of 0.1%.

15.19 Guignardia Fruit Rot

Guignardia fruit rot of guava was recorded on cultivar 'Beaumont' in transit as well as in the field from Bangalore, India in 1980 (Ullasa and Rawal, 1984). Zang *et al.* (2011) reported black spot of guava due to *Guignardia mangiferae* (strain pathogenic to guava) from Ningbo, China.

15.19.1 Pathogen (*Guignardia psidii* Ullasa & Rawal; *Guignardia mangiferae* Roy)

A fruit rot is caused by *Guignardia psidii* (Ullasa and Rawal, 1984; Misra, 1987). Colonies on PDA medium are greenish grey and become bluish black with abundant aerial mycelium. Dark grey to black, submerged mycelium consists of green to brownish black hyphae. On PDA the fungus produces both ascogenous as well as pycnidial stages after 10 days of incubation under laboratory conditions (Ullasa and Rawal, 1984).

G. mangiferae (strain bl2) colonies on PDA medium are dark green or olive, with a granular surface. Later the colony gradually become darker in colour, nearly black olive. The ascoma is spherical or pear-shaped, with a papillary orifice; the surface is often covered with irregular hyphal outgrowths, 125.7–553.3 μm × 125.7–272.2 μm in size; the asci are fasciculate, bitunicate, clavate, eight spored, 12.3–15.9 μm × 4.0–4.7 μm. The ascospores are single, hyaline, short cylindrical and swollen in the middle, slightly curved, 12.3–15.9 μm × 4.0–4.7 μm in size. Conidia are ovoid to almost pyriform, hyaline, 8.1–11.5 μm × 3.9–6.2 μm at the apex, with a flexible, 3.9–6.2 μm long mucoid apical appendage, and the entire conidium wall is surrounded by a stout, rigid, 1.1–2.5 μm thick mucoid sheath (Zang *et al.*, 2011).

15.19.2 Symptoms

Symptoms develop as minute depressed or flattened spots on the ripening fruits. In these spots, the fungus develops in a concentric manner. Several spots later coalesce and form bigger lesions. Ullasa and Rawal (1984) reported that no fungal fruiting structure appeared on fruit, while Misra and Prakash (1986) noted the formation of circular black spots on the surface of ripe fruits due to this fungus.

The typical symptoms due to *G. mangiferae* disease include black spots on fruits, which gradually expand. The spots are round or irregular in shape, slightly sunken, dark brown first and then black, surrounded by a light brown halo. Later, the spots become contiguous and the diseased part becomes lignified and shrunk, causing reduced fruit quality and decreased economic value (Zang *et al.*, 2011).

15.20 Pestalotia Fruit Rot

Pestalotia fruit rot of guava is caused by two species of *Pestalotia*: *Pestalotia psidii* and *Pestalotia olivacea.*

15.20.1 *Pestalotia psidii* Pat.

The disease is reported from Allahabad, Lucknow and Hisar (Srivastava and Tandon, 1969a; Misra and Prakash, 1986; Kaushik *et al.*, 1970).

Symptoms

The disease begins in the form of a brownish discoloration which becomes russet coloured after a week. As the spots enlarge, their central region becomes slightly depressed and subsequently minute, single or gregarious black acervuli with viscoid spore masses appear on the infected region. In partially rotted fruits, the margin of infected tissue appears ochraceous buff in colour (Srivastava and Tandon, 1969a).

Disease management

If fruits are stored at lower temperature and under drier conditions, they can be protected from *Pestalotia psidii* fruit rot. Postharvest wash with aureofungin (200 ppm) can also protect guava fruits for 5 days. 'Allahabad Safeda' guava was classified as susceptible, while 'Chittidar' was moderately susceptible and 'Apple Guava' was resistant to *Pestalotia psidii* (Singh and Bhargava, 1977a,b).

15.20.2 *Pestalotia olivacea* Guba

The disease was reported from Kurukshetra, India (Dhingra and Mehrotra, 1980).

Pathogen

The colony of the pathogen *P. olivacea* is white, hyphae hyaline, branched, septate, 2.5–3.5 µm wide; conidia four- or five-celled and varied coloured with the two end cells on either side hyaline and conical, the next cells light brown and the middle cells dark brown with thick wall and septa. Conidia 2.5–32.5 µm × 4.5–6.0 µm. Setulae, two or three on either side of the conidia attached to hyaline cells by a short pedicel, thread-like pointed ends, 23–25 µm long (Dhingra and Mehrotra, 1980).

Symptoms

In this rot, infection on fruits starts in the form of brownish coloured watery lesions, which later change to russet-coloured spots. White fluffy growth along with black pin-head-like acervuli appear after 1 week (Dhingra and Mehrotra, 1980).

15.21 Stylar End Rot

The stylar end rot of guava fruits is caused by *Phomopsis* sp. and has been recorded in India (Rai, 1956; Rawal, 1993). Later, it was considered to be caused by *Phomopsis psidii* de Camara (Srivastava and Tandon, 1969a).

15.21.1 Pathogen (*Phomopsis psidii* de Camara)

The disease is considered to be caused by *Phomopsis psidii* (Srivastava and Tandon, 1969a). The fungus produces dark carbonaceous pycnidia on rotting fruits and also in culture. Under moist conditions, masses of spores ooze out from these pycnidia through the ostiole. Pycnidia are ovoid and thick walled. They measure between 140 and 400 µm in diameter. Conidia are ovate to elongate, hyaline, in the range of 5–9 µm × 2.5–4.0 µm. Stylospores are long, slender, curved and of dimensions 16–30 µm × 0.8–1.4 µm. The fungus requires a minimum temperature for growth near 10°C, an optimum near 25°C and maximum of about 35°C (Rai, 1956).

15.21.2 Symptoms

The first visible symptom of the disease is the discoloration in the region lying just below and adjoining the persistent calyx (Fig. 15.18). The discoloration area gradually increases in size and turns dark brown. Later the affected area becomes soft. Along with the discoloration of the epicarp, the mesocarp tissue also shows discoloration and the diseased area is marked by being pulpy and light brown in colour, in contrast to the bright white colour of the healthy area of the mesocarp. At an advanced stage, due to disorganization of the inner affected tissues, the fruit shrinks and concentric wrinkles develop on the skin. Finally the whole fruit is affected and is covered with pycnidia. At all stages of development of the disease the affected tissue shows abundant fungal hyphae which are mostly intercellular (Rai, 1956).

15.21.3 Disease management

α-Naphthol (50 ppm) and guaicol (250 ppm) cause complete inhibition of mycelial growth of *Phomopsis psidii* and treated fruits showed no disease symptoms (Khare *et al.*, 1994).

15.22 Sour Rot

15.22.1 Pathogen (*Geotrichum candidum* Link ex Pers.)

A fruit rot of guava caused by *Geotrichum candidum* was first recorded in Nagpur,

Fig. 15.18. Stylar end rot of guava fruits.

India in October 1974. Mandarin orange, lemon, lime, sweet orange, banana, apple, tomato, grape vine, watermelon, muskmelon, cucumber fruits and tubers of potato were found to develop symptoms on artificial inoculation (Shankhapal and Hatwalne, 1976).

15.23 Soft Watery Rot

Soft watery rot is one of the most common, widely occurring diseases of guava in India. This disease was initially recorded from India (Edward *et al.*, 1964; Srivastava and Tandon, 1969a,b; Patel and Pathak, 1995). Adisa (1985) reported it in Nigeria, grouped it into a soft rot-causing organism and recognized the disease as of high occurrence.

15.23.1 Pathogen (*Botryodiplodia theobromae* Pat.)

Soft watery rot of guava is caused by *Botryodiplodia theobromae*. The optimum temperature for growth of the pycnidia of *B. theobromae* is 25°C; conidial germination is maximum at 30°C but at 10°C the spores do not germinate (Srivastava and Tandon, 1969a,b; Patel and Pathak, 1995). Germination of spores is highest at 100% RH and lowest at 30% RH (Patel and Pathak, 1995).

15.23.2 Symptoms

The infection starts as a brownish discoloration mostly at the stem end and gradually proceeds downwards in an irregular wavy manner. Finally, the whole fruit may be affected. In advanced cases numerous small pycnidia are produced over the entire surface of the fruit. The rot produced by the pathogen is soft and watery.

15.23.3 Disease management

Captan is found effective (Srivastava and Tandon, 1971). No rotting takes place for up to 10 days, if fruits are stored at 10–15°C (Srivastava and Tandon, 1969b). Storage at 10°C prevents decay; however, when guava stored at 10°C are brought back to room temperature, the decay is enhanced (Bhargava *et al.*, 1965).

Cultivars 'Safeda' and 'Apple Colour' were found susceptible, while 'Pear Shaped' was moderately susceptible (Srivastava and Tandon, 1969b).

15.24 Aspergillus Soft Rot

Aspergillus soft rot is caused by several species of *Aspergillus*, of which *Aspergillus awamori*, *Aspergillus wentii* and *A. niger* are important.

15.24.1 *Aspergillus awamori* Nakazawa

Soft rot of guava caused by *A. awamori* was reported from Allahabad, India (Lal *et al.*, 1980). Unlike other Aspergillus rots, which are mainly responsible for postharvest decay, this disease occurs on unripe fruits and results in 10–15% loss.

Symptoms

The disease spreads as small, circular, water-soaked spots, which enlarge and become russet brown in colour with age. The diseased lesions become soft in the middle and later develop warm-sepia to clove-brown coloured conidial heads surrounded by a white circular ring mixed with citron-yellow coloured fungus mycelia. The disease spreads on the whole fruit with a black mouldy growth. The fruit rots completely within 10–12 days and the soft pulpy tissue emits a fermented odour. High humidity and temperature greatly favour development of the disease. However, very young green fruits fail to show any soft rot symptoms.

Disease management

Fruits treated with Bavistin (carbendazim) and Saprol (triforine) at 1250 ppm completely check the rot (Arya *et al.*, 1981).

15.24.2 *Aspergillus wentii*

The infection starts as a discoloured area. The spot later turns pulpy and yellow brown. The infected portion shows maceration of the tissues. Optimum temperature and RH for development of the fungus are 35°C and 70%, respectively (Gupta *et al.*, 1979).

15.24.3 *Aspergillus niger*

The fungus develops brownish, soft spots which advance in size and depth, eventually becoming blackish in colour. Fruit shrink and consequently reduce in size. An intensive maceration of tissue results in oozing of yellow-brown watery secretion mixed with spores and emission of a disagreeable odour (Gupta *et al.*, 1979). With increase in the incubation period, the vitamin C content of both healthy and infected fruit tissue decreases but the rate of decline in healthy fruit is much less than in infected ones (Singh and Tandon, 1971). Adisa (1985) reported soft rot caused by *A. niger* from Nigeria also.

15.25 Pestalotiopsis Rot

15.25.1 Pathogen (*Pestalotiopsis psidii*)

Misra (1987) and Rawal (1993) reported fruit rot due to *Pestalotiopsis psidii* from India. It causes reddish or brown-yellow superficial spots which cover about 50% of the fruit surface and results in soft rotting.

15.26 Fusarial Rot

15.26.1 Pathogen (*Fusarium equiseti*; *Fusarium oxysporum*; *Fusarium moniliforme* var. *intermedium*)

Adisa (1985) reported rot of guava in Nigeria by *Fusarium equiseti* and *F. oxysporum*, while Misra (1987) reported *Fusarium moniliforme* var. *intermedium* causing large dark spots with cracking of epidermal tissue in Lucknow, India.

15.27 Rhizopus Fruit Rot

Fruit rot of guava caused by *Rhizopus stolonifer* was first reported from Kauai, Hawaii, USA by Ooka (1980). He observed soft rot affecting mature-green to fully ripe fruits in orchards of Kauai Island during 1978 and proved its pathogenicity. Adisa (1985) reported *R. stolonifer* and *Rhizopus oryzae* from Nigeria as causing soft rot.

15.27.1 Pathogen (*Rhizopus stolonifer* (Fr.) Lind.; *Rhizopus oryzae*)

Fruit rot of guava is caused by *R. stolonifer*. White sporangiophores and sporangia develop, which later turn black, where the aerial hyphae contact the surface. Sporangiophores and sporangia develop more densely at breaks that may occur in the epidermis of the affected area and at the point of infection.

15.27.2 Symptoms

In the early stages of disease development the lesions appear oily and water soaked. The lesion margins are distinct, and the lesions are slightly sunken at the margin. The rapidly extending lesions reduce the fruit flesh to a semi-solid state in a few days. The epidermis generally remains intact. Aerial hyphae develop at the point of infection. Although development is not extensive, it extends rapidly over the lesion and covers it with a sparse white to grey mycelium. Infected fruits remain attached to the tree until they are manually dislodged or fall in the course of natural maturation. Rhizopus rot is easily distinguished from Mucor rot in the field. Mucor rotted fruits are covered with abundant yellow mycelia and sporangia whereas Rhizopus rotted fruit show comparatively sparse aerial mycelium with dark grey to black sporangiophores and sporangia (Ooka, 1980).

15.28 Mucor Fruit Rot

15.28.1 Pathogen (*Mucor hiemalis* Wehmer)

A fruit rot of guava caused by *Mucor hiemalis* was first reported by Kunimoto *et al.* (1977) from the islands of Hawaii and Kauai. Up to 80% of mature-green fruits on trees were found to be infected.

15.28.2 Symptoms

The infected areas show water-soaked lesions, which develop within a week, and the entire fruit is covered with yellowish, fuzzy mass of fungal fruiting bodies and mycelia. Diseased fruits also give off a yeasty odour. The disease affects the yield of marketable fruits to a considerable degree. It was found that the disease develops only when guava fruits are wounded. *M. hiemalis* therefore is considered a wound parasite on guava fruits (Kunimoto *et al.*, 1977).

15.29 Controlling Fruit Rots

There are several options for controlling fruit rots that include preharvest fungicidal spray, fumigation of fruit by gases (Singh and Sharma, 1982), dips in antibiotics and other chemicals (Vir *et al.*, 1967), gamma radiation (Gupta and Chatrath, 1973b; Arya *et al.*, 1981), temperature treatment (Bhargava *et al.*, 1965), combination of heat and chemical treatment (Wells and Harvey, 1970; Wills *et al.*, 1982), waxing or oiling of fruits (Aulakh and Grover, 1969), wrapping of fruits and skin coatings (Kaul and Munjal, 1982), homeopathic medicine (Khanna and Chandra, 1977, 1989), and by some fungicides and hot water treatment (Pathak and Shekhawat, 1976).

The fruit skin of some guava cultivars contains antifungal substances which do not allow the germination of postharvest pathogens (Singh and Bhargava, 1976).

Weekly spraying of Bordeaux mixture (2:2:50) and copper oxychloride (0.2%) for the control of Phytophthora fruit rot of guava was recommended (Kothari, 1968). But Sohi and Sridhar (1969) observed a toxic effect of various copper fungicides on the fruits of cultivar 'Allahabad Safeda'. Sprays of Blitox, Fytolan and Fycol 8 E were found toxic to the fruits. Toxicity appears in the form of numerous small, dark brown specks, which later coalesce to form an extended russet area. However, excellent control of the disease was achieved with Dithane Z-78 (0.2%) and aureofungin (10 ppm).

Benlate in water at 3000 ppm and with 0.2% paraffinic mineral oil at 2000 ppm protects and checks various postharvest rots of guava fruits caused by *Pestalotia psidii*, *Phoma psidii* and *Gloeosporium psidii*. The loss in weight is minimized by Benlate treatments (Singh and Bhargava, 1977a). Dipping the fruit after harvest in benomyl solutions (temperature 48–50°C for 5 min) at 0.5–2.0 g/l was found effective for controlling fruit rot of guava (Wills *et al.*, 1982). Tetracycline as a post-infection dip at 500 ppm for 20 min and thiourea at 5.0% give protection against postharvest fruit rot (Gupta *et al.*, 1973). Postharvest fruit diseases of guava caused by fungi are controlled by the application of a Decco food-grade fruit coating. Application of coating at 4.04 ml per 100 g of fruits reduces the intensity of various diseases (Krishnaiah *et al.*, 1985). Treating guava fruits with Tecto 40 at 2000 and 2500 ppm concentration keeps the fruits completely free from various rots for up to 8 days and prolongs storage life (Bhargava and Singh, 1974).

Conidial germination and germ tube length of *C. gloeosporioides* (*Glomerella cingulata*) was decreased with increasing gamma ray dosage of 10 kr. Irradiation of fruits with 100 kr at 5, 18 and 24 h after inoculation gives good control of disease. Those fruits treated 42 h after inoculation rotted completely. Pre-inoculation irradiation was not effective (Gupta and Chatrath, 1973a).

15.30 Conclusion

There are a number of pathogens, mainly fungi, which affect guava crop at its different stages of growth. These pathogens cause various diseases on roots, stems, twigs, leaves and fruits in the field as well as on fruits after harvest during transit and storage. Out of all the diseases reported on guava, wilt of guava is the most important disease worldwide and extensive work has been done to understand and manage the disease. Various factors have been studied to understand the problem. More recently, association with root-knot nematodes has been added to this problem. Under pre- and postharvest conditions, pathogens cause various types of manifestation other than wilt, namely anthracnose, rots, rust, scab, sooty mould, stylar end rot, canker, dieback, leaf spot and damping off. Wilt and fruit rots are the most destructive diseases of guava and losses due to these are substantial.

Prevention is the only effective means of reducing losses from most guava diseases. The use of chemicals to control diseases is justified only if significant economic losses are anticipated and no practical alternative is available. Chemicals having less residual toxicity, protectant action or which can inhibit inoculum production for a long period are practical for controlling diseases.

Incidence of postharvest diseases is a major factor contributing to decrease fruit quality during storage and increase postharvest losses. Most of the postharvest diseases originate in the field as quiescent infections in immature fruit prior to harvest. Proper field management, careful harvesting and proper handling are required to combat the problems of postharvest diseases.

References

Abdel-Monaim, M.F., El-Morsi, M.E.A. and Hassan, M.A.E. (2014) Control of root rot and wilt disease complex of some evergreen fruit transplants by using plant growth promoting rhizobacteria in the New Valley Governorate, Egypt. *Journal of Phytopathology and Pest Management* 1, 23–33.

Adisa, V.A. (1985) Fruit rot diseases of guava (*Psidium guajava*) in Nigeria. *Indian Phytopathology* 38, 427–430.

Agarwal, G.P. and Ganguli, G. (1959) A leaf spot disease of *Anogeissus latifolia* Wall, due to *Pestalotiopsis versicolor* (Speg.) Steyaert. *Current Science* 28, 295–296.

Alam, K.M., Alam, M.M., Islam, M., Momotaz, R., Arifunnahar, M. *et al.* (2019) First report of *Nalanthamala psidii* causing wilt disease of guava in Bangladesh. *Plant Disease* 103, 1042.

Alfenas, A.C., Zauza, E.A.V. and Assis, T.F. (2003) First record of *Puccinia psidii* on *Eucalyptus globulus* and *E. viminalis* in Brazil. *Australasian Plant Pathology* 32, 325.

Alfieri, S.A. Jr, Langdon, K.R., Wehlberg, C. and Kimbrough, J.W. (1984) *Index of Plant Diseases in Florida.* Bulletin No. 11 (revised). Division of Plant Industry, Florida Department of Agriculture and Consumer Services, Gainesville, Florida, p. 389.

Amorim, E.P.R., Pio Ribeiro, G., Menezes, M. and Coelho, R.S.B. (1993) The pathogenicity and hyperparasitic action of *Fusarium decemcellulare* on *Puccinia psidii* in guava (*Psidium guajava*). *Fitopatologia Brasileira* 18, 226–229.

Amusa, N.A., Ashaye, O.A., Oladapo, M.O. and Oni, M.O. (2005) Guava fruit anthracnose and the effects on its nutritional and market values in Ibadan, Nigeria. *World Journal of Agricultural Sciences* 1, 169–172.

Andrade, A.C. (1951) Guava rust control by means of spraying. *Arquivos do Instituto Biológico, São Paulo* 20, 127–146.

Andrade, A.C. (1959) New fungicides for the control of guava rust. *Biologíca* 25, 178–179.

Anonymous (1949a) *Annual Administrative Report for the Year 1947.* Department of Agriculture, United Provinces, India, p. 48, 89.

Anonymous (1949b) *Distribution Maps of Plant Diseases – Maps 169–192.* Commonwealth Mycological Institute, Kew, UK, p. 38.

Anonymous (1950) *Annual Administrative Report for the Year Ending 30th June 1948.* Department of Agriculture, United Provinces, India, p. 125.

Anonymous (1953) *Report for the Period 1946–1951*. Directorate of Plant Protection, Quarantine and Storage, India, p. 44.
Anonymous (1965) *CMI Descriptions of Pathogenic Fungi and Bacteria – Set 6, Sheets 51–60*. Commonwealth Mycological Institute, Kew, UK.
Anonymous (1974) *Annual Report*. Indian Institute of Horticultural Research, Hessarghatta, India, p. 467.
Ansar, M., Saleem, A. and Iqbal, A. (1994) Cause and control of guava decline in the Punjab (Pakistan). *Pakistan Journal of Phytopathology* 6, 41–46.
Antu, S.K. (2013) Studies on canker disease of guava (*Psidium guajava*) caused by *Pestalotiopsis psidii* Pat. M.Sc. thesis, Mahatma Phule Krishi Vidyapeeth, Rahuri, India, p. 78.
Aparecido, C.C., Figueiredo, M.B. and Furtado, E.L. (2003a) Grupos de variabilidade fisiológica em populações de *Puccinia psidii*. *Summa Phytopathologica* 29, 234–238.
Aparecido, C.C., Figueiredo, M.B. and Furtado, E.L. (2003b) Effect of temperature on infection, teliospore formation and basidiospore production for *Puccinia psidii* (Uredinales). *Summa Phytopathologica* 29, 239–243.
Ariosa, T. (1982) New diseases of guava (*Psidium guajava* L.) in Sancti Spiritus Province. *Centro Agricola* 9, 3–7.
Arthur, J.C. (1915) Uredinales of Porto Rico based on collections by F.L. Stevens. *Mycologia* 7, 227–255.
Arthur, J.C. (1922) Uredinales. *North American Flora* 7, 481–540.
Arya, A., Dwivedi, D.K., Pandey, R.S., Shukla, D.N., Bhargava, S.N. and Lal, B. (1981) Chemical control of Aspergillus rot of guava. *Indian Phytopathology* 34, 359–360.
Athipunyakom, P. and Luangsaard, J.J. (2008) *Nalanthamala psidii*, cause of guava wilt disease in Thailand. In: *Proceedings of the 46th Kasetsart University Annual Conference, Kasetsart, Thailand, 29 January–1 February 2008*, pp. 504–512.
Athipunyakom, P. and Manoch, L. (1998) Guava wilt disease. In: *Proceedings of the 36th Kasetsart University Annual Conference, Kasetsart, Thailand, 3–5 February 1998*, p. 41 (poster in Thai).
Aulakh, K.S. and Grover, R.K. (1969) Use of oils for controlling ripe fruit rots of tomato caused by *Phoma destructum* and *Curvularia lycopersici*. *FAO Plant Protection Bulletin* 17, 90–91.
Beenken, L. (2017) *Austropuccinia*: a new genus name for the myrtle rust *Puccinia psidii* placed within the redefined family Sphaerophragmiaceae (Pucciniales). *Phytotaxa* 297, 53–61.
Bhargava, A.K., Sobti, A.K. and Ghasolia, R.P. (2003) Studies on guava (*Psidium guajava* L.) drying/wilt disease in orchards of Pushkar Valley. *Journal of Phytological Research* 16, 81–84.
Bhargava, S.N. and Singh, A.P. (1974) Thiabendazole storage of guava fruit. *Indian Phytopathology* 27(4), 613–615.
Bhargava, S.N., Ghosh, A.K., Srivastava, M.P., Singh, R.H. and Tandon, R.N. (1965) Studies on fungal diseases of some tropical fruits VII. Effect of temperature on the decay of mango, banana and guava caused by some important pathogens. *Proceedings of the National Academy of Sciences, India, Section B* 35, 393–398.
Bilgrami, K.S. and Purohit, D.K. (1971) A new pathogenic species of *Pestalotia*. *Indian Phytopathology* 24, 211–213.
Booth, T.H., Old, K.M. and Jovanovic, T.A. (2000) Preliminary assessment of high risk areas for *Puccinia psidii* (Eucalyptus rust) in the Neotropics and Australia. *Agriculture, Ecosystems & Environment* 82, 295–301.
Bose, S.K. and Muller, E. (1967) Central Himalayan fungi. *Indian Phytopathology* 20, 124–138.
Carnegie, A.J., Lidbetter, J., Walker, J., Horwood, M.A., Tesoriero, M. *et al.* (2010) *Uredo rangelii*, a taxon in the guava rust complex, newly recorded on Myrtaceae in Australia. *Australasian Plant Pathology* 39, 463–466.
Chandra Mohan (1985) Studies on guava decline in Punjab with special reference to wilt. PhD thesis, Punjab Agricultural University, Ludhiana, India.
Chandra Mohan, Jhooty, J.S. and Chand, T. (1986) Prevalence of guava decline in Punjab. *Plant Disease Research* 1, 77–78.
Chattopadhyay, S.B. and Bhattacharjya, S.K. (1968a) Investigation on wilt disease of guava (*Psidium guajava* L.) in West Bengal, I. *Indian Journal of Agricultural Sciences* 38, 65–72.
Chattopadhyay, S.B. and Bhattacharjya, S.K. (1968b) Investigation on wilt disease of guava (*Psidium guajava* L.) in West Bengal, II. *Indian Journal of Agricultural Sciences* 38, 176–183.
Chattopadhyay, S.B. and Sengupta, S.K. (1955) Studies on wilt of guava, in West Bengal. *Indian Journal of Horticulture* 12, 76–79.
Chibber, H.M. (1911) A working list of diseases of vegetable pests of some of the economic plants, occurring in the Bombay Presidency. *Poona Agricultural College Magazine* 2, 180–198.
Coelho, L., Alfenas, A.C., Ferreira, F.A. (2001) Variabilidade fisiológica de *Puccinia psidii* – Ferrugem do eucalipto. *Summa Phytopathologica* 27, 295–300.

Commonwealth Mycological Institute (1987) *Puccinia psidii* Winter. In: *Distribution Maps of Plant Diseases*, 4th edn. CAB International, Wallingford, UK, Map 181.

Cooke, T. and Drenth, A. (2009) Other fruit crops. In: Cooke, T., Persley, D. and House, S. (eds) *Diseases of Fruit Crops in Australia*. CSIRO Publishing, Collingwood, Australia, pp. 251–264.

Coutinho, T.A., Wingfield, M.J., Alfenas, A.C. and Crous, P.W. (1998) Eucalyptus rust: a disease with the potential for serious international implications. *Plant Disease* 82, 819–825.

Das, M. and Bose, K.N. (1993) *Colletotrichum acutatum* – a new fruit rotting pathogen of guava (*Psidium guajava* L.) in storage. *Indian Journal of Mycology and Plant Pathology* 23, 331.

Das Gupta, M.K. and Ghosal, P.K. (1977) It is possible to control guava wilt through oil cake amendments. *Science & Culture* 43, 131–133.

Das Gupta, S.N. and Rai, J.N. (1947) Wilt disease of guava (*Psidium guajava* L.). *Current Science* 16, 256–258.

De Castro, H. (1983) Padronização de metodologia de inoculação e availiação de resistência de *Eucalyptus* spp. á ferrugem causada por *Puccinia psidii* Winter. PhD thesis, ESALQ/USP, Piracicaba, Brazil.

De Castro, H.A., Krügner, T.L. and Filho, A.B. (1985) Especialização fisiológica no sistema *Eucalyptus grandis* W. Hill ex Maiden–*Puccinia psidii* Winter. *Ciência e Prática Lavras* 9, 80–92.

Dey, P.K. (1948) Plant pathology. In: *Administrative Report 1945–46*. Agriculture Department, Uttar Pradesh, India, pp. 43–46.

Dhingra, R. and Mehrotra, R.S. (1980) A few unrecorded post harvest diseases of fruits and vegetables. *Indian Phytopathology* 33, 475–476.

Edward, J.C. (1960) Wilt disease of guava. *The Allahabad Farmer* 34, 289–293.

Edward, J.C. (1961) Root stock for guava wilt control. *The Allahabad Farmer* 35, 5–10.

Edward, J.C. and Gaurishankar (1964) Root stock trial for guava (*Psidium guajava* L.). *The Allahabad Farmer* 38, 249–250.

Edward, J.C. and Srivastava, R.N. (1957) Studies on guava wilt. *The Allahabad Farmer* 31, 144–146.

Edward, J.C., Naim, Z. and Gaurishankar (1964) Canker and fruit rot of guava (*Psidium guajava* L.). *The Allahabad Farmer* 38, 1–3.

Ferrari, J.T., Nogueira, E.M.C. and Santos, A.J.T. (1997) Control of rust (*Puccinia psidii*) in guava (*Psidium guajava*). *Acta Horticulturae* 452, 55–58.

Ferreira, F.A. (1983) Eucalyptus rust. *Revista Arvore* 7, 91–109.

Ferreira, F.A. (1989) Ferrugem do eucalipto. In: *Patologia Florestal – Principais Doenças Florestals no Brasil*. Sociedade de Investigaçoes Florestais/Universidade Federal de Viçosa, Viçosa, Brazil, pp. 129–152.

Figueiredo, M.B. (2001) Life cycle and ecology of *Puccinia psidii*. *O Biológico* 63, 69–71.

Furtado, G.Q., Castro, H.A. and Pozza, E.A. (2005) Variabilidade fisiológica de *Puccinia psidii* Winter em *Eucalyptus grandis* e no hibrido urograndis. *Summa Phytopathologica* 31, 227–231.

Ganeshan, K., Poornima, K., Renukadevi, P. and Nakkeeran, S. (2019) Characterisation of antinemic and antimicrobial compounds of *Bacillus amyloliquefacien* (vb7), against nematode–disease complex causing guava decline in Tamil Nadu. *International Journal of Agriculture Sciences* 11, 9191–9196.

Glen, M., Alfenas, A.C., Zauza, E.A.V., Wingfield, M.J. and Mohammed, C. (2007) *Puccinia psidii*: a threat to the Australian environment and economy – a review. *Australasian Plant Pathology* 36, 1–16.

Goes, A., Martins, R.D. and Reis, R.F. (2004) Efeito de fungicidas cúpricos, aplicados isoladamente ou em combinação com mancozeb, na expressão de sintomas de fitotoxicidade e controle da ferrugem causada por *Puccinia psidii* em goiabeira. *Revista Brasileira de Fruticultura* 26, 237–240.

Gomes V.M., Souza, R.M., Mussi-Dias, V.D.A., Silveira, S.F. and Dolinski, C. (2011) Guava decline: a complex disease involving *Meloidogyne mayaguensis* and *Fusarium solani*. *Journal of Phytopathology* 159, 45–50.

Gomes, V.M., Souza, R.M., Almeida, A.M. and Dolinski, C. (2015) Relationships between *M. enterolobii* and *F. solani*: spatial and temporal dynamics in the occurrence of guava decline. *Nematoda* 1, e01014.

Grech, N.M. (1985) First report of guava rapid death syndrome caused by *Septofusidium* sp. in South Africa. *Plant Disease* 69, 726.

Grech, N.M. (1990) Guava wilting disease in Levubu. *CSFRI Information Bulletin* 218, 8.

Grech, N.M. (1994) Survey of guava plantations in Mexico and Honduras. Trip report, University of California.

Gupta, J.H. (1978) Damping off, a new disease of guava. *Indian Journal of Mycology and Plant Pathology* 8, 224.

Gupta, J.H. (1979) Control of damping off of guava by seed treatment with systemic and non-systemic fungicides. *Progressive Horticulture* 10, 53–55.

Gupta, J.P. and Chatrath, M.S. (1973a) Radiation induced biochemical mutation in *Colletotrichum gloeosporioides* Penz. causing anthracnose of guava (*Psidium guajava* L.). *Mycopathologia et Mycologia Applicata* 51(2/3), 217–221.

Gupta, J.P. and Chatrath, M.S. (1973b) Gamma radiation for the control of post harvest fruit rot of guava (*Psidium guajava*). *Indian Phytopathology* 26(3), 506–509.

Gupta, J.P., Chatrath, M.S. and Khan, A.M. (1973) Chemical controls of fruit rot of guava caused by *Colletotrichum gloeosporioides*. *Indian Phytopathology* 26(4), 650–653.

Gupta, P.C., Madaan, R.L. and Suhag, L.S. (1977) Varietal reaction of guava fruits to *Phytophthora nicotianae* var. *parasitica*. *Indian Journal of Mycology and Plant Pathology* 7, 177.

Gupta, V.K. and Misra, A.K. (2012) *Fusarium chlamydosporum*, causing wilt disease of guava (*Psidium guajava* L.) in India. *Archives of Phytopathology and Plant Protection* 20, 2425–2428.

Gupta, V.K., Misra, A.K. and Gaur, R.K. (2010a) Growth characteristics of *Fusarium* spp. causing wilt disease in *Psidium guajava* L. in India. *Journal of Plant Protection Research* 50, 430–440.

Gupta, V.K., Misra, A.K., Gaur, R.K., Jain, P.K., Gaur, D. and Sharma, S. (2010b) Current status of *Fusarium* wilt disease of guava (*Psidium guajava* L.) in India. *Biotechnology* 9, 176–195.

Gupta, Y.K., Roy, A.N., Yadav, S. and Gupta, M.N. (1979) Investigations on post harvest diseases of guava fruits. *Indian Phytopathology* 32, 623–624.

Haider, M., Bukhari, S.A.A., Binyamin, R. and Habib, A. (2016) Fungi associated with guava anthracnose and management of *Colletotrichum gloeosporioides* through biological and chemical means. *Pakistan Journal of Phytopathology* 28, 153–160.

Hamiduzzaman, M.M., Meat, M.B. and Ahmad, M.U. (1997) Effect of *Fusarium oxysporum* and nematode interaction on guava wilt. *Bangladesh Journal of Plant Pathology* 13, 9–11.

Hernández, J.R. (2006) Invasive Fungi. *Puccinia psidii*. Systematic Mycology and Microbiology Laboratory, Agricultural Research Service, US Department of Agriculture, Beltsville, Maryland. Available at: https://nt.ars-grin.gov/taxadescriptions/factsheets/index.cfm?thisapp=Pucciniapsidii (accessed 15 May 2020).

Hong, C.F., Hsieh, H.Y., Chen, K.S. and Huang, H.C. (2015) Importance of root infection in guava wilt caused by *Nalanthamala psidii*. *Plant Pathology* 64, 450–455.

Hsieh, S.P.Y., Liang, W.J., Kao, C.W. and Leu, L.S. (1976) Morphological and physiological characters of *Myxosporium psidii*, the causal organism of guava wilt. *Plant Protection Bulletin* 18, 309–317.

Hussain, M.Z., Rahman, M.A. and Bashar, M.A. (2014) Incidence of guava (*Psidium guajava* L.) wilt caused by *Fusarium oxysporum* Sch. f.sp. *psidii* in Bangladesh. *Journal of the Asiatic Society of Bangladesh, Science* 40, 97–105.

Islam, M.N., Chowdhury, M.S.M., Rahman, M.M., Alam, K.M., Arifunnahar, M. and Alam, M.M. (2015) Occurrence of anthracnose disease incidence and its severity on guava seedlings at different locations of Bangladesh. *International Journal of Sustainable Crop Production* 10, 21–31.

Jain, S.S. (1956) A preliminary note on the inactivation of *Fusarium oxysporium* f. *psidii* in guava plants by chemotherapeutic treatment. *Indian Journal of Horticulture* 13, 102–104.

Jhooty, J.S., Chand, J.N. and Krishnamurty, V. (1984) *Report of Committee constituted by ICAR on Guava Decline in Punjab and Haryana*. Submitted to ICAR, New Delhi.

Joubert, M.H. and Frean, R.T. (1993) An *in vitro* evaluation of fungicides against guava wilt. *Inligtingsbulletin, Instituut Vir Tropiese en Subtropiese Gewasse* 246, 3.

Junqueira, N.T.V., Andrade, L.R.M.d., Pereira, M., Lima, M.M. and Chaves, R.d.C. (2001) Diseases of guava (*Psidium guajava* L.) cultivated in Brazilian cerrados. *Circular Tecnica Embrapa Cerrados* 15, 31.

Kamle, M., Kalim, S., Bajpai, A., Chandra, R. and Kumar, R. (2012) *In vitro* selection for wilt resistance in guava (*Psidium guajava* L.) cv. Allahabad Safeda. *Biotechnology* 11, 163–171.

Katyal, S.L. (1972) Twenty-five years of research on fruit crops. *Indian Farming* 22, 14–16.

Kaul, J.L. and Munjal, R.L. (1982) Apple losses in Himachal Pradesh due to post harvest fungal pathogens. *Indian Journal of Mycology and Plant Pathology* 12, 209–210.

Kaushik, C.D., Chand, J.N. and Thakur, D.P. (1970) Fungi associated with decay of certain fruits and potato tubers in Haryana markets. *Journal of Research Ludhiana* 7, 648–650.

Kaushik, C.D., Thakur, D.P. and Chand, J.N. (1972) Parasitism and control of *Pestalotia psidii* causing cankerous disease of ripe guava fruits. *Indian Phytopathology* 25, 61–64.

Kawanishi, T., Uemastu, S., Kakishima, M., Kagiwada, S., Hamamoto, H. *et al.* (2009) First report of rust disease on ohia and the causal fungus in Japan. *Journal of Genetic and Plant Pathology* 75, 428-431.

Keith, L.M. and Zee, F.T. (2010) Guava diseases in Hawaii and the characterization of *Pestalotiopsis* spp. affecting guava. *Acta Horticulturae* 849, 269–276.

Keith, L.M., Velasquez, M.E. and Zee, F.T. (2006) Identification and characterization of *Pestalotiopsis* spp. causing scab disease of guava, *Psidium guajava*, in Hawaii. *Plant Disease* 90, 16–23.

Khan, R.M. and Misra, A.K. (2003) Influence of co-cultivation of marigold, garlic and turmeric on nemic population in guava cropping system. In: *Proceedings Souvenir & Abstract, Zonal Conference (East*

Zone), *Indian Society of Mycology and Plant Pathology & Seminar on Plant Diseases of National Importance with Special Reference to Guava Wilt and Mango Malformation, 4–5 April 2003*. CISH and IISR, Lucknow, India, pp. 24–25.

Khanna, K.K. and Chandra, S. (1977) Control of guava fruit rot caused by *Pestalotia psidii* with homeopathic drugs. *Plant Disease Reporter* 61, 362–366.

Khanna, K.K. and Chandra, S. (1989) Further investigation on the control of storage rot of mango, guava and tomato with homoeopathic drugs. *Indian Phytopathology* 42(3), 436–440.

Khare, V., Mehta, P., Kachhwaha, M. and Mehta, A. (1994) Role of phenolic substances in pathogenesis of soft rot diseases. *Journal of Basic Microbiology* 34, 323–328.

Killgore, E.M. and Heu, R.A. (2005) *A Rust Disease on Ohia*. New Pest Advisory No. 05-04. State of Hawaii Department of Agriculture, Honolulu, Hawaii.

Ko, W.H., Kunimoto, R.K. and Nishijima, W.T. (1982) Fruit rot of guava caused by *Phytophthora citricola*. *Plant Disease* 66, 854–855.

Kothari, K.L. (1968) Controlling fruit rot of guava. *Indian Horticulture* 12(3), 9–10.

Krishnaiah, J., Satyaprasad, C.H., Singh, T.G. and Thirupathaiah, V. (1985) Post harvest protection of guava fruits using Decco food grade coatings. *Indian Botanical Report* 4(2), 151–153.

Kunimoto, R.K., Ito, P.J. and Ko, W.H. (1977) *Mucor* rot of guava fruits caused by *Mucor hiemalis*. *Tropical Agriculture* 54, 185–187.

Kurosawa, E. (1926) Guava tachigare byo. *Transactions of the Natural History Society of Formosa* 83, 47–61.

Lal, B., Rai, R.N., Arya, A. and Tiwari, D.K. (1980) A new soft rot of guava. *National Academy Science Letters* 3, 259–260.

Leu, L.S. and Kao, C.W. (1979) Artificial inoculation of guava with *Myxosporium psidii*. *Plant Disease Reporter* 63, 1077–1079.

Leu, L.S., Kao, C.W., Wang, C.C., Liang, W.J. and Hsieh, S.P.Y. (1979) *Myxosporium* wilt of guava and its control. *Plant Disease Reporter* 63, 1075–1077.

Lim, T.K. and Khoo, K.C. (1990) *Guava in Malaysia: Production, Pests and Diseases*. Tropical Press, Kuala Lumpur.

Lim, T.K. and Manicom, B.Q. (2003) Diseases of guava. In: Ploetz, R.C. (ed.) *Diseases of Tropical Fruit Crops*. CAB International, Wallingford, UK, pp. 275–289.

Lin, C.C. (2005) Algal spot of guava. In: *Plant Protection Atlas Series 15: Guava Protection*. Bureau of Animal and Plant Health Inspection and Quarantine, Taipei, Taiwan, pp. 84–87. Available at: https://www.baphiq.gov.tw/publish/plant_protect_pic_15/P_pdf/03-08.pdf (accessed 20 January 2021).

Lui, L.J. (1972) Identification and occurrence of perfect stage and cultural and morphological variants of *Colletotrichum gloeosporioides* from guava in Puerto Rico. *Journal of Puerto Rico* 56, 171–180.

Madhu, M.R., Verma, K.K. and Vinod Kumar (2019) Distribution, prevalence and intensity of guava decline in western Haryana. *Journal of Entomology and Zoology Studies* 7, 521–524.

Manicom, B.Q. (1980) *Disease Management Progress Report*. CSFRI, Nelspruit, South Africa.

Marlatt, R.B. and Campbell, C.W. (1980a) Susceptibility of *Psidium guajava* selections to injury by *Cephaleuros* sp. *Plant Disease* 64, 1010–1011.

Marlatt, R.B. and Campbell, C.W. (1980b) Incidence of algal disease (*Cephaleuros* sp.) in selections of guava (*Psidium guajava*). *Proceedings of the Florida State Horticultural Society* 93, 109–110.

Marlatt, R.B. and Kimbrough, J.W. (1979) *Puccinia psidii* on *Pimenta dioica* in South Florida. *Plant Disease Reporter* 63, 510–512.

Martins, M.V.V., Silveira, S.F., Maffia, L.A., Rocabado, J.M.A. and Dias, V.M. (2011) Chemical control of guava rust (*Puccinia psidii*) in the Northern Region of Rio de Janeiro State, Brazil. *Australasian Plant Pathology* 40, 48–54.

Martins, M.V.V., da Silveira, S.F. and Maffia, L.A. (2014) Guava fruit loss caused by rust. *Summa Phytopathologica* 40, 107–113.

Mathur, R.S. (1956) Guava disease in India. *Indian Journal of Horticulture* 13, 26–29.

Mathur, R.S., Jain, S.S. and Swarup, J. (1964) Chemical treatment for guava wilt. *Proceedings of the National Academy of Sciences, India, Section B* 34, 33–36.

Mehta, N. (1987) Distribution of guava wilt in relation to age, soil type, management practices and varieties grown in Haryana. *Plant Disease Research* 2, 116–119.

Mehta, P.R. (1951) Observations on new and known diseases of crop plants of Uttar Pradesh. *Plant Protection Bulletin* 3, 7–12.

Midha, S.K. and Chohan, J.S. (1968) Chemical basis for incipient infection caused by *Gloeosporium psidii* in guava fruits. *Journal of Research* 5, 395–400.

Misra, A.K. (1987) *Studies on diseases of fruit crops. Annual Report*. CIHNP, Lucknow, India, pp. 124–125.
Misra, A.K. (1995) Guava wilt. In: Singh, S.J. (ed.) *Advances in Diseases of Fruits in India*. Kalyani Publishers, Ludhiana, India, pp. 183–190.
Misra, A.K. (1998–1999) Screening against wilt. *Annual Report*. CISH, Lucknow, India.
Misra, A.K. (2001) Diseases of guava and their management. In: Thind, T.S. (ed.) *Diseases of Fruits and Vegetables and Their Management*. Kalyani Publishers, Ludhiana, India, pp. 128–138.
Misra, A.K. (2003) Guava diseases – their symptoms, cause and management. In: Naqvi, S.A.M.H. (ed.) *Diseases of Fruits and Vegetables: Diagnosis and Management*, Vol. 2. Kluwer Academic Publishers, Dordrecht, The Netherlands, pp. 81–119.
Misra, A.K. (2004) *Networking Project on Investigation on Guava Wilt with Special Reference to Etiology and Management (UPCAR/UPDASP Word Bank Project)*. Final report submitted to UPCAR, Lucknow, India.
Misra, A.K. (2005) Important diseases of guava in India with special reference to wilt. In: *Souvenir of the 1st International Guava Symposium, CISH, Lucknow, India, 5–8 December 2005*, pp. 75–90.
Misra, A.K. (2006) Wilt of guava – a disease of national importance. *Indian Phytopathology* 59(3), 269–280.
Misra, A.K. (2008) *Networking Project on Wilt of Crops with Special Reference to Cultural, Morphological, Molecular Characterization and Pathogenic Variability of Isolates of India (Guava Wilt) (ICAR project)*. Final report submitted to Networking PI–IIPR, Kanpur and ICAR, New Delhi, p. 70.
Misra, A.K. (2017) Progressive steps in understanding and solving guava wilt – a national problem. *Indian Phytopathology* 70, 1–11.
Misra, A.K. (2018) Understanding and solving the problem of guava wilt for guava growers. In: *Proceedings of the Special National Symposium on Extension Plant Pathology: Technological Backstopping to the Farmers/Other Stakeholders, 25–26 September 2018*. Indira Gandhi Krishi Vishwavidyalaya, Raipur and IPS, New Delhi, pp. 124–125.
Misra, A.K. (2019a) Integrated management of guava wilt for sustainable agriculture. In: *Proceedings of the 71st Annual Meeting, Indian Phytopathological Society and National Symposium: Recent Challenges and Opportunities in Sustainable Plant Health Management, Institute of Agricultural Sciences, BHU, Varanasi, India, 26–28 February 2019*, p. 29.
Misra, A.K. (2019b) Ecofriendly management of guava wilt for sustainable horticulture. In: *Proceedings of the Annual Meeting, Indian Phytopathological Society of Mid Eastern Zone and National Conference on Plant Health Management for Ecofriendly and Sustainable Agriculture, Chandra Shekher Azad University of Agricultural & Technology, Kanpur, India, 25–26 November 2019*, p. 14.
Misra, A.K. and Gupta V.K. (2010) Relative pathogenicity of Fusarium wilt isolates to guava (*Psidium guajava* L.). *Journal of Mycology and Plant Pathology* 40, 72–77.
Misra, A.K. and Pandey, B.K. (1994–1995) Studies on guava wilt. *Annual Report*. CIHNP, Lucknow, India, p. 28.
Misra, A.K. and Pandey, B.K. (1999a) Pathogenicity and evaluation of fungicides against guava wilt pathogens. *Journal of Mycology and Plant Pathology* 29, 274–275.
Misra, A.K. and Pandey, B.K. (1999b) Guava wilt disease – a challenge for the coming millennium. In: *Proceedings of the National Symposium on Challenges & Prospects of Plant Pathology in the Coming Millennium, 9–11 December 1999*. NBRI, Lucknow, India, p. 22.
Misra A.K. and Pandey, B.K. (1999c) Natural wilting of guava plants during different months. *Indian Phytopathology* 52, 312.
Misra, A.K. and Pandey, B.K. (2000a) Pathogenicity and symptom production of wilt disease of guava by a new potent pathogen *Gliocladium roseum*. In: *Proceedings of the Indian Phytopathological Society Golden Jubilee International Conference on Integrated Disease Management for Sustainable Agriculture*, Vol. II. Indian Phytopathological Society, New Delhi, pp. 749–750.
Misra, A.K. and Pandey, B.K. (2000b) Progressive natural wilting of guava plants during different months. *Indian Phytopathology* 53, 423–427.
Misra, A.K. and Prakash, Om (1986) Studies on diseases of fruit crops. *Annual Report*. CIHNP, Lucknow, India, pp. 67–68.
Misra, A.K. and Prakash, Om (1990) *Guava Diseases – An Annotated Bibliography 1907–1990*. Bishen Singh Mahendra Pal Singh, Dehradun, India, p. 132.
Misra, A.K. and Prasad, B. (2003) Relative efficacy of different bio-agents for the control of guava wilt. *Journal of Mycology and Plant Pathology* 33, 494.
Misra, A.K. and Prasad, D. (2004) *Aspergillus niger* strain AN 17 potent bioagent to control wilt disease and its easy multiplication. In: *Proceedings of the Symposium on Recent Advances in Fungal Bioagents and Their Social Benefits, 10 September 2004*. NBRI, Lucknow, India, p. 12.

Misra, A.K. and Shukla, S.K. (2002) Assessment of loss due to guava wilt around Lucknow. In: *Proceedings of the National Seminar on Production and Post-Harvest Technology of Guava, Chandra Shekhar Azad University of Agriculture and Technology, Kanpur, India, 9–10 January 2002*, pp. 34–35.

Misra, A.K. and Singh, V.K. (2005) The CISH, Lucknow, marches towards ... solving guava wilt and mango malformation. *Indian Horticulture* 50, 25–27.

Misra, A.K., Om Prakash and Sen, B. (2000) Biological control of guava wilt by *Aspergillus niger* strain AN17 (Pusa Mrida). In: *Proceedings of the National Seminar on Hi-tech Horticulture, Bangalore, India, 26–28 June 2000*, p. 149.

Misra, A.K., Pandey, B.K. and Om Prakash (2003a) Wilt of guava, a disease of national importance and possible solutions. In: *Proceedings Souvenir & Abstract, Zonal Conference (East Zone), Indian Society of Mycology and Plant Pathology & Seminar on Plant Diseases of National Importance with Special Reference to Guava Wilt and Mango Malformation, 4–5 April 2003*. CISH and IISR, Lucknow, India, p. 23.

Misra, A.K., Rajan, S., Babita Prasad, Shukla, S.K. and Prasad, D. (2003b) Resistant source of wilt disease of guava. In: *Proceedings Souvenir & Abstract, Zonal Conference (East Zone), Indian Society of Mycology and Plant Pathology & Seminar on Plant Diseases of National Importance with Special Reference to Guava Wilt and Mango Malformation, 4–5 April 2003*. CISH and IISR, Lucknow, India, pp. 51–52.

Misra, A.K., Om Prakash, Pandey, B.K., Babita Prasad, Shukla, S.K. and Prasad, D. (2003c) Biological control of guava wilt. In: *Souvenir & Abstract, National Symposium on Organic Farming in Horticulture for Sustainable Production, 29–30 August 2003*. CISH, Lucknow, India, p. 65.

Misra, A.K., Om Prakash, Pandey, B.K., Babita Prasad, Shukla, S.K. and Prasad, D. (2004a) Biological control of guava wilt. In: Pathak, R.K., Ram Kishun, Khan, R.M. and Ram, R.A. (eds) *Organic Farming in Horticulture*. CISH, Lucknow, India, pp. 302–305.

Misra, A.K., Prasad, D., Babita Prasad and Shukla, S.K. (2004b) Effective management of wilt disease of guava. In: *Proceedings Souvenir & Abstract, National Symposium on Crop Surveillance: Disease Forecasting and Management, Indian Phytopathological Society, 19–21 February 2004*. Division of Plant Pathology, IARI, New Delhi, pp. 92–93.

Mitra, M. (1929) *Phytophthora parasitica* Dast. causing damping off disease of cotton seedlings and fruit rot of guava in India. *Transactions of the British Mycological Society* 14, 249–254.

Moustafa, M.S.H., Hala, A.M.E.D. and Asmaa, M.A.A. (2015) Pestalotia leaf spot a new disease affect guava trees in Egypt. *International Journal of Scientific & Engineering Research* 6, 1306–1312.

Nakasone, H.Y. (1998) Guava. In: Nakasone, H.Y. and Paull, R.E. (eds) *Tropical Fruits*. CAB International, Wallingford, UK, pp. 149–172.

Naresh, M., Madaan, R.L. and Daulta, B.S. (1987) Evaluation of guava hybrids to anthracnose fruit rot under field conditions. *Haryana Journal of Horticultural Sciences* 16, 82–85.

Narsimhan, M.J. (1938) *Pestalotia psidii* on guava in India. In: *Annual Administrative Report for the Year 1936–37*. Agriculture Department, Mysore, India, pp. 169–173.

Narsimhan, M.J. (1940) *Annual Administrative Report for the Year 1938–39*. Agriculture Department, Mysore, India, pp. 96–97.

Nelson, S.C. (2008) *Cephaleuros Species, the Plant-parasitic Green Algae*. Plant Disease No. PD-43. University of Hawaii, Honolulu, Hawaii.

Normand, F. (1994) Strawberry guava, relevance for Réunion. *Fruits* 49, 217–227.

Ooka, J.J. (1980) Guava fruit rot caused by *Rhizopus stolonifer* in Hawaii. *Plant Disease* 64, 412–413.

Opina, O.S. (1995) Epidemic development of Acremonium wilt of guava in the Philippines. *Philippine Phytopathology* 31, 127–131.

Pandey, R.R. and Dwivedi, R.S. (1985) *Fusarium oxysporum* f.sp. *psidii* as a pathogen causing wilt of guava in Varanasi district, India. *Phytopathologische Zhournal* 114, 243–248.

Patel, K.D. and Pathak, V.N. (1995) Development of Botryodiplodia rot of guava fruits in relation to temperature and humidity. *Indian Phytopathology* 48, 86–89.

Patel, M.K., Kamat, M.N. and Hingorani, G.M. (1950) *Pestalotia psidii* Pat. on guava. *Indian Phytopathology* 3, 165–176.

Pathak, V.N. and Shekhawat, P.S. (1976) Efficacy of some fungicides and hot water in control of anthracnose and *Aspergillus* rot of mango fruits. *Indian Phytopathology* I29(3), 315–317.

Piza, S.M.d.T. and Ribeiro, I.J.A. (1988) Influence of light and temperature on uredospore germination of *Puccinia psidii* Winter. *Bragantia* 47, 75–78.

Ploetz, R.C. (2007) Diseases of tropical perennial crops, challenging problems in diverse environments. *Plant Disease* 91, 644–663.

Poornima, K., Suresh, P., Kalaiarasan, P., Subramanian, S. and Ramaraju, K. (2016) Root knot nematode, *Meloidogyne enterolobii* in guava (*Psidium guajava* L.) a new record from India. *Madras Agricultural Journal* 103, 359–365.

Prakash, Om, Misra, A.K. and Shukla, S.K. (2002) *Penicillium citrinum* a potent pathogen against wilt disease of guava. In: *Proceedings of the Asian Congress of Mycology and Plant Pathology and Symposium on Plant Health for Food Security, 1–4 October 2002*. University of Mysore, Mysore, India, p. 180 (abstract no. PP-299).

Prasad, D., Shukla, S.K., Babita Prasad and Misra, A.K. (2003) Effect of intercrop and different doses of NPK on the incidence of guava wilt. In: *Proceedings Souvenir & Abstract, Zonal Conference (East Zone), Indian Society of Mycology and Plant Pathology & Seminar on Plant Diseases of National Importance with Special Reference to Guava Wilt and Mango Malformation, 4–5 April 2003*. CISH and IISR, Lucknow, India, p. 53.

Prasad, N., Mehta, P.R. and Lal, S.B. (1952) *Fusarium* wilt of guava (*Psidium guajava* L.) in Uttar Pradesh, India. *Nature* 4305, 753.

Quimio, A.J., de Villa, E.A. and Sumabat, R.S. (1984) Acremonium wilt of guava in the Philippines. *Philippine Phytopathology* 20, 4.

Quimio, T. and Quimio, A.J. (1975) Notes on Philippine grape and guava anthracnose. *Plant Disease Reporter* 59, 221–224.

Raghunathan, V. and Prasad, N.N. (1969) Occurrence of *Cercospora sawadae* on *Psidium guajava*. *Plant Disease Reporter* 53, 455.

Rahman, M.A., Ansari, T.H., Meah, M.B. and Yoshida, T. (2003) Prevalence and pathogenicity of guava anthracnose with special emphasis on varietal reaction. *Pakistan Journal of Biological Sciences* 6, 234–241.

Rai, J.N. (1956) Stylar end rot of the guava fruit. (*Psidium guajava* Linn.). *Proceedings of the Indian Academy of Sciences, Section B* 43, 55–61.

Rajgopalan, B. and Wilson, K.I. (1972a) *Diplodia natalensis* causing dry rot of guava fruits in South India. *Plant Disease Reporter* 56, 323–324.

Rajgopalan, B. and Wilson, K.I. (1972b) Field evaluation of fungicides against *Diplodia* rot of guava fruits. *Agricultural Research Journal of Kerala* 10, 194–195.

Ramaswamy, G.R., Sohi, H.S. and Govindu, H.C. (1984) Studies on spore germination in *Pestalotia psidii*, the causal organism of guava canker. *Indian Journal of Mycology and Plant Pathology* 14, 289.

Rana, O.S. (1981) *Diplodia* stem canker, a new disease of guava in Tarai regions of UP. *Science & Culture* 47, 370–371.

Rao, D.P.C. and Agarwal, S.C. (1976a) Efficacy of antibiotics against *Phomopsis destructum* causing fruit rot of guava. *Hindustan Antibiotic Bulletin* 18, 108–110.

Rao, D.P.C. and Agarwal, S.C. (1976b) Efficacy of fungicides against *Phomopsis* fruit rot of guava. *Indian Phytopathology* 29, 345–346.

Rao, D.P.C., Agarwal, S.C. and Saksena, S.B. (1976) *Phomopsis destructum* on *Psidium guajava* fruits in India. *Mycologia* 68, 1132–1134.

Rao, V.G. (1966) An account of market and storage diseases of fruits and vegetables in Bombay, Maharashtra (India). *Mycopathologia et Mycologia Applicata* 28, 165–176.

Rawal, R.D. (1993) Yield loss in guava by different root rots. *International Journal of Tropical Plant Diseases* 11, 69–72.

Rhoads, A.S. (1927) *Clitocybe* root rot of trees and other woody plants in Florida. *Phytopathology* 17, 56–57.

Rhoads, A.S. (1942) Notes on *Clitocybe* root rot of bananas and other plants in Florida. *Phytopathology* 32, 487–496.

Ribeiro, I.J.A. and Pommer, C.V. (2004) Breeding guava (*Psidium guajava*) for resistance to rust caused by *Puccinia psidii*. *Acta Horticulturae* 632, 75–78.

Rocabado, J.M.A. (2003) Epidemiologia e patogênese da ferrugem-da-goiabeira, causada por *Puccinia psidii*. PhD thesis, Universidade Estadual do Norte Fluminense Darcy Ribeiro, Campos, Brazil.

Rodrigues, N.J., Robbs, C.F. and Yamashiro, T. (1987) A bacterial disease of guava (*Psidium guajava*) caused by *Erwinia psidii* sp. nov. *Fitopatologia Brasileira* 12, 345–350.

Rodriguez, F.M.E. and Landa, J.B. (1977) Chemical soil disinfection against parasitic nematode in guava nurseries. *Centro Agricola de la Facultad de Ciencias Agricolos* 4, 57–77.

Ruehle, G.D. (1941) Algal leaf and fruit spot of guava. *Phytopathology* 31, 95–96.

Ruiz, R.A.R., Alfenas, A. and Ferreira, F.A. (1989a) Effect of temperature, light and inoculum source on teliospore and urediniospore production of *Puccinia psidii*. *Fitopatologia Brasileira* 14, 70–73.

Ruiz, R.A.R., Alfenas, A.C., Ferreira, F.A. and do Vale, F.X.R. (1989b) Influência da temperature, do tempo molhamento foliar, fotoperíodo e da intensidade de luz sobre a infeccao de *Puccinia psidii* em eucalipto. *Fitopatologia Brasileira* 14, 55–64.

Ruiz, R.A.R., Alfenas, A.C., Maffia, L.A. and Barbosa, M.B. (1989c) Progress of the eucalypt rust, caused by *Puccinia psidii* in the field. *Fitopatologia Brasileira* 14, 73–81.

Ruiz, R.A.R., Alfenas, A.C. and Demuner, N.L. (1991) Efficiency of fungicides for the control of rust (*Puccinia psidii*) on guava (*Psidium guajava*). *Summa Phytopathologica* 17, 147–153.

Schoeman, M.H. (1997) *Verslag oor 'n bsoek aan Maleisie om die voorkoms van koejawelverwelksiekte daar te ondersoek*. ITSC Report, Nelspruit, South Africa.

Schoeman, M.H. and Labuschagne, N. (2014) Preliminary evaluation of guava selections for guava wilt disease resistance in South Africa. *South African Journal of Plant and Soil* 31, 109–112.

Schoeman, M.H., Benade, E. and Wingfield, M.J. (1997) The symptoms and cause of guava wilt in South Africa. *Journal of Phytopathology* 145, 37–41.

Schoeman, M.H., Botha, F.A. and Manicom, B.Q. (2012) Guava wilt disease – the South African perspective. *Acta Horticulturae* 959, 67–72.

Schoeman, M.H., Labuschagne, N. and Calitz, F.J. (2017) Efficacy of fungicides, plant resistance activators and biological control agents against guava wilt disease caused by *Nalanthamala psidii*. *South African Journal of Plant and Soil* 34, 119–124.

Schroers, H.J., Geldenhuis, M.M., Wingfield, M.J., Schoeman, M.H., Wingfield, B.D. and Subraumanian, C.V. (2003) Introduction of *Nalanthamala* for the guava wilt fungus and palm pathogen *Gliocladium vermoesenii*. In: *Proceedings of the 8th International Congress of Plant Pathology, Christchurch, New Zealand, 2–7 February 2003*.

Schroers, H.J., Geldenhuis, M.M., Wingfield, M.J., Schoeman, M.H., Yen, Y.F. *et al.* (2005) Classification of the guava wilt fungus *Myxosporium psidii*, the palm pathogen *Gliocladium vermoesenii* and the persimmon wilt fungus *Acremonium diospyri* in *Nalanthamala*. *Mycologia* 97, 375–395.

Shankhapal, K.B. and Hatwalne, V.G. (1976) Sour rot of guava in India. *Current Science* 45, 565–566.

Sharma, S.K. (1981) Control of anthracnose of guava fruits caused by *Glomerella cingulata* (Stonem.) Spauld. and Schrenk. MSc thesis, Harayana Agricultural University, Hisar, India, p. 157.

Sharma, S.K., Singh, J.P. and Chand, J.N. (1983) Chemical control of anthracnose of guava caused by *Glomerella cingulata*. *Haryana Agricultural University Journal of Research* 13, 325–326.

Shukla, S.K., Babita Prasad, Prasad, D. and Misra, A.K. (2003) A cheap substrate for the mass multiplication of bio-agents of guava wilt. In: *Proceedings Souvenir & Abstract, Zonal Conference (East Zone), Indian Society of Mycology and Plant Pathology & Seminar on Plant Diseases of National Importance with Special Reference to Guava Wilt and Mango Malformation, 4–5 April 2003*. CISH and IISR, Lucknow, India, pp. 54–55.

Shukla, P.K., Fatima, T. and Rajan, S. (2019) Research on Fusarium wilt disease of guava. *Indian Phytopathology* 72, 629–636.

Simpson, J.A., Thomas, K. and Grgurinovic, C.A. (2006) Uredinales species pathogenic on species of Myrtaceae. *Australasian Plant Pathology* 35, 549–562.

Singh, A.P. and Bhargava, S.N. (1976) Presence of some antifungal substance in the skin of guava fruits (*Psidium guajava* L.). *Proceedings of the National Academy of Sciences, India, Section B* 46(3), 451–452.

Singh, A.P. and Bhargava, S.N. (1977a) Benlate as an effective post harvest fungicide for guava fruit. *Indian Journal of Horticulture* 34, 309–312.

Singh, A.P. and Bhargava, S.N. (1977b) Storage and transit studies in apple guavas. *Indian Journal of Horticulture* 34, 362–363.

Singh, B. and Lal, S.B. (1953) Wilt of guava. *Agriculture and Animal Husbandry* 3, 78–79.

Singh, G., Chohan, J.S. and Mann, S.K. (1976) Fruit rot of guava – a new disease problem in Punjab. *Indian Journal of Mycology and Plant Pathology* 6, 77.

Singh, J.P. and Sharma, S.K. (1982) Controlling anthracnose of guava caused by *Glomerella cingulata* by fumigation. *Indian Phytopathology* 35, 273–276.

Singh, N. (2020) Emerging problem of guava decline caused by *Meloidogyne enterolobii* and *Fusarium oxysporum* f. sp. *psidii*. *Indian Phytopathology* 73, 1–2.

Singh, R.H. and Tandon, R.N. (1971) Vitamin C content of guava fruits infected with *Aspergillus niger*. *Indian Phytopathology* 24, 807–809.

Singh, Y.H., Singh, L.G., Singh, T.S., Taibanganbidevi, N.G., Monteshori, S. and Yumlembam, R.A. (2017) *In vitro* efficacy of plant extract and bioagents against *Pestalotiopsis versicolor*, the incitant of leaf spot of guava in Manipur. *Plant Disease Research* 32, 86–90.

Sohi, H.S. (1983a) Studies on wilt disease of guava. *Annual Report*. Indian Horticulture Research Institute, Bangalore, India, p. 102.
Sohi, H.S. (1983b) Diseases of tropical and subtropical fruits and their control. In: Husain, A., Singh, K., Singh, B.P. and Agnihotri, V.P. (eds) *Recent Advances in Plant Pathology*. Print House, Lucknow, India, p. 73–86.
Sohi, H.S. and Sridhar, T.S. (1969) Injurious effect of copper sprays on guava fruits. *Indian Journal of Horticulture* 26, 155.
Sohi, H.S. and Sridhar, T.S. (1971) Controlling fruit rot of guava. *Indian Horticulture* 16, 9–10.
Solarte, F., Muñoz, C.G. and Maharachchikumbura, S.S.N. (2018) Diversity of *Neopestalotiopsis* and *Pestalotiopsis* spp., causal agents of guava scab in Colombia. *Plant Disease* 102, 49–59.
Sridhar, T.S. and Ullasa, B.A. (1978) Leaf blight of guava, a new record. *Current Science* 47, 442.
Sridhar, T.S., Ullasa, B.A. and Sohi, H.S. (1975) Occurrence of a new disease on grape seedlings caused by *Phytophthora nicotianae* var. *parasitica* (Dasture) Waterhouse from India. *Current Science* 44, 406.
Srivastava, A.K., Ahmad, R., Kumar, S. and Sukhada Mohandos (2001) Role of VA -mycorrhiza in the management of wilt disease of guava in the alfivols of Chotanagpur. *Indian Phytopathology* 54, 78–81.
Srivastava, H.P. (1963) Some leaf spot fungi. *Proceedings of the National Academy of Sciences, India* 34, 188–198.
Srivastava, M.P. and Tandon, R.N. (1969a) Post harvest diseases of guava in India. *Plant Disease Reporter* 53, 206–208.
Srivastava, M.P. and Tandon, R.N. (1969b) Studies on *Botryodiplodia* rot of guava. *Indian Phytopathology* 22, 268–270.
Srivastava, M.P. and Tandon, R.N. (1971) Efficacy of certain fungicides and an antibiotic against four isolates of *Botryodiplodia theobromae*. *Indian Phytopathology* 24, 396–397.
Stevenson, J.A. (1975) *Fungi of Puerto Rico and the American Virgin Islands*. Contribution No. 23. Reed Herbarium, Baltimore, Maryland.
Suhag, L.S. (1976) Observations on guava decline in Haryana and its control. *Pesticides* 10, 42–44.
Sydow, P. and Sydow, H. (1904–1924) *Monographia Uredinearum seu specierum omnium ad hunc usque diem cognitarum descriptio et adumbratio systematica*, Vol. 1 (1904), Vol. 3 (1915) and Vol. 4 (1924). Borntraeger Bros., Leipzig, Germany.
Tandon, I.N. (1961) A new seedling blight of guava and its control. *Indian Phytopathology* 14, 102–103.
Tandon, I.N. and Singh, B.B. (1969) Studies on anthracnose of guava and its control. *Indian Phytopathology* 22, 322–326.
Tandon, R.N. and Agarwala, R.K. (1954) Pathological studies of *Gloeosporium psidii* causing die back of guava. *Proceedings of the Indian Academy of Sciences, Section B* 40, 102–109.
Thirumalachar, M.J. (1945) An Ascomycetous parasite of *Cephaleuros*. *Proceedings of the Indian Academy of Sciences, Section B* 22, 374–377.
Tokeshi, H., Valdebenito, R.M. and Dias, A.S. (1980) Occurrence of a bacterial disease of guava in Sao Paulo state. *Summa Phytopathologica* 6, 85–87.
Uchida, J., Zhong, S. and Killgore, E. (2006) First report of a rust disease on Ohia caused by *Puccinia psidii* in Hawaii. *Plant Disease* 90, 524.
Ullasa, B.A. and Rawal, R.D. (1984) *Guignardia* fruit rot of guava – a new disease from Bangalore. *Current Science* 53, 435–436.
Uppal, B.N. (1936) *Physalospora psidii* on guava – a serious disease of guava in Bombay. *International Bulletin of Plant Protection* 10, 99.
Venkatakrishniah, N.S. (1952) *Glomerella psidii* (Del.) Sheld. and *Pestalotia psidii* Pat. associated with cankerous disease of guava. *Proceedings of the Indian Academy of Sciences, Section B* 36, 129–134.
Verma, B.R. and Sharma, S.L. (1976) Seasonal variation in symptoms caused by *Pestalotia psidii* on guava fruits. *Indian Journal of Mycology and Plant Pathology* 6, 97–98.
Vestal, E.F. (1941) *A Text Book of Plant Pathology*. Kitabistan, Allahabad, India/Karachi, Pakistan.
Vir, D., Raychaudhuri, S.P. and Thirumalachar, M.J. (1967) Aureofungin as fruit dip and wrap treatment for control of Diplodia rot of mango and Alternaria rot of tomato fruits during transit. *Indian Phytopathology* 20, 301–303.
Vos, J.E., Schoeman, M.H., Berjak, P., Watt, M.P. and Toerien, A.J. (2000) *In vitro* selection and commercial release of guava wilt resistant rootstocks. *Acta Horticulturae* 513, 69–79.
Wahid, O.A.A. (2001) Occurrence of Colletotrichum anthracnose disease of guava in Egypt. *International Journal of Pest Management* 47, 147–152.
Wang, C.L. and Hsieh, H.Y. (2005) Occurrence and pathogenicity of stem canker of guava in Taiwan caused by *Botryosphaeria rhodina*. *Plant Pathology Bulletin* 15, 219–230.

Webber, G.F. (1928) Plant pathology. In: *Report for the Fiscal Year Ending June 30, 1928*. University of Florida, Florida Agricultural Experiment Station, Gainesville, Florida, pp. 65R–78R.
Wells, J.M. and Harvey, J.M. (1970) Combination of heat and 2,6-dichloro-4-nitroaniline treatments for control of *Rhizopus* and brown rot of peaches, plums and nectarines. *Phytopathology* 60, 116–120.
Wills, R.B.H., Brown, B.I. and Scott, K.J. (1982) Control of ripe fruit rots of guavas by heated benomyl and guazatine dips. *Australian Journal of Experimental Agriculture and Animal Husbandry* 22(118/119), 437–440.
Winter, G. (1884) Repertorium. Rabenhorstii fungi europaei et extraeuraopaei. Cent. XXXI et XXXII. *Hedwigia* 23, 164–172.
Yadav, A.S. (1953) Some new hosts of *Cephaleuros* from Bihar. *Current Science* 22, 280.
Yang, H.R. and Chuang, T.Y. (1994) Pathogenicity and zymogram of anthracnose fungi isolated from some fruits. *Memoires of the College of Agriculture, National Taiwan University* 34, 1–8.
Zakir Hussain, M., Rahman, M.A. and Bashar, M.A. (2014) Incidence of guava (*Psidium guajava* L.) wilt caused by *Fusarium oxysporum* Sch. f.sp. *psidii* in Bangladesh. *Journal of Asiatic Society of Bangladesh, Science* 40, 97–105.
Zang, H.-L., Xu, Y., Duan, W.-J., Wang, J.-L. and Gu, J.-F. (2011) Isolation and identification of black spot pathogen in *Psidium guajava*. *Agricultural Science & Technology* 12, 1199–1200, 1212.
Zhong, S., Yang, B. and Alfenas, A.C. (2008) Permanent genetic resources: development of microsatellite markers for the guava rust fungus, *Puccinia psidii*. *Molecular Ecology Resources* 8, 348–350.

16 Postharvest Physiology and Storage

Margo Sulistio and Chun-Ta Wu*
National Taiwan University, Taiwan

16.1 Introduction

Guava fruit, indeed, is an intriguing plant material for ripening research, because this species contains both climacteric and non-climacteric cultivars. Guava on ripening is highly perishable. The shelf-life of guava under ambient conditions is around 3–5 days only, which is the major obstacle to commercializing guava fruit. The main factors affecting deterioration of guavas after harvesting are senescence, including softening after ripening, water loss, chilling injury, pathological breakdown and mechanical damage. Postharvest handling, which concerns the biology and techniques used for quality maintenance of fresh horticultural commodities during the time from harvesting through to ultimate utilization, is a vital phase of the guava industry. In this chapter, we provide an overview of the postharvest biology and technology of guava fruit.

16.2 Maturity Indices

Maturity at harvest is the most important factor that determines storage life and final fruit quality of guava. Harvesting fruits at appropriate stage of maturity is critical in maintaining postharvest quality of guava fruits (Tandon *et al.*, 1989; Azzolini *et al.*, 2004; Patel *et al.*, 2015; Deepthi, 2017). The advancement of fruit maturity, leading to the accumulation of sugars, increase in ascorbic acid, decrease in phenols and biosynthesis of aroma volatiles, influences the fruit flavour (Bashir and Abu-Goukh, 2003; Soares *et al.*, 2007). Late harvesting of guava results in reduction of its storage ability (Tandon *et al.*, 1989). Fruits at mature-green stage show longer shelf-life than those at colour-turning stage of maturity (Tandon *et al.*, 1989; Deepthi, 2017). Maturity indices are important for deciding when fruit should be harvested to provide some marketing flexibility and to ensure the attainment of acceptable eating quality to the consumer. Several attributes have been tried for judging guava fruit maturity. Two characteristics, fruit size and change in peel colour, were recommended as the best maturity indices (Mercado-Silva *et al.*, 1998; Singh and Pal, 2008a,b). In fact, peel colour, changing from dark green to light green at mature-green stage followed by de-greening and development of yellow colour when ripe, is commonly used to measure maturity and ripeness in guava. Maturity stages of climacteric 'Li-Tzy Bar' guava are shown in

*Email: wuct@ntu.edu.tw

DOI: 10.1079/9781789247022.0016

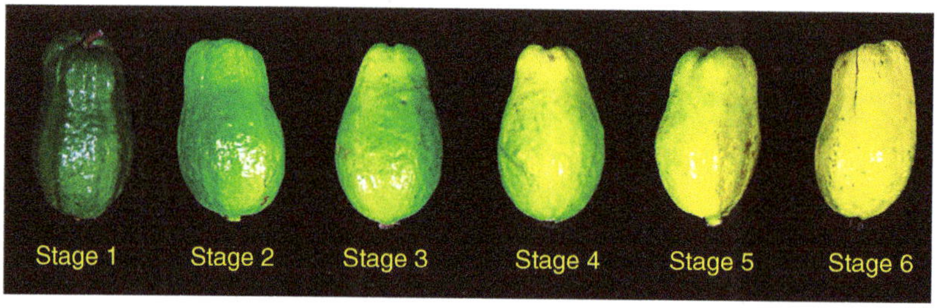

Fig. 16.1. Different stages of fruit maturity in 'Li-Tzy Bar' guava based on skin colour. Stages 1, 2, 5 and 6 represent immature-green, mature-green, fully ripe and overripe, respectively.

Fig. 16.1. For example, Cavalini *et al.* (2006) suggested that the most adequate maturity indices for 'Kumagai' guava were skin colour and pulp firmness, while for 'Paluma' fruit, the best indices were skin colour, pulp firmness and total titratable acidity (TTA).

Additionally, specific gravity, chemical attributes and fruit detachment force are useful for determining the time of harvesting (Rathore, 1976; Paull and Goo, 1983; Tandon *et al.*, 1989; Mercado-Silva *et al.*, 1998). Specific gravity generally declines during late stage of fruit maturation. Better fruit quality and consumer acceptability were found when fruit were harvested at specific gravity of <1.0 compared with >1.0 (Tandon *et al.*, 1989). Seasonal variation and inconvenient procedures for grading according to specific gravity, however, confine its utility (Mercado-Silva *et al.*, 1998). Chemical attributes including total soluble solids (TSS), TTA and tannin content may also be used to determine fruit maturity (Yusof and Mohamed, 1987; Cavalini *et al.*, 2006). Nevertheless, seasonal and cultivar variation limit their application to indicate fruit maturity (Rathore, 1976). 'Paluma' on ripening showed 11.1°Brix and soluble sugar content of 7.41 g 100 g^{-1} (Patricio *et al.*, 2012).

The harvest maturity of guava can also be estimated by days from anthesis or days from fruit set. Mostly, guava tree blooms approximately 60 days after pruning and fruit can be harvested from 100 to 150 days after blooming, depending on the season of the year (Nakasone and Paull, 1998). The number of days required for the cultivar 'Media China' to reach maturity was 130 days during spring–summer and 190 days in autumn–winter (Mercado-Silva *et al.*, 1998). For 'Allahabad Safeda', 165 days are required in Bangalore whereas 120 days are required under north Indian conditions (Selvaraj *et al.*, 1999). In term of days from fruit set, guava generally takes about 17–20 weeks to reach maturity (Bakshi *et al.*, 2014). Fully mature fruit of 'Hisar Srukha', 'Pant Prabhat' and 'Hisar Safeda' were attained at 69, 70 and 72 days after fruit set, respectively. However, a longer time after fruit set, 91 and 105 days, is needed by 'Lucknow-49' and 'Arka Amulya' to reach fully mature stage, respectively (Bakshi *et al.*, 2014). Coupled with fruit age, detachment force, texture and skin colour are also useful maturity indices to determine harvest maturity (Paull and Goo, 1983).

16.3 Postharvest Physiology

16.3.1 Respiration and ethylene production

Rates of respiration and ethylene production in guava fruits vary with cultivar, temperature, season and maturity stage (Osman and Ayub, 1996; Mercado-Silva *et al.*, 1998; Liu *et al.*, 2012; Gill, 2018). The respiration rates of guava fruit were 8 to 60 mg CO_2 kg^{-1} h^{-1} at 10°C and 18 to 130 mg CO_2 kg^{-1} h^{-1} at 20°C, respectively (Paull and Chen, 2016). Pink-pulped cultivars showed higher respiration rate than the white-pulped ones (Bashir and Abu-Goukh, 2003; Singh and Pal, 2008a). Ethylene production ranges from 1 to 20 µl kg^{-1} h^{-1} at 20°C (Kader, 2009). 'Jen-Ju Bar', also known as 'Pearl' guava, which is the leading guava cultivar for the fresh fruit market in Taiwan, released much less ethylene (C_2H_4) and carbon dioxide (CO_2) than 'Li-Tzy Bar' during postharvest storage at 20°C (Liu *et al.*, 2012) (Fig. 16.2). Guava cultivar 'Lucknow-49' produced less ethylene compared with 'Allahabad Safeda' and 'Hisar Safeda' guava (Singh and Pal, 2008a; Gill, 2018). Less ethylene production and higher levels of polyamines in 'Lucknow-49' contribute to better storability than 'Hisar Safeda' (Mondal *et al.*, 2008; Singh and Pal, 2008a). Investigation on different temperature storages in 'Kampuchea', 'Vietnam' and 'Taiwan' guava fruits showed that both respiration and ethylene emission were significantly higher in fruit stored at 27–33°C than those stored at 20 and 0–10°C (Osman and Ayub, 1996). Likewise, ethylene evolution and carbon dioxide production of 'Paluma' guava fruit showed similar trends to 'Kampuchea', 'Vietnam' and 'Taiwan' guavas (Bron *et al.*, 2005). Peak of respiration and massive ethylene production during postharvest storage were recorded 4–5 and 7–8 days for spring–summer fruit and autumn–winter fruit, respectively (Mercado-Silva *et al.*, 1998). The rate of ethylene production increases as the fruits ripen and reaches a peak level that may or may not necessarily coincide with the respiratory peak (Singh, 2011). Fruits harvested at advanced maturity (i.e. full size and colour-turning stage) had higher and earlier climacteric peaks than those harvested at small size and hard-green stage (Brown and Wills, 1983). Ethylene production from fruit of 'Pedro Sato' guava in the first four days of ripening at room temperature was practically constant (approximately 0.1 μl C_2H_4 kg^{-1} h^{-1}) and then progressed following a quadratic equation and increased to about 0.6 μl C_2H_4 kg^{-1} h^{-1} on the sixth day of ripening (Abreu *et al.*, 2012).

Guava is generally considered as a climacteric fruit (Akamine and Goo, 1979; Ali and Lazan, 1997; Singh, 2011). However, different from the vast majority of guava cultivars, certain cultivars behave in a non-climacteric manner during the ripening phase (Azzolini *et al.*, 2005; Liu *et al.*, 2012; Sulistio, 2019). Reports suggest that this characteristic of guava is varietal in nature (Porat *et al.*, 2009; Chen *et al.*, 2017). The Philippine cultivar 'Queso de Bola', which exhibits relatively low and steady carbon dioxide and ethylene production, is considered as a non-climacteric type (Rodeo *et al.*, 2018) since it behaves similarly to other non-climacteric cultivars such as 'Omni' from Israel (Porat *et al.*, 2009) and 'Jen-Ju Bar' from Taiwan (Liu *et al.*, 2012; Chen *et al.*, 2017; Sulistio, 2019).

Guava fruits exposed to ethylene display a distinct response on ripening physiology, according to cultivar and maturity stage of fruit (Reyes and Paull, 1995; Azzolini *et al.*, 2005; Liu *et al.*, 2012). Peel colour development and softening in immature-green fruits

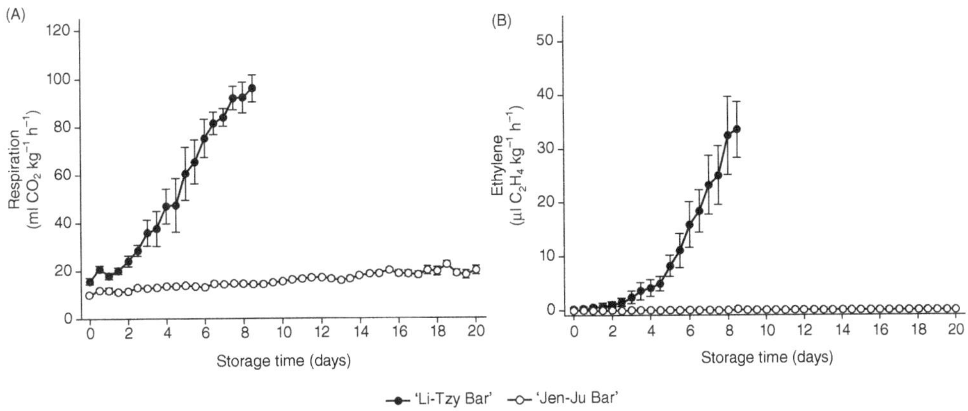

Fig. 16.2. Changes in the rate of (A) respiration and (B) ethylene production of mature-green 'Li-Tzy Bar' and 'Jen-Ju Bar' guava fruit at 20°C. Data shown are means with standard errors represented by vertical bars (n = 3). Modified from Liu *et al.* (2012).

were triggered by exogenous ethylene application; however, no such effect was observed on mature-green and quarter-yellow 'Beaumont' guava fruits (Reyes and Paull, 1995). Postharvest ethylene application on light-green mature 'Pedro Sato' guavas did not enhance endogenous massive ethylene production compared with the control group (Azzolini *et al.*, 2005). Increase of endogenous ethylene release was detected on mature-green 'Li-Tzy Bar' guavas treated with propylene, whereas 'Jen-Ju Bar' guavas harvested at the same maturity stage exhibited no significant enhancement of ethylene production in this experiment (Liu *et al.*, 2012). Nevertheless, postharvest fumigation with 1-methylcyclopropene (1-MCP), an ethylene action inhibitor, impedes ethylene biosynthesis and retards fruit ripening in both climacteric and non-climacteric guava cultivars (Azzolini *et al.*, 2005; Bassetto *et al.*, 2005; Singh and Pal, 2008b; Harb and Hasan, 2010; Phebe and Ong, 2010; Kumar *et al.*, 2014).

An interesting question raised is, what is the physiological mechanism(s) leading to a non-climacteric ripening manner of certain guava cultivars? Liu *et al.* (2012) found that although obvious respiration upsurge, peel pigmentation and pulp softening were observed, there was no significant increase of ethylene release from non-climacteric 'Jen-Ju Bar' guava fruits after a treatment with propylene, an ethylene analogue, at 1000 μl l^{-1} for 72 h (Fig. 16.3). This result implies that the failure of triggering System 2 massive ethylene biosynthesis, rather than defect of ethylene perception and/or signal transduction, is the underlying cause of the non-climacteric trait. In higher plants, ethylene is synthesized from methionine as the precursor, which is converted to *S*-adenosyl-L-methionine (AdoMet) and subsequently to 1-aminocyclopropane-1-carboxylic acid (ACC) via catalysis by AdoMet synthetase and ACC synthase (ACS), respectively (Yang and Hoffman, 1984). In the final step, ACC, the immediate precursor of ethylene, is oxidized by ACC oxidase (ACO) to form ethylene (Yang and Hoffman, 1984). The major difference between climacteric 'Li-Tzy Bar' and non-climacteric 'Jen-Ju Bar' guava fruits is that the latter variety cannot switch on *PgACS1*, a System 2 ACS gene, during ripening (Liu *et al.*, 2012; Chen *et al.*, 2017). Therefore, not enough ACC is produced by ACS to support massive ethylene production in mature-green 'Jen-Ju Bar' guava fruit. As a consequence, the fruit stays at mature-green phase and exhibits non-climacteric characteristics, including low and steady respiration and ethylene production rates after harvest, as well as maintenance of green colour in the peel and texture firmness in the pulp (Liu *et al.*, 2012; Chen *et al.*, 2017). In other words, the non-climacteric ripening trait of 'Jen-Ju Bar' is a consequence of limited ACC availability. This hypothesis is further strengthened by the results of *in vivo* feeding experiments using ethylene biosynthetic precursors and mature-green 'Jen-Ju Bar' fruit. 'Jen-Ju Bar' fruit behaved like a climacteric fruit, namely with increases

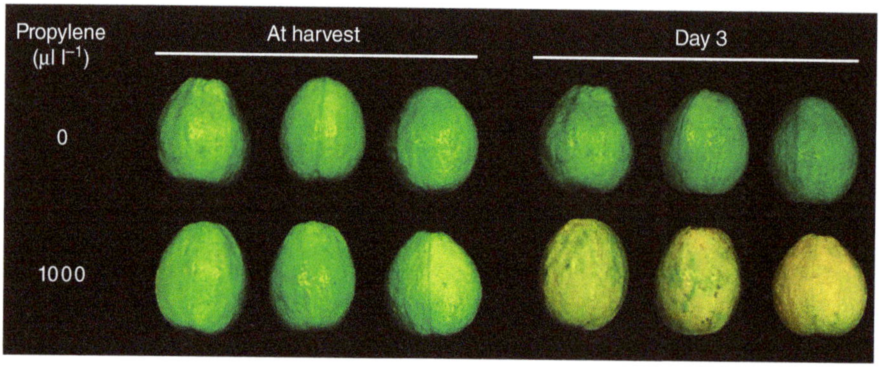

Fig. 16.3. Appearance of 'Jen-Ju Bar' guava fruit treated with or without propylene for 3 days at 20°C.

of respiration and ethylene production rates, fruit softening, skin coloration and aroma emission, when an ACC solution, but not AdoMet solution or water, was provided at the stem-end cavity (Liu *et al.*, 2012). These results directly demonstrate that ACS, rather than ACO, is the step in autocatalytic ethylene production that is blocked in 'Jen-Ju Bar' fruit. However, the importance of ACO activity in the ripening stage of guavas is not excluded. The increase in ethylene has been correlated with ACO activity during fruit ripening, indicating this enzyme may be a limiting step in the ethylene biosynthetic pathway in guava fruit since ACC level rose in line with fruit ripening (Mondal *et al.*, 2008).

16.3.2 Physicochemical changes during ripening

Change in colour

Fruit ripening is the physiological outcome of complex and interrelated biochemical changes resulting in the development of texture, taste, colour and flavour (Gill, 2018). In guava, degradation of chlorophyll coincidence with increasing carotenoid content leads to peel colour change from green to yellow (Jain *et al.*, 2003). However, several guava cultivars show exceptions, the skin colour stays green during ripening (Lim and Khoo, 1990; Liu *et al.*, 2012). In terms of pulp colour, it may change from white to creamy white, yellowish pink, deep pink or salmon red, depending upon cultivar (Singh, 2011; Mitra *et al.*, 2012; Gill, 2018). The presence of carotenoids may contribute to the intensity of the pulp colour of fruit during ripening (Wilberg and Rodriguez-Amaya, 1995). The main carotenoid in ripe guava is lycopene, followed by β-carotene and traces of other pigments (Padula and Rodriguez-Amaya, 1986; Wilberg and Rodriguez-Amaya, 1995; Mercadante *et al.*, 1999). In contrast, lycopene and β-carotene contents of guava cultivars from Indonesia were found to be similar (1150 and 984 μg 100 g^{-1}, respectively), and β-cryptoxanthin was at trace level (Setiawan *et al.*, 2001). In other studies, lycopene, β-carotene and phytofluene were identified as the predominant carotenoids in guavas from Cuba (Ventosa *et al.*, 2008).

Change in texture

Alteration of cell-wall carbohydrates in guava occurs with fruit softening during ripening (Singh, 2011). Substantial declines in pectin, cellulose, hemicellulose, lignin and starch during ripening render guava fruit softening (Jain *et al.*, 2001, 2003). The content of soluble pectin rises as the 'Pedro Sato' fruit ripening progresses, ranging from 0.13% (w/w) to 0.55% galacturonic acid. Throughout ripening, there was an increase in soluble pectin and a decrease in protopectin (Braga *et al.*, 2018). In 'Beaumont', 'Banaras' and 'Baladi' guavas, loss of tissue firmness during ripening is accompanied by a decrease in the level of total pectin, expressed either as ethanol-insoluble residue or on a fresh weight basis (Pal and Selvaraj, 1979; El-Zoghbi, 1994). The rate of fruit softening differs with cultivar; 'Beaumont' guava required about 1.5 days to reach 50% of initial firmness at harvest, whereas 'Kampuchea' took longer, about 24 days (Ali *et al.*, 2004). A reduction of 70% in the firmness of 'Pedro Sato' was observed during 4 days of storage at room temperature (Abreu *et al.*, 2012; Botelho *et al.*, 2016). Other guava cultivars also exhibited similar firmness loss, ranging among 85% for 'Paluma' (Cavalini *et al.*, 2015), 78% for 'Cortibel' (Werner *et al.*, 2009) and 83% for 'Allahabad Safeda' (Gill, 2018). Firmness variation of 34% was found between the third and seventh day of ripening for the cultivar 'Media China' (Mercado-Silva *et al.*, 1998). On the other hand, 'Kumagi' guava, which has a long shelf-life, showed only 25% variation between the first and tenth day of ripening (Carvalho *et al.*, 2001). The loss of tissue firmness in guava was correlated with the solubilization and depolymerization of pectin by pectic enzymes and the modification of cell-wall components during ripening (Ali and Lazan, 1997; Abu-Goukh and Bashir, 2003). Cell-wall constituents, namely pectin, hemicellulose, cellulose and lignin, decreased in the ripening guava (Jain *et al.*, 2001, 2003).

Remarkable textural changes in guava fruit are caused by simultaneous action of cell-wall hydrolysing enzymes, polygalacturonase (PG), pectin methylesterase (PME), β-galactosidase, (1→4)-β-glucanase and cellulase (Singh, 2011). Rising of PG activity was observed during fruit ripening, whereas PME increased at the beginning and declined subsequently (Abu-Goukh and Bashir, 2003; Jain *et al.*, 2003). Not only PG, increased activity of pectin esterase, β-galactosidase and cellulase were also detected with ripening of 'Kampuchea' guava (Chin *et al.*, 1994). Activity of PG, β-D-galactosidase and cellulase in 'Hisar Surkha' and 'Hisar Safeda' guava fruits rose progressively and markedly with the advancement of ripening on-tree as well as in storage (Sharma *et al.*, 2012). Maximum PME activity was found in 'L-49' (56.25 units g^{-1} fresh weight) at half-ripe stage then decreased at full-ripe stage (52.25 units g^{-1} fresh weight), followed by 'Allahabad Safeda' and 'Lalit' (Dolkar *et al.*, 2017). The cell-wall modification was also influenced by upsurge in activities of other enzymes, such as β-galactosidase, (1→4)-β-glucanase and cellulase (Abu-Goukh and Bashir, 2003; Ali *et al.*, 2004). In 'Pedro Sato' guava, the activity of β-D-glucosidase and esterase increased until the fourth day of ripening, which coincides with the decrease in firmness over the same period (Braga *et al.*, 2018).

Changes in starch, soluble solids content and total sugars

Sugar content is one of the important compositions present in the fruit determining the flavour characteristics and commercial assessment of guavas (Ali and Lazan, 1997). Guava fruit accumulates a very small amount of starch throughout its development. The starch content of guava fruit at mature-green stage is around 2–7% of fresh weight (Jain *et al.*, 2001, 2003; Liao, 2019). Bashir and Abu-Goukh (2003) observed a greater increase in total sugars in pulp than in peel after fruit firmness reached 1.21 kg cm^{-2}, which coincided with the climacteric peak of respiration. Soluble solids content (SSC) and total sugar content generally increase during ripening in guava fruit (El Bulk *et al.*, 1997; Bashir and Abu-Goukh, 2003; Singh and Pal, 2008a,b). In the Brazilian guava 'Pedro Sato', which behaved in a climacteric manner for changes of colour and loss of firmness, TSS and sugar contents showed no variation with the advancement of ripening and presented an average of 7.6°Brix, although the ethylene level increased starting from the fourth day of ripening (Abreu *et al.*, 2012). In ripe guava fruits, fructose is the main sugar, followed by glucose and sucrose (Singh, 2011; Liao, 2019). SSC, which is mainly associated with sugars and organic acids, usually rises during ripening, though seldom more than 3°Brix (Lazan and Ali, 1998; Mercado-Silva *et al.*, 1998). Rising SSC from 10 to 13°Brix was observed in two Indian guava cultivars, 'Allahabad Safeda' and 'Sardar' (Tandon *et al.*, 1989; Selvaraj *et al.*, 1999). Liu *et al.* (2012) reported that 'Li-Tzy Bar' guava showed increasing SSC from 7.9 to 8.2°Brix during ripening. The small change in TSS of the fruit during ripening was attributed to low starch content in pulp tissues at harvest.

Change in acidity

Citric acid is the predominant organic acid in guava fruit, followed by ascorbic, malic, glycolic, and fumaric acids (Wilson *et al.*, 1982). TTA initially rises with progressive fruit growth and then declines at full ripeness after the climacteric peak (Ali and Lazan, 1997). In climacteric fruits, the TTA usually decreases during ripening due to the use of organic acids, besides sugars, as a respiration substrate (Chitarra and Chitarra, 2005). In 'Pedro Sato' guava the TTA shows an average level of 0.5 citric acid equivalents per 100 g of pulp, which is similar to characteristics of non-climacteric fruits (Abreu *et al.*, 2012). The fruits of 'Sardar' had higher citric acid than those of 'Safeda', together with malic, tartaric, pyruvic, succinic, fumaric, oxaloacetic, α-ketoglutaric and malonic acids at various ripening stages (Selvaraj *et al.*, 1999). Titratable acidity is cultivar-dependent, either increasing (Lazan and Ali, 1998; Liao, 2019) or decreasing during fruit ripening (Paull and Goo, 1983; Liao, 2019).

Guava fruit is renowned for its rich content of vitamin C, also known as ascorbic acid (Paull and Duarte, 2012). Ascorbic acid concentration increases substantially at the beginning stage of ripening and then declines with senescence (Rathore, 1976; El Bulk *et al.*, 1997; Soares *et al.*, 2007; Gomez and Lajolo, 2008). The highest ascorbic acid content was reported in the peel of guava fruit (Wilson *et al.*, 1982). At the ripe stage (pulp firmness 0.3 kg cm^{-2}), the amount of ascorbic acid retained was 86.3% in the pulp and 85.6% in the peel of white-pulped guava fruits compared with 76.6 and 78.1% in the pulp and peel of pink-pulped guavas, respectively (Bashir and Abu-Goukh, 2003).

Change in polyphenols

The presence of polyphenols contributes to an astringent taste of guava fruits and their content declines markedly with ripening (El Bulk *et al.*, 1997). Young guava fruits had a total phenol content of 620 mg 100 g^{-1} tissue, with about 65% present in the form of condensed tannins. As fruit matures, the tannin content decreases considerably (Itoo *et al.*, 1987). Decrease of phenolic compounds in the white- and pink-pulped guava types was seven- and threefold, respectively (Bashir and Abu-Goukh, 2003). The decline in astringency was associated with increased polymerization of leucoanthocyanidins and hydrolysis of the arabinose ester of hexahydroxydiphenic acid during guava fruit ripening (Misra and Seshadri, 1968). Also the action of polyphenol oxidase contributed to the reduction in astringency of ripe guavas as its activity increased with ripening (Mowlah and Itoo, 1982).

Change in volatile compounds

About 270 volatile compounds have been identified in guava pulp and peel, including hydrocarbons, alcohols, aldehydes, ketones, acids, esters, basic sulfur compounds, ether, phenols, furans and epoxides (Simrat, 2009). The unique flavour of guava fruit is associated with the presence of C_6 aldehydes, C_6 alcohols, ethyl hexanoate, (*Z*)-3-hexenyl acetate, terpenes and 1,8-cineole (Chen *et al.*, 2006). Immature guava fruits contain mostly aldehyde compounds, while the content of esters increases progressively with maturity (Soares *et al.*, 2007). In mature fruits of guava, however, caryophyllene is the major sesquiterpene hydrocarbon in the volatile compounds (Chyau *et al.*, 1992). In another study, (*Z*)-3-hexenal, 3-sulfanyl-1-hexanol, 4-hydroxy-2,5-dimethyl-3(2*H*)-furanone, 3-sulfanylhexyl acetate, hexanal, ethyl butanoate, cinnamyl acetate and methional were identified as key aroma compounds of pink-pulped guavas (Steinhaus *et al.*, 2009).

Non-climacteric 'Jen-Ju Bar' fruits emit fewer aroma volatiles than those of climacteric guava cultivars. Ethylene appears to play a key regulatory role in the aroma production of guava fruits. Since the System 2 ethylene production is almost absent, 'Jen-Ju Bar' provided a good material for exploring the aroma synthesis controlled by ethylene during guava ripening. Zhang (2018) investigated the effects of various treatments with propylene on composition of the aroma volatiles in peel and pulp tissues of 'Jen-Ju Bar'. Thirteen main aroma compounds were identified in 'Jen-Ju Bar' fruit after propylene treatment, including nine esters, one alcohol, one aldehyde and two terpenoids. Among the volatile compounds, 3-hexenyl acetate, hexyl acetate, ethyl hexanoate and ethyl butanoate, the main aroma component of guava fruits, showed significant increase in production with the increase in propylene dosage and duration. The results strongly suggested that ethylene is the major elicitor of aroma synthesis in guava fruit (Zhang, 2018). Moreover, high lipoxygenase (LOX) activity in the peel tissue appears to be important for ethylene to regulate aroma production in guava fruits (Zhang, 2018).

16.4 Storage

16.4.1 Low-temperature storage

Cold storage is the most common practice for preservation in fresh produce. Fruits are normally pre-cooled immediately after harvest before transport, marketing or storage (Campbell, 1994). Low temperature may delay and/or retard ripening and may reduce spoilage. But the problem with guavas, as with most other tropical fruits, is that they are highly sensitive to chill temperature (Wang, 1993). These fruits have a higher critical temperature for the induction of chilling injury (CI) relative to the less sensitive or insensitive temperate species (Ali and Lazan, 1997). In general, a temperature of 8–10°C is regarded to be the critical limit for several guava cultivars (Tandon *et al.*, 1989; Reyes and Paull, 1995; González-Aguilar *et al.*, 2004; Singh and Pal, 2008a,b; Kader, 2009; Mahajan *et al.*, 2009; Rai *et al.*, 2010).

Storage behaviour of 'Queso de Bola' guava (classified as a non-climacteric cultivar) at different temperatures was studied by Rodeo *et al.* (2018). Fruit had a storage life of 7, 11 and 19 days at ambient (32 ± 2°C), 20°C and 15°C, respectively. Storage at 15°C for 2 weeks maintained sensory quality, slowed down ascorbic acid degradation, reduced the weight loss, prevented shrivelling and delayed the development of diseases. 'Kampuchea' guava remained in good condition for up to 3.6 weeks at 10°C (Silip and Hajar, 2007).

The susceptibility of guava fruit to CI may vary with the cultivar and maturity at harvest. The cultivar 'Criolla Roja' can be stored at 12 ± 2°C (Suárez *et al.*, 2009) while 'Kampuchea' guava remained good quality for over 3 weeks at 10°C (Silip and Hajar, 2007). A storage temperature of 8°C for 'Kumagai' guava facilitates its storability for up to 3 weeks and 4 days after transferring to 25°C (Gaspar *et al.*, 1997). Fruit maturity at harvest may influence the susceptibility of guava to CI; fruits harvested at colour-turning stage can be stored for 3 weeks at 7°C, have good appearance and suffer less decay than the fruits that are stored at mature-green stage (Vazquez-Ochoa and Colinas-Leon, 1990).

16.4.2 Controlled-atmosphere storage

Controlled atmosphere (CA) has been tested with success in many tropical fruits; however, it is not commonly used in storage of guava (Teixeira *et al.*, 2007). CA storage may assist the extension of many tropical and subtropical fruits by the precise regulation of low oxygen (O_2) and high carbon dioxide levels along with maintenance of optimum temperature (Gill, 2018). Previous studies showed that short-term exposure to low O_2 (<1–10 kPa) and high CO_2 (5–40 kPa) atmospheres prolonged the shelf-life of guava fruit (Gill, 2018). The recommended CA storage conditions for guava fruits are 2–5% O_2 and 0–1% CO_2 at temperature of 5–15°C (Kader, 2001). The shelf-life of guava under ambient conditions was lengthened by 2–3 days after keeping the fruit at very low O_2 (<1 kPa) and high CO_2 (40 kPa) at 40°C for 12 h (Singh and Pal, 2007).

CA is used to prolong storage life and reduce CI symptoms in guava fruits (Wang, 1993; Bautista and Silva, 1997). Storing guava for 30 days at 8°C under 5% O_2 + 2.5% CO_2, 5% O_2 + 5% CO_2 and 8% O_2 + 5% CO_2 delayed the appearance of CI symptoms in 'Lucknow-49', 'Allahabad Safeda' and 'Apple Colour' guava fruits, respectively (Singh and Pal, 2008a). No CI symptoms were observed in 'Pedro Sato' guava kept at 12.5°C for 28 days under 1–5% O_2 (Teixeira *et al.*, 2016). Keeping guava fruit at 10°C under 5% CO_2 for 3 weeks alleviated CI prevalence and increased postharvest life of 'Media China' guava fruits (Alba-Jiménez *et al.*, 2018).

Under CA storage, reduction of respiration rate, ethylene biosynthesis, colour development and pulp softening, maintenance of nutritional components, prevention of physiological disorders and decreased disease incidence have been reported (Teixeira *et al.*, 2007; Singh and Pal, 2008a,b; Teixeira and Durigan, 2010). Pretreatment of fruits in 1–3% calcium chloride ($CaCl_2$) solution and subsequent storage at 5% CO_2 at 10 ± 1°C for 24 days retained higher total phenolics content, antioxidant activity, citric acid and ascorbic acid contents of 'Safeda' guava fruits compared with those stored at room temperature (Javed *et al.*, 2018). Low O_2 concentrations (1 and 5 kPa) significantly reduced respiratory rates and delayed the ripening process of 'Pedro Sato' guava during storage at 12.5°C for 28 days, featuring maintenance of fruit green colour, higher TSS and reducing sugar contents (Teixeira and Durigan, 2010).

16.4.3 Modified-atmosphere storage

Storage under modified atmosphere (MA), as in polybags, or packaging (MAP) or coating (MAC) in polymeric films, may prolong the shelf-life of a wide variety of fruits (Pal *et al.*, 2004; Sivakumar and Korsten, 2006; Dhalsamant *et al.*, 2017). The MA treatments would ultimately result in a build-up of carbon dioxide and a depletion in the level of oxygen within the internal atmosphere surrounding the fruit. In many cases, respiration, as well as ethylene production and perception by the fruit, are reduced, ripening is delayed and hence the shelf-life of the fruit is extended markedly. In climacteric fruits like guava, where ripening is characterized by an increase in respiration and ethylene production, such MA treatments are likely to be beneficial (Ali and Lazan, 1997).

Application of MAP technology can reduce water loss and metabolic processes, resulting in an extension of guava fruit storability (Gaspar *et al.*, 1997; Mangaraj *et al.*, 2014; Antala *et al.*, 2015; Kumar *et al.*, 2017; Rana and Siddiqui, 2018). The storage conditions and thickness of the films influence the results of the MAP. Storing of 'Kumagai' guava fruit under MAP condition at 8°C for 3 weeks in 24.7 μm thick, low-density polyethylene (LDPE) film resulted in reduced weight loss, lower acidity and SSC, and lower CI symptoms (Gaspar *et al.*, 1997). The modified atmosphere of 5% O_2 and 4% CO_2 was reported to be suitable for preservation of 'Baruipur' guava fruits for 26 days of storage at 10°C, featuring the reduction of weight loss, pectin solubilization, retention of colour and firmness as compared with control fruits (Mangaraj *et al.*, 2014). Minimum weight loss, pulp/peel ratio, firmness, SSC, total sugars, ascorbic acid and titratable acidity were observed in 'Lucknow-49' guava fruits packed in 50 μm LDPE bags at 3% O_2 and 5% CO_2 gas composition at 5°C for 42 days of storage compared with controls (Antala *et al.*, 2015). 'Allahabad Safeda' guava fruits, when packed in 100-gauge polypropylene bags with four pores, showed significantly reduced loss of weight, lower CI, electrolyte leakage and titratable acidity, and markedly increased fruit firmness and organoleptic score, TSS, Brix/acid ratio and ascorbic acid contents compared with control fruits, allowing prolonged storage up to 25.63 days at 6 ± 1°C (Kumar *et al.*, 2017). Vacuum packing and MAP of 'Hisar Safeda' guava fruits in 200-gauge LDPE showed minimum weight loss, decay loss and ripening during storage at 37°C for 8 days (Rana and Siddiqui, 2018). MA conditions using unperforated LDPE of thickness 300 gauge effectively reduced decay loss and infection of 'Hisar Safeda' and 'Lucknow-49' guava fruits from different types of fungi during storage at 8°C for 8 days (Bishnoi and Sharma, 2015). Individually packaging 'Hisar Safeda' fruits in LDPE of 200-gauge thickness by shrink and cling wrapping and storing at 7 ± 3°C substantially reduced the magnitude of changes during storage, including ripening processes, as evident from lower SSC, higher ascorbic acid and polyphenol contents with lower polyphenol oxidase activity and physiological loss of weight (Rana *et al.*, 2015).

Several studies have reported the effectiveness of active MAP for guava storability. Active MAP by flushing with 5 kPa O_2 and 2.5 kPa CO_2 in 25 μm LDPE bags was more effective than passive MAP with the same film in maintaining the fruit quality (Singh, 2011). The combination of active MAP, an ethylene scavenger containing 3 g potassium permanganate and a moisture scavenger containing 46 g coarse silica gel extended the shelf-life of 'Baruipur' guava fruits up to 32 days. Fruits were less affected by CI, had brighter colour, lower firmness, lower physiological weight loss, and higher retention of total phenol and ascorbic acid contents compared with controls (Murmu and Mishra, 2018).

16.4.4 Edible coatings

Edible coatings may offer several beneficial effects including decreased rate of water loss, retarded fruit ripening and maintaining the texture of guava fruits (Kester and Fennema, 1986). MAC of mature-green guavas in cellulose- or carnauba-based emulsions delayed colour development and suppressed the increase in the level of TSS (McGuire and Hallman, 1995). Reduction in weight loss and ethylene production were observed in 'Lucknow-49' guava fruits treated with Sta-fresh™, a carnauba wax-based coating, during storage under normal conditions and in an evaporative cool chamber (8–12°C) for 7 and 14 days, respectively (Pal *et al.*, 2004). Glucomannan-based edible coating incorporated with beeswax extended the postharvest storability of a guava cultivar in Thailand for 13 days by reduction of weight loss and firmness without spoilage during storage at 29°C (Sothornvit, 2012). Olive oil and aloe vera composite coatings were also reported effective in maintaining physiological loss in weight, fruit firmness, TSS, titratable acidity, ascorbic acid and chlorophyll contents in 'Sardar' guava fruits during the storage at 27–29°C for 15 days (Kohli *et al.*, 2019).

Moreover, coating materials have been reported to enhance shelf-life and reduce decay incidence in guava fruits. Carnauba wax coating supplemented with 5% oil and 2% sucrose enhanced the storability of 'Media China' under refrigeration (Espinoza-Zamora *et al.*, 2008). 'Pearl' guava coated with 2% chitosan solution and stored at 11°C exhibited delayed fruit ripening, increased activities of antioxidant enzymes and maintenance of membrane integrity through reducing oxidative stress (Hong *et al.*, 2012). Carboxy methyl cellulose and cashew gum-based coating retarded colour change and decay incidence resulting in extended shelf-life of 'Kumagai' guava fruits stored at room temperature (Forato *et al.*, 2015). The combination of chitosan 2% (w/v) and jackfruit seed starch-based coating effectively delayed 'Paluma' guava fruit ripening and maintained firmness and colour as well as overall acceptance under 10 days of storage at room temperature and another 6 days of refrigeration (Rodrigues *et al.*, 2018). Similarly, application of 2% arrowroot starch plus 0.3% pomegranate seed oil substantially reduced the respiration rate and provided a better shelf-life in 'Paluma' guavas for 20 days under storage at 10 ± 2°C (de Medeiros Teodosio *et al.*, 2018).

16.4.5 Hot-water treatment

Heat treatment including hot-water immersion, hot air and hot vapour has been utilized as quarantine treatment for guavas (McGuire, 1997; Wang and Lin, 2009). Treatment of fruits in hot water for 35 min at 46.1°C controlled Caribbean fruit fly infestation and minimized overall loss of guava quality during storage (McGuire, 1997). Disinfection from oriental fruit fly is possible by vapour heat treatment at 46.5°C for 15 or 35 min (Wang and Lin, 2009). Combination of heat treatment with other postharvest approaches including CA storage (Murray *et al.*, 2007) or chemical treatment (Supapvanich and Promyou, 2017; Supapvanich *et al.*, 2019) has been applied for postharvest quality maintenance of guava.

Hot-water treatment at 40°C for 30 min followed by 0.1 mM methyl jasmonate dipping treatment maintained visual appearance and colour, retarded softening by inhibition of increase of soluble pectin and decrease of insoluble pectin, enhanced both antioxidant and free radical scavenging activities, increased bioactive compounds such as ascorbic acid, total phenols, flavonoids and peroxidase activity, and decreased catalase activity of 'Kim Ju' guava fruit during storage at 12 ± 1°C for 18 days (Supapvanich *et al.*, 2019).

16.4.6 Calcium treatment

Vacuum infiltration of wrapped fruits in 10% $CaCl_2$ at ambient temperature retarded softening and suppressed the increase in soluble pectin and titratable acid levels but had no effect on the incidence of disease compared with non-calcium treated fruit (Ali and Lazan, 1997). Fruits of 'Sardar' guava, when treated with calcium nitrate ($Ca(NO_3)_2$) at 1% and stored at 6–8°C, remained in marketable quality for up to 30 days (Mandal *et al.*, 2010). Application of 12% $CaCl_2$ solution was suggested for storing of 'Piracunga Vermelha' guava at 10°C for 3 weeks (Silva *et al.*, 2000). Postharvest treatment with $Ca(NO_3)_2$ at 1% extended the shelf-life up to 12 days at room temperature without adversely affecting fruit quality (Bhooriya *et al.*, 2018). Deepthi *et al.* (2016) suggested use of $Ca(NO_3)_2$ at 2% for storage of 'L-49' guava in cold storage (10 ± 1°C and 90 ± 5% relative humidity (RH)) that showed a potential shelf-life of 23.83 days. They observed longer shelf-life and better retention of quality when fruits were treated with calcium at mature-green stage rather than colour-turning stage. $Ca(NO_3)_2$-treated fruits showed delayed loss in skin greenness and their appearance was better, being free from shrivelling and dark spots, while $CaCl_2$-treated fruits had a dull appearance due to physiological weight loss and firmness compared with $Ca(NO_3)_2$ treatment.

16.4.7 Treatment with 1-methylcyclopropene

1-MCP, an ethylene action inhibitor, inhibits ethylene perception in plant tissues by competitively binding to ethylene receptors and controlling physiological responses such as ripening (Watkins, 2006, 2008). Application of 1-MCP to guava fruits resulted in suppression of respiration and ethylene production, delayed fruit softening and skin colour changes, enhanced cold storage life and alleviation of CI (Azzolini *et al.*, 2005; Bassetto *et al.*, 2005; Singh and Pal, 2008b). However, the effectiveness of 1-MCP in prolonging the postharvest life of guava varies with cultivar, ripening stage, exposure temperature, concentration and duration (Mitra *et al.*, 2012). The concentration of 1-MCP used in guava for extending the shelf-life varies from 300 to 1000 nl l^{-1} (Bassetto *et al.*, 2005; Singh and Pal, 2008b; Kumar *et al.*, 2014) and the duration of exposure varies from 3 to 24 h (Bassetto *et al.*, 2005; Singh and Pal, 2008b). One-day extension of storage period was found by exposing 'Pedro Sato' guava to 1-MCP at 300 nl l^{-1} for 6 or 12 h and at 900 nl l^{-1} for 3 h (Bassetto *et al.*, 2005). Using 1-MCP at 300 nl l^{-1} for 12 and 24 h or 600 nl l^{-1} for 6 h prolonged the storability of 'Allahabad Safeda' guava up to 5 days under low-temperature (10°C) storage (Singh and Pal, 2008b). 1-MCP treatment to 'Kampuchea' guava fruit at a dose of 600 nl l^{-1} for 6 h retained fruit colour, flesh firmness and delayed disease development during storage at 27°C and 70% RH for 5 days (Phebe and Ong, 2010). Application of 1-MCP at a concentration of 650 nl l^{-1} combined with cold storage at 12°C significantly delayed the ripening process, retarded peel colour change, maintained firmness, reduced biosynthesis of aroma volatiles and altered the odour volatiles composition, leading to extended postharvest life of 'Ben-Dov' guava (Harb and Hasan, 2010). Delaying colour development and respiration peak, as well as reducing the fruit softening and loss of fruit acidity were observed for 12 days in storage of 'Allahabad Safeda' guava by treatment with 1-MCP at 1000 nl l^{-1} for 4 h at 20°C (Kumar *et al.*, 2014). Similarly, postharvest treatment of guava with 500–1000 nl l^{-1} concentration of 1-MCP significantly retained high fruit size, fruit weight, palatability value, ascorbic acid content and titratable acidity up to 9 days of storage under ambient conditions (Singh *et al.*, 2017). Exposing physiologically mature 'Gola' and 'Surahi' guava fruits to 1-MCP at 500 nl l^{-1} for 12 h at 15°C, before storage at 10°C and 80% RH, markedly reduced the weight loss, reduced the decay incidence and preserved firmness and colour (Iqbal *et al.*, 2018).

16.4.8 Irradiation treatment

Irradiation of fruit after harvest was reported to enhance the shelf-life and can also be used for postharvest disinfection (Mitra *et al.*, 2012). The beneficial effects of irradiation in extending storage life and reducing decay incidence in guava fruit have been reported (Baghel *et al.*, 2005; Singh and Pal, 2009). The physiological response of fruit to γ-radiation varies with maturity, radiation dose and postharvest handling conditions (Thomas and Diehl, 1988). Ionizing radiation doses of less than 1 kGy are appropriate for guava fruits (Kader, 1986). Irradiation retarded physical and biochemical changes associated with ripening of fruits such as firmness, total acids, SSC and vitamin C (Baghel *et al.*, 2005; Singh and Pal, 2009). Postharvest exposure to 0.3 kGy delayed fruit ripening, reducing water loss and decay development in guava stored at room temperature (Thomas and Diehl, 1988). Irradiation of guava fruits with 0.1 kGy γ-radiation increased the post-harvest life of guava fruits by 8 days over control (Baghel *et al.*, 2005). Similarly, ionizing radiation treatment of 0.25 kGy prolonged the storage life and retarded the respiration and ethylene production of 'Lucknow-49' and 'Allahabad Safeda' guavas during 8 days of storage at 27 ± 2°C (Singh and Pal, 2009). Higher dose of 2 kGy γ-radiation was effective for disinfestation of 'Kadaro' white guava fruits from fruit flies, *Ceratitis* spp. and *Bactrocera* sp., and maintained the firmness after 4 days of storage at ambient temperature (Kabbashi *et al.*, 2012).

Combinations of irradiation with other postharvest treatments have been documented. Application of irradiation at 0.1 kGy combined with 6% waxol to 'Lalit' guava fruits caused minimum physiological loss in weight and maintained the fruit quality during storage for 16 days at ambient conditions (Yadav *et al.*, 2010). The combination of polystyrene packaging and 0.2 kGy γ-radiation on 'Pedro Sato' guava maintained higher quality and acceptability compared with an untreated group during cold storage at 10 ± 1°C and 90–95% RH for 28 days (de Campos *et al.*, 2011). The 'Kumagai' guavas treated with ultraviolet-C (UV-C) radiation and kept under low-temperature storage (8 ± 0.2°C) for 20 days maintained better quality indices than controls and the treatment completely inhibited pupation of the fruit fly, *Ceratitis capitata* (JanuáriaVieira *et al.*, 2014). Postharvest treatment with 0.3% potassium sorbate and 1.0 kGy radiation delayed the growth of microbes and ripening, resulting in extended shelf-life of guava (Hossain *et al.*, 2014). Reduction of decay incidence and water loss and maintenance of fruit quality in 'Baladi' guava were reported when treated with 0.2 kGy γ-radiation before storage and storage at 8 ± 1°C and 90% RH for 20 days (Hassanein *et al.*, 2018).

The positive effect of 0.1 kGy γ-radiation in maintaining vitamin C content was observed in winter guava fruits during storage at ambient conditions for 8 days (Pandey and Joshua, 2010). Pre-storage treatment of 'Lohan' guava with UV-C irradiation for 30 and 60 min resulted in increased contents of ascorbic acid, total phenolics and total antioxidant capacity of the fruit when stored at ambient temperature (Razali *et al.*, 2015). Exposure of 'Khaza' guava fruits to 0.2 kGy γ-radiation significantly increased the post-harvest life without any negative impacts on fruit quality (firmness, titratable acidity, SSC and vitamin C) as well as sensory quality parameters (appearance, taste, texture and flavour) as compared with non-irradiated fruits under ambient storage conditions and 85 ± 4% RH for 9 days (Sau *et al.*, 2018). Although some reports suggest that high irradiation doses lead to a significant decline in the vitamin C content of fruit at storage (Baghel *et al.*, 2005; Singh and Pal, 2009), Pandey and Joshua (2010), Razali *et al.* (2015) and Sau *et al.* (2018) reported either an increase or maintaining the vitamin C content during storage with 0.1 to 0.2 kGy treatment.

16.5 Postharvest Pathology

Incidence of postharvest disease is a major factor contributing to decrease the quality of fruits, alter the physical and chemical properties, reduce shelf-life and increase the postharvest losses in guava (Ali and Lazan, 1997; Martins *et al.*, 2007). Most of the postharvest diseases originate in the field as quiescent infections in immature fruit prior to harvest. Proper field management, including good sanitation and bagging of fruits in paper impregnated with fungicides, are some of the measures that can be taken to check earlier field infections that might develop into serious problems at storage (Mitra *et al.*, 2012).

Anthracnose, a fungal decay caused by *Colletotrichum gloeosporioides* Penz, is the most common and destructive fungal decay causing severe postharvest losses in guava (Lim and Manicom, 2003). The symptoms of anthracnose appear in the form of small brown lesions on the fruit surface, which later grow larger and develop into sunken patches (Singh, 2011). The infection of *C. gloeosporioides* was reported as causing significant postharvest losses of guava fruits in many countries, such as India, Taiwan, Brazil, Nigeria, Malaysia and Egypt (Rawal, 1993; Wahid, 2001; Amusa *et al.*, 2005; Martins *et al.*, 2007; Yeh and Shiesh, 2017). Hot-water treatment (47°C for 20 min) and also hot water combined with phosphite or hot water with 1-MCP displayed good control of postharvest anthracnose infection (Cruz *et al.*, 2015).

Guava black spot (GBS) and stylar end rot are also important postharvest diseases of guava fruit (Lim and Khoo, 1990; Fischer *et al.*, 2011; Yeh and Shiesh, 2017; Arafat, 2018). Similar to anthracnose, GBS is a latent infection in immature, developing fruits. Visible symptoms of the disease on guava fruit showed sunken lesions with concentric development, variation in colour ranging from greenish black to black and spread in severely affected fruit (Arafat, 2018). *Guignardia psidii* has been reported to cause black spot of guava fruits in the field and during transportation (Ullasa and Rawal, 1984), whereas *G. psidii* infection occurs in young fruits and stays quiescent until maturity (Tozetto and Ribeiro, 1993). Similarly, *Phyllosticta psidiicola* was reported as a potential cause of GBS disease in Taiwan and Venezuela (Lin *et al.*, 2003; González and Rondón, 2005). Unlike the previous studies, black spot disease on guava fruits in Egypt was caused by a novel isolate of *Phyllosticta capitalensis* (Arafat, 2018).

Stylar end rot caused by *Phomopsis psidii* may also emerge in the field and its symptoms appear as very small, discoloured, necrotic lesions on the surface. These lesions subsequently develop into a rough surface ring towards the stylar end of the fruit. The flesh and the seed cavity may disintegrate in severely infected fruits (Lim and Khoo, 1990). In Brazil, however, *Botryosphaeria dothidea*, *Neofusicoccum parvum* and *Neofusicoccum ribis* were identified as causal agents of stylar end rot disease (Júnior *et al.*, 2016). In another study, infection of guava cultivar in Egypt by *Botryosphaeria theobromae* and *Phomopsis* sp. displayed stylar end rot symptoms (Ammar and El-Naggar, 2014). Postharvest diseases are also dealt with in Chapter 15 of this volume.

16.6 Conclusion

Guava is an exquisite, nutritionally and economically valuable fruit crop of the tropics. It is considered an excellent source of vitamin C. Although most guava fruits have been placed in the climacteric category, the fruits of some guava cultivars display non-climacteric features. Climacteric guava fruits undergo rapid ripening changes, namely texture softening, peel pigmentation and odour emission, within a week under ambient conditions. Due to their highly perishable characteristics, mature guava fruits are difficult to handle for distant marketing and are more appropriate for processing rather than for fresh consumption. However, non-climacteric guava fruits show very low respiration and ethylene production rates as well as retain a slight green peel colour and crispy texture; therefore, having longer storability than their climacteric counterparts. These properties are important to facilitate international trade for fresh guavas. Moreover, since guava fruits tolerate low oxygen concentration, atmosphere modification technologies, such as CA and MAP, can potentially prolong postharvest life and maintain the quality. These handling techniques were recently applied in Taiwan for exporting non-climacteric 'Jen-Ju Bar' guavas to Canadian and US markets.

Possessing both climacteric and non-climacteric cultivars, guava becomes an intriguing plant material for ripening research. However, the underlying mechanism of this horticultural trait is still unclear. Elaborating the differences in gene structure, expression patterns and enzyme activities of ACS and ACO between climacteric and non-climacteric guavas could contribute comprehensively to increasing our understanding of guava ripening. In addition, the information may be beneficial to guava breeding projects for selecting cultivars with better longevity for the export market.

References

Abreu, J.R.d., Santos, C.D.d., Abreu, C.M.P.d., Pinheiro, A.C.M. and Corrêa, A.D. (2012) Ripening pattern of guava cv. Pedro Sato. *Ciência e Tecnologia de Alimentos* 32, 344–350.

Abu-Goukh, A.-B.A. and Bashir, H.A. (2003) Changes in pectic enzymes and cellulase activity during guava fruit ripening. *Food Chemistry* 83, 213–218.

Akamine, E.K. and Goo, T. (1979) Respiration and ethylene production in fruits of species and cultivars of *Psidium* and species of *Eugenia*. *Journal of the American Society for Horticultural Science* 104, 632–635.

Alba-Jiménez, J., Benito-Bautista, P., Nava, G., Rivera-Pastrana, D., Vázquez-Barrios, M.E. and Mercado-Silva, E. (2018) Chilling injury is associated with changes in microsomal membrane lipids in guava fruit (*Psidium guajava* L.) and the use of controlled atmospheres reduce these effects. *Scientia Horticulturae* 240, 94–101.

Ali, Z.M. and Lazan, H. (1997) Guava. In: Mitra, S.K. (ed.) *Postharvest Physiology and Storage of Tropical and Subtropical Fruits*. CAB International, Wallingford, UK, pp. 145–165.

Ali, Z.M., Chin, L.H. and Lazan, H. (2004) A comparative study on wall degrading enzymes, pectin modifications and softening during ripening of selected tropical fruits. *Plant Science* 167, 317–327.

Ammar, M. and El-Naggar, M. (2014) Screening and characterization of fungi and their associated mycotoxins in some fruit crops. *International Journal of Advanced Research* 2, 1216–1227.

Amusa, N., Ashaye, O., Oladapo, M. and Oni, M. (2005) Guava fruit anthracnose and the effects on its nutritional and market values in Ibadan, Nigeria. *World Journal of Agricultural Sciences* 1, 169–172.

Antala, D., Varshney, A., Davara, P. and Sangani, V. (2015) Modified atmosphere packaging of guava fruit. *Packaging Technology and Science* 28, 557–564.

Arafat, K. (2018) A novel isolate of *Phyllosticta capitalensis* causes black spot disease on guava fruit in Egypt. *Asian Journal of Plant Pathology* 12, 27–37.

Azzolini, M., Jacomino, A.P. and Spoto, M.H.F. (2004) Maturation stage and postharvest quality of 'Pedro Sato' guavas. *Revista Brasileira de Fruticultura* 26, 29–31.

Azzolini, M., Jacomino, A.P., Bron, I.U., Kluge, R.A. and Schiavinato, M.A. (2005) Ripening of 'Pedro Sato' guava: study on its climacteric or non-climacteric nature. *Brazilian Journal of Plant Physiology* 17, 299–306.

Baghel, B., Gupta, N., Khare, A. and Tiwari, R. (2005) Effect of different doses of gamma radiation on shelf-life of guava. *Indian Journal of Horticulture* 62, 129–132.

Bakshi, P., Wali, V., Sharma, A., Kour, G. and Bhat, D. (2014) *Maturity Indices of Guava Booklet*, 1st edn. Division of Fruit Science, Sher-e-Kashmir University of Agricultural Sciences and Technology, Jammu, India, pp. 4–6.

Bashir, H.A. and Abu-Goukh, A.B.A. (2003) Compositional changes during guava fruit ripening. *Food Chemistry* 80, 557–563.

Bassetto, E., Jacomino, A.P., Pinheiro, A.L. and Kluge, R.A. (2005) Delay of ripening of 'Pedro Sato' guava with 1-methylcyclopropene. *Postharvest Biology and Technology* 35, 303–308.

Bautista, P. and Silva, E. (1997) Effects of CA treatments on guava fruit quality. In: *Proceedings of the 7th International Controlled Atmosphere Conference, University of California, Davis, California*, p. 113.

Bhooriya, M., Bisen, B. and Pandey, S. (2018) Effect of postharvest treatments on shelf life and quality of guava (*Psidium guajava* L.) fruits. *International Journal of Chemical Studies* 6, 2559–2564.

Bishnoi, C. and Sharma, R. (2015) Influence of storage temperature on decay loss and microbial quality of stored guava (*Psidium guajava* L.). *International Journal of Agriculture, Environment and Biotechnology* 8, 621–624.

Botelho, L.N.S., Rocha, D.A., Braga, M.A., Silva, A. and Abreu, C.M.P.d. (2016) Quality of guava cv. Pedro Sato treated with cassava starch and cinnamon essential oil. *Scientia Horticulturae* 209, 214–220.

Braga, M.A., Marques, T.R., Simao, A.A., Botelho, L.N.S., Oliveira, L.S.d. and Abreu, C.M.P.d. (2018) Mechanism of firmness loss in guava cv. Pedro Sato during ripening at room temperature. *Food Science and Technology* 38, 26–32.

Bron, I.U., Ribeiro, R.V., Cavalini, F.C., Jacomino, A.P. and Trevisan, M.J. (2005) Temperature-related changes in respiration and Q10 coefficient of guava. *Scientia Agricola* 62, 458–463.

Brown, B. and Wills, R. (1983) Post-harvest changes in guava fruit of different maturity. *Scientia Horticulturae* 19, 237–243.

Campbell, C.A. (1994) Handling of Florida-grown and imported tropical fruits and vegetables. *HortScience* 29, 975–978.

Carvalho, H.d., Chitarra, M., Chitarra, A. and Menezes, J. (2001) Eficiência da concentração de cloreto de cálcio e do tempo de imersão no tratamento pós-colheita de goiaba branca cv. Kumagai. *Revista Brasileira de Fruticultura* 20, 375–381.

Cavalini, F.C., Jacomino, A.P., Lochoski, M.A., Kluge, R.A. and Ortega, E.M.M. (2006) Maturity indexes for 'Kumagai' and 'Paluma' guavas. *Revista Brasileira de Fruticultura* 28, 176–179.

Cavalini, F.C., Jacomino, A.P., Trevisan, M.J. and Miguel, A.C.A. (2015) Harvest time and quality of 'Kumagai' and 'Paluma' guavas. *Revista Brasileira de Fruticultura* 37, 64–72.

Chen, H.C., Sheu, M.J. and Wu, C.M. (2006) Characterization of volatiles in guava (*Psidium guajava* L.) cv. Chung-Shan-Yueh-Pa fruit from Taiwan. *Journal of Food and Drug Analysis* 14, 398–402.

Chen, K.E., Liu, T.C., Liu, Y.C. and Wu, C.T. (2017) 'Jen-Ju Bar' guava exhibited a non-climacteric ripening behavior resulting from a defect in the expression of System-2 ACC synthase *PgACS1*. *Acta Horticulturae* 1166, 63–70.

Chin, L., Ali, Z. and Lazan, H. (1994) Comparative softening of guava fruits: solubilization and depolymerization of cell wall carbohydrates during ripening. *Proceedings of the Malaysian Biochemical Society Conference* 19, 147–150.

Chitarra, M.I.F. and Chitarra, A.B. (2005) *Pós-colheita de Frutas e Hortaliças: Fisiologia e Manejo*, 2nd edn. Universidade Federal de Lavras, Lavras, Brazil.

Chyau, C.C., Chen, S.Y. and Wu, C.M. (1992) Differences of volatile and nonvolatile constituents between mature and ripe guava (*Psidium guajava* Linn.) fruits. *Journal of Agricultural and Food Chemistry* 40, 846–849.

Cruz, A.F., Medeiros, N.L., Benedet, G.L., Araújo, M.B., Uesugi, C.H. *et al.* (2015) Control of post-harvest anthracnose infection in guava (*Psidium guajava*) fruits with phosphites, calcium chloride, acetyl salicylic acid, hot water, and 1-MCP. *Horticulture, Environment, and Biotechnology* 56, 330–340.

de Campos, A.J., Fujita, E., Neves, L.C., Vieites, R.L. and Chagas, E.A. (2011) Gamma radiation and passive modified atmosphere on the quality of guavas 'Pedro Sato'. *Revista Brasileira de Fruticultura* 33, 350–356.

Deepthi, V.P. (2017) Physiological and biochemical changes during fruit growth, maturity and ripening of guava: a review. *Journal of Postharvest Technology* 5, 1–16.

Deepthi, V.P., Chandra Sekhar, R., Srihari, D. and Siva Sankar, A. (2016) Guava fruit quality and storability as influenced by harvest maturity and postharvest application of calcium salts. *Plant Archives* 16(1), 174–182.

de Medeiros Teodosio, A.E.M., Onias, E.A., de Oliveira, L.M., Rodrigues, M.H.B.S., Ribeiro, J.A. *et al.* (2018) Influence of different coatings on quality and shelf-life of guava under different storage temperature. *Journal of Experimental Agriculture International* 26, 1–10.

Dhalsamant, K., Mangaraj, S. and Bal, L.M. (2017) Modified atmosphere packaging for mango and tomato: an appraisal to improve shelf life. *Journal of Packaging Technology and Research* 1, 127–133.

Dolkar, D., Bakshi, P., Gupta, M., Wali, V., Kumar, R. *et al.* (2017) Biochemical changes in guava (*Psidium guajava*) fruits during different stages of ripening. *Indian Journal of Agricultural Sciences* 87, 257–-260.

El Bulk, R.E., Babiker, E.F.E. and Tinay, A.H.E. (1997) Changes in chemical composition of guava fruits during development and ripening. *Food Chemistry* 59, 395–399.

El-Zoghbi, M. (1994) Biochemical changes in some tropical fruits during ripening. *Food Chemistry* 49, 33–37.

Espinoza-Zamora, J., Baez-Sañudo, R., Saucedo-Veloz, C. and Mercado-Silva, E. (2008) Effect of application of waxes with vegetable oil and sucrose on the quality of Mexican guava cv. 'Media China'. *Acta Horticulturae* 849, 393–400.

Fischer, I.H., Almeida, A.M.d., Arruda, M.C.d., Bertani, R.M.d.A., Garcia, M.J.d.M. and Amorim, L. (2011) Postharvest damages in guavas from the Midwest region of the State of São Paulo. *Bragantia* 70, 570–576.

Forato, L.A., de Britto, D., de Rizzo, J.S., Gastaldi, T.A. and Assis, O.B. (2015) Effect of cashew gum–carboxymethylcellulose edible coatings in extending the shelf-life of fresh and cut guavas. *Food Packaging and Shelf Life* 5, 68–74.

Gaspar, J., Couto, F., Salomão, L.C.C., Finger, F.L. and Cardoso, A. (1997) Effect of low temperature and plastic films on post-harvest life of guava (*Psidium guajava* L.). *Acta Horticulturae* 45, 107–114.

Gill, K. (2018) Techniques for extending shelf life of guava fruits: a review. *Acta Horticulturae* 1205, 959–969.

Gomez, M.L.P. and Lajolo, F.M. (2008) Ascorbic acid metabolism in fruits: activity of enzymes involved in synthesis and degradation during ripening in mango and guava. *Journal of the Science of Food and Agriculture* 88, 756–762.

González-Aguilar, G., Tiznado-Hernandez, M., Zavaleta-Gatica, R. and Martínez-Téllez, M. (2004) Methyl jasmonate treatments reduce chilling injury and activate the defense response of guava fruits. *Biochemical and Biophysical Research Communications* 313, 694–701.

González, M. and Rondón, A. (2005) First report of *Cuignardia psidii*, an ascigerous state of *Phyllosticta psidiicola*, causing fruit rot on guava in Venezuela. *Plant Disease* 89, 773–773.

Harb, J. and Hasan, N. (2010) 1-MCP prolongs the shelf-life and changes the aroma profile of guava (*Psidium guajava* L.) fruit. *Acta Horticulturae* 934, 271–276.

Hassanein, R.A., Salem, E.A. and Zahran, A.A. (2018) Efficacy of coupling gamma irradiation with calcium chloride and lemongrass oil in maintaining guava fruit quality and inhibiting fungal growth during cold storage. *Folia Horticulturae* 30, 67–78.

Hong, K., Xie, J., Zhang, L., Sun, D. and Gong, D. (2012) Effects of chitosan coating on postharvest life and quality of guava (*Psidium guajava* L.) fruit during cold storage. *Scientia Horticulturae* 144, 172–178.

Hossain, F., Parvez, A., Munshi, M.K., Khalil, I. and Huque, R. (2014) Effect of radiation and chemical treatments on guava (*Psidium guajava* L.) to delay ripening in relation to organoleptic biochemical and microbiological properties. *International Journal of Current Microbiology and Applied Sciences* 3, 19–36.

Iqbal, Z., Randhawa, M.A., Zahoor, T., Asghar, M. and Beaudry, R. (2018) Influence of 1-methylcyclopropene on physico-chemical properties of 'Gola' and 'Surahi' guava (*Psidium guajava* L.) under air storage. *Pakistan Journal of Agricultural Sciences* 55, 389–396.

Itoo, S., Matsuo, T., Ibushi, Y. and Tamari, N. (1987) Seasonal changes in the levels of polyphenols in guava fruit and leaves and some their properties. *Journal of the Japanese Society for Horticultural Science* 56, 107–113.

Jain, N., Dhawan, K., Malhotra, S. and Singh, R. (2003) Biochemistry of fruit ripening of guava (*Psidium guajava* L.): compositional and enzymatic changes. *Plant Foods for Human Nutrition* 58, 309–315.

Jain, N., Dhawan, K., Malhotra, S.P., Siddiqui, S. and Singh, R. (2001) Compositional and enzymatic changes in guava (*Psidium guajava* L.) fruits during ripening. *Acta Physiologiae Plantarum* 23, 357–362.

JanuáriaVieira, S.M., Raga, A., Benedetti, B.C., de Oliveira, R.A., Di Marco, P.G. and de Toledo Scarponi, A.P. (2014) Effect of ultraviolet-C radiation on 'Kumagai' guavas infested by *Ceratitis capitata* (*Diptera—Tephritidae*) and on physical parameters of postharvest. *Scientia Horticulturae* 165, 295–302.

Javed, M.S., Randhawa, M.A., Ahmad, Z., Sajid, M.W., Nasir, M.A. and Tariq, M.R. (2018) Effect of $CaCl_2$ and controlled atmosphere storage on phytochemical attributes of guava. *Food Science and Technology* 38, 356–362.

Júnior, A.F.N., Fischer, I.H., Bragança, C.A., Júnior, N.S.M. and Amorim, L. (2016) Identification of Botryosphaeriaceae species that cause stylar-end rot of guavas and characterisation of the disease monocycle. *European Journal of Plant Pathology* 144, 271–287.

Kabbashi, E., Nasr, O.E., Musa, S.K. and Roshdi, M.A. (2012) Use of gamma irradiation for disinfestation of guava fruits from fruit flies [*Ceratitis* spp. & *Bactrocera* sp. (*Diptera: Tephritidae*)] in Khartoum State, Sudan. *Agricultural Science Research Journal* 2, 17–23.

Kader, A. (2009) Guava: recommendations for maintaining postharvest quality. Available at http://postharvest.ucdavis.edu/Commodity_Resources/Fact_Sheets/Datastores/Fruit_English/?uid=26&ds=798 (accessed 1 December 2019).

Kader, A.A. (1986) Potential applications of ionizing radiation in postharvest handling of fresh fruits and vegetables. *Food Technology* 40, 117–121.

Kader, A.A. (2001) A summary of CA requirements and recommendations for fruits other than apples and pears. *Acta Horticulturae* 600, 737–740.

Kester, J. and Fennema, O. (1986) Edible films and coatings: a review. *Food Technology* 60, 47–59.

Kohli, K., Tripathi, S., Kumar, A., Kumar, A., Singh, O. *et al.* (2019) Studies on postharvest changes in physico-biochemical properties of guava (*Psidium guajava* L.) cv. Sardar influenced by different composite coatings. *International Journal of Current Microbiology and Applied Sciences* 8, 1417–1424.

Kumar, K., Bhagwan, A., Kumar, A., Venkatlakxmi, K. and Madhavi, K. (2017) Effect of modified atmosphere packaging (MAP) on chilling injury and storage life of guava cv. Allahabad Safeda stored at 6 ± 1°C. *International Journal of Chemical Studies* 5, 771–776.

Kumar, R., Thakur, A. and Sharma, P. (2014) Effect of 1-methylcyclopropene (1-MCP) on quality and storage life of winter guava. In: *Proceedings of the 2nd International Conference on Agricultural and Horticultural Sciences*, Vol. 2, p. 216.

Lazan, H. and Ali, Z.M. (1998) Guava. In: Shaw, P.E., Chan, H.T. and Nagy, S. (eds) *Tropical and Subtropical Fruits*. AgScience Inc., Auburndale, Florida, pp. 446–485.

Liao, Y.S. (2019) The study of fruit development and ripening characteristics in 'Shiang Bar' and 'Jen-Ju Bar' guavas (*Psidium guajava* L.). Master thesis, National Taiwan University, Taipei, Taiwan.

Lim, T. and Manicom, B. (2003) Diseases of guava. In: Ploetz, R. (ed.) *Diseases of Tropical Fruit Crops*. CAB International, Wallingford, UK, pp. 275–289.

Lim, T.K. and Khoo, K.C. (1990) *Guava in Malaysia: Production, Pests and Diseases*. Tropical Press, Kuala Lumpur.

Lin, C., Lai, C. and Tsai, S. (2003) Ecological survey of guava new fruit rot – *Phyllosticta* rot (black spot) and other fruit rots. *Plant Protection Bulletin* 45, 263–270.

Liu, T.C., Liu, Y.C., Chen, K.E., Chao, C.W. and Wu, C.T. (2012) The nonclimacteric guava cultivar 'Jen-Ju Bar' is defective in system 2 1-aminocyclopropane-1-carboxylate synthase activity. *Postharvest Biology and Technology* 67, 10–18.

McGuire, R.G. (1997) Market quality of guavas after hot-water quarantine treatment and application of carnauba wax coating. *HortScience* 32, 271–274.

McGuire, R.G. and Hallman, G.J. (1995) Coating guavas with cellulose-or carnauba-based emulsions interferes with postharvest ripening. *HortScience* 30, 294–295.

Mahajan, B., Sharma, S. and Dhall, R. (2009) Optimization of storage temperature for maintaining quality of guava. *Journal of Food Science and Technology* 46, 604–605.

Mandal, G., Dhaliwal, H. and Mahajan, B. (2010) Effect of pre-harvest calcium sprays on post-harvest life of winter guava (*Psidium guajava* L.). *Journal of Food Science and Technology* 47, 501–506.

Mangaraj, S., Goswami, T., Giri, S. and Joshy, C. (2014) Design and development of modified atmosphere packaging system for guava (cv. Baruipur). *Journal of Food Science and Technology* 51, 2925–2946.

Martins, M.C., Amorim, L., Lourenço, S.A., Gutierrez, A.S.S. and Watanabe, H.S. (2007) Incidence of post harvest damages in guavas at the wholesale market of São Paulo and its relationship to pre harvest bagging. *Revista Brasileira de Fruticultura* 29, 245–248.

Mercadante, A.Z., Steck, A. and Pfander, H. (1999) Carotenoids from guava (*Psidium guajava* L.): isolation and structure elucidation. *Journal of Agricultural and Food Chemistry* 47, 145–151.

Mercado-Silva, E., Benito-Bautista, P. and de los Angeles García-Velasco, M. (1998) Fruit development, harvest index and ripening changes of guavas produced in central Mexico. *Postharvest Biology and Technology* 13, 143–150.

Misra, K. and Seshadri, T. (1968) Chemical components of the fruits of *Psidium guajava*. *Phytochemistry* 7, 641–645.

Mitra, S., Devi, H., Chakraborty, I. and Pathak, P. (2012) Recent development in postharvest physiology and storage of guava. *Acta Horticulturae* 959, 89–95.

Mondal, K., Singh, A., Saxena, N., Malhotra, S., Dhawan, K. and Singh, R. (2008) Possible interactions of polyamines and ethylene during ripening of guava (*Psidium guajava* L.) fruits. *Journal of Food Biochemistry* 32, 46–59.

Mowlah, G. and Itoo, S. (1982) Quantitative changes in guava polyphenols and the polyphenoloxidase (PPO) at different stages of maturation, ripening and storage. *Nippon Shokuhin Kogyo Gakkaishi* 29, 413–417.

Murmu, S.B. and Mishra, H.N. (2018) Selection of the best active modified atmosphere packaging with ethylene and moisture scavengers to maintain quality of guava during low-temperature storage. *Food Chemistry* 253, 55–62.

Murray, R., Lucangeli, C., Polenta, G. and Budde, C. (2007) Combined pre-storage heat treatment and controlled atmosphere storage reduced internal breakdown of 'Flavorcrest' peach. *Postharvest Biology and Technology* 44, 116–121.

Nakasone, H.Y. and Paull, R.E. (eds) (1998) Guava. In: *Tropical Fruits.* CAB International, Wallingford, UK, pp. 149–172.

Osman, A. and Ayub, M. (1996) Effects of different postharvest treatments on the respiration patterns of guava (*Psidium guajava* L.). *Acta Horticulturae* 464, 502–509.

Padula, M. and Rodriguez-Amaya, D.B. (1986) Characterisation of the carotenoids and assessment of the vitamin A value of Brazilian guavas (*Psidium guajava* L.). *Food Chemistry* 20, 11–19.

Pal, D. and Selvaraj, Y. (1979) Changes in pectin and pectin esterase activity in developing guava fruits. *Journal of Food Science and Technology* 16, 115–116.

Pal, R., Ahmad, M., Roy, S.K. and Singh, M. (2004) Influence of storage environment, surface coating, and individual shrink wrapping on quality assurance of guava (*Psidium guajava*) fruits. *Plant Foods for Human Nutrition* 59, 67–72.

Pandey, S. and Joshua, J.E. (2010) Influence of gamma-irradiation, growth retardants and coatings on the shelf life of winter guava fruits (*Psidium guajava* L.). *Journal of Food Science and Technology* 47, 124–127.

Patel, R., Maiti, C., Deka, B.C., Deshmukh, N., Verma, V. and Nath, A. (2015) Physical and biochemical changes in guava (*Psidium guajava* L.) during various stages of fruit growth and development. *International Journal of Agriculture, Environment and Biotechnology* 8, 63–70.

Patricio, F., Batista, M., de Lima, A.C.C.G., da Trindade, D., Araujo, A.L.S. and Alves, R.E. (2012) Chemical characterization of guava fruit produced in submiddle of São Francisco Valley, Brazil. *Acta Horticulturae* 869, 53–54.

Paull, R. and Chen, C. (2016) Guava. In: Gross, K., Wang, C. and Saltveit, M. (eds) *The Commercial Storage of Fruits, Vegetables, and Florist and Nursery Stocks*. Agriculture Handbook No. 66. US Department of Agriculture, Washington, DC, pp. 356–357.

Paull, R. and Duarte, O. (eds) (2012) Guava. In: *Tropical Fruits*, Vol. 2. CAB International, Wallingford, UK, pp. 91–155.

Paull, R. and Goo, T. (1983) Relationship of guava (*Psidium guajava* L.) fruit detachment force to the stage of fruit development and chemical composition. *HortScience* 18, 65–76.

Phebe, D. and Ong, P. (2010) Extending 'Kampuchea' guava shelf-life at 27°C using 1-methylcyclopropene. *International Food Research Journal* 17, 63–69.

Porat, R., Weiss, B., Zipori, I. and Dag, A. (2009) Postharvest longevity and responsiveness of guava varieties with distinctive climacteric behaviors to 1-methylcyclopropene. *HortTechnology* 19, 580–585.

Rai, M.K., Asthana, P., Jaiswal, V. and Jaiswal, U. (2010) Biotechnological advances in guava (*Psidium guajava* L.): recent developments and prospects for further research. *Trees* 24, 1–12.

Rana, S. and Siddiqui, S. (2018) Comparative effect of different individual wrappings on shelf life of guava (*Psidium guajava*). *Journal of Food Science and Technology* 55, 2935–2944.

Rana, S., Siddiqui, S. and Goyal, A. (2015) Extension of the shelf life of guava by individual packaging with cling and shrink films. *Journal of Food Science and Technology* 52, 8148–8155.

Rathore, D. (1976) Effect of season on the growth and chemical composition of guava (*Psidium guajava* L.) fruits. *Journal of Horticultural Science* 51, 41–47.

Rawal, R. (1993) Yield loss in guava by different fruit rots. *International Journal of Tropical Plant Diseases* 11, 69–72.

Razali, Z., George, D.S. and Ibrahim, N.A.N. (2015) Improvements in quality and antioxidant capacity of guava (*Psidium guajava* L.) exposed to UV-C irradiation. In: *Proceedings of the 3rd International Conference on Chemical, Agricultural and Medical Sciences (CAMS-2015)*, pp. 10–11.

Reyes, M. and Paull, R. (1995) Effect of storage temperature and ethylene treatment on guava (*Psidium guajava* L.) fruit ripening. *Postharvest Biology and Technology* 6, 357–365.

Rodeo, A.J.D., Gonzales, D.C.H. and Esguerra, E.B. (2018) Physiological and physicochemical changes in guava (*Psidium guajava* L. cv. Queso de Bola) fruit stored at different temperatures. *Philippine Journal of Crop Science* 43, 19–28.

Rodrigues, A.A.M., Silva, S.d.M., Dantas, A.L., Silva, A.F.d., Santos, L.d.S. and Moreira, D.d.N. (2018) Physiology and postharvest conservation of 'Paluma' guava under coatings using jack fruit seed-based starch. *Revista Brasileira de Fruticultura* 40, 1–8.

Sau, S., Datta, P., Sarkar, T. and Sarkar, S. (2018) Impact of low doses of gamma irradiation on off-season guava at ambient storage condition. *International Journal of Current Microbiology and Applied Sciences* 7, 295–307.

Selvaraj, Y., Pal, D., Edward Raja, M. and Rawal, R. (1999) Changes in chemical composition of guava fruits during growth and development. *Indian Journal of Horticulture* 56, 10–18.

Setiawan, B., Sulaeman, A., Giraud, D.W. and Driskell, J.A. (2001) Carotenoid content of selected Indonesian fruits. *Journal of Food Composition and Analysis* 14, 169–176.

Sharma, A., Sharma, R., Siddiqui, S. and Sharma, B. (2012) Comparison of cell wall degrading enzyme activities during ripening of guava fruit on-tree and in-storage. *Indian Journal of Horticulture* 69, 409–415.

Silip, J. and Hajar, S. (2007) Relationship between precooling, storage temperature and storage duration to the quality characteristics of guava (*Psidium guajava* L. cv. Kampuchea). *Acta Horticulturae* 735, 535–546.

Silva, A.d., Bassetto, E., Daiuto, A. and Vieites, R. (2000) Utilization of calcium chloride in the postharvest conservation of guava (*Psidium guajava* L.) fruits. *Agro-Ciencia* 16, 189–195.

Simrat, H. (2009) Production and post-harvest technology of guava in India – an overview. *Crop Research, Hisar, India* 38, 113–134.

Singh, J., Singh, K. and Singh, S.K. (2017) Effect of postharvest treatment of guava fruits with 1-methylcyclopropene and gibberellin on storage life and fruit quality. *International Archive of Applied Sciences & Technology* 8, 35–40.

Singh, S. (2011) Guava (*Psidium guajava* L.). In: Yahia, E.M. (ed.) *Postharvest Biology and Technology of Tropical and Subtropical Fruits*. Woodhead Publishing, Cambridge, pp. 213–240.

Singh, S. and Pal, R. (2007) Postharvest fruit fly disinfestation strategies in rainy season guava crop. *Acta Horticulturae* 735, 591–596.

Singh, S. and Pal, R. (2008a) Controlled atmosphere storage of guava (*Psidium guajava* L.) fruit. *Postharvest Biology and Technology* 47, 296–306.

Singh, S. and Pal, R. (2008b) Response of climacteric-type guava (*Psidium guajava* L.) to postharvest treatment with 1-MCP. *Postharvest Biology and Technology* 47, 307–314.

Singh, S. and Pal, R. (2009) Ionizing radiation treatment to improve postharvest life and maintain quality of fresh guava fruit. *Radiation Physics and Chemistry* 78, 135–140.

Sivakumar, D. and Korsten, L. (2006) Evaluation of the integrated application of two types of modified atmosphere packaging and hot water treatments on quality retention in the litchi cultivar 'McLean's Red'. *The Journal of Horticultural Science and Biotechnology* 81, 639–644.

Soares, F.D., Pereira, T., Marques, M.O.M. and Monteiro, A.R. (2007) Volatile and non-volatile chemical composition of the white guava fruit (*Psidium guajava*) at different stages of maturity. *Food Chemistry* 100, 15–21.

Sothornvit, R. (2012) Effect of edible coating on the qualities of fresh guava. *Acta Horticulturae* 1012, 453–459.

Steinhaus, M., Sinuco, D., Polster, J., Osorio, C. and Schieberle, P. (2009) Characterization of the key aroma compounds in pink guava (*Psidium guajava* L.) by means of aroma re-engineering experiments and omission tests. *Journal of Agricultural and Food Chemistry* 57, 2882–2888.

Suárez, J., Camacaro, D., Pérez, M. and Giménez, A. (2009) Effect of temperature and maturation stage on the postharvest fruit quality of guava (*Psidium guajava* L.) from MERCABAR, Lara State, Venezuela. *Revista Científica UDO Agrícola* 9, 60–69.

Sulistio, M. (2019) The effect of *Copia* LTR retrotransposon insertion in nonclimacteric 'Jen-Ju Bar' guava on the expression of *PgACS1*, a System 2 ACC synthase gene. *Proceedings of the Annual Conference of Taiwan Society for Horticultural Science* 65, 53–54.

Supapvanich, S. and Promyou, S. (2017) Hot water incorporated with salicylic acid dips maintaining physicochemical quality of 'Holland' papaya fruit stored at room temperature. *Emirates Journal of Food and Agriculture* 29, 18–24.

Supapvanich, S., Kernprie, Y., Boonyaritthongchai, P., Techavuthiporn, C., Tepsorn, R. and Youryon, P. (2019) Physicochemical quality maintenance and bioactive compounds enhancement of Thai guava fruit cv. 'Kim Ju' by using combinative hot water and methyl jasmonate immersion. *Emirates Journal of Food and Agriculture* 31, 395–404.

Tandon, D., Singh, B. and Kalra, S. (1989) Storage behaviour of specific-gravity-graded guava fruits. *Scientia Horticulturae* 41, 35–41.

Teixeira, G., Durigan, F., Santos, L., Ogassavara, F., Martins, R. *et al.* (2007) Effect of controlled atmosphere with reducing levels of oxygen on incidence of postharvest diseases in guava (*Psidium guajava* L. cv. 'Pedro Sato'). In: *Proceedings of the International Congress on Novel Approaches for the Control of Postharvest Diseases and Disorders*. CRIOF, University of Bologna, Bologna, Italy, pp. 240–247.

Teixeira, G.H. and Durigan, J.F. (2010) Effect of controlled atmospheres with low oxygen levels on extended storage of guava fruit (*Psidium guajava* L. 'Pedro Sato'). *HortScience* 45, 918–924.

Teixeira, G.H., Júnior, L.C.C., Ferraudo, A.S. and Durigan, J.F. (2016) Quality of guava (*Psidium guajava* L. cv. Pedro Sato) fruit stored in low-O_2 controlled atmospheres is negatively affected by increasing levels of CO_2. *Postharvest Biology and Technology* 111, 62–68.

Thomas, P. and Diehl, J. (1988) Radiation preservation of foods of plant origin. Part VI. Mushrooms, tomatoes, minor fruits and vegetables, dried fruits, and nuts. *Critical Reviews in Food Science and Nutrition* 26, 313–358.

Tozetto, L. and Ribeiro, W. (1993) Ocorrência de podridão de frutos de goiaba (*Psidium guajava*) causada por *Phyllosticta* sp. *Fitopatologia Brasileira* 18, 160.

Ullasa, B. and Rawal, R. (1984) Guignardia fruit rot of guava – a new disease from Bangalore, India. *Current Science* 58, 435–436.

Vazquez-Ochoa, R.I. and Colinas-Leon, M.T. (1990) Changes in guavas of three maturity stages in response to temperature and relative humidity. *HortScience* 25, 86–87.

Ventosa, M., Rodríguez, J.L. and Zerqueira, O.L. (2008) Determination of major carotenoids in guava (*Psidium guajava* L.). *Ciencia y Tecnología de Alimentos* 18, 1–4.

Wahid, O. (2001) Occurrence of Colletotrichum anthracnose disease of guava fruit in Egypt. *International Journal of Pest Management* 47, 147–152.

Wang, C.Y. (1993) Approaches to reduce chilling injury of fruits and vegetables. *Horticultural Reviews* 15, 63–95.

Wang, M.H. and Lin, H.L. (2009) Effect of cold storage and vapor heat treatment on the quality of 'Di-Wang' guava (*Psidium guajava* L.) fruits. *Journal of the Taiwan Society for Horticultural Science* 55, 113–126.

Watkins, C.B. (2006) The use of 1-methylcyclopropene (1-MCP) on fruits and vegetables. *Biotechnology Advances* 24, 389–409.

Watkins, C.B. (2008) Overview of 1-methylcyclopropene trials and uses for edible horticultural crops. *HortScience* 43, 86–94.

Werner, E.T., Junior, O., Bona, A.P.d., Cavati, B. and Gomes, T.D.U.H. (2009) Calcium chloride application in the post-harvest of guavas 'Cortibel'. *Bragantia* 68, 511–518.

Wilberg, V.C. and Rodriguez-Amaya, D.B. (1995) HPLC quantitation of major carotenoids of fresh and processed guava, mango and papaya. *Lebensmittel-Wissenschaft und -Technologie* 28, 474–480.

Wilson, C.W., Shaw, P.E. and Campbell, C.W. (1982) Determination of organic acids and sugars in guava (*Psidium guajava* L.) cultivars by high-performance liquid chromatography. *Journal of the Science of Food and Agriculture* 33, 777–780.

Yadav, M., Saurabh, S., Singh, D. and Singh, G.K. (2010) Effect of post harvest treatment with γ-irradiation on storage behaviour of guava (*Psidium guajava*) fruits. *Indian Journal of Agricultural Sciences* 80, 389–393.

Yang, S.F. and Hoffman, N.E. (1984) Ethylene biosynthesis and its regulation in higher plants. *Annual Review of Plant Physiology* 35, 155–189.

Yeh, S.-T. and Shiesh, C.C. (2017) The investigation and occurrence of guava disease in central part of Taiwan. *Acta Horticulturae* 1166, 115–124.

Yusof, S. and Mohamed, S. (1987) Physico-chemical changes in guava (*Psidium guajava* L.) during development and maturation. *Journal of the Science of Food and Agriculture* 38, 31–39.

Zhang, M.D. (2018) Effect of ethylene on aroma volatiles production and key aroma-related enzymes in guava fruit. Master thesis, National Taiwan University, Taipei, Taiwan.

Index

Note: Page numbers in **bold** type refer to **figures**
Page numbers in *italic* type refer to *tables*